TETRAHEDRON ORGANIC CHEMISTRY SERIES
Series Editors: **J E Baldwin, FRS & P D Magnus, FRS**

VOLUME 14

Principles of

Asymmetric Synthesis

Related Pergamon Titles of Interest

BOOKS

Tetrahedron Organic Chemistry Series:
CARRUTHERS: Cycloaddition Reactions in Organic Synthesis
DEROME: Modern NMR Techniques for Chemistry Research
HASSNER & STUMER: Organic Syntheses based on Name Reactions and Unnamed Reactions
PAULMIER: Selenium Reagents & Intermediates in Organic Synthesis
PERLMUTTER: Conjugate Addition Reactions in Organic Synthesis
SIMPKINS: Sulphones in Organic Synthesis
WILLIAMS: Synthesis of Optically Active Alpha-Amino Acids 2nd Edn*
WONG & WHITESIDES: Enzymes in Synthetic Organic Chemistry

JOURNALS

BIOORGANIC & MEDICINAL CHEMISTRY
BIOORGANIC & MEDICINAL CHEMISTRY LETTERS
TETRAHEDRON
TETRAHEDRON: ASYMMETRY
TETRAHEDRON LETTERS

Full details of all Elsevier Science publications/free specimen copy of any Elsevier Science journal are available on request from your nearest Elsevier Science office

* In Preparation

Principles of
Asymmetric Synthesis

ROBERT E. GAWLEY

University of Miami

and

JEFFREY AUBÉ

University of Kansas

PERGAMON

U.K.	Elsevier Science Ltd, The Boulevard, Langford Lane, Kidlington, Oxford, OX5 1GB, U.K.
U.S.A.	Elsevier Science Inc., 660 White Plains Road, Tarrytown, New York 10591-5153, U.S.A.
JAPAN	Elsevier Science Japan, Higashi Azabu 1-chome Building 4F, 1-9-15 Higashi Azabu, Minato-ku, Tokyo 106, Japan

First Edition 1996

Library of Congress Cataloging in Publication Data
A catalog record for this book is available from the
Library of Congress

British Library Cataloguing in Publication Data
A catalogue record for this book is available from the
British Library

ISBN 0 08 0418767 Hardcover
ISBN 0 08 0418759 Flexicover

Printed and bound in Great Britain by BPC Wheatons Ltd, Exeter

Table of Contents

Foreword

The development, over the last two decades of the area of asymmetric synthesis has massively increased the power of organic chemists to make molecules of defined handedness. The invention of chiral reagents which have permitted this great development is set out in this monograph. The whole gamut of these, even more efficient and selective, reagents described and most importantly the correct understanding of how these results are achieved, at the mechanistic level. This book will be an essential aid for all synthetic organic chemists both in industry and academe.

J E Baldwin, FRS

Asymmetric Synthesis—a foreword

Asymmetric synthesis, the selective generation of new chirality elements (as one definition goes), has developed from a specialty pursued by outsiders to an art cultured by some learned ones, and now may be considered a standard laboratory methodology for everybody's use. This development has taken place exponentially (explosively!) in the last two decades, triggered by a number of circumstances. Also the practitioners of pharmaceutical, vitamin, and agro synthesis need to produce enantiopure, rather than racemic active compounds (for registration!).

For many chemists (and, too often, for those making decisions about funding research), the invention of new reactions, the development of synthetic methodology, the systematic (retrosynthetic) analysis of target structures, the investigation of reaction mechanisms, and the total synthesis of complex natural products have lost their glory. Chemists' attention has shifted to areas such as combinatorial synthesis (driven by robot, computer, and miniaturization), material sciences, supramolecular chemistry, the origin of life, the biological and even the medical sciences. Yet, in all these fields chirality plays a central role—the molecules of life are chiral.

Organic synthesis has merged with transition-metal chemistry and with catalysis (homogeneous, heterogeneous, enantioselective). The increasing demand for efficient enantioselective synthetic methods was accompanied by, and it has in fact necessitated, a similarly dramatic development of the analytical tools for the determination of enantiomer ratios. In a 1990 *Angewandte Chemie* article I have stated: 'The primary center of attention for all synthetic methods will continue to shift toward catalytic and enantioselective variants; indeed, it will not be long before such modifications will be available with every standard reaction for converting achiral educts into chiral products.' A demonstration of this ongoing process are the volumes of monographs dedicated to the subject (see also the introduction of the present book): 450 pages in 1962 (Eliel, 'Stereochemistry of Carbon Compounds') and 450 pages in 1971 (Morrison–Mosher, 'Asymmetric Organic Reactions'), 1800 pages in 1983/84 (Morrison, 'Asymmetric Synthesis'), 900 pages in 1993 (Brunner–Zottimeier, 'Handbook of Enantioselective Catalysis with Transition Metal Compounds'), 1250 pages in 1994 (Eliel–Wilen, 'Stereochemistry of Organic Compounds'), 6000 pages in 1995 (Helmchen *et al.*, Houben–Weyl, 'Stereoselective Synthesis'). Numerous smaller works, some specialized (some superficial), new journals ('Chirality', 'Tetrahedron: Asymmetry'), and treatises in review organs have appeared concomitantly.

The authors of 'Principles of Asymmetric Synthesis' have managed to cover the theme (on *ca.* 350 pages) in a most condensed and masterly way. In contrast to many a competitor they have chosen well-defined topics (enolate alkylations, direct and conjugate addition to carbonyl compounds with formation of one or two new chirality centers, rearrangements and cycloadditions, reductions and oxidations), they use clear-cut concepts and language to present not only synthetic results but also mechanistic considerations (with due care). They include a glossary of stereochemical terms (with some recommendations about which ones not to use: fight the brutalization of chemical language!), and they present a chapter on analytical methods. All of this is accompanied by extensive referencing (up to 217 per chapter, over 1300 total) to the very latest literature, and, in spite of the broad coverage, there are some insightful and profound sections (close to the authors heart of interest, *cf.* the discussion of the stereochemical course of the *Wittig* rearrangement).

The book meets the requirements for today's teaching of stereoselective reactions, and it is of a quality and competence which will make it a handy reference work even for those experts doing research in the various areas covered.

Dieter Seebach

Preface

The field of asymmetric synthesis evolved from the study of diastereoselectivity in reactions of chiral compounds, through auxiliary-based methods for the synthesis of enantiomerically pure compounds (diastereoselectivity followed by isomer separation and auxiliary cleavage), to asymmetric catalysis. In the former case, diastereomeric mixtures ensue, and an analytical technique such as chromatography is used for isomer purification. In the latter instance, enantiomers are the products, and chiral stationary phases can be used for chromatographic purification. Furthermore, many methods have now been developed that generate numerous stereocenters in a single step. Highly selective reactions that produce one or more stereocenters with a high degree of selectivity ($\geq 90\%$), along with modern purification techniques, allow the preparation - in a single step - of chiral substances in $\geq 98\%$ ee for many reaction types.

In this book, we introduce one new paradigm: the recognition of a distinction between interligand and intraligand asymmetric induction. Briefly (see Section 1.3), this distinction recognizes the intimate involvement of metals in nearly all highly stereoselective reactions, and reflects the evolutionary development of modern asymmetric synthesis from issues of simple stereoselectivity to asymmetric catalysis. We hope that readers – after they become cognizant of this distinction – will find it useful in their efforts to improve on the methods described in this book.

The field continues to grow at an exponential rate, so that comprehensive coverage in a monograph is no longer possible. To even address the topic in a significant way is a formidable task, and to render the subject manageable, decisions had to be made about what to include and what to leave out.

In the end, we have selected several reaction categories, which comprise many of the most useful synthetic tools. As the title implies, the focus is on the principles that govern relative and absolute configurations in transition state assemblies. There are only a few principles, but they recur constantly. For example, organization around a metal atom, $A^{1,3}$ strain, van der Waals interactions, dipolar interactions, etc., are factors affecting transition state energies, and which in turn dictate stereoselectivity *via* transition state theory. One might call these analyses *molecular recognition at a saddle point*.

In writing this book, we have adhered to fairly strict limitations regarding terminology (and ask others to do the same). Sloppy terminology (*e.g.* "optically active synthesis") serves no useful purpose and intended meanings are often obscured (especially to future generations). To help readers understand our wording and to guide proper usage, a detailed, annotated glossary of stereochemical terms is included (and highlighted at the page margin for easy reference). This glossary will be useful to readers who are unfamiliar with the precise definitions of stereochemical terms (or their source). Included also are a number of stereochemical terms that are not in current usage, but which are included for reference purposes.

The book comprises 8 chapters. The first provides background, introduces the topic of asymmetric synthesis, outlines principles of transition state theory as applied to stereoselective reactions, and includes the glossary. The second chapter details methods for analysis of mixtures of stereoisomers, including an important section on sample preparation. Then follow four chapters on carbon-carbon bond forming reactions, organized by reaction type and presented in order of increasing mechanistic complexity: Chapter 3 is about enolate alkylations, Chapter 4 nucleophilic additions to carbonyls. Chapter 5 is on aldol and Michael additions (2 new stereocenters), while Chapter 6 covers rearrangements and cycloadditions. The last two chapters cover reductions and oxidations.

In addition to tables of examples that show high selectivity, a transition state analysis is presented to explain - to the current level of understanding - the stereoselectivity of many of the reactions covered. Examination of these rationales often exposes the weaknesses of current theories, in that they cannot always explain the experimental observations. These shortcomings provide a challenge for future mechanistic investigations.

The decision has been made to omit details about auxiliary and/or catalyst preparation, auxiliary attachment and removal, and details of experimental procedures. Although we usually do not comment on the ease of auxiliary or catalyst preparation or removal, we have tried to focus on methods where this issue is not a problem – at least not on a laboratory scale. Also, when comparing selectivity data among various reagents for a given reaction, the reader should recognize that the indicated selectivities are for individually optimized experimental conditions (*i.e.,* solvent, temperature, time, etc.) that may not be the same from entry to entry, making relative selectivity comparisons tenuous. Examples included in the tables and schemes do not necessarily reflect the scope and limitations of a reaction; in some cases only the more stereoselective examples are included. We also do not include much detail on synthetic applications of the methods described. Figures that illustrate synthetic targets prepared using asymmetric synthesis as a key step – with the relevant stereocenters highlighted – are included, but outlines of the synthetic plans are not. Names of specific target molecules may be found in the index.

This book is intended for advanced undergraduate or graduate students, and other chemists needing a guide to the principles of asymmetric synthesis and stereo-selectivity, and for experts who seek leading references to the primary literature. The book could be used for a course in organic mechanisms, stereochemistry, reactions, or synthesis.

REG wishes to thank his coauthor for his help in this project, which has taken five years to complete, and which would never have seen the light of day had it not been for Jeff's cheerful encouragement and constructive criticism all along the way, and for his willingness to jump in and help finish it by contributing the last chapter and taking responsibility for the index.

Much of the work on this book was completed during a sabbatical leave for REG, at the Swiss Federal Institute of Technology (ETH), Zürich, which was funded in part by a Fogarty Senior International Fellowship from the National Institutes of Health, in part by a sabbatical leave from the University of Miami, and in part by the ETH. This financial support is warmly acknowledged, with thanks. Special thanks are also due to Professor Dieter Seebach for his generous hospitality during the sabbatical year, and to his colleagues, Professors Arigoni, Diederich, Dunitz, Prelog, and Vasella, who jointly contributed to making the year in Zürich a most enjoyable one.

Many of our friends and colleagues have contributed to this work with helpful discussions, or by reading and commenting on various portions of this book. Among these, Professor Vladimir Prelog deserves special thanks for his exhaustive critique of an early draft of the glossary. Others, in alphabetical order, are: Steve Buchwald, Luis Echegoyen, Dieter Enders, Jeff Evanseck, Carl Hoff, Ken Houk, Dieter Seebach, Bill Purcell, and Andrea Vasella. Additionally, we are grateful to Jennifer Badiang, Kristine Frank, Vijaya Gracias, Michael Hoemann, and Klaas Schildknegt for their valuable comments and hard work in helping with the index. The insights and criticisms of these colleagues have helped us greatly, but we retain responsibility for the mistakes that remain.

Finally, it has not been possible to describe all of the contributions that are relevant to each of the topics in the book. To those authors who feel that their work has not been given its due, or whose work has been overlooked, we offer our apologies. Comments are welcome.

Robert E. Gawley
Coral Gables, Florida

Jeffrey Aubé
Lawrence, Kansas

April 23, 1996

Reprint acknowledgements

Figure 2.12: Reprinted with permission from *J. Am. Chem. Soc.,* vol. 91; T. Williams; R. G. Pitcher; P. Bommer; J. Gutzwiller; M. Uskokovic, *"Diastereomeric Solute-Solute Interactions of Enantiomers in Achiral Solvents. Nonequivalence of the Nuclear Magnetic Resonance Spectra of Racemic and Optically Active Dihydroquinine"*, pages 1871-1872 (1969). Copyright 1969, American Chemical Society.

Figure 4.6b: Reprinted with permission from *J. Am. Chem. Soc.,* vol. 95; H. B. Bürgi; J. D. Dunitz; E. Schefter, *"Geometrical Reaction Coordinates. II. Nucleophilic Addition to a Carbonyl Group"*, pages 5065-5067 (1973). Copyright 1973 American Chemical Society.

Figure 4.7: Reprinted from *Tetrahedron,* vol. 30; H. B. Bürgi; D. Dunitz; J. M. Lehn; G. Wipff, *"Stereochemistry of Reaction Paths at Carbonyl Centres"*, pages 1563-1572 (1974), with kind permission from Elsevier Science Ltd., The Boulevard, Langford Lane, Kidlington OX5 1GB, UK.

Figure 4.9a: Reprinted with permission from *J. Am. Chem. Soc.,* vol. 109; E. P. Lodge; C. H. Heathcock, *"On the Origin of Diastereofacial Selectivity in Additions to Chiral Aldehydes and Ketones: Trajectory Analysis"*, 2819-2820 (1987). Copyright 1987 American Chemical Society.

Figure 6.14: Reprinted with kind permission of VCH Verlagsgesellschaft and Professor G. Helmchen from *"Concerning the Mechanism of the Asymmetric Diels-Alder Reaction: First Crystal Structure Analysis of a Lewis Acid Complex of a Chiral Dienophile"*, by T. Poll; J. O. Metter; G. Helmchen, *Angew. Chem. Int. Ed. Engl.* **1985**, *24*, 112-114.

Figure 7.12 (lower part) and Figure 7.13c: Reprinted from *Tetrahedron,* vol. 30; K. N. Houk; N. G. Rondan; Y.-D. Wu; J. T. Metz; M. N. Paddon-Row, *"Theoretical Studies of Stereoselective Hydroborations"*, pages 2257-2274 (1984), with kind permission from Elsevier Science Ltd., The Boulevard, Langford Lane, Kidlington OX5 1GB, UK.

Chapter 1

Introduction

1.1 Why we do asymmetric syntheses

The world is chiral [1-3]. Most organic compounds are chiral. Chemists working with perfumes, cosmetics, nutrients, flavors, pesticides, vitamins, and pharmaceuticals, to name a few examples [4-8], require access to enantiomerically pure compounds. In the pharmaceutical industry [8,9], more than half of the drugs available on the market in 1990 (worldwide) were chiral, and roughly half of those were sold as a single enantiomer. But of those drugs sold as a single enantiomer, roughly 90% were natural products or semisynthetic derivatives. By contrast, nearly 90% of all chiral synthetic drugs sold at that time were racemic [10].

As our ability to produce enantiomerically pure compounds grows, so does our awareness of the differences in pharmacological properties that a chiral compound may have when compared with its enantiomer or even its racemate [11,12]. We easily recognize that all biological receptors are chiral, and as such can distinguish between the two enantiomers of a ligand or substrate. Enantiomeric compounds often have different odors or tastes [13-15].[1] Thus, it is obvious that two enantiomers should be considered different compounds when screened for pharmacological activity [8,12,16]. The demand for enantiomerically pure compounds as drug candidates is not likely to let up in the foreseeable future.

How might we obtain enantiomerically pure compounds? Historically, the best answer to that question has been to isolate them from natural sources. Hence the dependence on natural product isolation for the production of enantiomerically pure pharmaceuticals. Derivatization of natural products or their use as synthetic starting materials has long been a useful tool in the hands of the synthetic chemist, but it has now been raised to an art form by practitioners of the "Chiron" approach to total synthesis, wherein complex molecules are dissected into chiral fragments that may be obtained from natural products [17-25].

So if the objective is to obtain an enantiomerically pure compound, one has a choice to make: synthesize the molecule in racemic form and resolve it [26], find a plant or bacterium that will make it for you, start with the appropriate chiron (but beware of racemic or partly racemic natural products), or plan an asymmetric synthesis. Among the factors to consider in weighing the alternatives are the amount of material required, the cost of the starting materials, length of synthetic plan, etc., factors which have long been important to synthetic design [27,28]. For the purposes of biological evaluation, it may be *desirable* to include a resolution so that one synthesis will provide both enantiomers. But for the production of a single

---

[1] For example, the enantiomers of limonene smell and taste like oranges or lemons, the enantiomers of phenylalanine taste bitter or sweet, the enantiomers of carvone taste like spearmint or caraway, all depending on the absolute configuration.

enantiomer, resolution will have a maximum theoretical yield of 50% unless the unwanted enantiomer can be recovered and recycled. In most cases, the chiron and enzyme/organism approaches will be restricted to the production of only one enantiomer by a given route, notwithstanding the talent of some investigators to produce both enantiomers of a target from the same chiral starting material. The chiron approach consumes the chiral natural product, while the asymmetric synthesis routes (historically) do not. However, the lines of distinction between these categories are fading.

1.2 What is an asymmetric synthesis?

The most quoted definition of an asymmetric synthesis is that of Marckwald [29]:

'Asymmetrische' Synthesen sind solche, welche aus symmetrisch constituirten Verbindungen unter intermediärer Benutzung optisch-activer Stoffe, aber unter Vermeidung jedes analytischen Vorganges, optisch-activ Substanzen erzeugen.[2]

In modern terminology, the core of Marckwald's definition is the conversion of an achiral substance into a chiral, nonracemic one by the action of a chiral reagent. By this criterion, the chiron approach falls outside the realm of asymmetric synthesis. Marckwald's point of reference of course, was biochemical processes, so it follows that modern enzymatic processes [30-32] are included by this definition. Marckwald also asserted that the nature of the reaction was irrelevant, so a self-immolative reaction or sequence[3] such as an intermolecular chirality transfer in a Meerwein-Pondorf-Verley reaction would also be included:

Interestingly, the Marckwald definition is taken from a paper that was rebutting a criticism [33] of Marckwald's claim to have achieved an asymmetric synthesis by a group-selective decarboxylation of the brucine salt of 2-ethyl-2-methylmalonic acid [34,35]:

Thus from the very beginning, the definition of what an asymmetric synthesis might encompass, or even if one was possible, has been a matter of debate. On the latter point, the idea that a chemist could synthesize something in optically active form from an achiral precursor was doubted in some circles, even in Marckwald's

---

2 'Asymmetric' syntheses are those which produce optically active substances from symmetrically constituted compounds with the intermediate use of optically active materials, but with the avoidance of any separations.

3 Self-immolative processes are those that generate a new stereocenter at the expense of an existing one, either in a single reaction or in a sequence whereby the controlling stereocenter is deliberately destroyed in a subsequent step.

time. That doubt, expressed in a published lecture in 1898 [36] was one of the last tenets of the vitalism theory to die.

When one considers that historically, isolated natural products were the major source of chiral nonracemic chemicals, and when the labor of extraction was also considered, it is no wonder that recovery of the chiral reagent was important. Nowadays, the source of a reagent is the nearest chemical catalog, and it makes little difference if the chiral substance, whether to be used as a chiron or as a recoverable or consumable reagent, is obtained by isolation, enzymatic synthesis, or resolution. For the purposes of synthetic planning, the most important variable is the cost of the process relative to the value of the product. Also, the scale of the planned synthesis must be considered: an affordable cost for the preparation of a few grams of product may not be feasible for the production of several hundred kilos. It may also be important to consider the availability of either enantiomer of the chiral reagent.

In 1974, Eliel proposed the following criteria for judging an asymmetric synthesis [37]:

1. The synthesis must be highly stereoselective.
2. If the chiral auxiliary (adjuvant) is an integral part of the starting material, the chiral center (or other chirality element) generated in the asymmetric synthesis must be readily separable from the auxiliary without racemization [of the new stereocenter].
3. The chiral auxiliary or reagent must be recoverable in good yield and without racemization.
4. The chiral auxiliary or catalyst should be readily and inexpensively available in enantiomerically pure form.

Several comments [38] are appropriate regarding these guidelines. The first is obviously the most important, and is universally applicable to all synthetic strategies. Chromatographic or other purification techniques often provide a practical solution to low selectivity in unfavorable cases, however. Points 2 and 3 address auxiliary-based techniques, and are predicated on the higher cost of chiral reagents. Condition 3 is less important when a chiral catalyst has a high turnover number or when the chiral auxiliary is very inexpensive. Point 4 also becomes less important in catalytic processes as the turnover number increases.

The simplicity of the Marckwald definition has been its most enduring feature, but our understanding of structure and mechanism has evolved since Marckwald's time,[4] and spectroscopic and chromatographic techniques have displaced polarimetry as the primary determinant of enantiomeric purity. In light of these

---

[4] As a point of reference, consider that in Marckwald's time the van't Hoff - le Bel theory of tetrahedral carbon was accepted, but what we now know as an sp^2 or trigonal carbon, was not. It was thought, at least by some, that the fourth site of a carbonyl carbon was an unoccupied site on a tetrahedron. For example, under the term 'asymmetric induction' in the first collective index of *Chemical Abstracts,* we find reference to a paper (**8**:3431[1]) entitled "Preparation of *l*-benzaldehyde through asymmetric induction . . ." (E. Erlenmeyer, F. Landesberger, G. Hilgendorff *Biochem. Z.* **1914**, *64,* 382-392). The formula for benzaldehyde was PhCHL.OL, where L indicates an unoccupied position.

developments, as well as the new applications of double asymmetric induction (*vide infra*) that are not addressed by the Marckwald definition, a broader definition is appropriate:

> *Asymmetric synthesis* is a reaction or reaction sequence that *selectively* creates one configuration of one or more new *stereogenic elements* by the action of a chiral reagent or auxiliary, acting on *heterotopic* faces, atoms, or groups of a substrate. The stereo-selectivity is primarily influenced by the chiral catalyst, reagent, or auxiliary, despite any stereogenic elements that may be present in the substrate.

In 1933, two short monographs [39,40] summarized virtually everything known about asymmetric synthesis and asymmetric induction. By 1971, the field was summarized in another short monograph of about 450 pages [41], but by then the art of organic synthesis was ready for a rapid advance: ten years later, a five volume treatise [42] was necessary (~1800 pages). The literature continues to grow at such a rate that comprehensive coverage is increasingly difficult. Although not restricted to asymmetric synthesis, the recent nine volume *Comprehensive Organic Synthesis* encyclopedia [43] subtitled 'Selectivity, Strategy, and Efficiency in Modern Organic Chemistry' is ~7000 pages. In 1995, a 6000-page treatise entitled *Stereoselective Synthesis* and advertised as "the whole of organic stereochemistry" appeared as part of the Houben-Weyl series [44].

It is the primary aim of the present work to provide a *concise* analysis of the stereochemical features of transition states in a variety of reaction types. These control elements are only partly understood at present, but as the intra- and intermolecular forces that govern transition state assemblies come further into focus, the principles outlined in this book will be refined and improved. The ultimate (attainable?) goal is clear: the production of any relative and absolute configuration of one or more stereogenic units through the use of chiral catalysts that do not require consideration of chirality elements extant in the substrate.

1.3 Stereoselectivity: intraligand *vs.* interligand asymmetric induction

It is the primary goal of this book to analyze the factors that influence stereoselectivity when one stereoisomer predominates over others. For illustrative purposes, consider the addition of a nucleophile to a carbonyl. The faces of unsymmetric carbonyls are heterotopic, either enantiotopic (if there are no stereo-centers in the molecule) or diastereotopic (if there are), as shown in Figure 1.1 (see also glossary, Section 1.6). In order to achieve a predominance of one stereoisomer (enantiomer or diastereomer) over the other, the transition states resulting from attack from the heterotopic *Re* or *Si* faces must be diastereomeric. This will be the case if either the carbonyl compound or the reagent (or both) is (are) chiral.

It is useful to classify stereoselective reactions as a preliminary step to identify-ing the factors influencing stereoselectivity. Metals are intimately involved in almost all highly stereoselective reactions,[5] so our classification will begin there.

---

[5] For a review of stereoselective reactions of free radicals, see ref. [45,46].

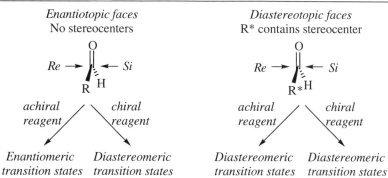

Figure 1.1. Additions to heterotopic faces of an aldehyde.

The first step is to examine the ligands on the metal. The most important ligand will be the substrate; other ligands may simply be solvent, monodentate or bidentate "spectator" ligands, or a ligand that is involved in the reaction (a "player"). Take the addition of a Grignard reagent to an aldehyde as an example (Scheme 1.1). The Grignard reagent (ignoring Schlenck processes) has an organic ligand (*e.g.*, Me), a halide (X), and several bound solvent molecules (L_n) in its coordination sphere. Addition to a carbonyl does not occur, however, until the aldehyde coordinates to the magnesium. Notice that four types of ligand are now apparent in the intermediate shown in brackets: the aldehyde substrate, the methyl group that adds to the carbonyl, the halide, and the bound solvent. The aldehyde and the methyl are the "players" while the halide and the solvent ligands are "spectators".

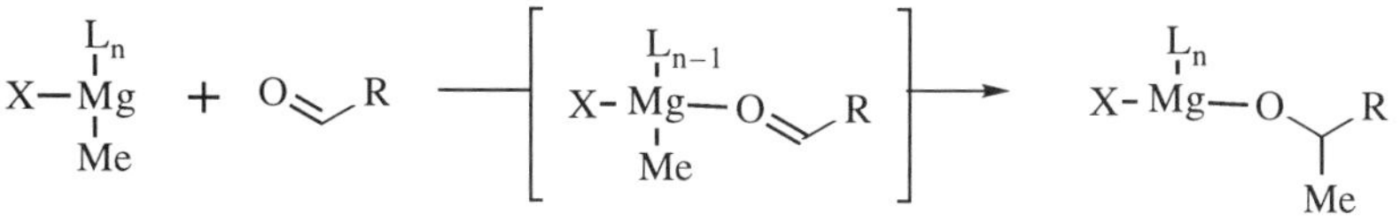

Scheme 1.1. The addition of a Grignard reagent to an aldehyde.

Several examples of carbonyl additions (see also Chapter 4) serve to illustrate a further ligand classification that has tremendous bearing on stereoselectivity, and which illustrates in a nutshell the history of asymmetric synthesis while also pointing us toward its future. In the 1950s, Cram examined the influence of an adjacent stereocenter on the stereoselectivity of nucleophilic additions to carbonyls [47,48]. In the example illustrated (Scheme 1.2a, taken from Eliel's later work, [49], one diastereomer is formed to the near exclusion of the other. The chirality sense (*R/S*) of the new stereocenter is determined by the chirality center adjacent to the carbonyl. Both the "old" and the "new" stereocenters are within the same ligand on the metal, however, so the asymmetric induction is *intraligand*. Later, the auxiliary shown in Scheme 1.2b was developed for use in an asymmetric synthesis of α-hydroxy aldehydes. Here, the oxathiane fragment is *removed* after directing the selective formation of one of two possible diastereomers [50]. Note, however, that this example is also a case of *intraligand* asymmetric induction, and that both of these examples have diastereomeric transition states because the carbonyl-containing substrate is chiral. Scheme 1.2c shows the addition of a Grignard to an ketone, after modifying the reagent with a chiral diol. Here, since the ketone is achiral, the two

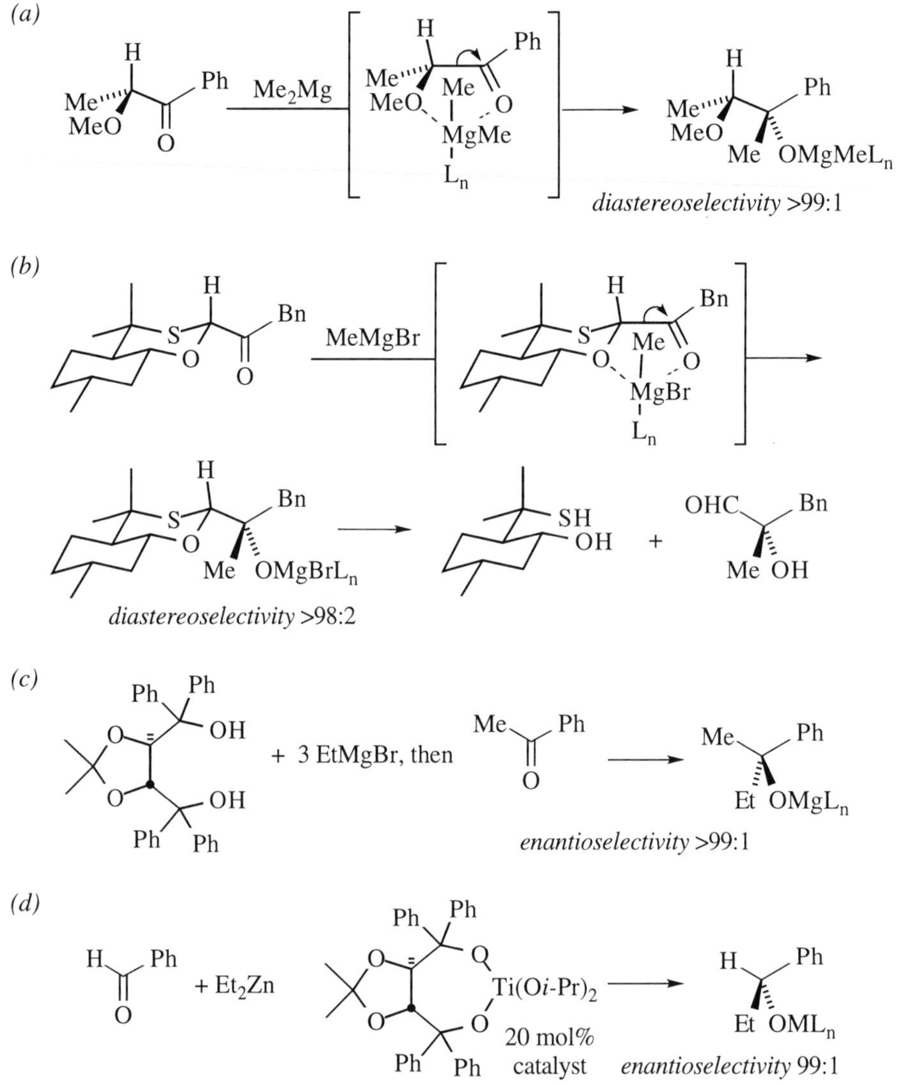

(a)

diastereoselectivity >99:1

(b)

diastereoselectivity >98:2

(c)

+ 3 EtMgBr, then

enantioselectivity >99:1

(d)

+ Et$_2$Zn

Ti(Oi-Pr)$_2$
20 mol%
catalyst   *enantioselectivity 99:1*

Scheme 1.2. Intraligand *vs.* interligand asymmetric induction: *(a)* Diastereoselective addition *via* Cram's cyclic model ([49], *cf.*, Section 4.2). *(b)* Asymmetric synthesis of a pure enantiomer *via* diastereoselective addition to a carbonyl with a chiral auxiliary [50]. *(c)* Enantioselective addition of ethyl Grignard to an aldehyde using a chiral ligand on magnesium [51]. *(d)* Catalytic enantioselective addition of diethylzinc to an aldehyde using a chiral ligand on titanium [52].

faces of the carbonyl are enantiotopic, and the transition states are rendered diastereomeric by the chiral diol ligand. Although the details of the reaction are unknown, the only chirality element present in the reactants are in the diol ligand, making this *interligand* asymmetric induction.[6]

---

[6] In principle a metal atom may also be stereogenic (and therefore influence the stereoselectivity), but a separate category is not needed since chiral nonracemic metal complexes containing achiral ligands are rare in asymmetric synthesis [53].

The mental process of "removing" the stereocenter from the substrate and putting it on another ligand of the metal allows the introduction of the element of asymmetric catalysis [54,55], as shown by the diethylzinc addition in Scheme 1.2d [56]. Here, the chiral reagent is used in less than stoichiometric quantities. Catalytic processes are, of course, more cost-effective than stoichiometric processes, and have the added advantage of decreasing the environmental impact of disposing of (or recycling) byproducts produced in stoichiometric quantities.

These four examples illustrate the progress made in stereoselective reactions in the last few decades, and which has evolved through several distinct phases: (*i*) Diastereoselective synthesis by addition of nucleophiles to carbonyls having a neighboring stereocenter; (*ii*) the extension of the same notion to the synthesis of a single enantiomer *via* diastereoselection and auxiliary removal; and (*iii*) enantioselective addition to an achiral substrate by a stoichiometric reagent; and (*iv*) enantioselective addition mediated by a chiral catalyst. Extrapolation of the trend that is apparent in these simple examples points inexorably toward the goal stated at the end of the previous section: the production of new stereogenic units through the use of chiral catalysts that do not require consideration of existing chirality elements.

1.4 Selectivity: kinetic and thermodynamic control

The means by which stereoselectivity is achieved in various reactions and processes are widely variable. However, the asymmetric induction that results in any given process must fall into one of only two categories: thermodynamic or kinetic control, the latter being the more common.

Consider a starting material, **A**, that may give two possible products, **B** and **C**. Figure 1.2a illustrates how equilibration might occur to afford an equilibrium mixture of **B** and **C** by one of two possible routes. The reactions **A** → **B** and **A** → **C** might be reversible, or **A** and **C** could equilibrate by a route that does not involve **A**. Either way, the product ratio (**C/B**) is given by

$$\mathbf{C/B} = \frac{[\mathbf{C}]}{[\mathbf{B}]} = K = e^{-\Delta G/RT}, \qquad (1.1)$$

where ΔG is the free energy difference between **C** and **B**: $\Delta G = G_B - G_C$. Processes such as this are under *thermodynamic control*.

Under conditions of *kinetic control* (Figure 1.2b), the conversion of **A** into either **B** or **C** is irreversible, the relative rates of formation of each product determine the outcome, and the product ratio (**C/B**) is given by

$$\mathbf{C/B} = \frac{k_1}{k_2} = e^{-\Delta\Delta G^{\ddagger}/RT}, \qquad (1.2)$$

where k_1 and k_2 are the rate constants for the formation of **B** and **C**, respectively.

$\Delta\Delta G^{\ddagger}$ is the difference in the transition state energies for each process:

$$\Delta\Delta G^{\ddagger} = \Delta G_B^{\ddagger} - \Delta G_C^{\ddagger}, \qquad (1.3)$$

where $\Delta G_B^{\ddagger}$ and $\Delta G_C^{\ddagger}$ are the free energies of activation for the formation of **B** and **C**, respectively.

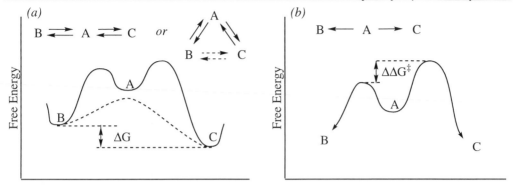

Figure 1.2. *(a)* Conversion of **A** into a mixture of **B** and **C** under *thermodynamic control.* Note that **B** and **C** may equilibrate *via* **A** or by another route (dashed line). *(b)* Conversion of **A** into **B** and **C** under *kinetic control.*

Two types of selectivity may used to establish new stereogenic units in a molecule: *diastereoselectivity* and *enantioselectivity.* In *diastereoselective* reactions, either kinetic or thermodynamic control are possible, but in *enantioselective* reactions, the products are isoenergetic and only kinetic control is possible.[7]

Equations 1.1 and 1.2 establish the exponential dependence of selectivity on free energy and temperature. Figure 1.3 shows plots of these equations at three different temperatures.[8] The curves of Figure 1.3 illustrate a number of points:

1. The steepest part of the curves occurs in the region where the selectivity (K or k_1/k_2) is ≤10. Because of the exponential relationship, a doubling of the free energy difference at 10:1 will increase the selectivity to 100:1.

2. The total energy differences that afford 100:1 selectivity are not great. For comparison, recall that ΔG for the cis and trans isomers of the dimethyl-cyclohexanes or between the axial and equatorial conformations of methyl-cyclohexane is 1,600-1,700 cal/mole.

3. In the "flat" part of the curves, small differences in energy will produce large differences in selectivity. For example at 0°, an increase in ΔG or ΔΔG‡ of 873 cal/mole increases selectivity from 20:1 to 100:1. By comparison, ΔG for the gauche and anti forms of butane is 900 cal/mole.

4. At lower temperatures, the selectivity curves "flatten out" more quickly. Thus for a given process, subtle changes in the stereochemical control elements will usually have a greater influence if the reaction is carried out at low temperature. Note that this does not necessarily mean that lowering the temperature of a reaction will result in increased selectivity (*vide infra*).

5. Selectivities may be expressed in any of several ways: as K or k_1/k_2, % enantiomer excess (ee), % diastereomer excess (de), or the percentage of the major enantiomer (% es) or diastereomer (% ds).[9] It is worth keeping these

---

[7] Unless the reaction is conducted in a nonracemic chiral solvent.

[8] Strictly speaking, these equations and the curves in Figure 1.3 are valid only for a unimolecular reaction in the gas phase, but to a first approximation they serve as useful tools for our purposes.

[9] Use of enantiomer excess (the excess of one enantiomer over its racemate) to describe *selectivity* is inappropriate since it implies that the minor enantiomer and an equal amount of the major enan-

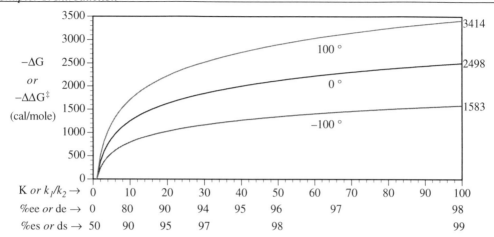

Figure 1.3. The relationship between selectivity and free energy (for the competitive formation of two products) at −100°, 0°, and 100° C. The free energy values for product ratios of 100:1 are labeled on the right ordinate (cal/mole).

parallel scales in mind when evaluating selectivities, as all are used interchangeably in the literature. Since routine purification techniques can often remove 5 to 10% of isomeric impurities, selectivities higher than 95% (ds or es) may not be required from a practical standpoint.

Regarding the effect of temperature on selectivity, reliance on equations such as 1.1 and 1.2 can be misleading, since free energy itself is temperature dependent:

$$G = H - T\Delta S. \tag{1.4}$$

Combination of equations 1.2 and 1.4 [38] gives

$$\frac{k_1}{k_2} = \left(e^{-\Delta\Delta H^{\ddagger}/RT}\right)\left(e^{\Delta\Delta S^{\ddagger}/R}\right), \tag{1.5}$$

where $\Delta\Delta H^{\ddagger}$ and $\Delta\Delta S^{\ddagger}$ are the differences in enthalpy and entropy of activation for the formation of **B** and **C**, defined as was $\Delta\Delta G^{\ddagger}$ in equation 1.3.[10] Equation 1.5 shows that only the enthalpy term is temperature dependent. An example of the effect this can have is shown in the additions of organolithiums to lactaldehyde shown in Scheme 1.3 [57]. The addition of methyllithium has $\Delta\Delta H^{\ddagger} = -260$ cal/mole and $\Delta\Delta S^{\ddagger} = 0$. Since $\Delta\Delta H^{\ddagger}$ is negative, the exponent of the first term is positive and lowering the temperature from 308° to 208° results in an increase in k_1/k_2. In contrast, the addition of phenyllithium has $\Delta\Delta H^{\ddagger} = +340$ cal/mole and $\Delta\Delta S^{\ddagger} = +28$ e.u. With a positive $\Delta\Delta H^{\ddagger}$ and $\Delta\Delta S^{\ddagger}$, **C** is favored by enthalpy and **B** is favored by entropy. In this case, the reaction is entropy controlled: since the exponent of the first term is negative, lowering the temperature decreases the preference for **B**;

---

tiomer are formed in a random process. In this book, percent enantioselectivity (% es) and percent diastereoselectivity (% ds) will be used to describe selectivity. These terms suffer the drawback that a process that forms two stereoisomers randomly is nevertheless 50% "selective." On the other hand, few of the processes discussed in this book are stereorandom, and the correlation with the familiar concept of "% yield" has obvious advantages.

10 $\Delta\Delta H^{\ddagger} = \Delta H^{\ddagger}_{\mathbf{B}} - \Delta H^{\ddagger}_{\mathbf{C}}$, negative if **B** is favored; $\Delta\Delta S^{\ddagger} = \Delta S^{\ddagger}_{\mathbf{B}} - \Delta S^{\ddagger}_{\mathbf{C}}$, positive if **B** is favored.

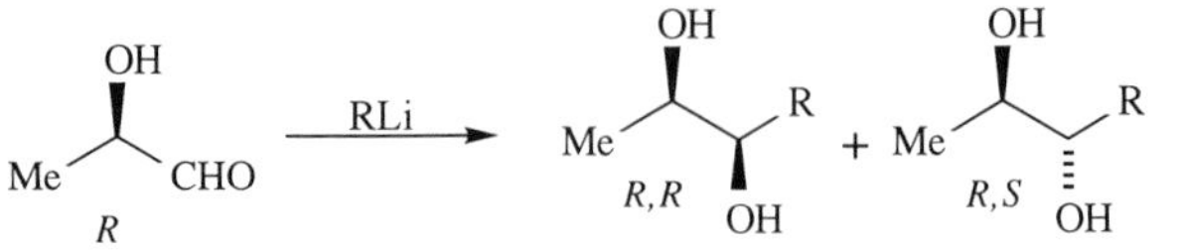

R	R,R/R,S (°K)
Me	60:40 (308)
Me	65:35 (208)
Ph	73:27 (308)
Ph	68:32 (213)

Scheme 1.3. Effect of temperature on addition of organolithiums to *R*-lact-aldehyde [57].

nevertheless, the entropic preference prevails and **B** is still the major product, albeit in lower amount [57]. Thus, although lowering the temperature often increases selectivity, it does not necessarily do so in all cases.

1.5 Single and double asymmetric induction

For the purposes of illustration, Figure 1.4 illustrates two reaction types in generic form. Single asymmetric induction occurs if a single chirality element directs the selective formation of one stereoisomer over another by selective reaction at one of the heterotopic (*Re/Si*) faces of a trigonal atom.

*Nucleophilic addition
to carbonyls*

*Electrophilic addition
to enolates*

Figure 1.4. Two types of reactions that distinguish hetero-topic faces.

An interesting circumstance develops when two of these techniques are combined in the same reaction, such as when the second reactant also contains a chirality element (*e.g.*, when a chiral nucleophile reacts with a chiral carbonyl compound): the chirality elements of each reactant may influence stereoselectivity either in concert or in opposition. This phenomenon is known [58,59] as double asymmetric induction. A simple illustration is shown in Scheme 1.4 and involves the reaction of

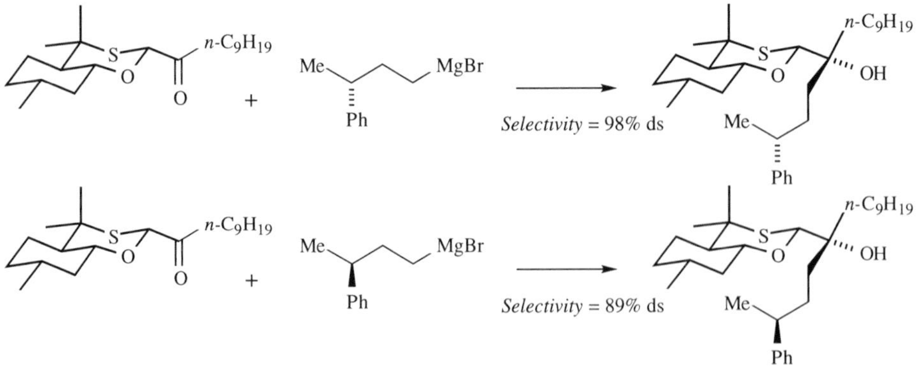

Selectivity = 98% ds

Selectivity = 89% ds

Scheme 1.4. Double asymmetric induction: changing the absolute configuration of a chiral nucleophile affects the stereoselectivity of addition to a chiral ketone [60].

the two enantiomers of a chiral Grignard reagent with a chiral ketone [60].[11] Note the difference in diastereoselectivity observed for the two reactions, clearly resulting from the change in absolute configuration of the remote stereocenter of the Grignard.

In order to understand the phenomenon of double asymmetric induction, we need to have a clear picture of the inherent selectivities of each of the chiral partners in closely related single asymmetric induction processes. Consider for example the kinetically controlled aldol addition reactions shown in Scheme 1.5 [58].[12] The first two illustrated reactions are examples of single asymmetric induction with inherently low selectivities. Scheme 1.5a is the reaction of a chiral Z(O)-enolate with an achiral aldehyde [61], and illustrates the *Si*-facial preference of the *S* enantiomer of the enolate of 78:22. In Scheme 1.5b, an achiral enolate that is structurally similar to the chiral enolate of Scheme 1.5a is allowed to react with a chiral aldehyde [62]. The 73:27 product ratio reflects the *Re*-facial preference of the aldehyde.[13] Note that the absolute configuration of the new stereocenters in the

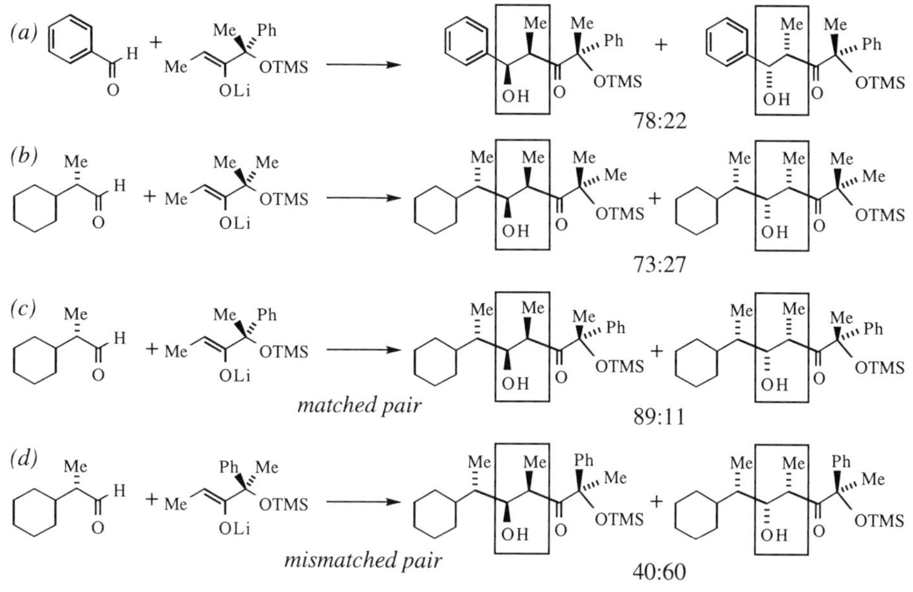

Scheme 1.5. Examples of single and double asymmetric induction in the aldol addition reaction. *(a)* Reaction of a chiral enolate and an achiral aldehyde; *(b)* Reaction of an achiral enolate with a chiral aldehyde; *(c)* Matched pair double asymmetric induction with a chiral enolate and a chiral aldehyde; *(d)* Mismatched pair double asymmetric induction with a chiral enolate and the aldehyde enantiomeric to that shown in *(a)*. (After ref. [58]).

---

[11] The selectivity of the reactions illustrated in Scheme 1.4 are rationalized by Cram's cyclic model, discussed in Section 4.2.

[12] In these examples, two stereocenters are created, but only two of the four possible stereoisomers are formed. As will be explained in detail in Chapter 5, the two *syn* isomers are produced stereoselectively from the Z(O)-enolate. (See glossary, section 1.6, for definition of this term.)

[13] Obviously if the absolute configuration of either of the chiral reactants in Schemes 1.5a or 1.5b were reversed, the absolute configuration of the new stereocenters would also be reversed *i.e.*, the enantiomers of the illustrated products would be produced in the same ratio.

major products are the same. Since both chiral reactants, the enolate of the first reaction and the aldehyde of the second both prefer the same absolute configuration in the addition product, we may expect that reaction of the chiral enolate with the chiral aldehyde would afford product having the same absolute configuration.

When the *S*-enolate and the *S*-aldehyde (Scheme 1.5c) were allowed to react, the expected product was indeed formed, but the selectivity was higher (89:11) than in either of the previous examples because the inherent selectivities of the two chiral species are mutually reinforcing [58]. This is an example of *matched pair* double asymmetric induction. A *mismatched* double asymmetric induction would result from reversing the absolute configuration of either of the two chiral reactants. For example, when the *S*-aldehyde and now the *R*-enolate (Scheme 1.5d) were allowed to react, the two products were formed in a ratio of 40:60 [58]. The higher selectivity of the enolate (78% ds) over the aldehyde (73% ds) is manifested in the absolute configuration obtained as the major isomer in Scheme 1.5d.

With this example in mind, let us reexamine the principles of selectivity presented earlier and apply them to the case of double asymmetric induction. In Figure 1.2b, two possible products are formed under kinetic control. This reaction diagram is applicable to the examples of Schemes 1.5a and 1.5b, in that *a single chirality element* operates to render the two transition structures diastereomeric. Now imagine what the effect of a second chirality element might be. Figure 1.5a illustrates the case of a matched pair: the second chirality element increases $\Delta\Delta G^{\ddagger}$ by lowering the energy of the already favored transition state (A $\rightarrow$ B) and/or raising the energy of the disfavored one. The previously favored isomer is formed with increased selectivity. Figure 1.5b illustrates the mismatched case, wherein the second chirality element decreases $\Delta\Delta G^{\ddagger}$ by increasing the energy of the favored transition state and/or decreasing the energy of the disfavored one. In this example, the second chirality element decreases $\Delta\Delta G^{\ddagger}$ but does not change its sign. Obviously, additional perturbation of the transition states could reduce $\Delta\Delta G^{\ddagger}$ to zero, or reverse the selectivity by changing the sign of $\Delta\Delta G^{\ddagger}$.

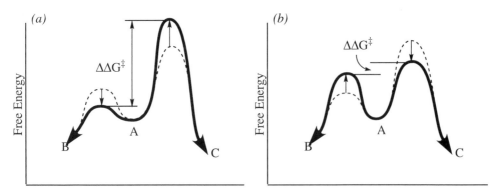

Figure 1.5. Double asymmetric induction. The dashed lines represent a hypothetical case of single asymmetric induction. *(a)* Matched pair: $\Delta\Delta G^{\ddagger}$ is increased by the influence of a second chirality element; *(b)* Mismatched pair: $\Delta\Delta G^{\ddagger}$ is decreased by the influence of a second chirality element.

There are two important lessons here. The first is that a matched pair will afford higher selectivities than either chiral reactant would afford on its own. The second lesson is more subtle. In considering the two single asymmetric induction reactions, suppose that one of the chiral reagents is much more selective than the other. In this instance, the mismatched pair may still be a highly selective reaction. Figure 1.6 illustrates an energy diagram wherein the stereoselectivity due to the second chirality element completely overwhelms that of the first. The dotted lines indicate a preference, in single asymmetric induction, for product B. Under the influence of a much more highly selective reagent, the double asymmetric induction (bold line) favors A by "changing the sign" of $\Delta\Delta G^{\ddagger}$. Even though this is a mismatched pair, it still may be very selective. In such cases, the chiral reagent is the primary determinant of the absolute configuration of the new stereocenter(s) in the product!

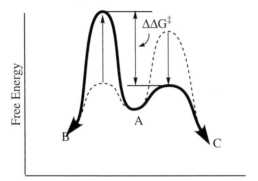

Figure 1.6. Reagent-based stereocontrol in double asymmetric induction.

In a general sense, if the species having an overwhelmingly higher "inherent selectivity" is the chiral auxiliary, chiral reagent or chiral catalyst, and the less selective species is a synthetic intermediate being carried on to a target, then *the reagent can be used to determine the absolute configuration of the product, independent of the chirality sense and bias of the substrate.* This concept is known as *"Reagent-Based Stereocontrol"* [58,59]. As we will see throughout this book, a number of reagents deliver high enough selectivities to achieve this important goal.

Two examples of such processes are shown in Scheme 1.6. One is the titanium TADDOLate-catalyzed addition of diethylzinc to myrtenal (see Section 4.3, [52]; the other is the Sharpless asymmetric epoxidation (see Section 8.2.2, [58,63]). In both cases, the diastereoselectivity for the reaction of the substrate with an achiral reagent is low (65-70% ds), while the catalysts have enantioselectivities of >95% with achiral substrates. In these cases of double asymmetric induction, the catalyst completely overwhelms the facial bias of the chiral substrate.

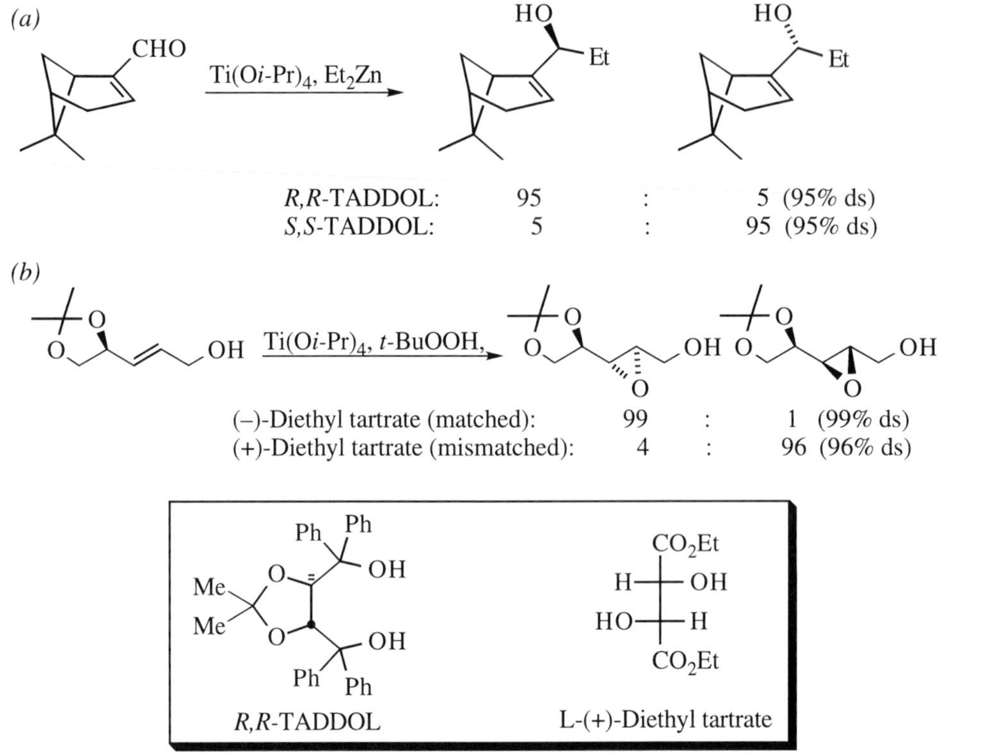

(a)

CHO

Ti(O*i*-Pr)$_4$, Et$_2$Zn →

HO
Et

HO
Et

R,R-TADDOL: 95   :   5 (95% ds)
S,S-TADDOL: 5   :   95 (95% ds)

(b)

OH   Ti(O*i*-Pr)$_4$, *t*-BuOOH, →   OH   OH

(−)-Diethyl tartrate (matched):   99   :   1 (99% ds)
(+)-Diethyl tartrate (mismatched):   4   :   96 (96% ds)

Ph Ph
O   OH
Me
Me   O
OH
Ph Ph

R,R-TADDOL

CO$_2$Et
H —— OH
HO —— H
CO$_2$Et

L-(+)-Diethyl tartrate

Scheme 1.6. Matched and mismatched double asymmetric induction demonstrating "Reagent-Based Stereocontrol": (a) The diethylzinc addition catalyzed by titanium TADDOLates (Chapter 4, [52]). (b) The Sharpless asymmetric epoxidation (Chapter 8, [58,63].

1.6 Glossary of stereochemical terms[14]

A values: The free energy difference $(-\Delta G°)$ between equatorial and axial conformations of a substituted cyclohexane, positive if equatorial is preferred. For a compilation of values, see ref. [64], and references cited therein.

$A^{1,2}, A^{1,3}$ *strain:* see *allylic strain.*

Absolute asymmetric synthesis: A synthesis in which *achiral* reactants are converted to *nonracemic, chiral* products, and where the *enantioselectivity* is induced only by an external force such as circularly polarized light in a photochemical reaction [65].

Absolute configuration: The arrangement in space of the ligands of a *stereogenic unit,* which may be specified by a stereochemical descriptor such as *R* or *S, D* or *L, P* or *M.* See also *chirality sense, chirality element, stereogenic element*

Achiral: See *chiral.*

Achirotopic: See *chirotopic.*

Allylic strain: The destabilization of a molecule, or an individual conformation, by *van der Waals* repulsion between substituents on a double bond and those in an allylic position [66]. Two types have been identified (see bold bond in figure): $A^{1,2}$ strain occurs between substituents on an allylic carbon and the adjacent sp^2 carbon. $A^{1,3}$ strain occurs between substituents on an allylic carbon and the distal sp^2 carbon. The latter effect can be quite strong [67]. Originally [66], the terms were defined in the context of cyclohexane derivatives, but more recently the effects have been recognized as important factors in conformational dynamics of acyclic systems [67].

$A^{1,2}$ *strain:* $A^{1,3}$ *strain:*

R   R   R   R

Alternating symmetry axis (S_n): An axis about which a rotation by an angle of $360/n$, followed by a reflection across a plane perpendicular to the axis results in an entity that is indistinguishable from (superimposable on) the original. Also called a rotation-reflection axis. See also *symmetry axis.*

Alpha (α), beta (β): Stereodescriptors used commonly in carbohydrate [68] and steroid [69] nomenclature to describe *relative configuration.* In steroids, "any [substituent] that lies on the same side of the ring plane as the C_3-hydroxyl group of cholesterol [see illustration] is described as β-oriented, and the carbon to which the group is joined has the β-configuration. The opposite orientations and configurations are designated α" [69]. The α, β nomenclature is often extended to other ring systems, but a reference stereocenter must be

---

[14] Note that other terms defined in this glossary are italicized.

specified, either explicitly or by convention (see for example ref. [70]). Often, reference is made to a 2-dimensional drawing in which a reference plane is specified. If the reference plane is horizontal, β is above and α is below the plane, as illustrated below. If the plane is vertical, β is toward the viewer.

In carbohydrates, the β-anomer has the C_1-hydroxyl or alkoxyl group on the opposite side of a *Fischer projection* as the substituent (∗) that defines D or L (see *Fischer-Rosanoff convention*); this need not be the position at which the ring is closed. The α-anomer has the two on the same side [68].

steroids

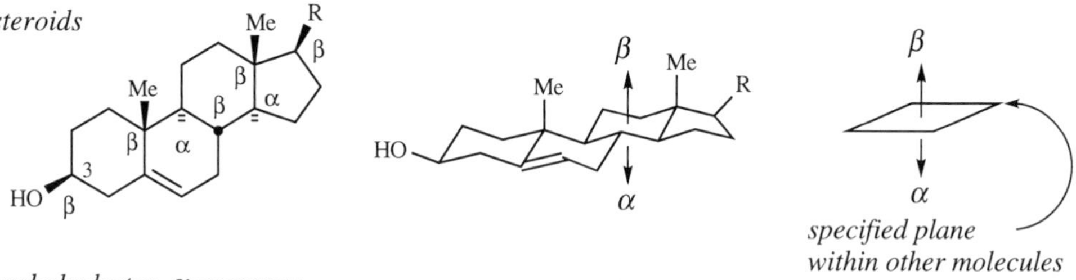

specified plane within other molecules

carbohydrates, α anomers:

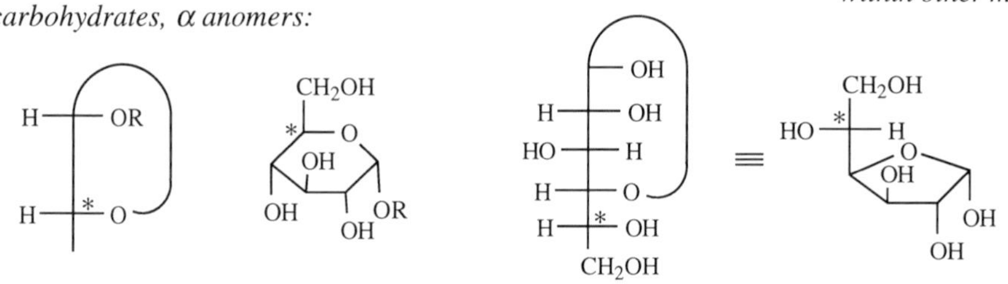

carbohydrates, β anomers:

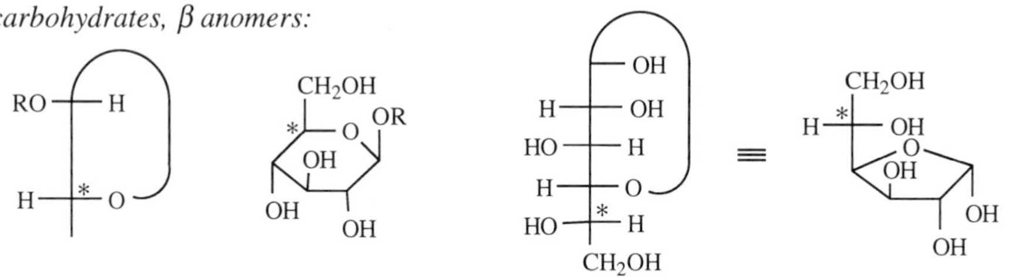

Angle strain: Destabilization of a molecule due to a variation of bond angles from "optimal" values (109° 28' for a tetrahedral atom). Also called *Baeyer strain*.

Anomeric effect: Originally, the unexpected stability of a C-1 alkoxy group of a glycopyranoside occupying the *axial* position. This effect is now more generally considered to be a conformational preference of an X–C–Y–C moiety for a *synclinal (gauche) conformation* (where X and Y are hetero atoms, and at least one is a nitrogen, oxygen, or fluorine). See illustration, below [71,72].

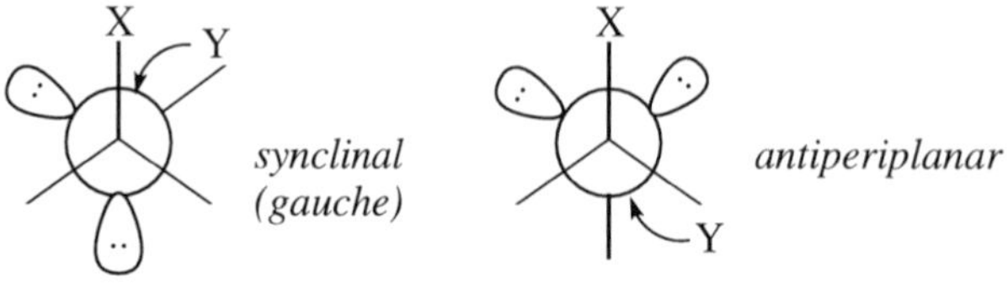

synclinal (gauche)

antiperiplanar

Antarafacial, suprafacial: In a reaction where a molecule undergoes two changes in bonding (either making or breaking), the relative spatial arrangement is suprafacial if the changes occur on the same face of the molecular fragment and antarafacial if on opposite faces [73].

Anti: See *torsion angle; syn, anti.* Also used to describe *antarafacial* addition or elimination reactions [74]. Formerly used to describe the configuration of azomethines such as oximes and hydrazones (See *E, Z*).

Anticlinal: See *torsion angle.*

Antiperiplanar: See *torsion angle.*

Aracemic: Synonym for *nonracemic* [75]. See also *scalemic.*

Asymmetric: Lacking all symmetry elements, *i.e.,* belonging to symmetry point group C_1.

Asymmetric carbon atom: van't Hoff's definition for a carbon atom having four different ligands (*i.e.* Cabcd). See also *stereogenic center, stereogenic element.*

Asymmetric center: See *stereogenic center.*

Asymmetric destruction: See *kinetic resolution.*

Asymmetric induction: The preferential formation of one *enantiomer* or *diastereomer* over another, due to the influence of a *stereogenic element* in the substrate, reagent, catalyst, or environment (such as solvent). Also, the preferential formation of one configuration of a stereogenic element under similar circumstances. When two reactants of a reaction are stereogenic, the stereogenic elements of each reactant may operate either in concert (matched pair) or in opposition (mismatched pair). This phenomenon is known [58,59] as double asymmetric induction, or double diastereoselection. See Section 1.5.

Asymmetric synthesis: A reaction or reaction sequence that *selectively* creates one configuration of one or more new *stereogenic elements* by the action of a chiral reagent or auxiliary, acting on *heterotopic* faces, atoms, or groups of a substrate. The stereoselectivity is primarily influenced by the chiral catalyst, reagent, or auxiliary, despite any stereogenic elements that may be present in the substrate. See Section 1.2.

Asymmetric transformation: The conversion of a mixture (usually 1:1) of stereoisomers into a single stereoisomer or a mixture in which one isomer predominates. An *"asymmetric transformation of the first kind"* involves such a conversion without separation of the stereoisomers. An *"asymmetric transformation of the second kind"* also involves separation, such as an equilibration accompanied by selective crystallization of one stereoisomer [76]. The terms "first- and second-order asymmetric transformations" to describe these processes are inappropriate. See also *stereoconvergent.*

Atropisomers: Stereoisomers arising from *restricted rotation* around a single bond (*i.e., conformers*), with a high enough rotational barrier that the isomers can be isolated (16 - 20 kcal/mole at room temperature), such as *ortho-*

G
L
O
S
S
A
R
Y

disubstituted biaryls [77]. The chirality sense of a conformation may be described using the *P, M* system.

Axial, equatorial: Bonds or ligands of a cyclohexane (or saturated 6-membered heterocycle) chair conformation. The axial bonds are parallel to the C_3 (S_6) axis of cyclohexane (or the corresponding position of a heterocycle), and each equatorial bond is parallel to two of the ring bonds. In a cyclohexene, the corresponding allylic bonds or ligands are called pseudoaxial (ax') and pseudo-equatorial (eq'). In a trigonal bipyramidal structure, the three ligands in a plane with the central atom are also known as equatorial.

Axis of chirality: See *chirality element, stereogenic axis, stereogenic element.*

Baeyer strain: See *angle strain.*

Bisecting and eclipsing conformations: In a structure with the grouping R₃C–C=X, the conformation in which a torsion angle R-C-C=X is *antiperiplanar*, and the torsion angles to the other two R groups is equal or nearly so is the *bisecting* conformation. The conformation in which a torsion angle R-C-C=X is *synperiplanar* is called *eclipsing*.

bisecting conformation

eclipsing conformation

Boat: See *chair, boat, twist;* and *half-chair, half-boat.*

Bond opposition strain: See *eclipsing strain.*

Bowsprit, Flagpole: In the cyclohexane boat conformation the ligands on the two carbons that are out of the plane of the other four. Endocyclic ligands are *flagpole,* exocyclic ligands are *bowsprit.*

Bürgi-Dunitz trajectory: The angle of approach of a nucleophile toward a carbonyl carbon, 107° (probably more accurately 105±5°) [78-80]. See Section 4.1.

The Bürgi-Dunitz trajectory

Cahn-Ingold-Prelog method: See *CIP method.*

CDA, chiral derivatizing agent: A reagent of known enantiomeric purity that is used for derivatization and analysis of enantiomer mixtures by spectroscopic or chromatographic means. See Section 2.3.1.

Center of chirality: See *stereogenic center.*

Center of symmetry, center of inversion (i): A point in an object that is the origin of a set of Cartesian axes, such that when all coordinates describing the object (x, y, z) are converted to $(-x, -y, -z)$, an identical entity is obtained. Equivalent to a two-fold *alternating axis* (S_2).

Centers of inversion (•)

Chair, boat, twist-boat: The cyclohexane *conformation* (point group D_{3d}) in which carbons 1, 2, 4, and 5 are coplanar and atoms 3 and 6 are on opposite sides of the plane is the *chair.* When atoms 3 and 6 are on the same side of the '1-2-4-5' plane, and also lie in a mirror plane, the conformation (point group C_{2v}) is called a *boat.* If atoms 3 and 6 are moved to either side of the boat's '3-6' mirror plane, the conformation (point group D_2) is the *twist-boat.* The chair and twist-boat conformations are at energy minima ($\Delta G = 5.6 - 8.5$ kcal/mole for cyclohexane [81]) while the boat is at a higher energy saddle-point. The *twist-boat* is sometimes called the *skew* conformation (however, see *torsion angle*). These terms are also applied to similar conformations of substituted cyclohexanes and to heterocyclic analogs. See also *axial, equatorial.*

chair	*boat*	*twist-boat*

Chair-chair inversion: See *ring reversal*

Chiral: A geometric figure, or group of points is chiral if it is nonsuperimposable on its mirror image [82]. A *chiral* object lacks all of the second order (improper) symmetry elements, σ (*mirror plane*), i (*center of symmetry*), and S (*rotation-reflection axis*). In chemistry, the term is (properly) only applied to entire molecules, not to parts of molecules. A chiral compound may be either *racemic* or *nonracemic.* An object that has any of the second order symmetry elements (*i.e.,* that is superimposable on its mirror image) is *achiral.* It is inappropriate to use the adjective *chiral* to modify an abstract noun: one cannot have a chiral opinion and one cannot execute a chiral resolution or synthesis.

Chiral auxiliary: A chiral molecule that is covalently attached to a substrate so as to render enantiotopic faces or groups in the substrate diastereotopic. After the diastereoselective reaction, the auxiliary should be removable and recoverable intact. See Section 1.2.

Chirality: The property that is responsible for the nonsuperimposabllility of an object, or a group of points, with its mirror image.

Chirality axis: See *chirality element.*

Chirality center: See *chirality element.*

Chirality element, element of chirality: A *stereogenic axis, center,* or *plane* that is reflection variant. See also *stereogenic element.*

Chirality plane: See *chirality element.*

Chirality sense: The property that distinguishes enantiomorphs such as a right or left threaded screw. For molecules, the chirality sense may be described by *R, S;* or *P, M.* See also *absolute configuration.*

Chirality transfer: Asymmetric induction in which one stereogenic element is sacrificed as another is created.

Chiroptic: Referring to the optical properties of chiral substances, such as optical rotation, circular dichroism, and optical rotatory dispersion.

Chirotopic: The property of "any atom, and, by extension, any point or segment of the molecular model, whether occupied by an atomic nucleus or not, that resides in a chiral environment" [83]. *Achirotopic* is the property of any atom or point that does not reside in a chiral environment (see also [84]). "Chirotopic atoms located in chiral molecules are enantiotopic by external comparison between enantiomers. Chirotopic atoms located in achiral molecules are enantiotopic by internal and therefore also by external comparison.... All enantiotopic atoms are chirotopic" [83].

CIP (Cahn, Ingold, Prelog) method, CIP system: The CIP sequencing rules establish the conventional ordering of ligands for the unambiguous description of absolute configuration by descriptors such as *R, S; P, M; E, Z.*

There are several steps in the method, which is abbreviated as follows (for the definitive rules, see ref. [85,86]:

Ligancy complementation: All atoms other than hydrogen are complemented to quadriligancy by providing one or two duplicate representations of any ligands which are doubly or triply bonded, respectively, and then adding the necessary number of phantom atoms of atomic number zero [85]. For example, the representation of a carbonyl is expanded as follows:

where 0 denotes a phantom atom.

Sequence rules:

0. Nearer end of axis or side of plane precedes further.

1. Higher atomic number precedes lower.

2. Higher atomic mass precedes lower.

3. Cis (*Z*) precedes trans (*E*). Some special cases require the following qualification [86]: when two ligands (indistinguishable by rules 1 and 2) differ by one having the ligand of higher rank in a cis position (*Z*) to the core of the stereogenic unit, and the other in a trans position (*E*), the former takes precedence.

4. Like pair precedes unlike pair. (For a listing of like and unlike pairs, see *l,u* in this glossary).

5. *R* precedes *S;* and *M* precedes *P*

To implement the sequence rules, it is useful to construct a digraph of the ligands to be compared, as shown below. The ligands of the proximal atoms (1 and 2) are placed in the digraph such that 11 has precedence over 12, 12 over 13, 21 over 22, etc. Another layer of ligands, labeled 111, 112, ... 233, could be constructed if necessary (only one such set is shown below). In implementing the sequence rules, 1 is first compared with 2. If there is no difference, 11 is compared with 21, then 12 with 22, etc., until a decision is reached. If comparison of ligands in the next sphere is necessary, the branches of highest priority are followed [85,86].

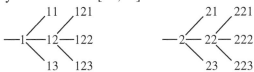

For the vast majority of cases, *CIP* rank can be determined using only ligancy complementation and sequence rule 1. The rules result in the following (descending) sequence of CIP rank for several common functional groups [87]: $COOCH_3$, COOH, COPh, CHO, $CH(OH)_2$, *o*-tolyl, *m*-tolyl, *p*-tolyl, Ph, C≡CH, *t*-Bu, cyclohexyl, vinyl, isopropyl, benzyl, allyl, *n*-pentyl, ethyl, methyl, D, H.

For the assignment of CIP descriptors, see *R, S; P, M;* and *E, Z.*

cis, trans: A stereochemical prefix to describe the relationship between two ligands on a double bond or a ring: cis if on the same side, trans if on opposite sides. For alkenes, the cis-trans nomenclature can be ambiguous and the *E, Z* descriptor is preferred. In a ring, the reference conformation (real or hypothetical) is planar, and approximates a circle or an oval, not a kidney. See *cis-trans isomers.*

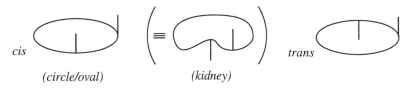

cis

(circle/oval)   *(kidney)*   *trans*

cisoid conformation (usage discouraged): See *s-cis, s-trans*.

cis-trans isomers: Stereoisomeric alkenes or cycloalkanes (or heterocyclic analogs), that differ in the position of ligands relative to a reference plane: cis if on the same side, *trans* if on opposite sides.

| *trans* | *cis* | *cis* | *cis* | *trans* |

Clinal: See *torsion angle*.

Configuration: The arrangement of atoms in space that distinguishes *stereoisomers*, excluding *conformational isomers*. *Atropisomers* are a special case of conformational isomers that, because they are isolable at room temperature, may have an absolute configuration descriptor assigned to the *stereogenic axis*. See also *absolute configuration, chirality sense, relative configuration*.

Conformation: In a molecule of a given *constitution* and *configuration,* the spatial array of atoms affording distinction between *stereoisomers* that can be interconverted by rotation around single bonds. The *chirality sense* of conformations may be specified using the *P, M* nomenclature.

Conformational analysis: The analysis of the chemical and physical properties of different conformations of a molecule.

Conformational isomers (conformers): Stereoisomers at potential energy minima (local or global) having identical *constitution* and *configuration,* which differ only in *torsion angles.*

Constitution: The description of the number and kind of atoms in a molecule and their bonding (including bond multiplicities, but not *relative* or *absolute con-figuration,* or *conformation*).

Constitutional isomers: Isomers that differ in connectivity, such as CH_3CH_2OH and CH_3OCH_3.

Cram's rule (cyclic model): A model for predicting the major stereoisomer resulting from nucleophilic addition to an aldehyde or a ketone having an adjacent stereocenter that is capable of chelation (especially 5-membered ring chelation). After chelate formation, the nucleophile adds from the side opposite the larger of the remaining substituents on the α-stereocenter [48]. See Section 4.2.

Cram's rule (open chain model): A model to predict the major stereoisomer resulting from nucleophilic addition to a ketone or aldehyde having an adjacent *stereocenter*. The rule originally formulated by Cram in 1952 [47] has evolved into the current Felkin-Anh formulation [79,88,89], illustrated below. In the transition structure, the largest substituent of the stereocenter, or the substituent having the lowest-lying σ* orbital (L) is perpendicular to the carbonyl, and the nucleophile attacks from the opposite side, on a trajectory that places it approximately 107° away from the carbonyl (the *Bürgi-Dunitz trajectory*). The favored transition structure (a), has this trajectory nearly eclipsing the site of the smaller of the two remaining substituents. See Section 4.1.

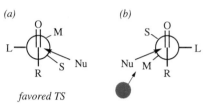

favored TS

CSA (Chiral solvating agent): A diamagnetic additive of known enantiomeric purity used to induce anisochrony in *enantiomers* of a *racemate* for NMR analysis. See Section 2.3.4.

CSP (Chiral stationary phase): A *nonracemic* chiral stationary phase for the chromatographic separation of *enantiomers*. See Section 2.4.

CSR (Chiral shift reagent): A paramagnetic lanthanide complex of known enantiomeric purity used to induce anisochrony in *enantiomers* of a *racemate* for NMR analysis. See Section 2.3.3.

D, L: See *Fischer-Rosanoff convention*

d, l, dl: Obsolete alternatives for (+)- and (−)- used to designate the sign of rotation of *enantiomers* at 589 nm (the sodium D line), and (±)- for a *racemate*. Sometimes used as arbitrary descriptors for a single enantiomorph.

Diastereoisomers: See *diastereomers*.

Diastereomer excess (percent diastereomer excess, % de): In a reaction in which two (and only two) diastereomeric products are possible, the percent diastereomeric excess, % de is given by:

$$\% \text{ de} = \frac{|D_1 - D_2|}{D_1 + D_2} \cdot 100 = |\%D_1 - \%D_2| \ .$$

where D_1 and D_2 are the mole fractions of the two diastereomeric products. If a reaction can produce more than two diastereomers, the ratio should itself be reported, or the selectivity reported as % *ds, i.e.,* with the product ratio(s) normalized to 100%. See *diastereoselectivity*.

Diastereomers (diastereoisomers): Stereoisomers that are not enantiomers (including alkene *E, Z* isomers).

Diastereoselectivity (percent diastereoselectivity, % ds): In a reaction in which more than one diastereomer may be formed (with mole fractions D_1, D_2, ... D_n produced) the diastereoselectivity is the mole fraction formed of the major product (or the desired product), expressed as a percent:

$$\% \ ds = \frac{D^*}{D_1 + D_2 \ ... \ + D_n} \cdot 100 \ ,$$

where D^* is the mole fraction of the desired isomer [90]. See Section 1.4. See also *enantioselectivity*.

Diastereotopic: The relationship of two ligands of an atom that are constitutionally equivalent, but in positions that are not symmetry related. Replacement of either ligand yields a pair of diastereomers. Also, faces of a trigonal atom that are not symmetry related, such that addition to either face gives a pair of dia- stereomers. Reflection variant faces may be specified as *Re* or *Si*, and ligands, L, may be specified as L_{Re} or L_{Si}, by noting on which face of a triangle the ligand in question sits (see *heterotopic*). Note that addition of a ligand to the *Re* face of a trigonal atom affords a tetrahedral array with the new ligand in the L_{Re} position. Reflection invariant descriptors are *re, si*, as illustrated below [91,92]. See also *Re, Si, homotopic,* and *enantiotopic.*

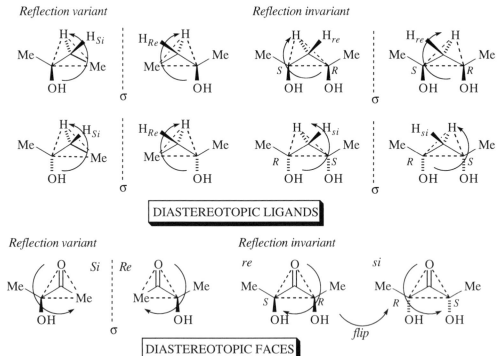

Dihedral angle: The angle between two defined planes. The term is most commonly applied to vicinal bonds on a *Newman projection*. See also *torsion angle.*

Dissymmetric: Obsolete synonym for *chiral*. Not equivalent to *asymmetric*, since chiral substances may have symmetry. See also *asymmetric.*

Double asymmetric induction: See *asymmetric induction.*

Dunitz angle: See *Bürgi-Dunitz trajectory.*

E, Z: Descriptors for the arrangement of ligands around double bonds. On either end of the double bond, the group of highest *CIP* rank is identified. If the two higher-ranking groups are on the same side of the double bond, the descriptor of the stereoisomer is Z (zusammen = together); if on opposite sides, E (entgegen = apart). See also *cis, trans* isomers. For enolates, some authors modify this rule such that the OM ligand (anionic oxygen with its metal) takes the highest priority. See *E(O), Z(O).*

E isomers: Z isomers:

E(O), Z(O): Descriptors for the arrangement of ligands around enolate double bonds. The standard *E/Z* stereochemical descriptor is modified such that the OM group is given priority over the carbonyl substituent, independent of the metal and the other substituent [93]. The priority descriptors for the α-carbon are maintained, as illustrated by the following examples:

Z(O)-enolates:

Eclipsed, Eclipsing: Two ligands on adjacent atoms are *eclipsed* if their torsion angle is near 0° (*i.e., synperiplanar*). See also *bisecting conformation, eclipsing strain, torsion angle.*

Eclipsing conformation: See *bisecting conformation.*

Eclipsing strain: See *torsional strain.*

Element of chirality: See *chirality element*

Enantioconvergent: See *stereoconvergent*

Enantiomer: A *stereoisomer* that is not superimposable on its mirror image. See also *enantiomorphous.*

Enantiomer excess, ee (percent enantiomer excess, % ee): For a mixture of a pure *enantiomer* and its *racemate*, the percent excess of the pure enantiomer over the racemate. % ee is given by:

$$\% \ ee = \frac{|E_1 - E_2|}{E_1 + E_2} \cdot 100 = |\%E_1 - \%E_2| \ ,$$

where E_1 and E_2 are the mole fractions of the two enantiomers. See *enantiomer purity.*

Enantiomer purity: A description of the enantiomer composition of a sample, historically expressed as *% ee*. Because this term implies that the impurity is the racemate (not the minor enantiomer), many authors prefer to use *enantiomer ratio, er,* normalized to 100%.

Enantiomer ratio, er: The ratio of two *enantiomers*. When used as an expression of enantiomer purity, this ratio is often normalized to 100% (*i.e.,* 99:1, 80:20).

Enantiomerically enriched (enantioenriched): A sample that has one *enantiomer* in excess.

Enantiomerically pure, enantiopure: A sample which contains (within the limits of detection) only one enantiomer. Note that this is not synonymous with *homochiral* [94].

Enantiomorphous: Not superimposable on its mirror image.

Enantioselectivity (percent enantioselectivity, % es): In a reaction or reaction sequence in which one enantiomer (E_1) is produced in excess, the enantioselectivity is the mole fraction formed of the major enantiomer, expressed as a percent:

$$\% \text{ es} = \frac{E_1}{E_1 + E_2} \cdot 100 \ .$$

See also *diastereoselectivity.*

Enantiotopic: The relationship of two ligands of an atom that are related by a *mirror plane, center of symmetry,* or *alternating axis,* but not by a simple (proper) *symmetry axis.* Replacement of either ligand yields a pair of *enantiomers.* Also, faces of a trigonal atom that are not symmetry related, such that addition to either face gives a pair of enantiomers. Note that addition of a ligand to the *Re* face affords a tetrahedral array with the new ligand in the L_{Re} position. The faces may be specified as *Re* or *Si,* and reflection variant ligands are best specified as L_{Re} or L_{Si}, as illustrated below [91,92]. See also *Re, Si, homotopic, heterotopic,* and *diastereotopic.*

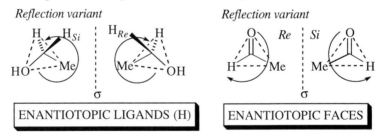

endo, exo: The stereochemical prefix that describes the relative configuration of a substituent on a bridge (not a bridgehead) of a bicyclic system. If the substituent is oriented toward the larger of the other bridges, it is *endo;* if it is oriented toward the smaller bridge, it is *exo.*

ent: A prefix to the name of a chiral molecule to indicate its *enantiomer.*

Envelope: The conformation of a five-membered ring in which four atoms are coplanar, and the fifth (the flap) is out of the plane.

Epimerization: The interconversion of *epimers.*

Epimers: Diastereomers that differ in *configuration* at one of two or more *stereogenic units.*

Equatorial: See *axial, equatorial.*

erythro, threo: Terms used to describe relative configuration at adjacent stereocenters. Originally, the term was derived from carbohydrate nomenclature (*cf.* erythrose, threose). In this sense, if the molecule is drawn in a *Fischer projection,* the *erythro* isomer has identical or similar substituents on the same side of the vertical chain and the *threo* isomer has them on opposite sides. In the early 1980s, proposals appeared to redefine these terms based on *zig-zag projections* [95] and *CIP priority* [96], but the latter usages are now discouraged [97]. See *l, u; pref, parf; syn, anti.*

exo: See *endo, exo.*

Felkin-Anh model: See *Cram's rule (open chain model).*

Fischer Projection (or Fischer-Tollens projection): A planar projection formula in which the vertical bonds lie behind the plane of the paper and the horizontal bonds lie above the plane. Used commonly in carbohydrate structures, where each carbon in turn is placed in the proper orientation for planar projection.

Fischer projection   hash/wedge view   ball and stick stereo view

Fischer-Rosanoff convention: A method for the specification of absolute configuration, still in common use for amino acids and sugars. When drawn in a *Fischer projection* with C_1 at the top, if the functional group of the specified stereocenter is on the right, the absolute configuration is D, if on the left, it is L. For amino acids, the reference stereocenter is C_2; for sugars it is the highest numbered stereocenter [98].

D-glucose   L-glucose   D-serine   L-serine

* Reference stereocenter

Flagpole: See *bowsprit, flagpole*

Free rotation, restricted rotation: In the context of an experimental observation, *free rotation* is sufficiently fast (*i.e.,* the rotational barrier is sufficiently low) that different *conformations* are not observable. Conversely, *restricted rotation* is sufficiently slow (the barrier is sufficiently high) that conformational isomers can be observed.

Gauche: Synonomous with a *synclinal* alignment of groups attached to adjacent atoms (*i.e.,* a torsion angle of near +60° or –60°). See *torsion angle.*

Geometric isomers: Synonym for *cis-trans* double bond isomers.

Half-boat: See *half-chair, half-boat.*

Half-chair, half-boat: Terms used most commonly to describe conformations of cyclohexenes in which four contiguous carbon atoms atoms lie in a plane. If the other two atoms lie on opposite sides of the plane, the conformation is a half-chair; if they are on the same side, it is a half-boat, as shown below. Also used for 5-membered rings, where three adjacent atoms define the plane.

cyclohexene half-chair   cyclopentane half-chair

Helicity: The chirality sense of a helix. May be specified by *P, M.*

Heterochiral: See *homochiral.*

Heterotopic: Either *diastereotopic* or *enantiotopic.* Refers to either the *Re* or *Si* half space of a two-dimensionally chiral triangle, as shown below [91,92]. See also *Re, Si, enantiotopic, diastereotopic.*

CIP rank a>b>c

Homochiral: A descriptor for objects and molecules having the same *chirality sense* [99]. Thus, "two equal and similar right hands are homochirally similar. Equal and similar right and left hands are heterochirally similar..." [82]. A set of right shoes, or an assembly of molecules (such as a mixture of amino acids) that have the same *relative configuration* or *chirality sense* are homochiral [99,100]. This term should not be used to describe enantiomerically pure compounds, since the term homochiral describes a <u>relationship</u>, not a <u>property</u> [94].

Homofacial, heterofacial: The *relative configuration* of stereocenters (in different molecules) having three identical ligands and one different is homofacial if the fourth ligand is on the same *heterotopic* face in both, and heterofacial if on opposite faces, as shown below. [91,92,101]. See also *relative configuration.*

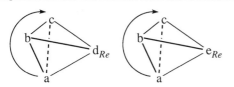 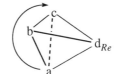 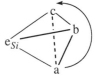

homofacial: *d* and *e* both reside on the *Re* face of the *abc* triangle

heterofacial: *d* is on the *Re* side of the *abc* triangle, whereas *e* is on the *Si* face.

Homotopic: Ligands that are related by an *n*-fold rotation axis. Similarly, faces of a trigonal atom that are related by an *n*-fold rotation axis. Replacement of any of the ligands or addition to either of the faces gives an identical compound. See also *heterotopic, enantiotopic,* and *diastereotopic*.

Inversion: See *Walden inversion, pyramidal inversion,* and *ring inversion*.

Isomers: Compounds that have the same molecular formula but which have different *constitutions* (*constitutional isomers*), *configurations* (*enantiomers, diastereomers*), or *conformations* (*conformational isomers*), and therefore have different chemical and/or physical properties.

Kinetic resolution: The separation (or partial separation) of *enantiomers* due to a difference in the rate of reaction of the two enantiomers in a *racemic mixture* with an *nonracemic chiral* reagent.

l, u: Descriptors for the specification of *relative configuration*. A pair of stereo-genic units has the relative configuration *l* (for *like*) if the descriptor pairs are *RR, SS, RRe, SSi, ReRe, SiSi, MM, PP, RM, SP, ReM,* or *SiP*. The pair is specified as *u* (*unlike*) if they have descriptor pairs *RS, RSi, ReS, ReSi, MP, RP, SM, ReP,* and *SiM* [86]. Reflection invariant descriptors (*r, s, re, si, p,* and *m*) may be substituted in place of the reflection variant descriptors above. Note the use of lower case *l* and *u* letters, implying a *reflection invariant* relationship.

lk, ul: An extension of the *l, u* nomenclature to describe topicity. If a reagent of configuration *R* (or the *Re* face of a trigonal atom) preferentially approaches the *Re* face of a trigonal atom, the topicity is *lk* (*like*); it is *ul* (*unlike*) if it approaches the *Si* face. Similarly, the approach of an achiral reagent to diastereotopic faces of a trigonal atom is *lk* if the *Re* face is preferred in the *R* enantiomer, and vice versa; the topicity is *ul* if the *Si* face is preferred in the *R* enantiomer, and vice versa. In short, if the first letters of the two *stereo-chemical descriptors* are the same, the topicity is *lk*. If they are different, it is *ul* [102]. See the more complete listing of like and unlike pairs under *l, u*. Note the use of lower case *l* and *u* letters, implying a *reflection invariant* relationship. *Lk* and *Ul* would be used if the topicity were *reflection variant*, which would occur if one of the components was reflection invariant.

M, P: See *P, M*.

meso: A stereoisomer that has two or more *stereogenic units*, but which is *achiral* because of a *symmetry plane*. The plane reflects *enantiomorphic* groups.

Newman projection: A projection formula that represents the spatial arrangement of the ligands on two adjacent atoms as viewed down the bond joining them.

2-butanol:

zig-zag projection   Newman projection   ball and stick stereo view
               of one conformer

Nonbonded interactions: Attractive or repulsive "through space" forces between atoms or groups in a molecule (intermolecular or intramolecular) that are not directly bonded to each other.

Nonracemic: Not *racemic*.

Optical activity: The property of a substance to rotate plane polarized light. See Section 2.2

Optical purity (op, % op): The ratio of the observed *specific rotation* of a substance to the maximum possible rotation of the substance, expressed as a percent:

$$\% \text{ op} = \frac{[\alpha]}{[\alpha]_{\text{max}}} \cdot 100 \ .$$

Usually (but not always) it is equal to *enantiomer excess,* or *% ee.* This term is less frequently used now, as enantiomer ratios are often determined by non-polarimetric methods. See Section 2.2.

Optical yield: For a chemical reaction, the *enantiomer excess* of the products relative to that of the starting material, expressed as a percent. In asymmetric synthesis, the denominator may be the ee of the chiral reagent or catalyst.

P, M: Descriptors of chirality sense of a helix. Once the axis of the helix is identified, one chooses the ligands of the highest *CIP rank.* If the smallest angle between the ligands (*i.e.,* ≤ 180°) in a projection is clockwise going from front to rear, the chirality sense is *P* (plus), if counterclockwise, it is *M* (minus) [85,103]. Additionally, *P, M* may be used to describe the chirality sense of a helix of any sequence of atoms as long as they are explicitly identified. Note that it does not matter which end of the helix is viewed.

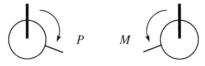

These descriptors can be used to specify enantiomeric *conformers,* such as gauche butane, and the *absolute configuration* of *stereogenic axes* and *planes.* Prelog and Helmchen have recommended the use of *P,M* instead of *R,S* for specifying the absolute configuration of planes and axes of chirality [86]. See also ref. [101], Ch. 14.

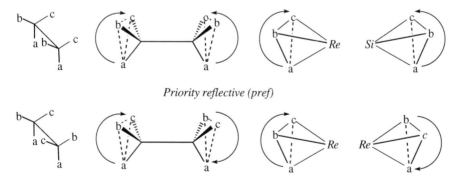

P-butane   *M*-2-hydroxyparacyclophane

M-2,3-pentadiene   *P*-binaphthol

Pitzer strain: See *torsional strain*.

Planar chirality: See *stereogenic element*.

Point group (symmetry point group): The symmetry classification of a molecule based on its symmetry elements (axes, planes, etc.).

Pref, parf: Descriptors of relative configuration based on *CIP priority*. The relationship between two chirality centers is pref (priority reflective) if the order of decreasing priority of the three remaining groups at one chirality center is a reflection of the order of decreasing priority of the groups at the other center. When the orders of decreasing priority are not reflective of each other, the relative configuration is parf (priority antireflective). If the chirality centers are not adjacent, the intervening bonds are neglected and the two centers are treated as if they were directly linked. If the two centers are part of a ring, they are treated as if connected by a bond that replaces the shorter path [104].

Priority reflective (pref)

Priority antireflective (parf)

Prochiral: Tetrahedral atoms having *heterotopic* ligands, or heterotopic faces of trigonal atoms, may be described as being prochiral. Note that it is inappropriate to describe an entire molecule as being prochiral [105]. Heterotopic faces are described using *Re, Si* if *reflection variant*, and *re, si* if reflection invariant [91]. If the *CIP priority* of the three ligands is clockwise, the face (toward the observer) is *Re;* if counterclockwise, it is *Si* [105]. For heterotopic ligands, two conventions have been used to describe prochirality. Both use the CIP rank of the ligands to specify the "prochirality sense" of each ligand. The broader rule is that of Prelog and Helmchen [91,92]. In this method, a tetrahedron is constructed of the four ligands around the prochiral center. If the ligand 'L' is sitting on the *Re* face of the triangle formed by the other three ligands, it is specified L_{Re} (or L_{re} if *reflection invariant*); similarly, the ligand would be specified L_{Si} or L_{si} if on the *Si* face. See also *enantiotopic, diastereotopic, heterotopic,* and *relative configuration.*

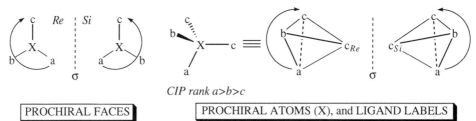

CIP rank a>b>c

| PROCHIRAL FACES | PROCHIRAL ATOMS (X), and LIGAND LABELS |

Another convention, used in biochemistry to specify the hydrogen atoms of a prochiral methylene, replaces a hydrogen with a deuterium. If such replacement results in the *R* configuration, the ligand position is *pro-R*. If the *S* configuration is obtained, it is *pro-S* [105]. If *reflection invariant*, the descriptors are *pro-r* and *pro-s*.

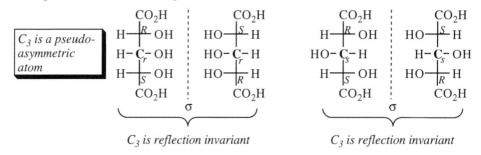

Pseudoasymmetric atom: A *stereogenic* atom of a stereoisomer that has two *enantiomorphic* ligands (reflection invariant), and two other different ligands. Exchange of any two ligands generates a *diastereomeric* compound. The *CIP* descriptors for pseudoasymmetric atoms are *r, s*. Use of this term is discouraged in favor of *stereogenic center*.

| C_3 is a pseudo-asymmetric atom | | | | |

C_3 is reflection invariant C_3 is reflection invariant

Pseudoaxial, pseudoequatorial: See *axial, equatorial.*

Pseudorotation: Term used by some authors to describe the out of plane motion of the ring atoms in cyclopentane during fast conformational interchange of the many envelope and twist conformers. Usage is discouraged.

Pyramidal inversion: The change of bond directions in a trivalent central atom having a pyramidal (tripodal) arrangement with the central atom at the apex of the pyramid. The inversion appears to move the central atom to a similar position on the other side of the pyramid. If the central atom is *stereogenic*, pyramidal inversion reverses its *absolute configuration*.

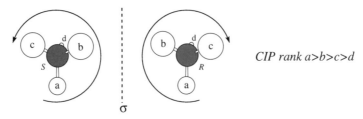

r, s: See *pseudoasymmetric atom.*

R, S: CIP descriptors for the specification and description of absolute configuration, as follows:

Stereogenic center: After the CIP rank of ligands is determined (see *CIP method*), the tetrahedron is arranged so that the ligand of lowest priority is to the rear. If the order of the other three ligands (highest to lowest) is clockwise, the *absolute configuration* is *R* (latin *rectus*, right); if counterclockwise, it is *S* (latin *sinister*, left).

CIP rank a>b>c>d

Stereogenic axis: The descriptors may be modified to R_a, S_a when applied to a stereogenic axis [85], although it is usually more convenient to use the *P, M* system to specify the configuration of stereogenic axes and planes [86].

Stereogenic plane: The descriptors may be modified to R_p, S_p when applied to a stereogenic plane [85], although it is usually more convenient to use the *P, M* system to specify the configuration of stereogenic axes and planes [86].

The symbols *R** and *S** may be used to describe *relative configuration.* Thus, *R*,R** describes a racemate of *l* configuration and *R*,S** describes a racemate of *u* configuration. See *l,u.*

rac: A prefix to the name of a chiral molecule to indicate that it is the *racemate.*

Racemate (racemic mixture): An equimolar mixture of two *enantiomers,* whose physical state is unspecified [26]. Some authors restrict the term 'racemate' to a crystalline compound whose unit cell contains equal numbers of enantiomeric molecules and 'racemic mixture' to a mechanical mixture of two crystals that form a eutectic of two enantiomers. The latter is now referred to as a conglomerate [26].

Racemization: The conversion of a nonracemic substance into its racemate.

Re, Si (re, si): Stereochemical descriptors for *heterotopic* faces. If the *CIP priority* of the three ligands is clockwise, the face (toward the observer) is *Re* (latin *rectus,* right); if counterclockwise, it is *Si* (latin *sinister,* left) [105].

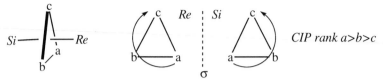

The descriptors may be used to describe the faces of trigonal atoms,

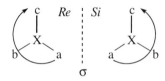

or the ligand position of a tetrahedral stereogenic unit,

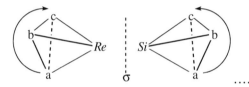

Lower case descriptors *(re, si)* are used for the rare cases that are *reflection invariant* [91]:

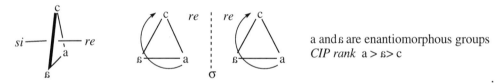

For examples of reflection invariant stereogenic centers and faces, see *diastereotopic,* and *pseudoasymmetric atom.*

Reflection variant, reflection invariant: The terms used to describe an object and its relationship with its mirror image. If the two are identical, the object is reflection invariant. If the object is *enantiomorphous* to its mirror image, it is reflection variant.

Relative configuration: The *configuration* of any *stereogenic* element with respect to another. Relative configuration is *reflection invariant.* The relative configuration of pairs of *stereogenic* units in the same molecule may be described as *R*, R** or *l* if they have the same *CIP* descriptor, and *R*, S** or *u* if they are different. (See *l, u* for a complete list of like and unlike descriptors.) The term can also be used in an intermolecular sense as follows: if the two molecules contain stereogenic units abcd and abce, and if e and d both sit on the same *heterotopic face,* the two stereogenic units have the same relative configuration. If not, they have the opposite relative configuration. The term may be applied to starting material and products of a reaction sequence. See also *homofacial, heterofacial.*

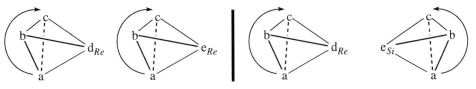

Same relative configuration   *Opposite relative configuration*

Resolution: The separation of a *racemic mixture* into (at least one of) its component enantiomers. See also *kinetic resolution.*

Restricted rotation: See *free rotation.*

Retention of configuration: The product of a chemical reaction has retained its configuration if the product has the same *relative configuration* as the starting material. See also *Walden inversion, relative configuration.*

Ring inversion (ring reversal): The interconversion of cyclohexane *conformations* having similar shapes (chair - chair), accompanied by interchange of the *equatorial* and *axial* substituents. Similarly, the interchange of any such similarly shaped conformations in a cyclic molecule.

Rotamers: Stereoisomers of the same constitution and configuration, that differ only by torsion angles.

Rotation angle (α): The rotation of the plane of polarized light after passing through an *optically active* sample. If the angle of rotation is clockwise, the sample is dextrorotatory and the sign of rotation is positive (+). If the angle is counterclockwise, the sample is levorotatory, and the sign of rotation is (−). See also *optical activity, specific rotation,* and Section 2.2.

Rotation-reflection axis: See *alternating symmetry axis* .

Rotational barrier: The energy barrier between two *conformers.*

s-cis, s-trans: Conformational descriptors for the single bond linking two double bonds (darkened below). The *synperiplanar* conformation is *s*-cis, and the *antiperiplanar* conformation is *s*-trans. See *torsion angle.*

Sawhorse formula: A perspective drawing that indicates the spatial arrangements of the ligands on two adjacent tetrahedral atoms. The bond between the two atoms is a diagonal line, with the nearer atom at the bottom.

        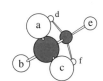

sawhorse projection   *ball and stick view*

Scalemic: Not *racemic* [106,107]. Synonomous with *aracemic, nonracemic.*

Sense of chirality: See *chirality sense.*

Sequence rules: See *CIP method.*

Si, si: See *Re, Si.*

Skew: See *chair, boat, twist-boat.*

Specific rotation: The specific rotation of a sample, $[\alpha]$, is defined as:

$$[\alpha]_\lambda^t = \frac{100\alpha}{l \cdot c} ,$$

where t is temperature, λ is wavelength of the light, α is the observed *rotation*, l is the sample path length (in dm), and c is the concentration (in g/100 mL). $[\alpha]$ is normally reported without units, but the concentration and the solvent are usually specified in parentheses after the value of $[\alpha]$. See Section 2.2.

Staggered conformation: The *conformation* of two tetrahedral carbons is staggered if the *torsion angle* between the ligands is approximately ±60°.

Stereocenter: See *stereogenic center.*

Stereochemical descriptor: A letter-symbol or prefix to specify configuration or conformation, such as *R, S, E, Z, P, M, cis, trans,* etc.

Stereoconvergent: A reaction or reaction sequence is stereoconvergent if stereo-isomerically different starting materials yield the same stereoisomeric product. The sequence may be more specifically labeled either *enantioconvergent* or *diastereoconvergent.*

Stereoelectronic effect: An effect on structure and reactivity due to the orientation and alignment of bonded or nonbonded electron pairs [108].

Stereogenic axis: A set of two pairs of tetrahedrally arranged bonding positions (D_2 or C_{2v} point symmetry), each occupied by two different ligands. Exchange of the ligands of either pair reverses the *absolute configuration.* Examples include unsymmetrically substituted allenes and 2,6,2',6'-tetrasubstituted biphenyls. If the axis is *reflection variant,* it may be called a *chirality axis.* The absolute configuration may be described by either *P, M* or *R, S.* (Note that for a stereogenic axis, $R \equiv M$ and $S \equiv P$). See also *stereogenic element.*

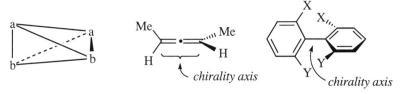

Stereogenic center (stereocenter): An atom in a molecule (or a focal set of atoms) with four equivalent tetrahedral bonding positions (T_d point symmetry), occupied by four different ligands. Exchange of any two ligands reverses the *absolute configuration.* If the center is reflection variant, it may be called a *chirality center.* If it is reflection invariant, it is sometimes called a *pseudoasymmetric atom,* although usage of this term is discouraged. *Stereogenic center* is thus an extension of the 'asymmetric carbon atom' of van't Hoff and LeBel, and now includes species such as N^+abcd and the sulfur atom of unsymmetric sulfoxides (where the fourth 'ligand' is a lone pair), as well as tetrahedral arrays of ligands with T_d symmetry. The *absolute configuration* may be described by the *CIP method. See R,S.*

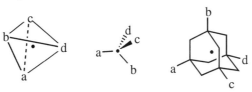

Stereogenic element, stereogenic unit: A *center, axis,* or *plane* in a molecule in which exchange of two ligands leads to a new *stereoisomer.* If the stereogenic element is reflection variant, the elements are *chirality center, chirality axis,* and *chirality plane.* The bonding positions of stereogenic <u>centers</u> have point symmetry T_d; the bonding positions of stereogenic <u>axes</u> have point symmetry D_2 or C_{2v}; the bonding positions of stereogenic <u>planes</u> have point symmetry C_s. As a result, there must be four different ligands (abcd) on a T_d bonding center to create stereogenicity. On an axis, only the two ligands of each pair need be different (ab/ab), the two pairs may be the same. In a stereogenic plane, only one of the ligands in the plane need be different. See also *stereogenic axis, stereogenic center, stereogenic plane.*

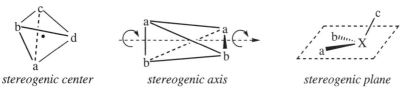

| stereogenic center | stereogenic axis | stereogenic plane |

Stereogenic plane: A planar structural fragment that, because of *restricted rotation* or structural requirements, cannot lie in a symmetry plane. If the stereogenic plane is reflection variant, the element may be called a *chirality plane.* For example with a monosubstituted paracyclophane, the stereogenic plane includes the plane of the benzene ring. For a 1,2-disubstituted ferrocene, the disubstituted cyclopentadiene lies in a chirality plane. The *absolute configuration* may be specified by either *R, S* or *P, M.* See also *stereogenic element.*

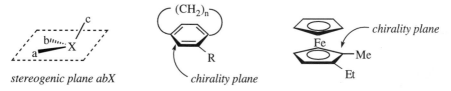

| stereogenic plane abX | chirality plane | |

Stereoheterotopic: Either *enantiotopic* or *diastereotopic.*

Stereoisomers: Isomers of the same *constitution* that differ only in the position of atoms and ligands in space (*i.e., enantiomers* and *diastereomers*).

Stereoselectivity: In a reaction, the preferential formation of one *stereoisomer* over another (or others). See also *diastereoselectivity, enantioselectivity.*

Stereospecific: A pair of reactions are stereospecific if *stereoisomeric* educts afford stereoisomeric products. A stereospecific process is necessarily 100% *stereoselective,* but the converse is not necessarily true, even if the stereoselectivity is 100%. Use of the term to describe a reaction that is merely highly stereoselective is discouraged.

Structure: The *constitution, configuration,* and *conformation* of a molecule. Formerly, the term was used as a synonym for *constitution* alone.

Structural isomers: Obsolete term for *constitutional isomers.*

Superimposable, superposable: Two objects are superimposable if they can be brought into coincidence by translation and rotation. For chemical structures, *free rotation* around single bonds is permissible. Thus, two molecules of *R*-2-butanol are considered superimposable independent of their conformations.

Suprafacial: See *antarafacial, suprafacial.*

syn, anti: Prefixes that describe the relative configuration of two substituents with respect to a defined plane or ring (syn if on the same side, anti if opposite). Such planes may be defined arbitrarily, but some that are in common usage are illustrated below. Formerly, these terms were used to describe the configuration of oximes, hydrazones, etc. (see *E, Z*). See also *torsion angle.*

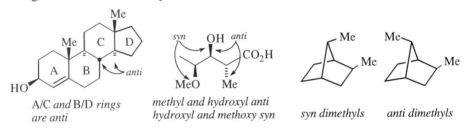

| A/C and B/D rings are anti | methyl and hydroxyl anti hydroxyl and methoxy syn | syn dimethyls | anti dimethyls |

Synclinal: See *torsion angle.*

Symmetry axis (Cn): An axis of an object, about which a rotation by an angle of 360/n gives an entity that is superimposable on the original. See also *alternating symmetry axis.*

Symmetry elements: Axes, centers, or planes of symmetry.

Symmetry plane (σ): A mirror plane which bisects an object, such that reflection of one half produces a fragment that is superimposable on the other half.

Thorpe-Ingold effect: The original phenomenon observed by Thorpe and Ingold was an accelerating effect on cyclizations [109,110]. They attributed the effect to a bond angle compression, as shown below, whereby geminal substituents enlarge bond angle α by *van der Waals* repulsion, and thereby compress bond angle β. (It is likely that this explanation is an oversimplification. For a recent study on geminal effects on ring closure rates, see ref. [111]).

$$\alpha_1 \begin{pmatrix} H \\ & \ddots R \\ H & R \end{pmatrix} \beta_1 \qquad \boxed{\begin{array}{c} \alpha_1 < \alpha_2 \\ \beta_1 > \beta_2 \end{array}} \qquad \alpha_2 \begin{pmatrix} Me \\ & \ddots R \\ Me & R \end{pmatrix} \beta_2$$

Threo: See *erythro, threo.*

Torsion angle: The angle, in a molecular fragment A–B–C–D (having ABC and BCD bond angles ≤ 180°), between the planes ABC and BCD (see the *Newman projection,* below), always defined such that the absolute value is less than 180°. If (looking from either direction) the turn from A to D or D to A is clockwise, the torsion angle is positive; if it counterclockwise, it is negative (see also *P, M*). If the torsion angle is 0° to ±90°, the angle is *syn;* if between ±90° and 180°, it is *anti.* Similarly, angles from 30 to 150° and –30 to –150° are *clinal.* Combination gives *synperiplanar* for angles between 0° and ±30°; 30° to 90° and –30° to –90° are *synclinal;* 90° to 150° and –90° to –150° are *anticlinal;* and ±150° to 180° are *antiperiplanar* [103]. Often the *synperiplanar* conformation is called eclipsed, the *antiperiplanar* conformation anti, and the *synclinal* conformation gauche or skew.

synperiplanar
0°
–30°   +30°
synclinal   *synclinal*
–90°   +90°
anticlinal   *anticlinal*
–150°   +150°
180°
antiperiplanar

Torsional strain: Destabilization of a molecule due to a variation of a *torsional angle* from an optimal value (*e. g.,* 60° in a saturated molecule). Also called *Pitzer strain, eclipsing strain.*

Torsional isomers: See conformational isomers.

trans: See *cis, trans isomers.*

Transannular interaction: Literally: cross-ring interactions. Non-bonded interaction between ligands attached to nonadjacent atoms in a ring, for example in a cyclohexane *boat* or in medium-sized rings.

transition state, transition structure: In a chemical reaction, the transition *state* is the ensemble of molecular structures that are at the free energy saddle point between reactants and products. The transition *structure* corresponds to the single set of atomic coordinates at the saddle point of the potential energy surface (internal, or enthalpic energy at 0° K). Thus, coordinates of the transition state vary with temperature, whereas those of the transition structure do not. In a practical sense, a structure that is drawn on a piece of paper (whether derived from a computation or not) should be referred to as a transition structure, since it is static. The transition state is an ensemble of similar structures undergoing translational, vibrational, and rotational motion.

transoid conformation (usage discouraged): See *s-cis, s-trans*.

twist-boat: See *chair, boat, twist-boat*.

u: See *l, u*.

ul: See *lk, ul*.

van der Waals interactions: Attractive or repulsive interactions resulting from close approach of two molecules [112-114]. Modern usage (especially in molecular mechanics calculations) also uses the term van der Waals interactions to describe the attractive and repulsive interactions created by intramolecular approach of molecular fragments [115]. See also *nonbonded interactions*.

Walden inversion: Conversion of Xabcd into Xabcd (for an identity reaction) or Xabce, of opposite *configuration*. Synonymous with *inversion of configuration*.

Walden inversion

Z: See *E, Z*.

Z(O): See *E(O), Z(O)*.

Zig-zag projection: A stereochemical projection in which the main chain of an acyclic compound is drawn in the plane of the paper with 180° torsion angles, with substituents above the plane drawn with bold or solid wedges, and hashed lines for substituents behind the plane.

zig-zag projection

1.7 References

1. M. Gardner *The Ambidexterous Universe: Mirror Asymmetry and Time-Resolved Worlds, 2nd Ed*; Charles Scribner's Sons: New York, 1979.
2. *Chirality. From Weak Bosons to the alpha Helix*; R. Janoschek, Ed.; Springer: New York, 1991.
3. R. A. Hegstrom; D. K. Kondpudi *Scientific American* **1990**, *262*, 108-115.
4. S. C. Stinson *Chem. Eng. News* **1994**, *May 16*, 10-25.
5. *Chirality in Industry: The Commercial Manufacture and Applications of Optically Active Compounds*; A. N. Collins; G. N. Sheldrake; J. Crosby, Eds.; Wiley: New York, 1992.
6. R. A. Sheldon *Chirotechnology. Industrial Synthesis of Optically Active Compounds*; Marcel Dekker: New York, 1993.
7. J. W. Scott In *Topics in Stereochemistry*; E. L. Eliel, N. L. Allinger, Eds.; Wiley-Interscience: New York, 1991; Vol. 19, p 209-226.
8. R. Crossley *Tetrahedron* **1992**, *48*, 8155-8178.
9. S. C. Stinson *Chem. Eng. News* **1993**, *Sep 27*, 38-65.
10. E. J. Ariens, quoted in *Chem. Eng. News* **1990**, *Mar 19*, 40.
11. E. J. Ariens *Trends. Pharm. Sci.* **1986**, *7*, 200-205.
12. *Stereochemistry and Biological Activity of Drugs*; E. J. Ariens; W. Soudijin; P. B. M. W. M. Timmermans, Eds.; Blackwell: Oxford, 1983.
13. G. F. Russel; J. I. Hills *Science* **1971**, *172*, 1043-1044.
14. L. Friedman; J. G. Miller *Science* **1971**, *172*, 1044-1045.
15. J. Solms; L. Vuataz; R. H. Egli *Experientia* **1965**, *21*, 692-694.
16. C. H. Easson; E. Stedman *Biochem. J.* **1933**, *27*, 1257-1266.
17. S. Hanessian *Total Synthesis of Natural Products*; Pergamon: Oxford, 1983.
18. G. M. Coppola; H. F. Schuster *Asymmetric Synthesis. Construction of Chiral Molecules Using Amino Acids*; Wiley-Interscience: New York, 1987.
19. *Carbohydrates as Raw Materials*; W. Lichtenthaler, Ed.; VCH: New York, 1991.
20. T.-L. Ho *Enantioselective Synthesis: Natural Products from Chiral Terpenes*; Wiley: New York, 1992.
21. J. W. Scott In *Asymmetric Synthesis*; J. D. Morrison, Ed.; Academic: Orlando, 1984; Vol. 4, p 1-226.
22. G. W. J. Fleet *Chem. Br.* **1989**, *25*, 287-292.
23. T. Money *Nat. Prod. Rep.* **1985**, 253-289.
24. T. Money In *Studies in Natural Products Chemistry*; Atta-Ur-Rahman, Ed.; Elsevier: Amsterdam, 1989; Vol. 4, p 625-697.
25. S. Hanessian; J. Franco; B. Larouche *Pure Appl. Chem.* **1990**, *62*, 1887-1910.
26. J. Jacques; A. Colbert; S. H. Wilen *Enantiomers, Racemates and Resolutions*; Wiley-Interscience: New York, 1981.
27. R. E. Ireland *Organic Synthesis*; Prentice-Hall: Englewood Cliffs, NJ, 1969.
28. T.-L. Ho *Tactics of Organic Synthesis*; Wiley-Interscience: New York, 1994.
29. W. Marckwald *Chem. Ber.* **1904**, *37*, 1368-1370.
30. C.-H. Wong; G. M. Whitesides *Enzymes in Synthetic Organic Chemistry*; Pergamon: Oxford, 1994.
31. H. L. Holland *Organic Synthesis with Oxidative Enzymes*; VCH: New York, 1991.
32. A. J. Pratt *Chem. Brit.* **1989**, *25*, 282-286.
33. J. B. Cohen; T. S. Patterson *Chem. Ber.* **1904**, *37*, 1012-1014.
34. W. Marckwald *Chem. Ber.* **1904**, *37*, 349-354.
35. E. Erlenmeyer; F. Landesberger *Biochem. Z.* **1914**, *64*, 366-381.
36. F. R. Japp *Nature* **1898**, *58*, 452-460.
37. E. L. Eliel *Tetrahedron* **1974**, *30*, 1503-1513.

38. E. L. Eliel In *Asymmetric Synthesis*; J. D. Morrison, Ed.; Academic: Orlando, 1983; Vol. 2, p 125-155.
39. P. D. Richie *Asymmetric Synthesis and Asymmetric Induction*; Oxford: London, 1933.
40. G. Kortum *Sammlung Chimie U. Chemische Technology*; F. Enke: Stuttgart, 1933.
41. J. D. Morrison; H. S. Mosher *Asymmetric Organic Reactions*; Prentice-Hall: Englewood Cliffs, NJ, 1971.
42. *Asymmetric Synthesis*; J. D. Morrison, Ed.; Academic: Orlando, 1983-85.
43. *Comprehensive Organic Synthesis. Selectivity, Strategy, and Efficiency in Modern Organic Chemistry*; B. M. Trost; I. Fleming, Eds.; Pergamon: Oxford, 1991.
44. *Stereoselective Synthesis*; G. Helmchen; R. W. Hoffmann; J. Mulzer; E. Schaumann, Eds.; Georg Thieme: Stuttgart, 1995; Vol. E21.
45. N. A. Porter; B. Giese; D. P. Curran *Acc. Chem. Res.* **1992**, *24*, 296-304.
46. G. S. Miracle; S. M. Cannizzaro; N. A. Porter *Chemtracts-Org. Chem.* **1993**, *6*, 147-171.
47. D. J. Cram; F. A. A. Elhafez *J. Am. Chem. Soc.* **1952**, *74*, 5828-5835.
48. D. J. Cram; K. R. Kopecky *J. Am. Chem. Soc.* **1959**, *81*, 2748-2755.
49. X. Chen; E. R. Hortelano; E. L. Eliel; S. V. Frye *J. Am. Chem. Soc.* **1992**, *114*, 1778-1784.
50. S. V. Frye; E. L. Eliel *Tetrahedron Lett.* **1985**, *26*, 3907-3910.
51. B. Weber; D. Seebach *Angew. Chem. Int. Ed. Engl.* **1992**, *31*, 84-86.
52. D. Seebach; A. K. Beck; B. Schmidt; Y. M. Wang *Tetrahedron* **1994**, *50*, 4363-4384.
53. H. Brunner *Acc. Chem. Res.* **1979**, *12*, 250-257.
54. *Catalytic Asymmetric Synthesis*; I. Ojima, Ed.; VCH: New York, 1993.
55. R. Noyori *Asymmetric Catalysis in Organic Synthesis*; Wiley-Interscience: New York, 1994.
56. D. Seebach; D. A. Plattner; A. K. Beck; Y. M. Wang; D. Hunziker; W. Petter *Helv. Chim. Acta* **1992**, *75*, 2171-2209.
57. C. Zioudrou; P. Chrysochou *Tetrahedron* **1977**, *33*, 2103-2108.
58. S. Masamune; W. Choy; J. S. Petersen; L. R. Sita *Angew. Chem. Int. Ed. Engl.* **1985**, *24*, 1-76.
59. K. B. Sharpless *Chemica Scripta* **1985**, *25*, 71-77.
60. T. Kogure; E. L. Eliel *J. Org. Chem.* **1984**, *49*, 576-578.
61. S. Masamune; S. A. Ali; D. L. Snitman; D. S. Garvey *Angew. Chem. Int. Ed. Engl.* **1980**, *19*, 557-558.
62. C. T. Buse; C. H. Heathcock *J. Am. Chem. Soc.* **1977**, *99*, 8109-8110.
63. T. Katsuki; A. W. M. Lee; P. Ma; V. S. Martin; S. Masamune; K. B. Sharpless; D. Tuddenham; F. J. Walker *J. Org. Chem.* **1982**, *47*, 1373-1378.
64. J. March In *Advanced Organic Chemistry, 4th ed.*; Wiley-Interscience: New York, 1992, p 145.
65. G. Bredig *Angew. Chem.* **1923**, *36*, 456-458.
66. F. Johnson *Chem. Rev.* **1968**, *68*, 375-413.
67. R. W. Hoffmann *Chem. Rev.* **1989**, *89*, 1841-1860.
68. W. Pigman; D. Horton In *The Carbohydrates. Chemistry and Biochemistry, 2nd ed.*; Academic: New York, 1972, p 1-67.
69. L. F. Fieser; M. Fieser *Natural Products Related to Phenanthrene, 3rd ed*; Reinhold: New York, 1949.
70. D. H. R. Barton *Experientia* **1950**, *6*, 316-320.
71. *The Anomeric Effect and Associated Stereoelectronic Effects. ACS Symposium Series 539*; G. R. J. Thatcher, Ed.; American Chemical Society: Washington, 1993.
72. A. J. Kirby *The Anomeric Effect and Related Stereoelectronic Effects at Oxygen*; Springer: Berlin, 1983.
73. P. Muller, ed. *Pure Appl. Chem.* **1994**, *66*, 1077-1184.
74. V. Gold *Pure Appl. Chem.* **1983**, *55*, 1281-1371.
75. E. L. Eliel; S. H. Wilen *Chem. Eng. News* **1991**, *July 22*, 2.

76. M. M. Harris In *Progress in Stereochemistry*; W. Klyne, P. B. D. de la Mare, Eds.; Butterworths: London, 1958, p 157-195.
77. R. Kuhn In *Stereochemie. Eine Zusammenfassung der Ergebnisse, Grundlagen und Probleme*; K. Freudenberg, Ed.; Franz Deuticke: Leipzig, 1933, p 801-824.
78. H. B. Bürgi; J. D. Dunitz; E. Schefter *J. Am. Chem. Soc.* **1973**, *95*, 5065-5067.
79. N. T. Anh; O. Eisenstein *Nouv. J. Chimie* **1977**, *1*, 61-70.
80. H. B. Bürgi; D. Dunitz; J. M. Lehn; G. Wipff *Tetrahedron* **1974**, *30*, 1563-1572.
81. H. M. Pickett; H. L. Strauss *J. Am. Chem. Soc.* **1970**, *92*, 7281-7290.
82. L. Kelvin In *Baltimore Lectures*; C. J. Clay and Sons: London, 1904, p 436 and 619.
83. K. Mislow; J. Siegel *J. Am. Chem. Soc.* **1984**, *106*, 3319-3328.
84. R. Dagani *Chem. Eng. News* **1984**, *June 11*, 21-23.
85. R. S. Cahn; C. K. Ingold; V. Prelog *Angew. Chem. Int. Ed. Engl.* **1966**, *5*, 385-415, 511.
86. V. Prelog; G. Helmchen *Angew. Chem. Int. Ed. Engl.* **1982**, *21*, 567-583.
87. L. C. Cross; W. Klyne, collators *Pure Appl. Chem.* **1976**, *45*, 11-30.
88. M. Chérest; H. Felkin; N. Prudent *Tetrahedron Lett.* **1968**, 2199-2204.
89. N. T. Anh *Topics in Current Chemistry* **1980**, *88*, 145-162.
90. D. Seebach; R. Naef *Helv. Chim. Acta* **1981**, *64*, 2704-2708.
91. V. Prelog; G. Helmchen *Helv. Chim. Acta* **1972**, *55*, 2581-2598.
92. C. E. Wintner *J. Chem. Educ.* **1983**, *60*, 550-553.
93. S. Masamune; T. Kaiho; D. S. Garvey *J. Am. Chem. Soc.* **1982**, *104*, 5521-5523.
94. E. L. Eliel; S. H. Wilen *Chem. Eng. News* **1990**, *Sep 19*, 2.
95. C. H. Heathcock; C. T. Buse; W. A. Kleschick; M. C. Pirrung; J. E. Sohn; J. Lampe *J. Org. Chem.* **1980**, *45*, 1066-1081.
96. R. Noyori; I. Nishida; J. Sakata *J. Am. Chem. Soc.* **1981**, *103*, 2106-2108.
97. C. H. Heathcock In *Asymmetric Synthesis*; J. D. Morrison, Ed.; Academic: Orlando, 1984; Vol. 3, p 111-212.
98. M. A. Rosanoff *J. Am. Chem. Soc.* **1906**, *28*, 114-121.
99. E. Ruch *Theor. Chim. Acta (Berl.)* **1968**, *11*, 183-192.
100. E. Ruch *Acc. Chem. Res.* **1972**, *5*, 49-56.
101. E. L. Eliel; S. H. Wilen; L. N. Mander *Stereochemistry of Organic Compounds*; Wiley-Interscience: New York, 1994.
102. D. Seebach; V. Prelog *Angew. Chem. Int. Ed. Engl.* **1982**, *21*, 654-660.
103. W. Klyne; V. Prelog *Experientia* **1960**, *16*, 521-523.
104. F. A. Carey; M. E. Kuehne *J. Org. Chem.* **1982**, *47*, 3811-3815.
105. K. R. Hanson *J. Am. Chem. Soc.* **1966**, *88*, 2731-2742.
106. C. H. Heathcock; B. L. Finkelstein; E. T. Jarvi; P. A. Radel; C. R. Hadley *J. Org. Chem.* **1988**, *53*, 1922-1942.
107. C. H. Heathcock *Chem. Eng. News* **1991**, *Feb 4*, 3.
108. P. Deslongchamps *Stereoelectronic Effects in Organic Chemistry*; Pergamon: Oxford, 1983.
109. R. M. Beesley; C. K. Ingold; J. F. Thorpe *J. Chem. Soc.* **1915**, *107*, 1080-1106.
110. C. K. Ingold *J. Chem. Soc.* **1921**, *119*, 305-329.
111. M. E. Jung; B. T. Vu *Tetrahedron Lett.* **1996**, *37*, 451-454.
112. G. C. Maitland; M. Rigby; E. B. Smith; W. A. Wakeham *Intermolecular Forces. Their Origin and Determination*; Clarendon: Oxford, 1981.
113. T. Kihara *Intermolecular Forces*; Wiley: New York, 1976.
114. H. Margenau; N. R. Kestner *Theory of Intermolecular Forces*; Pergamon: Oxford, 1969.
115. U. Burkert; N. L. Allinger *Molecular Mechanics. ACS Monograph 177*; American Chemical Society: Washington, DC, 1982.

Chapter 2

Analytical Methods

2.1 The importance of analysis, and which method to choose

Central to the art and practice of stereoselective synthesis is the analysis of the outcome of the reaction. Given the results, we may compare them to our intent (or hope) and act accordingly. Additionally, we may try to understand the factors that governed the formation of the observed products.

The separation of the enantiomorphous crystals of racemic sodium ammonium tartrate by Pasteur in 1848, and his observation that the two forms were optically active in solution, linked the concept of molecular chirality to optical activity [1]. When Emil Fischer began the first serious attempts at asymmetric synthesis in the latter 19th century, the polarimeter was the most reliable tool available to evaluate the results of an enantioselective reaction (by determination of optical purity), and it remained the primary tool for nearly 100 years. Only recently has analytical chemistry brought us to the point where we can say that polarimetry has been superceded as the primary method of analysis in asymmetric synthesis.

It is apparent from the preceding chapter that the analysis of enantiomers (by whatever means) addresses only part of the problem: often, a stereoselective reaction produces a mixture of diastereomers, and polarimetry is an inappropriate technique. Thus, asymmetric synthesis requires the means for the analysis of both enantiomeric and diastereomeric mixtures. Ultimately, the *ratio* of isomers and the *configuration* of each new stereocenter should be determined.

In choosing a technique for the analysis of a stereoselective reaction, a number of questions must be addressed:

1. How much material is available for the analysis, and what limits of detection are desired?
2. Are the stereoisomers enantiomeric or diastereomeric, and how many possible stereoisomers are there?
3. Do the products have a chromophore that might aid analysis by CD/ORD or UV spectroscopy, or detectability by HPLC?
4. Do the products have a functional group available for derivatization by a chiral or achiral reagent, or for interaction with a stationary phase in chromatography or with a chiral agent in solution?
5. If polarimetry is to be used for the analysis of enantiomers, is the specific rotation of the pure enantiomer known with certainty, or will it have to be determined?
6. Once the ratio of stereoisomers is determined, how will the configuration of each new stereocenter be assigned? Can the same method be used for the determination of product ratio *and* the assignment of configuration?

It should be recognized from the outset that an important aspect of any analysis of stereoisomer distribution is that the analysis reflect the ratio in the crude product without unintended enrichment by chromatographic means during sample purification and preparation. Most readers are probably aware that chromatography can separate diastereomers, so that care must be taken to insure that the diastereomer mixture being analyzed is the same as that produced in a reaction. In 1983, Crooks reported that *enantiomer enrichment* occured upon chromatographic purification of a partly racemic nicotine sample [2]. Such enrichment has since been recorded for sulfoxides [3], amino acid derivatives [4-6], biaryls [5], alcohols [5,7], ketones such as the Wieland-Mischer ketone [8], and drugs such as chlormezanone and camazepam [5], and the list is growing (see Figure 2.1). The phenomenon is not observed with totally racemic samples. A likely explanation for this type of enrichment is that the chromatography is separating a heterochiral dimer of the analyte from the monomer (or a homochiral dimer). In support of this hypothesis, Matusch and Coors [5] showed that the phenomenon was more pronounced with higher column loadings (higher concentration). On the other hand, Dreiding and colleagues were unable to find any evidence for dimerization of the Wieland-Mischer ketone, either polarimetrically or spectroscopically [8]. In the latter case, it may be that the aggregation is taking place in the higher concentrations that occur at or near the surface of the stationary phase. *Thus, when preparing a sample for analysis, pooled fractions should include those eluting before and after the "main band" so as to minimize any adventitious stereoisomer enrichment.*

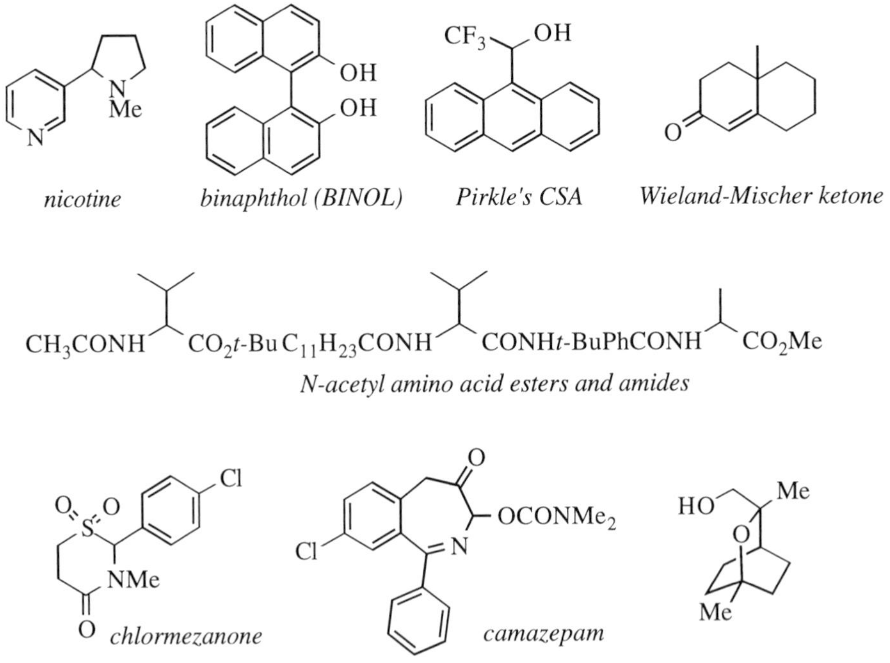

Figure 2.1. Compounds known to undergo enantiomeric enrichment upon chromatography on an achiral stationary phase such as silica gel [2-8].

Two types of analysis exist: those that separate the stereoisomers and those that do not. Polarimetry of course, fits the latter category. Separation is also not necessary for NMR analysis of a derivatized sample or with the aid of a chiral solvating agent or shift reagent. Chiral stationary phase (CSP) chromatographic techniques such as capillary GC or HPLC obviously *do* separate the analyte isomers, and may also facilitate the assignment of absolute configurations. The sections which follow describe the advantages and disadvantages of some of the more popular methods. If possible, more than one technique should be employed when a new synthetic method is being developed; use of procedures whose stereochemical consequences are well established may often rely on a single type of analysis.

2.2 Polarimetry, the old-fashioned way[1]

Polarimetry measures the rotation of a plane of monochromatic polarized light after having passed through a sample, as shown schematically in Figure 2.2.

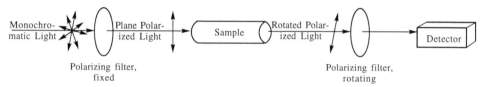

Figure 2.2. Schematic representation of a polarimeter.

It is not intuitively obvious why a chiral medium should have this effect, until the linearly polarized light beam, represented by the sine wave in Figure 2.3a, is broken down into the two circularly polarized components shown in Figures 2.3b and c.[2] When the linearly polarized beam passes through a perpendicular plane, the point of intersection moves along a line. When the circularly polarized beams pass

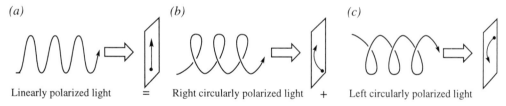

Figure 2.3. Representations of the waveforms of polarized light beams passing through a perpendicular plane.

through the same plane (the helices are moved without being rotated), the point of intersection describes a circle, moving either to the right or the left depending on the chirality sense of the helix. Note that the vector sum of the right- and left-circularly polarized beams equals the linearly polarized beam. The right- and left-circularly polarized beams are refracted equally by an achiral medium; that is, their refractive indices n_R and n_L (which measure change in velocity), are equal. As shown in Figure 2.4a, the vector sum of the two refracted circularly polarized beams remains in the plane of the incident polarized light, *i.e.*, the plane is not

1   Monographs: [9-15].

2   This analysis is an oversimplification. For a thorough treatment, see ref. [16].

rotated. In a medium where n_L and n_R are not equal, the two beams are shifted out of phase, and their vector sum is rotated out of plane by an angle, α, as shown in Figure 2.4b. A medium which produces this effect is *circularly birefringent*. A solution of a pure enantiomer is circularly birefringent. In contrast, an equimolar mixture of two enantiomers will have an equal number of refractions to the right and left, and the net result will be $\alpha = 0$. Thus, a polarimeter cannot distinguish an achiral compound from a racemate. It was Pasteur's discovery of circular bire-fringence in solutions of enantiomorphous crystals of racemic sodium ammonium tartrate [1] that set the stage for the development of stereochemical theory by establishing the presence of chiral molecules in an optically inactive compound.

Figure 2.4. The effect on the transmitted plane by refraction of circularly polarized light beams relative to the incident plane of polarized light (dashed line). *(a)* When $n_L = n_R$, the vector sum (solid arrows) of the two circularly polarized beams (dashed arrows) remains in the same plane as the incident beam. *(b)* When $n_L \neq n_R$, the vector sum of the two waves is rotated $\alpha°$ away from the plane of the incident light.

To summarize, the differential refraction of right- and left-circularly polarized light by a chiral nonracemic substance results in the rotation of the plane of the sine wave that is the vector sum of the two circularly polarized beams. That the two circularly polarized beams should be refracted differently by a chiral substance is apparent if one considers their helicity and imagines the interaction of a helix with a chiral substance in the context of double asymmetric induction explained in the previous chapter: any chiral entity will interact differently with the two enantiomeric forms of another chiral entity. On a macroscopic scale, we can easily perceive with our right hand the difference between a right- and left-handed screw, just as a chiral molecule may detect the difference between right- and left-circularly polarized light. On the molecular scale, whether n_R and n_L differ enough to be measured depends on the system. If the 'refractivity'[3] of the various ligands around a stereocenter in a chiral molecule are nearly the same, the difference between n_R and n_L may be too small to detect, and no rotation will be observed. From a practical standpoint, it may be possible to change the wavelength to increase the difference in n_R and n_L.

---

[3] This term is used loosely, and is related to the polarizability of each ligand. Interestingly, before the days of IR and NMR spectroscopy attempts were made to quantify the refractive index of individual functional groups as a means of deducing structure. For a summary of 'Molar Refraction,' see S. Glasstone *Physical Chemistry*, Van Nostrand: Princeton (1946) pp. 528-534, and other texts of the same period.

In addition to polarimetry, other chiroptical properties may be useful for the assignment of absolute configuration, although they are rarely used to determine enantiomeric purity [9-12,17,18]. Optical rotatory dispersion, ORD, measures the optical rotation of a compound as a function of wavelength, and its theory is the same as for simple polarimetry described above. Circular dichroism, CD, is similar, but differs in that the substrate must have a chromophore that absorbs at the wavelengths employed. In this special case, the molar absorptivities (extinction coefficients) of the right- and left-circularly polarized beams are different. Thus, in addition to being out of phase, the vectors of the transmitted beams are also of unequal magnitude. As a result, the emergent beam no longer traverses a line, but describes an ellipse, and the emergent light is *elliptically polarized.* In the region of such a CD band, the ORD exhibits 'anomalous' behavior (a Cotton effect) due to the absorption. The mean wavelength between an ORD peak and trough [9] is close to the λ_{max} of the chromophore absorbing the light. It is not unusual for the ORD curve to change sign in such a region. Because ORD measures a rotation, it is theoretically finite at all wavelengths, but since CD measures a difference in absorption, it only occurs in the vicinity of an absorption band.

The degree of rotation observed in a polarimeter, α, is dependent on the number of chiral species the light encounters on its passage through the sample chamber, as well as the wavelength of the light. Thus, analytical accuracy dictates strict control of a number of experimental parameters, such as temperature, concentration, light source, and path length. To minimize the effects of these variables and to increase the reproducibility, specific rotation, $[\alpha]$, is defined as:

$$[\alpha]_\lambda^T = \frac{100\alpha}{l \cdot c} , \qquad (2.1)$$

where T is the temperature, λ is the wavelength of the light (often the D lines of sodium at 589.0 and 589.6 nm and abbreviated simply 'D'), α is the observed rotation, l is the sample path length in decimeters, and c is the concentration in grams per 100 milliliters of solution. To insure reproducibility, it is common practice to report the concentration and solvent along with the specific rotation, and the units are understood.[4] For example, if a solution of 0.014 g in 1.0 mL of ethanol solution afforded a measured rotation of +1.375°, the specific rotation would be reported as:

$$[\alpha]_D^{25} +98 \ (c = 1.4, EtOH) .$$

This denotes a specific rotation of +98 deg·mL/g·dm measured at the D line of sodium, temperature 25 °C, at a concentration of 1.4 grams per 100 milliliters of ethanol. For pure liquids (or solids), the equation

$$[\alpha]_\lambda^T = \frac{\alpha}{l \cdot \rho} \qquad (2.2)$$

is used, where ρ is the density in grams per milliliter.

---

[4]   It is incorrect to report specific rotation in "degrees."

Specific rotation was defined over 150 years ago, which accounts for the unusual units of path length and concentration: decimeters were used because a long path length was needed to get an accurate measurement, and mass was used instead of molecular weight because molecular weights were uncertain in the early 19th century. The D line of sodium was chosen because it is easily produced in a flame and is nearly monochromatic. Now that molecular weights are no longer an unknown, molecular rotation, [Φ], may be used instead of [α]:

$$[\Phi]_\lambda^T = \frac{M[\alpha]}{100} , \qquad (2.3)$$

where M is the molecular weight. Molecular rotation is commonly used in ORD.

Sign of rotation reflects absolute configuration (and is often used to assign it), and the magnitude of the rotation is used to determine the optical purity, usually expressed as a percent:

$$\% \text{ optical purity} = \frac{100[\alpha]_\lambda^T}{[\alpha_0]_\lambda^T} , \qquad (2.4)$$

where $[\alpha]_\lambda^T$ is the observed specific rotation, and $[\alpha_0]_\lambda^T$ is the specific rotation of the pure enantiomer under identical conditions. The optical purity of an enantiomerically pure compound is 100%, and 0% for a racemate. Ideally, the specific rotation of a partly racemic mixture varies linearly with enantiomeric composition. Thus, a 3:1 mixture of a enantiomers whose $[\alpha_0] = +98$ should exhibit $[\alpha] = +49$, and the optical purity would be 50%.

For a chiral compound, percent enantiomer excess (ee) is defined as:

$$\% \text{ enantiomeric excess} = 100 \frac{R-S}{R+S} , \qquad (2.5)$$

where R and S represent the amounts of the two enantiomers. Thus, a 3:1 mixture of two enantiomers is 50% ee, which expresses the excess of one enantiomer over the racemate.

The optical purity is usually, *but not always,* equal to enantiomer excess. In order for the two to be equal, it is necessary that there be no aggregation. It is possible, for example, that a homochiral or heterochiral dimer (see Glossary, Section 1.6, for definitions) would refract the circularly polarized light differently than the monomer (or each other). In 1968 [19] Krow and Hill showed that the specific rotation of (S)-2-ethyl-2-methylsuccinic acid (85% ee) varies markedly with concentration, and even changes from levorotatory to dextrorotatory upon dilution. In 1969 [20], Horeau followed up on Krow and Hill's observation, and showed that the "optical purity" (at constant concentration) and enantiomer excess of (S)-2-ethyl-2-methylsuccinic acid were unequal except when enantiomerically pure or completely racemic. This deviation from linearity is known as *the Horeau effect,* and its possible occurence should be remembered when determining enantiomeric purity by polarimetry.

For optical purity to accurately reflect enantiomeric purity, it is obvious that the sample must be free of any chiral impurities. It may not be as obvious that achiral impurities can also cause significant error. For example, Yamaguchi and Mosher [21] showed that the specific rotation of enantiomerically pure 1-phenylethanol could be enhanced from $[\alpha]_D^{20}$ +43.1 (c = 7.19, cyclopentane) to $[\alpha]_D^{20}$ +58.3 (c = 2.64, cyclopentane) by the addition of 10.6 g/100 mL of acetophenone. Presumably, this enhancement is due to an interaction between the alcohol and the ketone, either through hydrogen bonding or hemiacetal formation.

In order to determine optical purity, it is necessary to know $[\alpha_o]$. In natural product synthesis, the rotation of the target is usually known, but the original authors may not have established that the isolated material was enantiomerically and chemically pure. One course of action is to resolve a sample and measure $[\alpha_o]$ yourself. This may prove tedious, but it has the advantage of eliminating any ambiguities between sample and standard. Another possibility is to calculate $[\alpha_o]$ by isotopic dilution or kinetic resolution.

In the isotopic dilution technique [22], an enantiomerically enriched sample of unknown %ee is diluted with a second, isotopically labelled sample of the same compound of known %ee (usually a racemate) and known isotope content. Measurement of the isotopic content after dilution, and comparison of rotations before and after dilution, allows extrapolation to rotation at 100% ee.

Kinetic resolution may be used to determine $[\alpha_o]$ in two ways [23,24]. The results of two kinetic resolutions of a racemic compound, allowed to go to different (known) extents of conversion, can be used to calculate the specific rotation of the enantiomerically pure compound. Alternatively, one may use two 'reciprocal' kinetic resolutions: racemic A is resolved by B *and* racemic B is resolved by A. If the racemate is used in large excess in both cases, and the stereoselectivities of the resolutions are not too high, $[\alpha_o]$ for A may be calculated if $[\alpha_o]$ for B is known, or vice versa.

The discovery of the anomalies mentioned above are partly responsible for the declining popularity of polarimetry for the determination of enantiomer ratios. Even if the experimentalist is alert to these sources of error, the possibility still exists that an early determination of specific rotation, against which a new value must be compared, is itself in error. Thus, caution is advised. Nevertheless, if used carefully, polarimetry can provide a simple, efficient, and inexpensive method for the analysis of enantiomeric purity.

2.3 Nuclear magnetic resonance

For the analysis of diastereomeric mixtures, NMR is an obvious choice, and derivatization of enantiomers with a chiral reagent can also be an excellent method of analysis (reviews: [25,26]). In the 1960s, a number of discoveries were made that facilitated the direct observation of diastereomeric and enantiomeric mixtures by NMR. The development of chiral derivatizing agents, lanthanide shift reagents, and chiral solvating agents made it possible to observe (and integrate) separate signals for enantiomers. The following discussions elaborate on each.

2.3.1 Chiral derivatizing agents (CDAs)

The derivatization of a mixture of enantiomers with a chiral reagent produces diastereomers that may be analyzed by NMR spectroscopy or by chromatography [27]. In order to be useful, a number of requirements must be met:

1. The CDA must be enantiomerically pure, or (less satisfactorily) its enantiomeric purity must be known accurately.
2. The reaction of the CDA with both enantiomers must go to completion under the reaction conditions, or (again less satisfactorily) the relative rate of reaction for each enantiomer must be known.
3. The CDA must not racemize under the derivatization or analysis conditions, and its attachment should be mild enough so that the substrate does not racemize either.
4. If analysis is by HPLC, the CDA should have a chromophore to enhance detectability. If analysis is by NMR, the CDA should have a functional group that gives a singlet and that is remote from other signals for easy integration.

The importance of the first point is evident if we consider the following reactions of an analyte with a CDA:

$$\text{Analyte } (R + S) + \text{CDA } (R') \rightarrow \text{Diastereomeric derivatives } R\text{–}R' + S\text{–}R' \quad (2.6)$$

$$\text{Analyte } (R) + \text{CDA } (R'+S') \rightarrow \text{Diastereomeric derivatives } R\text{–}R' + R\text{–}S' \quad (2.7)$$

Equation 2.6 illustrates the derivatization of a mixture of R and S enantiomers of analyte with an enantiomerically pure derivatizing agent R'. If the reaction is complete for both enantiomers of analyte, the ratio of diastereomeric derivatives R–R' and S–R' will equal the ratio of enantiomers R and S of the analyte. Equation 2.7 shows how a CDA that is not enantiomerically pure can cause problems. If there is no kinetic resolution of the CDA by the analyte, the ratio of diastereomeric derivatives R–R' and R–S' will reflect the diastereomer ratio of the CDA. Note that S–R' (Eq. 2.6) and R–S' (Eq. 2.7) are enantiomeric and standard methods of analysis will not distinguish them. If both the analyte and the CDA are 90% ee (95:5 ratio of enantiomers), the four possible diastereomers will have the statistical ratio of .9025/.4075/.4075/.0025 (Eq. 2.8). Since the products are two diastereomeric racemates, combination of the enantiomers yields a .9050/.0950 ratio, or 81% ee.

$$\begin{array}{cccccc}
.95R/.05S & +.95R'/.05S' & \rightarrow & R\text{–}R' & + \ S\text{–}R' \ + & R\text{–}S' \ + & S\text{–}S' \\
\text{Analyte} & \text{CDA} & & .9025 & .0475 & .0475 & .0025
\end{array} \qquad (2.8)$$

Although a number of CDAs have been developed over the years [28], by far the most popular is Mosher's acid [29-31], α-methoxy-α-(trifluoromethyl)phenylacetic acid, abbreviated MTPA. It is commercially available in either enantiomeric form, and is used for the derivatization of alcohols and amines. Two recent reports [32,33] indicate that the enantiomeric purity of commercially available material may vary from 94 to 99.8% ee, and one might expect similar levels of enantiomeric purity from the original preparation [29]. The enantiomeric purity of MTPA may be determined by esterification of diacetone glucose and examination of the NMR [33], or more accurately by chiral stationary phase gas chromatography of the MTPA methyl ester [32].

Derivatization of chiral alcohols and amides of general structure RCHZR' (Z = OH or NH$_2$) yields esters and amides that are frequently referred to as 'Mosher esters' or 'Mosher amides.' ^{1}H, ^{13}C, or ^{19}F NMR may be used to observe the diastereomeric derivatives [29,34]. Most commonly, the –OCH$_3$ is observed by ^{1}H NMR or the –CF$_3$ is observed by ^{19}F NMR. In most cases, one or the other of these nuclides will be well enough separated that accurate integration will be possible. In problematic cases, additional spectral dispersion may be obtained by adding a lanthanide shift reagent such as Eu(fod)$_3$ (fod = 6,6,7,7,8,8,8-heptafluoro-2,2-dimethyl-3,5-octanedionato ligand) to the NMR sample [21,35-37].[5]

Models have been proposed to correlate chemical shift data with absolute configuration [30]. In 1973, Mosher observed that derivatization of enantiomerically pure esters and amides with both enantiomers of MTPA and comparison of chemical shifts produced some interesting trends. With reference to the top structures in Figure 2.5[6] (L$_2$ is smaller than L$_3$), two experimental trends were observed:

1. When *R*-MTPA was used, the ^{19}F chemical shift was at lower field than when the *S* enantiomer was used [39].
2. When *R*-MTPA was used, the ^{1}H signals in L$_2$ were at higher field than the ^{1}H signals in L$_2$ when the *S*-enantiomer was used [30], a trend that holds except when the secondary alcohol or amine is significantly hindered [40,41].

Two situations may arise in practice: determination of absolute configuration of a pure enantiomer, and assignment of configuration of the major enantiomer of a mixture. When a steric difference exists between L$_2$ and L$_3$ (L$_3$ larger), the absolute configuration can be determined by correlation with the known examples. Assuming that the Cahn-Ingold-Prelog priority of the ligands around the unknown

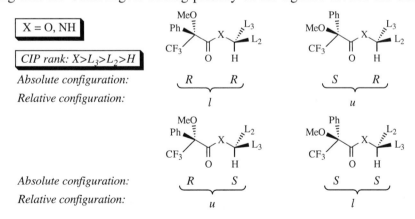

Figure 2.5. Diastereomeric MTPA derivatives. Note that *l* or *u* isomers may be obtained by derivatization of a single enantiomer with racemic MTPA (horizontal pairs) or by derivatization of a racemate with enantiomerically pure MTPA (vertical pairs).

---

[5] Note that this technique was applied in the mid 1970s using low-field spectrometers. It may not be useful on high-field spectrometers. See Section 2.3.3, especially pp. 56-57.

[6] A crystal structure of an *O*-methylmandelate ester has been obtained [38] that supports this shielding model.

center is X>L_3>L_2>H, derivatization with both enantiomers of MTPA will give *R-R* (relative configuration *l*) and *S-R* diastereomers (relative configuration *u*), illustrated by the top two structures in Figure 2.5. By derivatizing a pure enantiomer with racemic MTPA, the absolute configuration may be determined. From the data gathered by Mosher and Yamaguchi, the following empirical rules can be stated:

1. The ^{1}H L_2 signals of the *l* diastereomer will be upfield of the L_2 signals of the *u* isomer.
2. The ^{1}H L_3 signals and the ^{19}F CF_3 signals of the *l* diastereomer will be downfield of the corresponding signals of the *u* isomer.

If an unequal mixture of enantiomers is present, symmetry considerations dictate that derivatization with only one MTPA enantiomer is necessary, since (Figure 2.5) derivatization of a racemate with one enantiomer of MTPA also produces an *l/u* mixture. Thus either the two left structures or the two right ones could be used to establish configurations. Again assuming the Cahn-Ingold-Prelog priority of the ligands around the unknown center is X>L_3>L_2>H, the following empirical rules apply:

1. If *R*-MTPA is used, the *R* configuration at the unknown stereocenter will give L_2 signals upfield of the *S* diastereomer and L_3 and CF_3 signals downfield of the *S*-diastereomer.
2. If *S*-MTPA is used, the *S*-configuration at the unknown stereocenter will give L_2 signals upfield of the *R* diastereomer and L_3 and CF_3 signals downfield of the *R*-diastereomer.

As a model for assignment of absolute configuration of Mosher esters and amides in the presence of Eu(fod)$_3$, Yamaguchi proposed the equilibria shown in Figure 2.6. For both diastereomers, the europium is chelated by the MTPA carbonyl and methoxy oxygens. In the top case, the smaller carbinol ligand, L_2, is closest to the europium and is more deshielded than L_3. In the bottom case, (larger) L_3 is closer to the europium, causing steric repulsion and disfavoring the equilibrium as drawn. Because of this repulsive interaction, K_R is larger than K_S, and the top ester-europium complex is more abundant. This model therefore predicts that the sterically less demanding ligand should be more deshielded in the *R* configuration.

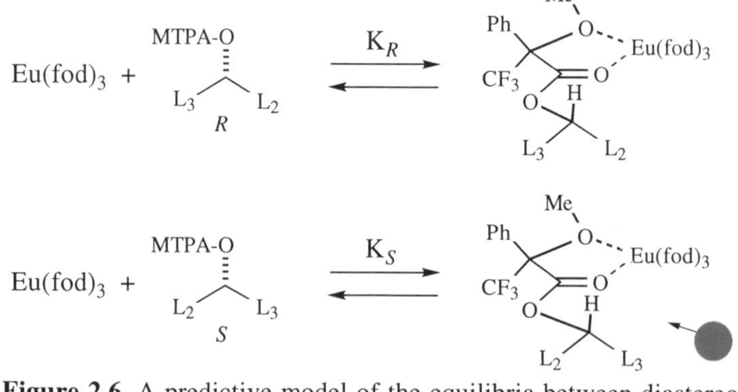

Figure 2.6. A predictive model of the equilibria between diastereo-meric Mosher esters and a europium shift reagent.

In conclusion, two points must be emphasized. First, the rationales presented in Figures 2.5 and 2.6 are only models, and do not necessarily represent preferred conformations.[6] Second, it should be restated that in order for the CDA method to be accurate, any adventitious kinetic resolution in the derivatization must be quantitated or eliminated. For example, Heathcock has noted that MTPA derivatization of a racemic alcohol (0% ee) afforded a 1.7:1 mixture of Mosher esters (26% de) and the % ee determinations had to be corrected accordingly [42]. More recently, Svatos used a five-fold excess to force a derivatization to completion [43]. If the appropriate control experiments are done, derivatization with Mosher's reagent can be a very reliable method for determination of enantiomer ratios and absolute configuration of amines and alcohols. For the derivatization of ketones, chiral diols may be used [44], but similar control experiments should be undertaken.

2.3.2 Achiral derivatizing agents

Imaginative tricks can also be used to analyze enantiomeric mixtures. For example an achiral, bifunctional derivatizing agent may be used to randomly dimerize a mixture of enantiomers. If a statistical ratio can be proven in control experiments, the ratio of the chiral to meso diastereomers can be used to calculate the enantiomer composition [45]. Figure 2.7 shows several derivatizing agents that are available. In principle, the ratio could be determined by chromatographic or spectroscopic methods. ^1H, ^{13}C and ^{31}P NMR provide a particularly facile method for the analysis of alcohols [45-47]. The following generic reaction indicates the process:

$$R + S + AX_2 \rightarrow RAR + SAS + RAS + SAR \quad , \qquad (2.9)$$

stoichiometry: 1   x   *probabilities:* $1 \cdot 1$ x^2 $1 \cdot x$ $1 \cdot x$

where R and S indicate the absolute configuration of the alcohols, AX_2 is the 'dimerization' reagent. RAR and SAS are a d,l pair, while RAS and SAR are meso. The latter may or may not be identical,[7] but it is necessary to recognize (statistically) that either an '$SAX + R$' or an '$RAX + S$' sequence would produce a meso isomer. If the S/R ratio is x, then the probability for the formation of the d,l pair is $(1 + x^2)$ and $2x$ for the meso isomer(s). Thus the d,l/meso ratio is given by

$$\frac{d,l}{meso} = y = \frac{1 + x^2}{2x} . \qquad (2.10)$$

Solving for x gives

$$x = \text{enantiomer ratio} = \frac{2y \pm \sqrt{4y^2 - 4}}{2} . \qquad (2.11)$$

There is an ambiguity in this determination in that the mathematics is oblivious to absolute configuration, hence the "±" in the quadratic formula. Thus, although x was defined as *S/R*, the solutions will be *S/R* and *R/S*.

---

[7]   If 'A' is stereogenic in the products, the two will be diastereomers. For example, derivatization of alcohols with PCl_3 affords phosphonates in which the phosphorus is stereogenic and two meso isomers are produced [46].

$$R*OH \ + \ AX_2 \quad \longrightarrow \quad \begin{array}{ll} (R*O)_2A & \textit{d,l pair} \\ \\ (R*O)_2A & \textit{meso} \end{array} \quad \longrightarrow \quad R*OH$$

$$AX_2 = COCl_2, \ CH_2(COCl)_2, \ C_6H_4(COCl)_2, \ Me_2SiCl_2, \ Ph_2SiCl_2, \ PCl_3$$

Figure 2.7. Achiral derivatizing agents for the determination of enantiomer ratio and partial resolutions [45-47].

It is also of interest to note that, on a preparative scale, this method may be used to enrich the enantiomer ratio of a partially resolved racemate [45]. For example, dimerization of a 9:1 mixture (80% ee), followed by separation and then cleavage of the *d,l* isomer affords an 81:1 mixture of enantiomers (98.8% ee) with an 82% theoretical yield.

2.3.3 Chiral shift reagents (CSR)

The notion that enantiotopic groups would be anisochronous in a chiral environment was first suggested by Mislow in 1965 [27], and is the basis for the analysis of enantiomer ratios by chiral shift reagents and chiral solvating agents. Figure 2.6 illustrated that two *diastereomeric* complexes between an achiral lanthanide shift reagent and a ligand may afford differential deshielding of nuclides in the ligand. In these examples, enantiotopic nuclides were rendered diastereotopic (*i.e.,* placed in a chiral environment) by virtue of the MTPA derivatization. A simpler alternative is to use a chiral additive that renders the nuclei diastereotopic in a supramolecular complex. Two types of additive will be discussed: chiral shift reagents (CSR) and chiral solvating agents (CSA). Advantages of the CSR method are:
1. The chiral shift reagent need not be enantiomerically pure.
2. There can be no accidental resolution, deresolution, or racemization during a derivatization (but beware of enantiomer enrichment during sample purification, *vide supra*).
3. A wide range of functional groups can be analyzed with this technique, since all that is required is a Lewis basic atom to coordinate to the lanthanide.

Disadvantages are that absolute configurations cannot generally be determined without reference to a known sample, and that both enantiomers must be available to insure peak separation. An additional disadvantage has developed with the advent of high-field NMR spectrometers: the technique is not as effective at high fields, as explained below.

The first CSR was introduced by Whitesides in 1970 [48]. A 1973 monograph covered the details of lanthanide shift reagents [49], and two reviews have since covered chiral lanthanide shift reagents [50,51]. For our purposes, there are a few things we ought to know about shift reagents themselves, and about how they interact with ligands. Lanthanide shift reagents are *tris* complexes of β-diketonate ligands. For symmetrical, achiral diketones, the complexes exist as an equilibrating mixture of two enantiomeric forms, the Δ and Λ. If the ligand is an unsymmetrical diketone, the Δ and Λ isomers each exist as cis and trans (fac, mer) isomers. If the diketone ligand is chiral, the Δ and Λ forms are diastereomers. Thus, a lanthanide (chiral) *tris*-(diketonato) complex is an equilibrating mixture of four diastereomers.

Assuming that coordination of an additional ligand is an outer sphere phenomenon, each face of each octahedron is a potential binding site. For reasons of symmetry, a cis complex (of an unsymmetrical diketone) has four unique binding sites and a trans complex has two. These are illustrated for the cis-Δ and trans-Δ isomers in Figure 2.8 (there are six more for the Λ isomer). Thus, the four equilibrating lanthanide isomers may produce as many as twelve 1:1 complexes. To complicate matters further, the lanthanide-ligand stoichoimetry may not be 1:1, the shift reagents may not be monomers, and their structure may change on complexation with the extra ligand. Fortunately, a detailed understanding of CSR-ligand binding is not necessary for their use, although equilibration has sometimes been observed during the first few minutes after mixing [51]. Because the equilibration is fast, we can consider the chiral shift reagent as a single species, although the experimentalist should be aware of the possible complexity of the equilibrating systems in the event they are observed at the spectrometer. The important point to remember is that the fast equilibria yield time-averaged spectra.

Because the phenomenon of lanthanide induced shift results from a fast exchange phenomenon, the linewidths of the peaks are governed by the fast exchange approximation, Equation 2.12 [52,53]:

$$\text{linewidth} = \delta v = \frac{\pi (\Delta \delta)^2}{2k} \quad , \tag{2.12}$$

where δv is the linewidth, $\Delta \delta$ is the chemical shift difference (in Hz) for the nuclide

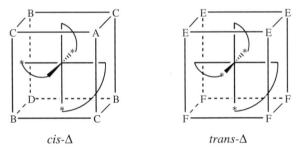

cis-Δ   *trans-Δ*

Figure 2.8 Schematic representation of the two Δ-diastereomers of an octahedral complex of an unsymmetrical bidentate ligand, showing the six (A - F) unique outer sphere coordination sites.

in question, and k is the rate constant for exchange. Since chemical shift anisochrony ($\Delta \delta$) is directly proportional to the field strength, H_0, it is clear that linewidth will increase with the square of the field strength; this problem is especially severe when $\Delta \delta$ is large, as is often the case in lanthanide shift reagents.[8] Since accurate integration is the objective of a CSR analysis, broad lines are undesirable and analysis is actually better accomplished at low field [54,55]! Since low field NMRs (≤100 MHz) are becoming increasingly rare, it is useful to note

---

[8]   This problem is exacerbated by the fact that the lanthanide atom is paramagnetic and its complexes have broad lines to start with.

that there is also a temperature dependence hidden in Equation 2.12. Since k is related to energy of activation, the following proportionality holds:

$$\text{linewidth} = \delta v \propto H_0^2 \, e^{\Delta G^{\ddagger}/RT} \quad , \tag{2.13}$$

which shows that linewidth is proportional (not only) to the square of the field strength, but also to the temperature. Although increasing the field of the spectro-meter broadens the lines, raising the temperature tends to counteract this effect. Operationally, it is wise to conduct a CSR analysis on the lowest field spectrometer available; if line broadening is a problem, warming the sample may help [55-57]. Failing that, spin-echo techniques may be used to eliminate broadened lines [54].

The interaction of a ligand with a lanthanide complex may result in a change in chemical shift, $\Delta\delta$, for some of the nuclides of the ligand, especially those that are in the spatial vicinity of the coordinating atom. If the shift reagent and the ligand are chiral, there may be different lanthanide-induced shifts for corresponding nuclei in the two enantiomers of the ligand (nuclides that are enantiotopic by external comparison). This induced anisochrony (chemical shift difference) is $\Delta\Delta\delta$. Equation 2.14 illustrates a simplified view of the equilibration of a racemate with a chiral shift reagent, in which the equilibria of the CSR are ignored so that the CSR may be considered as a single species.

$$(+)\text{-CSR}\cdot R \underset{}{\overset{K_R}{\rightleftarrows}} (+)\text{-CSR} + R,S \underset{}{\overset{K_S}{\rightleftarrows}} (+)\text{-CSR}\cdot S \tag{2.14}$$

Two possible mechanisms have been suggested as the source of $\Delta\Delta\delta$: $K_R \neq K_S$, or $(+)\text{-CSR}\cdot R$ and $(+)\text{-CSR}\cdot S$ have different geometries [58]. It is likely that both of these mechanisms operate to differing extents in various systems. Regarding K_R and K_S, note that nuclei that are enantiotopic by internal comparison, such as the methylene protons of benzyl alcohol or the methyls of dimethyl sulfoxide, can be differentiated by CSRs [59]. Clearly no stability difference is required for inducing anisochrony. An important consequence of this fact is that the enantiomeric purity of compounds that are chiral by virtue of isotopic substitution (*e.g.*, C_6H_5CHDOH) may be evaluated by this method (as well as by the CSA method described in the next section).

Since the spectrum observed is a time average of the free and CSR-bound ligand, the combination of the enantiomeric forms of a CSR with the enantiomeric forms of a ligand is a dynamic phenomenon. Because of this dynamic relationship, and in contrast to the 'static' derivatization discussed in the preceeding section, the CSR need not be enantiomerically pure. Consider the two extremes: the CSR is enantiomerically pure, and the CSR is racemic.[9] Equation 2.14 illustrates the binding of an enantiomerically pure $(+)$-CSR with the two enantiomers of a ligand.[10] The observed spectrum is a time average of the spectrum of each free

---

[9]   For simplicity, we will consider only homochiral *tris* complexes (*i.e.*, only complexes in which all three diketones have the same chirality sense). Dynamic exchange would in fact produce a number of heterochiral complexes, but on average their effects would cancel.

[10] A similar set of equilibria would result from analysis of an enantiomerically pure ligand by a racemic CSR (*cf.* Figure 2.5).

enantiomer and its (+)-CSR complex. On the other hand, if the CSR is racemic and both enantiomers of the ligand are present, Equation 2.15 applies. The spectrum of the *R* enantiomer would now be a time average of the free *R* enantiomer, the (+)-CSR·*R* complex, and the (–)-CSR·*R* complex. Likewise, the spectrum of the *S* enantiomer would be a time average of the free *S* enantiomer, the (+)-CSR·*S* complex, and the (–)-CSR·*S* complex. For reasons of symmetry (*e.g.*, (–)-CSR·*R* = (+)-CSR·*S*), the two time-averaged spectra will be identical, and the lanthanide-induced shifts will be the same (i.e., $\Delta\Delta\delta = 0$) if the CSR is 0% ee. An intermediate case, such as where the CSR is 80% ee, would produce a different time average, such that $\Delta\Delta\delta$ would decrease, but the integral of the peaks corresponding to the two enantiomers of the ligand would be the same.

$$(+)\text{-CSR}\cdot R + (-)\text{-CSR}\cdot R \; \rightleftarrows \; (\pm)\text{-CSR} + R,S \; \rightleftarrows \; (+)\text{-CSR}\cdot S + (-)\text{-CSR}\cdot S \qquad (2.15)$$

Lanthanides are 'hard' Lewis acids, and the best binding occurs with ligands that contain 'hard' Lewis basic atoms. Approximate binding affinities are primary amine > hydroxyl > ketone > aldehyde > ether > ester > nitrile [50]. Chiral shift reagents have also been used with sulfoxides, arsine sulfides, amino acids, and certain transition metal complexes [51]. Carboxylic acids decompose lanthanide diketonato complexes [58], and so they should be esterified before analysis. Other functional group interconversions may aid the analysis (increase $\Delta\Delta\delta$) by changing the binding characteristics [50]. Such a change might be desirable in several circumstances. For example, weak binding of a sterically hindered hydroxyl might be increased by acetylation (the binding site becomes the more accessible ester carbonyl). In multifunctional molecules, it might be worthwhile to block binding at one site in order to improve binding at another. This might be accomplished by trifluoro-acetylation of an alcohol or amine or by ketalization of a carbonyl.

Figure 2.9 illustrates the three ligands found in the most common and commercially available chiral shift reagents, and the abbreviations used for each. The tfc [60] and hfc [61] ligands are sold as the europium, ytterbium or praseodymium complexes, while the dcm ligand [58] is sold as the europium complex. Since $\Delta\Delta\delta$ is a function of concentration, temperature, and ligand, a comparison of "resolving power" among the different reagents is difficult. Nevertheless, for 1-phenylethanol and 1-phenylethyl amine, the largest $\Delta\Delta\delta$ for europium complexes was found for the dcm complex Eu(dcm)$_3$,[11] while Eu(tfc)$_3$ and Eu(hfc)$_3$ were about the same [51]. In choosing lanthanides, europium and ytterbium induce downfield shifts while praseodymium induces upfield shifts. Additionally, the three metals may also cause line broadening to differing extents [50]. For 1-phenylethanol and 1-phenylethyl amine, Pr(hfc)$_3$ induced larger shifts than Eu(hfc)$_3$, and did so at lower concentration [51]. Still, the concensus appears to be that no single CSR is superior with all possible ligands.

---

[11] Since the dcm ligand is a C_2-symmetric β-diketone, there is no *cis/trans (fac/mer)* isomerization in the complex. As a result, the number of outer sphere coordination sites is reduced from 12 to 4 (two for the Δ and two for the Λ isomers). Spectral averaging of fewer isomeric complexes may account for the larger differentiation by this ligand.

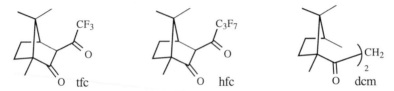

Figure 2.9 The most common ligands for chiral shift reagents: trifluoro-acetylcamphor (tfc) [60], heptafluorobutanoylcamphor (hfc) [61], and dicampholylmethane (dcm) [58].

Fraser recommends the following experimental protocol [51]:

1. Try as many as four CSRs, the approximate order of capacity being $Eu(dcm)_3 > Pr(hfc)_3 \approx Yb(hfc)_3 > Eu(hfc)_3$.
2. Try changing the temperature. Lower temperature can have a substantial influence on lanthanide-induced shifts [58,61,62], while warming may sharpen lines [55].
3. If still unsuccessful, try derivatizing the ligand to make it a stronger, harder, Lewis base.

Before conducting a CSR study, the experimentalist should consult Sullivan's review for detailed experimental guidelines [50]. Briefly, the guidelines suggest: dry the substrate, the solvent, and the CSR (by sublimation if prepared fresh or over phosphorous pentoxide *in vacuo* if purchased); keep the substrate concentration low (~0.1 - 0.25 M); add the CSR (either as a solid or as a concentrated solution) in small increments, and filter the solution after each addition (the molar ratio needed for a good induced shift is rarely >1:1, and too much lanthanide can broaden lines and even cause the induced shifts to decrease); re-shim the spectrometer after the CSR is added to compensate for the presence of the paramagnetic ions, and check for paramagnetic precipitates after the sample has been spinning for several minutes. Additionally, recall (*vide supra*) that the method is usually more effective at low field.

2.3.4 Chiral solvating agents (CSA)

Mislow's 1965 suggestion [27] that enantiotopic nuclides would be anisochronous in a chiral solvent was reduced to practice the next year by Pirkle [63], in the form of a chiral solvating agent.[12] In the intervening years, analysis of enantiomer ratios and the assignment of absolute configuration by chiral solvating agents has become a very useful tool. There are a number of features that distinguish the CSAs from the CSRs, and which highlight the complementarity of the two techniques [64]:

1. In contrast to the complex equilibria of the CSRs, CSAs are simple diamagnetic compounds. Since the dynamics of a CSA and its interaction with a solute may be reasonably well understood, deduction of absolute configuration is often possible.

---

[12] Note the distinction between the terms 'shift reagent' and 'solvating agent.' Because of the differences in the mechanism of binding and induced anisochrony, the former is reserved for lanthanide complexes and the latter for diamagnetic compounds.

2. Anisochrony in CSAs is usually induced in enantiotopic groups of a ligand by the presence of an anisotropic moiety in the CSA, such as an aromatic ring (as opposed to a paramagnetic metal atom).
3. The induced chemical shift changes of diamagnetic CSAs are not usually as large as with CSRs, and the range of structural types that can be addressed is not as broad. An advantage of smaller $\Delta\delta$ is that the effect of field strength on linewidth (*cf.* Equations 2.9 and 2.10) is not as problematic.
4. Because CSAs are diamagnetic, line broadening is not as much of a problem as with CSRs. Therefore, it is often possible to deduce enantiomer ratios by comparison of peak *height,* obviating the need for a complete separation of all lines in a shifted pair of multiplets.

As with CSRs, both enantiomers should be available to insure the presence of induced anisochrony.[13]

By far the most studied CSAs are the 1-(aryl)trifluoroethanols and the 1-(aryl)ethyl amines (aryl = phenyl, 1-naphthyl, 9-anthryl) that associate primarily through hydrogen-bonding mechanisms. Chiral acids are finding increased use for the analysis of amines as their diastereomeric salts,[14] although assignment of configuration is risky due to aggregation and other dynamic phenomena [64]. Figure 2.10 lists several of the readily available CSAs along with some of the structural types with which they have been used for determination of enantiomer excess and absolute configuration.[15]

The equilibria that describe the 1:1 interactions of a CSA and a pair of enantiomeric solutes (Equation 2.16) is similar to the one used to explain shift reagents (Equation 2.11). An important distinction is that we *assumed,* for the sake of simplicity, that the CSR was a single species. For chiral solvating agents, that assumption is not necessary, because it is fact. As a result, the analysis of the geometry of the diastereomeric solvates, (+)-CSA·*R* and (+)-CSA·*S*, often allows determination of absolute configuration.

$$(+)\text{-CSA·}R \underset{}{\overset{K_R}{\rightleftarrows}} (+)\text{-CSA} + R,S \underset{}{\overset{K_S}{\rightleftarrows}} (+)\text{-CSA·}S \qquad (2.16)$$

As was the case with chiral shift reagents, preferential population of one diastereomer over the other ($K_R \neq K_S$) is not a prerequisite for induced anisochrony of enantiotopic groups. Additionally, since the CSA is diamagnetic, it may be used in excess over the analyte. A five-fold excess is usually sufficient to drive the equilibria of Equation 2.16 to the "outside," such that the solute is present only as its two diastereomeric solvates. Since the observed spectra are time-averages of all the species in solution, this chemical trick simplifies analysis of absolute configuration by focussing on the diastereomeric solvates alone.

---

[13] In principle, induced anisochrony could also be established by studying an enantiomerically pure analyte with racemic CSA (*cf.* Figure 2.5).

[14] Chiral acids are often used for classical resolution of racemic amines. It is evident from this discussion that anisochrony in soluble diastereomeric salts might possibly be used to monitor the progress of such a classical resolution.

[15] For a more complete list organized by solute type, see ref. [65].

The first rationale for "recognition"[16] of enantiomers was the three-point model proposed by Easson and Stedman in 1933 to explain the interaction of racemic drugs with biological receptors [66]. A similar model was proposed by Ogston in 1948 to explain the enantioselectivity of enzyme reactions [67]. These simplistic models proposed three simultaneous *binding* interactions to explain enzyme enantio-specificity. Similarly, the best rationale for understanding induced anisochrony in enantiomers is based on three interactions, although all three do not have to be binding [64].

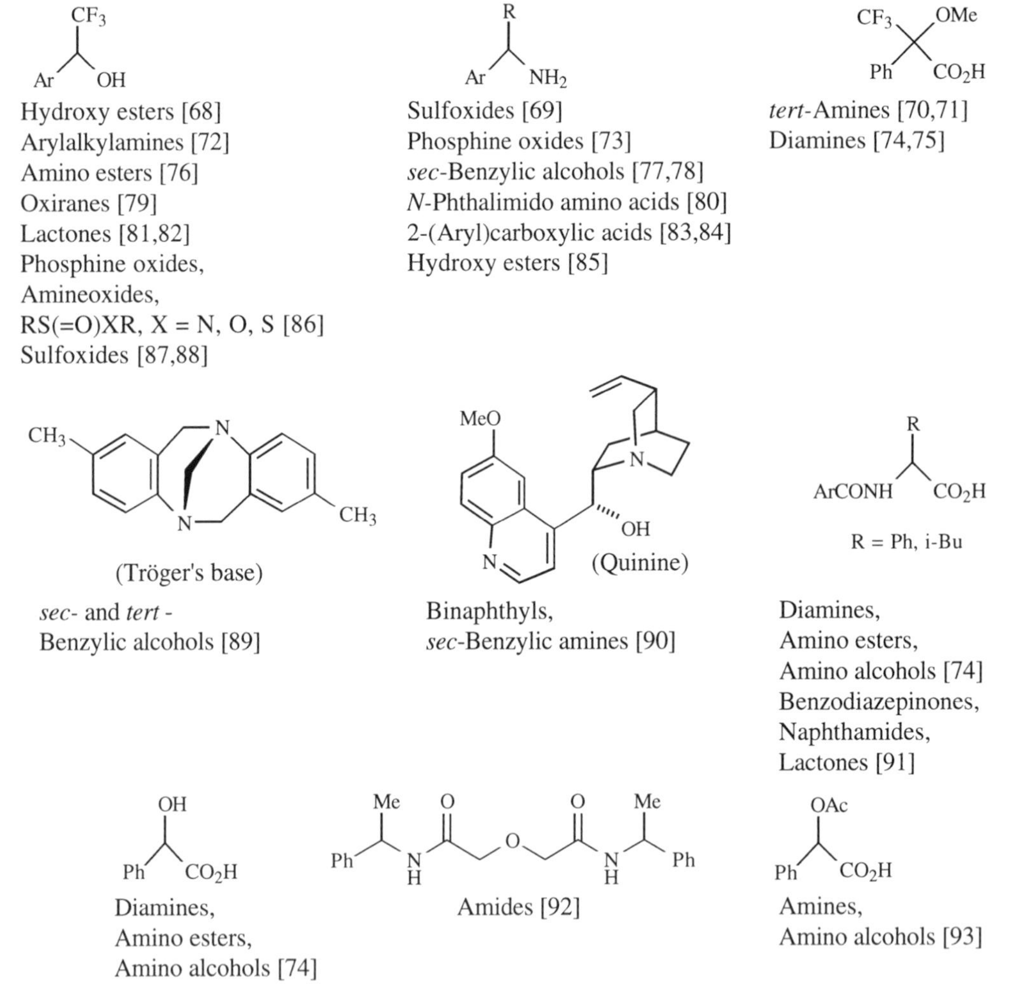

CF₃

Ar   OH

Hydroxy esters [68]
Arylalkylamines [72]
Amino esters [76]
Oxiranes [79]
Lactones [81,82]
Phosphine oxides,
Amineoxides,
RS(=O)XR, X = N, O, S [86]
Sulfoxides [87,88]

R

Ar   NH₂

Sulfoxides [69]
Phosphine oxides [73]
sec-Benzylic alcohols [77,78]
N-Phthalimido amino acids [80]
2-(Aryl)carboxylic acids [83,84]
Hydroxy esters [85]

CF₃   OMe

Ph   CO₂H

tert-Amines [70,71]
Diamines [74,75]

(Tröger's base)

sec- and *tert* -
Benzylic alcohols [89]

(Quinine)

Binaphthyls,
sec-Benzylic amines [90]

R

ArCONH   CO₂H

R = Ph, i-Bu

Diamines,
Amino esters,
Amino alcohols [74]
Benzodiazepinones,
Naphthamides,
Lactones [91]

OH

Ph   CO₂H

Diamines,
Amino esters,
Amino alcohols [74]

Me   O   O   Me

Ph   N   N   Ph
      H   H

Amides [92]

OAc

Ph   CO₂H

Amines,
Amino alcohols [93]

Figure 2.10 Common chiral solvating agents and some classes of compounds they have been used with. For a more complete listing, see ref. [65].

_____

16   The anthropomorphic notion that a chiral molecule can somehow 'recognize' or 'discriminate' the chirality sense of another chiral molecule is a convenience that is used commonly, realizing that it is the observer that does the recognizing, not the molecules [64].

Figure 2.11 illustrates the principle with two specific binding interactions and a third, which provides the anisotropy for enantiomer discrimination.[17] In this example, there are two hydrogen bond donors in the CSA: the hydroxyl and the benzylic hydrogen (which has been rendered acidic by the neighboring trifluoromethyl group). Suppose for example, that the solute is dibasic and binds preferentially such that OH and B_1 interact and CH and B_2 interact. Note that the solute substituent that is syn to the aryl group on the CSA will be shielded relative to the other (R_2 on the left and R_1 on the right). *This is the third point required for discrimination of enantiomers.* Because of the shielding cone above the aromatic ring, the time-averaged spectrum will have R_2 at higher field in the absolute configuration on the left. If the preference of the two bonding interactions between the CSA and the solute are known, then the absolute configuration of the CSA can be used to determine the absolute configuration of the solute. Once again, there need be no difference in stability between the two solvates, since protons that are enantiotopic by internal comparison can also be differentiated by CSAs [91]. Detailed models for the assignment of absolute configuration have only been made for the 1-(aryl)trifluoroethanols and the 1-(aryl)ethyl amines, and the reader is referred to Pirkle's review for further details [64].

Figure 2.11 The interaction of a 1-aryltrifluoroethanol chiral solvating agent and the two enantiomers of a dibasic solute.

Although much of the usefulness of CSAs is for analysis of enantiomeric excess, the principles described above may manifest themselves in other ways. For example in 1969, Williams *et al.* demonstrated that dihydroquinine can serve as a chiral solvating agent for itself [94]. Thus, the NMR spectrum of the racemate and the (–)-enantiomer were *not* superimposable, and a 3:1 mixture of enantiomers exhibited anisochronous signals for several enantiotopic protons. The spectra of racemic, 100% ee, and 50% ee dihydroquinine are shown in Figure 2.12. Clearly, the three spectra are different. Although the spectra of both the pure enantiomer and the racemate are similar and easily interpreted, they do not match. The spectrum of the partially enriched enantiomer is unexpectedly complex. Assuming binary association, the phenomenon can be understood in terms of the following equilibria,

$$S* + S* \rightleftarrows S*\cdot S* \qquad (2.17)$$

$$S* + R* \rightleftarrows S*\cdot R* \qquad (2.18)$$

$$R* + R* \rightleftarrows R*\cdot R* , \qquad (2.19)$$

where $R*$ and $S*$ represent the two enantiomers. Since the molecule also contains an anisotropic perturbing function, anisochrony of the homochiral ($S*\cdot S*$) and the

---

[17] This example illustrates two *binding* interactions, although one attractive and one repulsive interaction would also suffice, so long as a third is present as well.

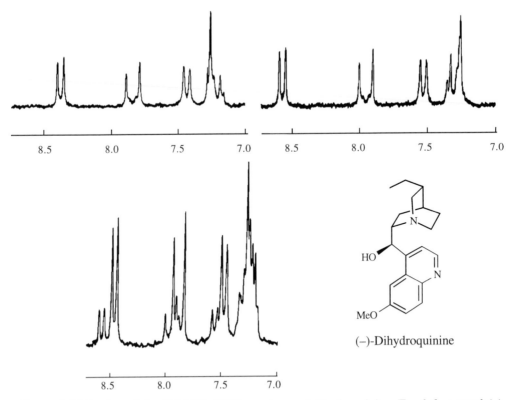

Figure 2.12 Portion of the 100 MHz NMR spectrum of dihydroquinine. Top left, natural (–)-enantiomer; top right, racemate; bottom left, 3:1 mixture of the two enantiomers (50% ee). Reprinted with permission from ref. [94], copyright 1969, American Chemical Society.

heterochiral ($S^*\cdot R^*$) dimers ensues. The observed spectrum is a time-average of the species from Equations 2.17-2.19 that are present in the samples. The pure enantiomer represents the time average of the species in Equation 2.17 (or 2.19) alone, whereas the racemate represents the time average of all three equilibria. In the present case, these time-averaged spectra are not identical. When the enantiomers are not present in equal amounts, the weighting factors for each equilibrium's contribution to the time averaged spectrum are unequal. For example, if the S^* enantiomer predominates, the primary interactions will be equilibria 2.17 and 2.19, and the time-averaged spectra for the enantiomers are different (S^*, $S^*\cdot S^*$, $S^*\cdot R^*$ vs. R^*, $S^*\cdot R^*$). In other words, the spectrum of enantiomerically enriched dihydroquinine is the result of 'self-induced nonequivalence' whereby the enantiomer in excess acts as the CSA for both enantiomers of the racemate.

Williams also noted [94] that the spectral anomalies of the dihydroquinine enantiomer, racemate, and various mixtures were solvent dependent (CH_3OD reduced the anisochrony, as did *O*-acylation) and became identical at high dilution. In synthesis, such anomalies should be remembered when interpreting spectra and when comparing spectra of synthetic materials with literature data if one is enantiomerically pure and the other is fully or partly racemic, or if the spectra were recorded at different solute concentrations.

2.4 Chromatography

Asymmetric reactions and processes give rise to two kinds of stereoisomeric products: diastereomers and enantiomers. The physical separation of these isomers with simultaneous analysis of isomer distribution (peak integration) is an excellent way to determine the selectivity of a reaction. For the analysis of *diastereomers*, standard chromatographic techniques suffice, although the chromatographic method should be accompanied by another technique that determines the configuration of the new centers. Diastereomer analysis also ensues in cases of double asymmetric induction, and the configuration of known centers in the reactants may be used as a point of reference for determination of the new stereocenter(s) by NMR or X-ray.

Early methods for chromatographic analysis of *enantiomers* called for derivatization with a chiral reagent. This is a method that is still used, although the problems of kinetic resolution discussed in Section 2.3.1 should be recalled when planning such an analysis. A much more appealing method is the direct separation of enantiomers by gas or liquid chromatography on a chiral stationary phase (CSP). The growing popularity of this method is evidenced by the number of monographs published on the subject in the last few years [95-102]. This method has a number of appealing features:

1. No kinetic resolution arises as a result of double asymmetric induction in a chiral derivatization scheme, although care must still be taken to avoid enantiomer enrichment (or depletion) during workup (*cf.* Figure 2.1 and accompanying discussion).
2. The order of enantiomer elution for a given class of compounds is often known, so that enantiomeric purity and absolute configuration can be determined simultaneously.
3. The sensitivity of GC or HPLC detectors are such that very small amounts of analyte, as little as a few micrograms under favorable circumstances, may be analyzed. This is far below the limits of detection in polarimetry and NMR.
4. Integration of chromatographic peaks is usually much more accurate than measurement of rotations or integration of NMR peaks. Therefore chromatography is the method of choice when accuracy is important, and is especially applicable to the analysis of samples of high enantiomeric purity.
5. Scale-up may allow for preparative purification to 100% ee.

Of course for new classes of compounds being studied on a chiral stationary phase, that a mixture of enantiomers are separable must be proven by analyzing a racemate, and the order of elution must be established by correlation with compounds of known configuration. In certain instances, derivatization may be necessary to improve chromatographic behavior and/or detectability.

In order to appreciate the forces that are responsible for chromatographic resolution, we need to review some of the principles of chromatography.[18] The elution of a sample through a chromatography column is accomplished by a partitioning of the sample between a stationary phase and mobile phase. In GC the

---

[18] For a comprehensive treatment, see ref. [103].

stationary phase is a liquid and the mobile phase is a gas, while in HPLC the stationary phase is a solid and the mobile phase is a liquid. In the extreme, a sample that does not interact with the stationary phase is eluted in the amount of time it takes the mobile phase to travel the column, t_0. Samples that *do* interact with the stationary phase will obviously take longer. Their retention may be expressed as t_R (retention time), V_R (retention volume), or κ' (capacity ratio).

The latter is defined as

$$\kappa' = \frac{A_s}{A_m} \quad , \qquad\qquad (2.20)$$

where A_s and A_m represent the amount of solute in the stationary and mobile phases, respectively. Thus the capacity ratio is the equilibrium constant for the partitioning of the analyte between the mobile and stationary phases. The capacity ratio can also be expressed in terms of retention times,

$$\kappa' = \frac{t_R - t_0}{t_0} \quad , \qquad\qquad (2.21)$$

where t_0 is the retention time of an unretained compound, usually visible as the solvent front (see Figure 2.13).

For two peaks to be 'resolved' chromatographically, the capacity ratios, κ_1' and κ_2', must be different. For analytical purposes, two interdependent chromatographic properties must be considered: the chromatographic separability factor, α, and the resolution, R_S. The chromatographic separability factor is defined as

$$\alpha = \frac{\kappa_2'}{\kappa_1'} \quad , \qquad\qquad (2.22)$$

where κ_1' and κ_2' are the capacity ratios of the first and second eluting peaks, respectively. Combination of Equations 2.18 and 2.19 gives

$$\alpha = \frac{t_{R2} - t_0}{t_{R1} - t_0} \quad , \qquad\qquad (2.23)$$

where t_{R1} and t_{R2} are the retention times of the first and second eluting peaks, respectively. Using Equations 2.21 and 2.22, capacity ratios and separability factors can be easily obtained from a chromatogram, as shown in Figure 2.13.

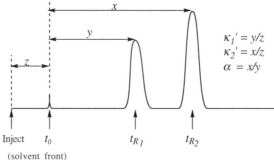

Figure 2.13 A hypothetical chromatogram, showing the retention time of an unretained compound (t_0), the retention times of two analytes, t_{R1} and t_{R2}, and the relationship of these quantities to the capacity ratios, κ_1' and κ_2', and the chromatographic separability factor, α.

Because the capacity ratios reflect the equilibrium between two analytes and the stationary phase, the separability factor, α, is directly related to the free energy difference between the analyte·stationary phase complexes, according to

$$\Delta\Delta G = -RT\ln\alpha. \qquad (2.24)$$

Rearrangement gives

$$\alpha = e^{-\Delta\Delta G/RT}, \qquad (2.25)$$

which is the same exponential relationship described in the previous chapter (*cf.* Equations 1.1 and 1.2). For the interaction of chiral stationary phases with enantiomers, the complexes are diastereomeric; in order for separation to occur, they must *not* be isoenergetic.[19] Separability factors of 1.1 are common in CSP chromatography, which translates to a free energy difference, $\Delta\Delta G$ (at 25° C), of only 56 cal/mole (*cf.* Figure 1.3)! It is the *amplification* of this difference during the chromatographic process that accounts for the separation.

Resolution, R_S, of chromatographic peaks is the ratio of the peak separation to the average peak width:

$$R_S = \frac{2(t_{R_2} - t_{R_1})}{w_1 + w_2}, \qquad (2.26)$$

where w_1 and w_2 are the widths of the first and second peaks, respectively. Thus, the resolution is dependent on both the separation factor, α, and the column efficiency (number of theoretical plates). As shown in Figure 2.14, the same resolution may give rise to two closely spaced narrow peaks or to two broader peaks that are more widely separated.

Racemization of either the analyte or the chiral stationary phase may give rise to peak coalescence, but the two are easily distinguished by their appearance, as shown in Figure 2.15. Over time, racemization of the CSP may occur, and this will reduce

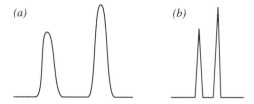

Figure 2.14 Hypothetical chromatograms of identical resolution. *(a)* Large α on a low efficiency column. *(b)* Smaller α on a high efficiency column.

the separation factor, α. If racemization of a CSP is possible (such as with an amino acid derived CSP), it is wise to periodically run a standard to check for peak coalescence. Another type of peak coalescence is due to racemization of the analyte on the column [104,105]. The appearance of such a phenomenon depends on the relative rates of racemization and separation [105]. The two extremes are fast and

---

[19] Recall that the diastereomeric complexes of enantiomers with CSRs or CSAs *may* be isoenergetic and still exhibit anisochrony (Sections 2.3 and 2.4).

slow racemization, relative to the separation. Fast racemization would yield a single sharp peak, and extremely slow racemization would go undetected. Intermediate cases might appear as a trough between the peaks, or a hump, as shown in Figure 2.15b and c [105].

The historical origins of CSP chromatography are early in the 20th century when it was observed that certain dyes were enantioselectively adsorbed onto biopolymers such as wool [106-108]. Although there were isolated instances of chromatographic resolutions earlier,[20] development of CSP chromatography as a useful tool did not take place until capillary gas chromatography (GC) and high

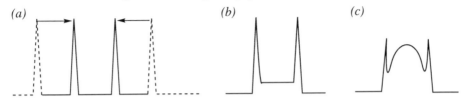

Figure 2.15 Hypothetical chromatograms showing peak coalescence due to racemization. *(a)* Racemization of the chiral stationary phase causes the peaks to move closer together (α decreases, ultimately to zero if the CSP is racemic). *(b)* and *(c)* Racemization of the analyte on the column may produce a trough between the enantiomer peaks, as in *(b)*, which may grow to a hump, as in *(c)*, or even a single peak, depending on the relative rates of racemization and separation [105].

performance liquid chromatography (HPLC) were popularized in the 1970s. The separation of enantiomers on a CSP requires the formation of diastereomeric adsorbate 'complexes' between the analyte and the CSP. A number of CSPs have come into use, but only in a few cases has detailed work been done to rationalize the relative stabilities of the diastereomeric adsorbates. Indeed, the energy differences that are required for enantiomer separation on an efficient column are so small (~50 cal/mole, *vide supra*) that caution is advised in overinterpreting enantiomer 'recognition' models.

In 1952, Dalgliesh [111] extended the 3-point model [66,67] to CSP chromatography, to explain the separation of amino acid enantiomers by paper chromatography. Dalgliesh postulated a 3-point *attraction*, which now seems to be somewhat oversimplified. More recently, Pirkle has argued that, although three points are required, all need not be attractive [112]. At least one, however, must be stereochemically dependent. A detailed study of chiral solvating agents (*vide supra*) has led to fairly exact models of 3-point solvation to explain the chemical shift effects of the CSA. Immobilization of one or the other of the CSA components on silica gel produced separations having the same order of elution as expected based on the selectively solvated species in the NMR experiment. For example, N-(3,5-dinitrobenzoyl)leucine amide bonded to silica gel shows a high degree of selectivity (α = 9.7) for the enantiomers of methyl N-(2-naphthyl)alaninate [113]. The converse is also true: N-(2-naphthyl)alanine ester as a CSP shows a high affinity for N-(3,5-dinitrobenzoyl)leucine derivatives [114]. The model for the complexation of

---

[20] For more detailed accounts of the early history of CSP chromatography see ref. [95,109,110].

these two species involves an aromatic π-stacking interaction and two hydrogen bonds, as indicated in Figure 2.16a. This model is supported by intermolecular NOE enhancements in CDCl₃ [113], and an X-ray crystal structure of the bimolecular complex [115]. The latter is illustrated in Figure 2.16b, which confirms the three interactions proposed earlier [113,114], and which provides strong experimental support for the model. Although detailed models of other CSP separations have not been as extensively studied as the Pirkle systems, it is likely that they also conform to some variant of the 3-point rule [112,116].

In CSP-GC and CSP-HPLC, there are only a few categories of chiral selectors used as stationary phases. There is a broader variety of CSP columns on the market for HPLC than for GC, but most types have been investigated in both media. For the most up to date information, literature from vendors of CSP columns should be

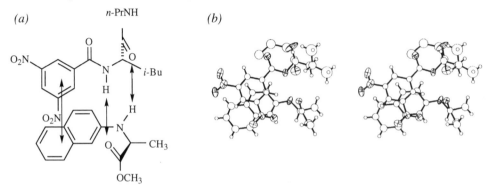

(a)　　　　*n*-PrNH　　　　　*(b)*

Figure 2.16 Supramolecular complex of *N*-(3,5-dinitrobenzoyl)leucine *n*-propyl amide and methyl *N*-(2-naphthyl)alaninate. *(a)* Schematic representation of the three recognition points deduced from NOE data [113]. *(b)* Stereoview of a bimolecular crystal. The orientation of the two species concurs with solution NOE data. Reprinted with permission from ref. [115].

consulted. In the US, Regis, J. T. Baker, and Daicel have a variety of HPLC columns to choose from, and they have published some useful tables on selecting the correct column for a given application [117,118]. A recent review also lists a number of compound types that are resolvable on various CSPs [119]. Similar information on GC columns and applications is available from Applied Science or Supelco, and Souter's monograph [95] also contains an extensive listing of CSPs for GC, and the types of compounds separated by each. An expedient method of finding a solution to a separation problem may be the use of computerized databases. In addition to the standard databases such as those offered by *Chemical Abstracts,* Roussel and Piras have constructed a database dedicated to the enantiomeric resolution of racemic mixtures by HPLC [120].

Here, only general categories of chiral stationary phases will be mentioned.[21] One of the more popular types of GC and HPLC columns use donor-acceptor interactions such as those illustrated in Figure 2.16 for enantiomer separation.

---

[21] Lough's monograph gives a particularly thorough coverage of CSP types for HPLC [97], while Souter's slightly older text is particularly good for GC [95].

Types of donor-acceptor interactions are hydrogen bonding, π-stacking, dipole stacking, etc. Derivatization is usually required for both GC and HPLC applications.

In HPLC, a large variety of analyte types have been resolved on columns packed with derivatized cellulose. Cellulose acetate, benzoate, and carbamate derivatives provide CSPs that will separate a very broad range of analyte types, although a single CSP may not have broad applicability to a wide variety of analytes, and derivatization may be required. Separation is achieved by donor-acceptor interactions, with inclusion phenomena sometimes playing a secondary role.

Microcrystalline cellulose triacetate, cyclodextrin- and crown ether-derived CSPs, as well as some chiral synthetic polymers, achieve enantiomer separation primarily by forming host-guest complexes with the analyte; in these cases, donor-acceptor interactions are secondary. Solutes resolved on cyclodextrins and other hydrophobic cavity CSPs often have aromatic or polar substituents at a stereocenter, but these CSPs may also separate compounds that have chiral axes. Chiral crown ether CSPs resolve protonated primary amines.

Chiral ligand exchange chromatography utilizes immobilized transition metal complexes that selectively bind one enantiomer of the analyte, which is usually an amino acid.

Proteins such as bovine serum albumin, immobilized to silica, achieve enantiomer separation primarily *via* hydrophobic and electrostatic interactions. Although the protein-based CSP columns have low capacity and preparative use is impossible, these phases offer the analyst the convenience of being able to resolve a broad spectrum of analytes with a single column.

2.5 Summary

Over seventy five years after the van't Hoff - Le Bel theory of the asymmetric carbon atom was introduced, Bijvoet and colleagues established for the first time the absolute configuration of a chiral molecule, sodium rubidium tartrate, using anomalous dispersion of X-rays [121].[22] Recall that Emil Fischer's assignment of the absolute configuration of D- and L-tartrate was arbitrary and for the first half of the twentieth century was hypothetical. Thus, the assignment of the absolute configuration of all known chiral molecules is predicated on this single method. More commonly, it is used to establish molecular constitution and relative configurations. This is a very powerful tool [122] that should not be overlooked by the synthetic practitioner, although its implementation is usually left to a specialist.

The simplest methods for analysis are polarimetry and NMR, when applicable. Advances in chromatographic science continue apace, and it is likely that further advancements will be made to ease the burden of analysis. Chromatography is the

---

[22] It is interesting that tartrate was the first resolved compound [1] as well as the first compound whose absolute configuration was established; it is fitting that this seminal work was done at the *van't Hoff* Laboratories of the University of Utrecht. Given the role tartaric acid played in the establishment of the field of stereochemistry, it was perhaps inevitable that it would also play a major role in asymmetric synthesis, as will be seen in a number of examples throughout this book.

method of choice for analysis of compounds of very high isomeric purity, since it is the only method currently available that can accurately detect and quantify <<1% contamination. The goal of asymmetric synthesis is to produce very highly selective reactions, but when this is achieved the job of identifying the chromatographic peaks due to minor stereoisomers becomes more difficult. It is tempting for the analyst to assume that the small peak(s) near the major one is the 'other' isomer, but this is a risky assumption. The safest bet is to synthesize the 'other' isomer(s) independently, but this may not be feasible. The next best thing is to couple the chromatograph to some sort of spectroscopic device such as a mass spectrometer or a diode array UV-VIS detector.[23]

For the analysis of compounds that are chiral by virtue of isotopic substitution, NMR is the method of choice, since energetic differences between diastereomeric complexes are not required for induced anisochrony. When it works, NMR is also one of the simplest and fastest techniques available. For monofunctional or weakly basic solutes, chiral shift reagents are more likely to succeed, whereas chiral solvating agents are simpler (when they work) and are better for the assignment of absolute configuration.

2.6 References

1. L. Pasteur *Ann. chim. et phys.* **1848**, *24[3]*, 442-459.
2. K. C. Cundy; P. A. Crooks *J. Chromatog.* **1983**, *281*, 17-33.
3. D. Piter; S. Daudien; O. Samuel; H. B. Kagan *J. Org. Chem.* **1994**, *59*, 370-373.
4. A. Dobashi; Y. Motoyama; K. Kinoshita; S. Hara *Anal. Chem.* **1987**, *59*, 2209-2211.
5. R. Matusch; C. Coors *Angew. Chem. Int. Ed. Engl.* **1989**, *28*, 626-627.
6. R. Charles; E. Gil-Av *J. Chromatog.* **1984**, *298*, 516-520.
7. R. M. Carman; K. D. Klika *Aus. J. Chem.* **1991**, *44*, 895-896.
8. W.-L. Tsai; K. Hermann; E. Hug; B. Rohde; A. S. Dreiding *Helv. Chim. Acta* **1985**, *68*, 2238-2243.
9. C. Djerassi *Optical Rotary Dispersion*; McGraw-Hill: New York, 1960.
10. P. Crabbe *Optical Rotary Dispersion and Circular Dichroism in Organic Chemistry*; Holden-Day: SF, 1965.
11. G. Snatzke *Optical Rotary Dispersion and Circular Dichroism in Organic Chemistry*; Heyden and Son: London, 1967.
12. S. F. Mason *Modern Optical Activity and the Chiral Discriminations*; Cambridge University: Cambridge, 1982.
13. D. J. Caldwell; H. Eyring *The Theory of Optical Activity*; Wiley: New York, 1971.
14. E. Charney *The Molecular Basis of Optical Activity: Optical Rotatory Dispersion and Circular Dichroism*; Wiley: New York, 1979.
15. L. D. Barron *Molecular Light Scattering and Optical Activity*; Cambridge University: Cambridge, 1983.
16. J. H. Brewster In *Topics in Stereochemistry*; E. L. Eliel, N. L. Allinger, Eds.; Wiley-Interscience: New York, 1967; Vol. 2, p 1-72.
17. J. B. Lambert; H. F. Shurvell; L. Verbit; R. G. Cooks; G. H. Stout *Organic Structural Analysis*; Macmillan: New York, 1976.

---

[23] The UV spectra of enantiomers are identical of course, and those of diastereomers are usually superimposable as well.

18. M. Legrand; M. J. Rougrer In *Stereochemistry. Fundamentals and Methods*; H. B. Kagan, Ed.; Georg Thieme: Stuttgart, 1977; Vol. 2, p 33-184.
19. G. Krow; R. K. Hill *Chem. Commun.* **1968**, 430-431.
20. A. Horeau *Tetrahedron Lett.* **1969**, 3121-3124.
21. S. Yamaguchi; H. S. Mosher *J. Org. Chem.* **1973**, *38*, 1870-1877.
22. K. K. Anderson; D. M. Gash; J. D. Robertson In *Asymmetric Synthesis*; J. D. Morrison, Ed.; Academic: Orlando, 1983; Vol. 1, p 45-58.
23. A. R. Schools; J. P. Guette In *Asymmetric Synthesis*; J. D. Morrison, Ed.; Academic: Orlando, 1983; Vol. 1, p 29-44.
24. A. Horeau In *Stereochemistry. Fundamentals and Methods*; H. B. Kagan, Ed.; Georg Thieme: Stuttgart, 1977; Vol. 3, p 51-94.
25. A. Gaudemer In *Stereochemistry. Fundamentals and Methods*; H. B. Kagan, Ed.; Georg Thieme: Stuttgart, 1977; Vol. 1, p 44-136.
26. D. Parker *Chem. Rev.* **1991**, *91*, 1441-1457.
27. M. Raban; K. Mislow *Tetrahedron Lett.* **1965**, 4249-4253.
28. S. Yamaguchi In *Asymmetric Synthesis*; J. D. Morrison, Ed.; Academic: Orlando, 1983; Vol. 1, p 125-152.
29. J. A. Dale; D. L. Dull; H. S. Mosher *J. Org. Chem.* **1969**, *34*, 2543-2549.
30. J. A. Dale; H. S. Mosher *J. Am. Chem. Soc.* **1973**, *95*, 512-519.
31. D. E. Ward; C. K. Rhee *Tetrahedron Lett.* **1991**, *32*, 7165-7166.
32. W. A. König; K.-S. Nippe; P. Mischnick *Tetrahedron Lett.* **1990**, *31*, 6867-6868.
33. W. R. Roush; L. K. Hoong; M. A. J. Palmer; J. C. Park *J. Org. Chem.* **1990**, *55*, 4109-4117.
34. D. Enders; B. B. Lohray *Angew. Chem. Int. Ed. Engl.* **1988**, *27*, 581-583.
35. S. Yamaguchi; J. A. Dale; H. S. Mosher *J. Org. Chem.* **1972**, *37*, 3174-3176.
36. S. Yamaguchi; F. Yasuhara; K. Kabuto *Tetrahedron* **1976**, *32*, 1363-1367.
37. S. Yamaguchi; K. Kabuto *Bull. Chem. Soc. Jpn.* **1977**, *50*, 3033-3038.
38. C. Siegel; E. R. Thornton *Tetrahedron Lett.* **1988**, *29*, 5225-5228.
39. G. R. Sullivan; J. A. Dale; H. S. Mosher *J. Org. Chem.* **1973**, *38*, 2143-2147.
40. T. Kusumi; Y. Fujita; I. Ohtani; H. Kakisawa *Tetrahedron Lett.* **1991**, *32*, 2923-2926.
41. T. Kusumi; Y. Fujita; I. Ohtani; H. Kakisawa *Tetrahedron Lett.* **1991**, *32*, 2939-2942.
42. J. S. Dutcher; J. G. Macmillan; C. H. Heathcock *J. Org. Chem.* **1976**, *41*, 2663-2669.
43. A. Svatos; J. Valterová; D. Saman; J. Vrkoc *Collect. Czech. Chem. Commun.* **1990**, *55*, 485-490.
44. H. Hiemstra; H. Wynberg *Tetrahedron Lett.* **1977**, 2183-2186.
45. J. P. Vigneron; M. Dhaenens; A. Horeau *Tetrahedron* **1973**, *29*, 1055-1059.
46. B. L. Feringa; A. Smaardijk; H. Wynberg *J. Am. Chem. Soc.* **1985**, *107*, 4798-4799.
47. X. Wang *Tetrahedron Lett.* **1991**, *32*, 3651-3654.
48. G. M. Whitesides; D. W. Lewis *J. Am. Chem. Soc.* **1970**, *92*, 6979-6980.
49. R. E. Sievers *Nuclear Magnetic Resonance Shift Reagents*; Academic: New York, 1973.
50. G. R. Sullivan In *Topics in Stereochemistry*; E. L. Eliel, N. L. Allinger, Eds.; Wiley-Interscience: New York, 1978; Vol. 10, p 287-329.
51. R. R. Fraser In *Asymmetric Synthesis*; J. D. Morrison, Ed.; Academic: Orlando, 1983; Vol. 1, p 173-196.
52. J. A. Pople; W. G. Schneider; H. J. Bernstein *High-Resolution Nuclear Magnetic Resonance*; McGraw-Hill: New York, 1959.
53. J. W. Emsley; J. Feeney; L. H. Sutcliffe *High Resolution Nuclear Magnetic Resonance Spectroscopy, vol. 1*; Pergamon: Oxford, 1965.
54. J. M. Bulsing; J. M. K. Sanders; L. D. Hall *J. Chem. Soc., Chem. Commun.* **1981**, 1201-1203.
55. T. J. Wenzel; C. A. Morin; A. A. Brechting *J. Org. Chem.* **1992**, *57*, 3594-3599.
56. D. H. G. Grout; D. Whitehouse *J. Chem. Soc., Perkin Trans. 1* **1977**, 544-549.

57. D. J. Pasto; J. K. Borchardt *Tetrahedron Lett.* **1973**, 2517-2520.
58. M. D. McCreary; D. W. Lewis; D. L. Wernick; G. M. Whitesides *J. Am. Chem. Soc.* **1974**, *96*, 1038-1054.
59. R. R. Fraser; M. A. Petit; M. Miskow *J. Am. Chem. Soc.* **1972**, *94*, 3253-3254.
60. H. L. Goering; J. N. Eikenberg; G. S. Koermer *J. Am. Chem. Soc.* **1971**, *93*, 5913-5914.
61. R. R. Fraser; M. A. Petit; J. K. Saunders *J. Chem. Soc., Chem. Commun.* **1971**, 1450-1451.
62. N. Ahmed; N. S. Bhacca; J. Selbin; J. D. Wander *J. Am. Chem. Soc.* **1971**, *93*, 2564-2565.
63. W. H. Pirkle *J. Am. Chem. Soc.* **1966**, *88*, 1837.
64. W. H. Pirkle; D. J. Hoover In *Topics in Stereochemistry*; E. L. Eliel, N. L. Allinger, Eds.; Wiley-Interscience: New York, 1982; Vol. 13, p 263-331.
65. G. R. Weisman In *Asymmetric Synthesis*; J. D. Morrison, Ed.; Academic: Orlando, 1983; Vol. 1, p 153-171.
66. C. H. Easson; E. Stedman *Biochem. J.* **1933**, *27*, 1257-1266.
67. A. G. Ogston *Nature* **1948**, *162*, 963.
68. L. Schjelderup; A. J. Aasen *Acta Chem. Scand., Ser. B* **1986**, *B40*, 601-603.
69. M. Desmukh; E. Duñach; S. Juge; H. Kagan *Tetrahedron Lett.* **1984**, *25*, 3467-3470.
70. B. E. Maryanoff; D. F. McComsey *J. Het. Chem.* **1985**, *22*, 911-914.
71. F. J. Villani, Jr.; M. J. Constanzo; R. R. Inners; M. S. Mutter; D. E. McClure *J. Org. Chem.* **1986**, *51*, 3715-3718.
72. W. H. Pirkle; T. G. Burlingame; S. D. Beare *Tetrahedron Lett.* **1968**, 5849-5852.
73. A. Tambuté; A. Begos; M. Lienne; M. Claude; R. Rosset *J. Chromotogr.* **1987**, *396*, 65-81.
74. S. C. Benson; P. Cai; M. Coilon; M. A. Tokles; J. K. Snyder *J. Org. Chem.* **1988**, *53*, 5335-5341.
75. C. A. R. Baxter; H. C. Richard *Tetrahedron Lett.* **1972**, 3357-3358.
76. W. H. Pirkle; S. D. Beare *J. Am. Chem. Soc.* **1969**, *91*, 5150-5155.
77. W. H. Pirkle; S. D. Beare *J. Am. Chem. Soc.* **1967**, *89*, 5485-5487.
78. W. H. Pirkle; M. S. Hoekstra *J. Magn. Reson.* **1975**, *18*, 396-400.
79. I. Moretti; F. Taddei; G. Torre; N. Spassky *J. Chem. Soc., Chem. Commun.* **1973**, 25-26.
80. A. Ejchart; J. Jurczak; K. Bankowski *Bull. Acad. Pol. Sci. Ser. Sci. Chim.* **1971**, *19*, 731-734.
81. W. H. Pirkle; D. L. Sikkenga; M. S. Pavlin *J. Org. Chem.* **1977**, *42*, 384-387.
82. W. H. Pirkle; D. L. Sikkenga *J. Org. Chem.* **1977**, *42*, 1370-1374.
83. J.-P. Guetté; L. Lacombe; A. Horeau *C. R. Hebd. Seances Acad. Sci. Ser. C* **1968**, *267*, 166-169.
84. A. Horeau; J.-P. Guetté *C. R. Hebd. Seances Acad. Sci. Ser. C* **1968**, *267*, 257-259.
85. W. H. Pirkle; S. D. Beare *Tetrahedron Lett.* **1968**, 2579-2582.
86. W. H. Pirkle; S. D. Beare; R. L. Muntz *J. Am. Chem. Soc.* **1969**, *91*, 4575.
87. W. H. Pirkle; S. D. Beare; R. L. Muntz *Tetrahedron Lett.* **1974**, 2295-2298.
88. W. H. Pirkle; S. D. Beare *J. Am. Chem. Soc.* **1968**, *90*, 6250-6251.
89. S. H. Wilen; J. Z. Qi; P. G. Williard *J. Org. Chem.* **1991**, *56*, 485-487.
90. C. Rosini; G. Uccello-Banetta; D. Pini; C. Abete; P. Salvadori *J. Org. Chem.* **1988**, *53*, 4579-4581.
91. W. H. Pirkle; A. Tsipouras *Tetrahedron Lett.* **1985**, *26*, 2989-2992.
92. B. S. Jursic; S. I. Goldberg *J. Org. Chem.* **1992**, *57*, 7370-7372.
93. D. Parker; R. J. Taylor *Tetrahedron* **1987**, *43*, 5451-5456.
94. T. Williams; R. G. Pitcher; P. Bommer; J. Gutzwiller; M. Uskokovic *J. Am. Chem. Soc.* **1969**, *91*, 1871-1872.
95. R. W. Souter *Chromatographic Separations of Stereoisomers*; CRC: Boca Raton, Fl, 1985.
96. S. G. Allenmark *Chromatographic Enantioseparation Methods and Applications, 2nd Ed*; Ellis Horwood: Chichester, 1991.
97. *Chiral Liquid Chromotography*; W. J. Lough, Ed.; Blackie: Glasgow and London, 1989.

98. *Chromatographic Chiral Separations*; M. Zief; L. J. Crane, Eds.; Marcel Dekker: New York, 1988.

99. *Chiral Separations*; D. Stevenson; I. D. Wilson, Eds.; Plenum: New York, 1988.

100. W. A. Koenig *The Practice of Enantiomer Separation by Capillary Gas Chromatography*; Huethig: Heidelburg, 1987.

101. W. A. König *Gas Chromatographic Enantiomer Separation with Modified Cyclodextrins*; Hüthig Buch Gmbh: Heidelberg, 1992.

102. *Chiral Separations by Liquid Chromatography. ACS Symposium Series 471*; A. Satinder, Ed.; American Chemical Society: Washington, 1991.

103. L. R. Snyder; J. J. Kirkland *Introduction to Modern Liquid Chromatography*; Wiley-Interscience: New York, 1974.

104. V. Schurig; W. Bürkle *J. Am. Chem. Soc.* **1982**, *104*, 7573-7580.

105. W. Bürkle; H. Karfunkle; V. Schurig *J. Chromatogr.* **1984**, *288*, 1-14.

106. R. Wilstätter *Chem. Ber.* **1904**, *37*, 3758-3760.

107. A. W. Ingersoll; R. Adams *J. Am. Chem. Soc.* **1922**, *44*, 2930-2937.

108. G. Konrad; H. Musso *Chem. Ber.* **1984**, *117*, 423-426.

109. B. Feibush; N. Grinberg In *Chromatographic Chiral Separations*; M. Zief, L. J. Crane, Eds.; Marcel Dekker: New York, 1988, p 1-14.

110. A. Pryde In *Chiral Liquid Chromotography*; W. J. Lough, Ed.; Blackie: Glasgow, 1989, p 23-35.

111. C. E. Dalgliesh *J. Chem. Soc.* **1952**, 3940-3942.

112. W. H. Pirkle; T. C. Pochapsky *Chem. Rev.* **1989**, *89*, 347-362.

113. W. H. Pirkle; T. C. Pochapsky *J. Am. Chem. Soc.* **1987**, *109*, 5975-5982.

114. W. H. Pirkle; T. C. Pochapsky; G. S. Mahler; D. E. Corey; D. S. Reno; D. M. Alessi *J. Org. Chem.* **1986**, *51*, 4991-5000.

115. W. H. Pirkle; J. A. Burke, III; S. R. Wilson *J. Am. Chem. Soc.* **1989**, *111*, 9222-9223.

116. V. A. Davankov *Chromatographia* **1989**, *27*, 475-482.

117. I. W. Wainer *A Practical Guide to the Selection and Use of HPLC Chiral Stationary Phases*; J. T. Baker: Phillipsburg, NJ, 1988.

118. *1989 Application Guide for Chiral Column Selectron*; Daicel Industries: New York; Tokyo; Los Angeles; Dusseldorf, 1989.

119. W. H. Pirkle; T. C. Pochapsky In *Advances in Chromatography*; J. C. Giddings, E. Grushka, P. R. Brown, Eds.; Marcel Dekker: New York, 1987; Vol. 27, p 73-127.

120. C. Roussel; P. Piras *Pure Appl. Chem.* **1993**, *65*, 235-244.

121. J. M. Bijvoet; A. F. Peerdeman; A. J. v. Bommel *Nature* **1951**, *168*, 271-272.

122. R. Parthasarathy In *Stereochemistry. Fundamentals and Methods*; H. B. Kagan, Ed.; Georg Thieme: Stuttgart, 1977; Vol. 1, p 181-234.

Chapter 3

Enolate, Azaenolate, and Organolithium Alkylations

Originally, the term 'carbanion' was used to refer to anionic reactive intermediates whose actual structure was rather poorly understood. In recent years, considerable advances have been made in developing the chemistry of carbanionic species and in understanding the structure of 'carbanions,' especially as regards the involvement of the metal [1-6]. In this chapter we will focus on three types of intermediate that fall into the category of 'carbanion.' Our discussion will be further limited to alkylations: carbon-carbon bond-forming reactions with electrophiles such as alkyl halides that produce only one stereocenter. Aldol and Michael additions are covered in Chapter 5, and reactions with heteroatom electrophiles that form carbon-oxygen or carbon-nitrogen bonds are discussed in Chapter 8.

Carbanions that have found use in asymmetric synthesis are stabilized by one or more substituents (Figure 3.1). By far the most common 'carbanion stabilizing' functional group is the carbonyl. Although early texts (*e.g.,* [7]) referred to the conjugate base of carbonyl compounds as carbanions, these species are now universally known as enolates (Figure 3.1a). Closely related to enolates are their nitrogen analogs, azaenolates (Figure 3.1b). As we will see, the fact that there is a substituent on the nitrogen is important to asymmetric synthesis because it provides a convenient foothold for attachment of a chiral auxiliary. In recent years, a new type of chiral 'carbanion' has emerged: organolithium species in which the carbon bearing the metal is stereogenic (Figure 3.1c,d). Again, the negative charge is stabilized in these species, but not by resonance as is the case with enolates and azaenolates. Instead, heteroatoms on the lithiated carbon provide inductive stabilization; in some instances chelation may be involved.

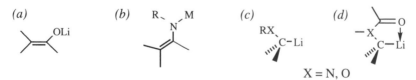

Figure 3.1. Enolates, azaenolates, and α-heteroatom organolithiums.

3.1 Enolates and azaenolates[1]

The deprotonation of a carbonyl gives a nucleophilic species that reacts with electrophiles such as alkyl halides to afford products of substitution at the α carbon. Because of this reactivity, the intermediate species used to be drawn with a negative

---

[1] For comprehensive coverage of enolate alkylations, see refs. [8,9].

charge on carbon (Figure 3.2a). Resonance considerations later suggested that the negative charge should be placed on the more electronegative oxygen (Figure 3.2b). When enolate reactions are carried out in aqueous or alcohol solution, the ionic species are separately solvated, and this type of representation is justified. Unfortunately, the same usage has persisted, even when the reactions are conducted in aprotic solvents where solvent-separated ions are not likely to exist. A more appropriate notation is to affix the metal to the oxygen (Figure 3.2c). Spectroscopic and X-ray data have revealed that metal enolates are usually (always?) aggregated, both in the solid state and in ethereal or hydrocarbon solution (Figure 3.2d). The illustrations in Figure 3.2 show the historical progression of these notations, which Seebach whimsically calls "the route of the sorcerer's apprentice" [5].

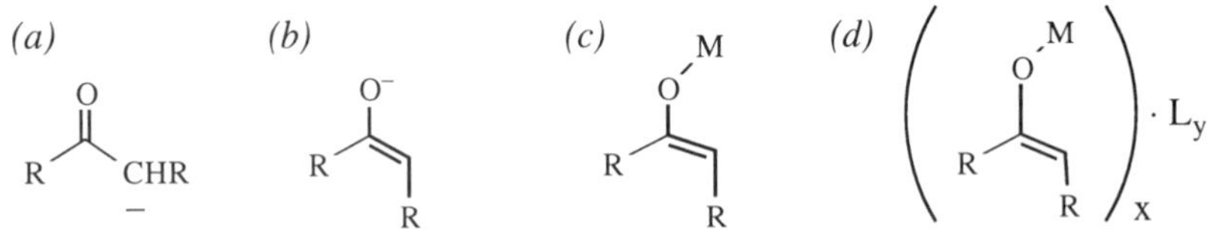

Figure 3.2. Various notations for an enolate, from a naked carbanion or enolate, via a metal enolate, to supramolecular aggregates (after ref. [5]).

Enolates may form supramolecular[2] species such as dimers, tetramers, or hexamers, and these species are often in equilibrium (Scheme 3.1). Enolates may also form mixed aggregates with added salts or with secondary or tertiary amines. The existence of such species has been proven in the solid state by X-ray crystallography, and colligative effects and NMR studies have confirmed their existence in solution (reviews: [5,6,8,12-14]; see also: [15-18]). Interestingly, dimers are even found in crystals of tetrabutylammonium malonates and cyanoacetates [19], indicating that a metal is not necessary for supramolecular organization!

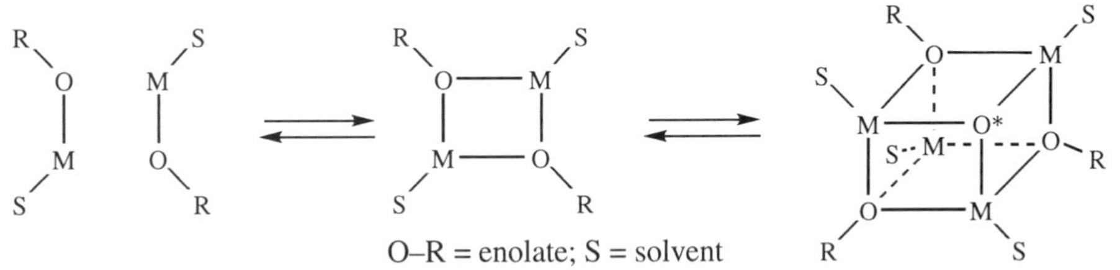

O–R = enolate; S = solvent

Scheme 3.1. Equilibrating dimeric and tetrameric enolate aggregates. Formal charges are not shown. There is probably more than one solvent molecule coordinated to the monomers (left) and dimer (middle). In the tetramer (right), the R moiety is deleted from the indicated oxygen (O*) for clarity.

Chemical evidence also confirms the presence of supramolecular complexes in enolate reactions. For example, added salts can affect the product ratio of enolate alkylations [20-25]. Evidence of mixed aggregation between enolates and secondary amines includes experiments such as those illustrated in Scheme 3.2 where quench-

---

2   The term supramolecular was coined by Lehn to refer to "organized entities ... that result from the association of two or more chemical species held together by intermolecular forces" [10,11].

Creger:

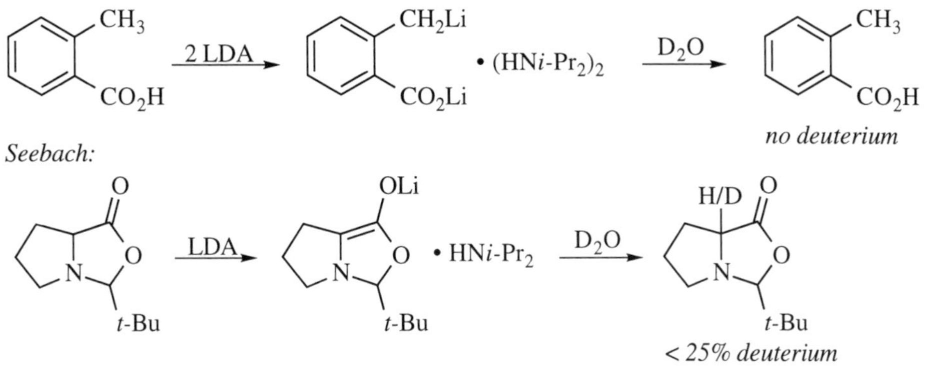

Seebach:

Scheme 3.2. Enolate·diisopropylamine complexes do not incorporate deuterium upon quenching with D$_2$O. *(a)* Creger demonstrated the phenomenon with *o*-toluic acid [26]. *(b)* Seebach showed that lactone enolates behave similarly [27].

ing with D$_2$O or MeOD produces little or no deuterium incorporation, indicating that the enolate is protonated by the secondary amine from within a supramolecular aggregate [26,28-32].

The phrase "conducted tour mechanism" was coined by Cram to describe the removal of a proton by a base and its subsequent return to a different face of the same molecule from which it was removed [1]. Originally, the conducted tour mechanism was postulated to explain the observation that rates of racemization of deuterated carbon acids were faster than hydrogen-deuterium exchange in solutions of potassium *tert*-butoxide/*tert*-butyl alcohol. Thus, "the basic catalyst takes hydrogen or deuterium on a 'conducted tour' of the substrate from one face of the molecule to the [other]" (ref. [1], p. 101). This process was envisioned as a rotation of the carbanion within the solvent cage. We now recognize that the secondary amine forms a mixed aggregate with the enolate, such that the reprotonation (and perhaps conformational motion) is 'intrasupramolecular.'

A complete understanding of enolate chemistry must include knowledge of the aggregation of the enolate species involved [5]. Consider the reaction of an enolate with an alkyl halide as it may have been depicted over the years (Scheme 3.3), progressing from the simple carbanion alkylation to the reaction of supramolecular aggregates.

The mechanism depicted in Scheme 3.3d may be the closest to reality for the reaction of an enolate with an alkyl halide, but this picture is dependent on the individual system under study. For our purposes, we can rationalize most enolate reactions by considering metal enolates as monomers (as in Scheme 3.3c), while realizing that the other coordination sites of the metal may be occupied by ligands that may be solvent molecules, additives such as HMPA, DMPU or TMEDA,[3]

---

[3]   HMPA: hexamethylphosphoramide. DMPU: *N,N'*-dimethylpropyleneurea. TMEDA: tetramethyl-ethylenediamine. Both are additives that coordinate metals and may inhibit aggregation. Note that mechanistic interpretation of the effect of additives, especially TMEDA in THF solvent, are risky [33].

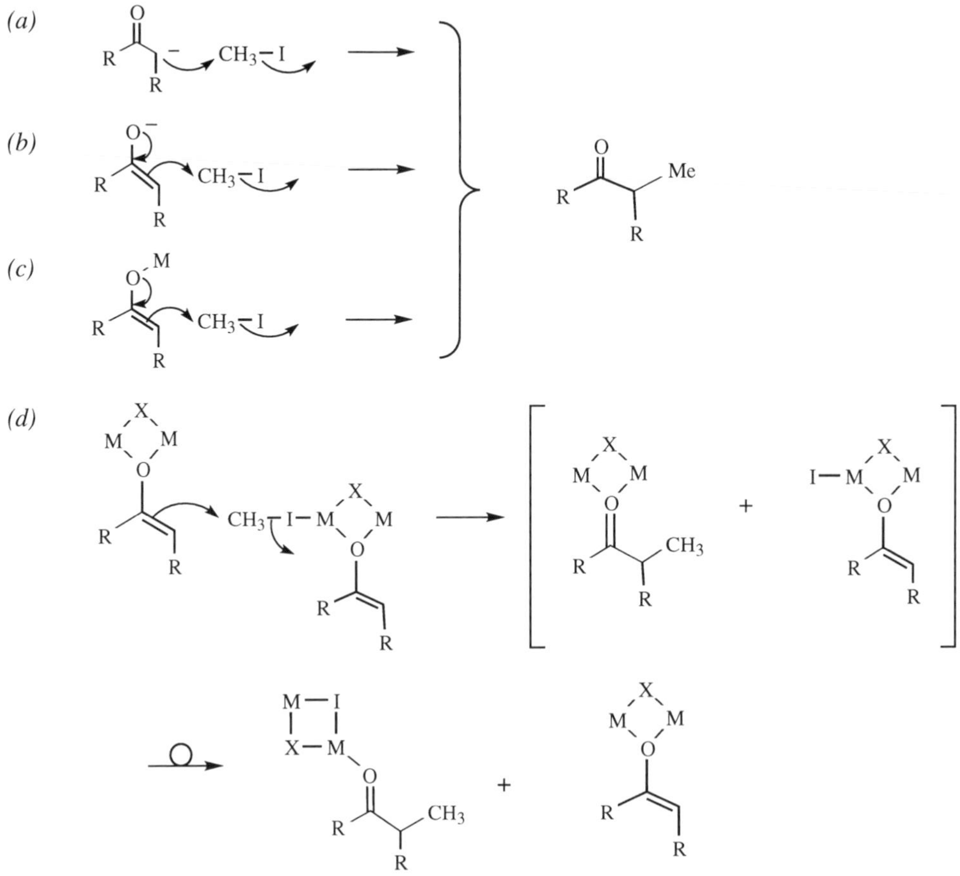

(a)

(b)

(c)

(d)

Scheme 3.3 Alkylation of an "enolate" *(a)* naked carbanion; *(b)* naked enolate; *(c)* metal enolate; *(d)* supramolecular alkylation and rearrangement (after ref. [5]).

anions of added salts, or another molecule of enolate. The interested reader is referred to Seebach's review to see the types of supramolecular complexes that may arise in the chemistry of lithium enolates [5].

3.1.1 Deprotonation of carbonyls[4]

A number of bases may be used for deprotonation, but the most important ones are lithium amide bases such as those illustrated in Figure 3.3.[5] Although other alkali metals may be used with these amides, lithium is the most common. Amide bases efficiently deprotonate virtually all carbonyl compounds, and do so regioselectively with cyclic ketones such as 2-methylcyclohexanone (*i.e.*, C_2 vs. C_6 deprotonation) and stereoselectively with acyclic carbonyls (*i.e.*, E(O)- vs. Z(O)-[6] enolates. If the carbonyl is added to a solution of the lithium amide, deprotonations are irreversible and kinetically controlled [36-38]. Under such conditions, the con-

---

4   For a review on enolate formation, see ref. [34].
5   LDA is stable in both ether and THF at room temperature for 24 hours, but that LTMP has a half-life of 12 hours in THF and only 4 hours in ether at room temperature [35].
6   See glossary, section 1.6, for the definition of the E(O)/Z(O) descriptors of enolate geometry.

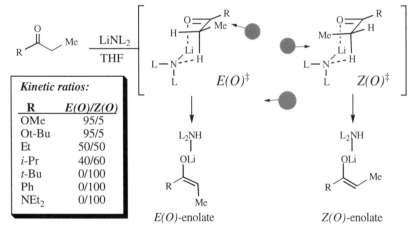

Figure 3.3. LDA = lithium diisopropyl amide, LICA = lithium isopropyl cyclohexyl amide, LTMP = lithium tetramethylpiperidide, LHDS = lithium hexamethyldisilyl amide.

figuration of an acyclic enolate is determined during the deprotonation step, and subsequent isomerization is probably not important [38]. The most commonly cited deprotonation model has the lithium amide base and the carbonyl reacting in a cyclic, 6-membered transition state such that the α-proton and the metal are transferred simultaneously [39]. This mechanism, proposed by Ireland in 1976, may be used to explain the preferred formation of $E(O)$- or $Z(O)$-enolates of acyclic esters, ketones and amides [14], at least in the absence of coordinating additives such as HMPA (*vide infra*). As shown in Scheme 3.4, the transition states for the deprotonation to the $E(O)$- and $Z(O)$-enolates are apparently controlled by a balance of 1,2-eclipsing interactions between the α-methyl group and the carbonyl substituent, R, and 1,3-diaxial interactions between the nitrogen ligand and α-methyl. For esters, the atom attached to the carbonyl is oxygen, and the alkyl group (even one as large as a *tert*-butyl) can rotate away from the α-methyl and have no effect on enolate geometry. In such cases, $E(O)^{\ddagger}$ is more stable than the $Z(O)^{\ddagger}$ due to the 1,3-diaxial interaction of the nitrogen ligand and the α-methyl in the latter. As the steric requirements of the carbonyl substituent increase, especially in the plane of the forming double bond, van der Waals repulsion increases the $A^{1,3}$ strain in the enolate and $E(O)^{\ddagger}$ is destabilized relative to $Z(O)^{\ddagger}$. Thus, for large substituents such as *tert*-butyl, and for substituents such as phenyl or dialkylamides that are coplanar with the carbonyl, the $Z(O)$-enolate is formed exclusively. Molecular mechanics calculations confirm the general validity of these arguments [40].

Scheme 3.4. Ireland model [39] transition structures for the deprotonation of acyclic carbonyls (after ref. [14]). The gray circles point out the sources of strain in the transition states: $A^{1,3}$ strain increasing as the enolate develops in $E(O)^{\ddagger}$ and 1,3-diaxial strain in $Z(O)^{\ddagger}$. For structural information from X-ray data, see ref. [29].

In the presence of coordinating additives such as HMPA, DMPU or TMEDA, the trend outlined in Scheme 3.4 may not hold [36,41-43]. For example, in the presence of HMPA, LDA deprotonation of 3-pentanone affords a 5:95 mixture of $E(O)$- and $Z(O)$-enolates under conditions of thermodynamic control (equilibration by reversible aldol addition) [39,41], but a 50:50 mixture under kinetic control [41,42].

For esters, thermodynamic equilibration of enolates is less likely, but additives can still affect the selectivity. Using LDA in THF for example, deprotonation of ethyl propionate is 94% $E(O)$-selective, but in THF containing 45% DMPU, deprotonation is 93-98% $Z(O)$-selective [36]. Ireland rationalizes this observation in terms of the transition states in Scheme 3.4 as follows: in the absence of additives, there is a close interaction between the metal, the carbonyl oxygen and the base which leads to a tight transition structure and $E(O)\ddagger$ is favored. In the presence of coordinating additives, there is more effective solvation for the lithium, and therefore weakened interaction between the lithium and the carbonyl oxygen. The cyclic transition structures will be expanded, and may even open to an acyclic transition structure. When the association between the base and the ester is diminished, the 1,3-diaxial strain in $Z(O)\ddagger$ is reduced, whereas $E(O)\ddagger$ (and acyclic structures with similar torsion angles) are still destabilized by $A^{1,3}$ strain [36].

For α-silyloxyacetates, Yamamoto reported the selective deprotonations shown in Scheme 3.5. These examples are consistent with the trend noted by Ireland, in that HMPA favors formation of the $Z(O)$-enolate, but other factors may also be involved. In fact, the transition structures proposed by Yamamoto (Scheme 3.5, inset) invoke chelation by the silyloxy group in one instance but not the other. Note however, that the Ireland argument could also be applied here: with LTMP as the

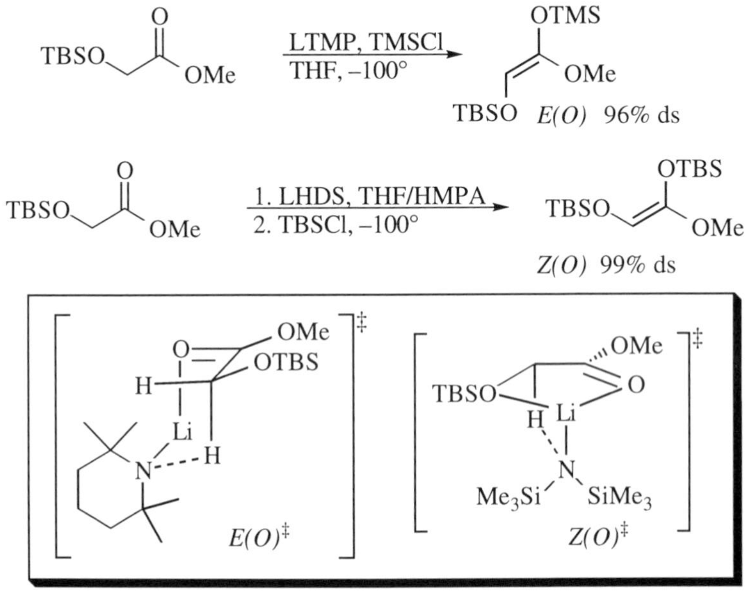

Scheme 3.5. Selective formation of either $E(O)$- or $Z(O)$-enolates of silyloxyacetates [43]. *Inset:* The authors suggest that the $E(O)$-enolate is formed according to the Ireland rationale (Scheme 3.4), but that the $Z(O)$-enolate is formed by chelation in the transition state.

base, the $Z(O)^{\ddagger}$ Ireland transition structure (Scheme 3.4) would be destabilized considerably by 1,3-diaxial interactions between the silyloxy and the bulky tetramethylpiperidine. With LHDS in THF/HMPA, the lithium would by solvated by the HMPA, and the 1,3-diaxial interactions would be attenuated as explained above, but they could also be diminished because of the longer Si–N bond distance (compared to C–N) in the amide. Independent of mechanism, *the bottom line is that both ester E(O)- and Z(O)-enolates can be produced selectively.*

The Ireland rationale also fails to account for phenomena such as changes in selectivity as the reaction proceeds and for the effect of added lithium salts. For example, the deprotonation of 3-pentanone by LTMP affords a 97:3 $E(O)/Z(O)$ selectivity at 5% conversion, but <90:10 selectivity at ≥80% conversion [38]. Moreover, the presence of 0.3-0.4 equivalents of lithium chloride or ≥1.0 equivalents of lithium bromide enhance the $E(O)/Z(O)$ selectivity (at complete conversion) to 98:2.[7,8] LTMP is one of the most sterically hindered lithium amides known, and there is some evidence that the formation of mixed aggregates is sterically driven: mixed dimerization with sterically unhindered LiX species provides a simple means to alleviate the steric demands of LTMP aggregates (Scheme 3.6). For example, a 50:50 mixture of cyclohexanone enolate and LTMP shows significant heterogeneous aggregation, whereas a similar mixture with LDA shows <5% mixed dimer [44]. The observation of decreased selectivity as enolate accumulates or with added lithium halide [38], as well as the observation of equilibrating mixed aggregates of LTMP, lithium enolates, and lithium halides [44] led to the conclusion that lithium salt dependent selectivities stem from the intervention of mixed aggregates in the

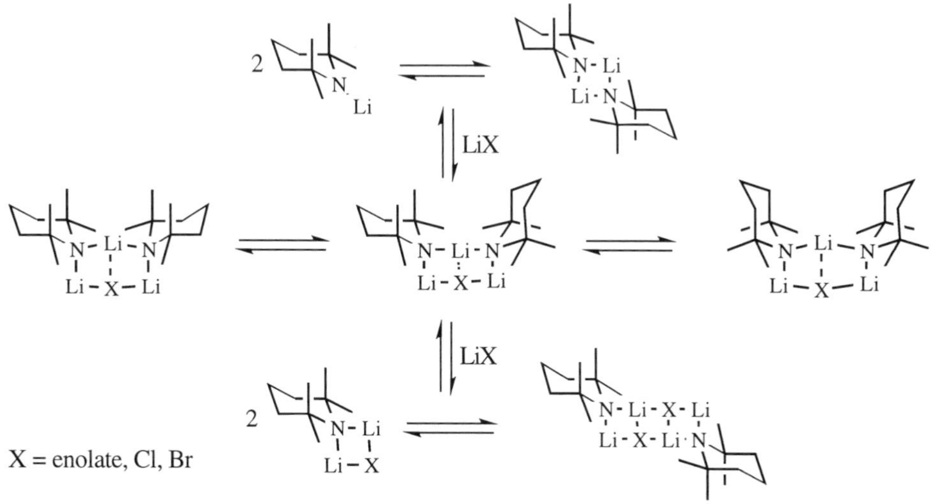

Scheme 3.6. Proposed dynamic equilibria of LTMP and added lithium salts (after ref. [44]).

---

7 For a simple protocol for the preparation of LTMP/LiBr solutions by deprotonation of TMP·HBr with butyllithium, which affords a 50:1 ratio of the $E(O)$ and $Z(O)$ enolates of 3-pentanone, see ref. [38].

8 Curiously, the presence of ≥1.0 equivalents of lithium chloride leaves the $E(O)/Z(O)$ ratio unchanged from the ratio in the absence of lithium halide [38].

product determining transition state(s) [38]. Lithium bis(2-adamantyl)amide, which is even more hindered than LTMP, forms mixed aggregates with ketone enolates but not with lithium halides, and enolizes ketones with a very high degree of *E(O)/Z(O)* selectivity [17]. The *E(O)/Z(O)* ratio of ketone enolates is also dependent on the amount of hexane in the THF solvent [45].

In light of the anomalies described above, it is apparent that the Ireland model is an oversimplification, but a clearer picture has not yet emerged. Indeed, expecting such a simple model to account for kinetic selectivites in a number of solvent systems with a variety of carbonyl substrates and amide bases is asking a lot. There is some evidence that an 8-membered ring transition structure may be involved (deprotonation by a lithium amide 'open dimer'), and computational studies indicate that neither 6- nor 8-membered ring transition structures bear much resemblance to carbocyclic 6- or 8-membered rings. For a detailed discussion of these issues, see ref. [46].

3.1.2 The transition state for enolate alkylations

The earliest work on the origin of stereoselectivity of enol and enolate reactions was done some forty years ago in the steroid arena [47,48], at the beginning of the modern era of stereochemistry. More recent efforts have focussed on the stereoelectronic effects exerted by the frontier orbitals on the trajectory of electrophilic attack [49]. Specifically, Agami suggested that the approach trajectory for the electrophile should be as shown in Figure 3.4a and b [50-52]. Using *ab initio* methods, Houk found a transition structure for the alkylation of acetaldehyde enolate with methyl fluoride which agreed with Agami's prediction of Figure 3.4a. An 'out of perpendicular' component (*à la* Figure 3.4b) was not found, but the methyl fluoride transition state is late relative to methyl iodide, and a structure associated with an earlier transition state (less bond making between nucleophile and electrophile) would probably exhibit this feature [53]. Note the pyramidalization of the α-carbon in Houk's transition structure, a feature that crops up in a number of calculated transition structures [54,55] and which appears to be important in other reaction types as well [29,56]. Often, pyramidal sp^2 atoms are found in X-ray crystal structure structures of ground state reactants such as enones.[9]

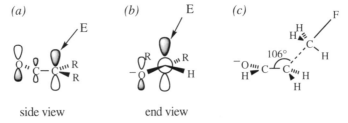

Figure 3.4. Theoretical approach trajectories (drawn in the plane of the paper) for electrophilic attack at an enolate carbon. *(a)* and *(b)* Agami's trajectory [50-52]; *(c)* Houk's trajectory [53].

---

[9] See the discussion in Section 4.4.3 (Figure 4.23) for a discussion of the phenomenon of pyramidalization in nucleophilic additions to trigonal atoms.

Studies on the stereoselective alkylation of conformationally rigid cyclohexanone enolates (summarized in ref. [14]) indicate that the transition state is early. In these systems, axial attack affords a product in a chair conformation while equatorial attack affords a twist-boat (Scheme 3.7). If the relative stability of these conformers were felt in the transition state, significant selectivity would ensue. However the low selectivities observed (55:45 for the reaction of the lithium enolate of 4-*tert*-butyl-cyclohexanone with methyl iodide [57]) suggest an early transition state according to the Hammond postulate. Somewhat higher selectivity for axial deuteration [57] is consistent with a less exothermic reaction. A higher propensity for axial alkylation (70:30) with a tetrabutylammonium cation [58] or when the *tert*-butyl group is in the 3-position (80:20) suggest that other factors (aggregation?) are also at work.

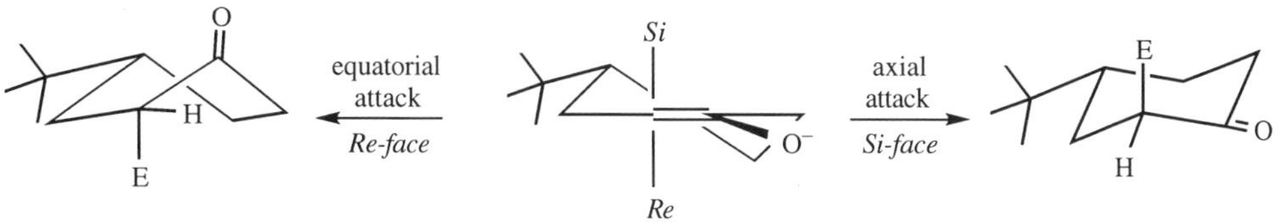

Scheme 3.7. Equatorial and axial approach of an electrophile to a cyclohexanone.

Stereocenters at the β-position can have an important effect on the differentiation of the enolate faces. The conformation of the 2-3 (allylic) bond of an acyclic enolate is governed primarily by $A^{1,3}$ strain (see glossary, section 1.6, and ref. [59]) such that the most stable conformation has the smallest substituent eclipsing the double bond, independent of enolate geometry (Figure 3.5).

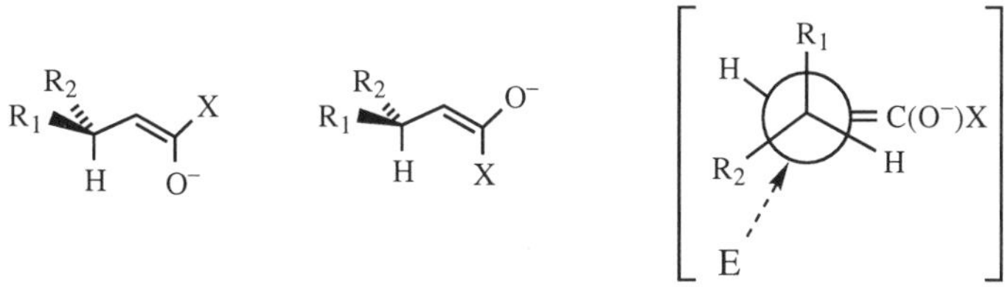

Figure 3.5. Ground state (left and center) and transition state conformations of β-substituted enolates.

In the transition state, the α-carbon is pyramidalized, and the substituents on the β-carbon are rotated such that the substituent that is the better σ-donor is perpendicular to the double bond [55,60,61]. The opposite face is then preferred by the approaching electrophile, as shown on the right in Figure 3.5. This 'antiperiplanar effect' is a phenomenon that occurs quite often in organic chemistry,[10] and arises because of the favorable overlap of an allylic σ-bond with the π-orbital of the enolate. The resulting perturbation raises the energy of the enolate HOMO and renders it more reactive [61]. For substituents (R_1 and R_2, Figure 3.5) that differ only in steric bulk, the selectivity is small (65:35), but the example in Scheme 3.8 illustrates how the electronic effect of an alkoxy substituent can profoundly influence face selectivity by proper alignment of its lone pairs.

---

[10] Another type is the Felkin-Anh descendant of Cram's rule, which is discussed in detail in Chapter 4 (Section 4.1).

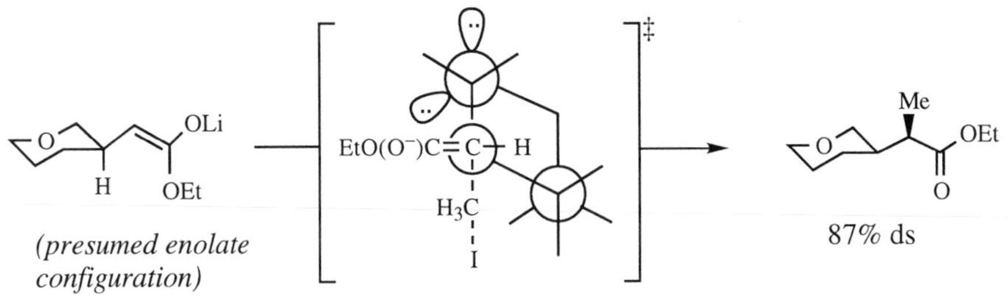

Scheme 3.8. Stereoselective alkylation of an ester enolate determined solely by stereoelectronic effects [61].

3.1.3 Enolate and azaenolate alkylations

In Section 1.3 (pp. 4-7), the relationship of extant chirality in a reacting system to any newly created stereocenters was categorized according to the relationship of the former and the latter in a metal complex in the transition state. Thus, *intraligand* asymmetric induction occurs when both the "old" and the "new" stereocenters are on the same ligand of the metal, and *interligand* asymmetric induction occurs when the existing stereocenters are on another ligand. Evans [14] had grouped chiral enolate systems into three categories, based on the location of the existing stereocenter relative to any rings present (intrannular if it is within a ring and extrannular if it is not). In the present context, these categories are sub-classes of intraligand asymmetric induction, as shown in Figure 3.6: *intraannular*, in which the existing stereocenter is contained in a ring that is bonded to the enolate at two points, *extraannular*, in which the moiety containing the stereocenter is bonded to the enolate at one point, or *chelate-enforced intraannular*, in which the stereocenter is contained in a chelate ring containing the enolate metal.

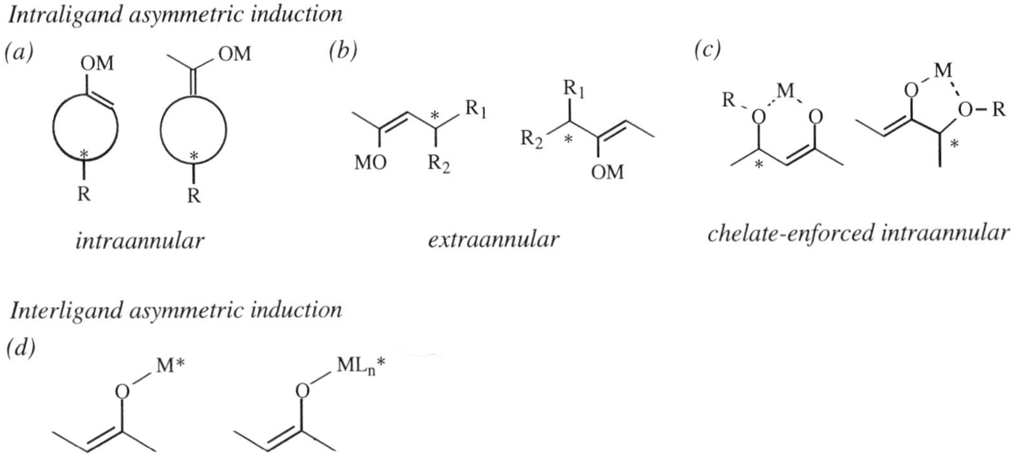

Intraligand asymmetric induction

Interligand asymmetric induction

Figure 3.6. Two categories of asymmetric induction are intraligand *(a-c)* and interligand *(d)*. The former may be subdivided [14] into *(a)* intraannular; *(b)* extraannular; and *(c)* chelate-enforced intraannular.

The widespread use of enolate alkylations for carbon-carbon bond formation has led to the development of a large number of methods for asymmetric synthesis, and the search goes on. The following discussion is intended to highlight enolate alkylation methods that seem to have broad applicability or which illustrate one of the categories mentioned above.

Intraligand asymmetric induction. An instructive introduction to intraannular alkylations is the 'self-regeneration of chirality centers' concept introduced by Seebach [62-66]. Scheme 3.9 illustrates the concept and Table 3.1 lists several representative examples. A chiral educt, such as an amino acid derivative, is condensed with pivaldehyde. This derivatization creates a new stereocenter selectively, and this second stereocenter then controls the selectivity of the subsequent alkylation by directing the electrophile to the face of the enolate opposite the *tert*-butyl group, a good example of intraannular 1,3-asymmetric induction. After purification of the alkylation product, hydrolysis affords enantiomerically pure products.

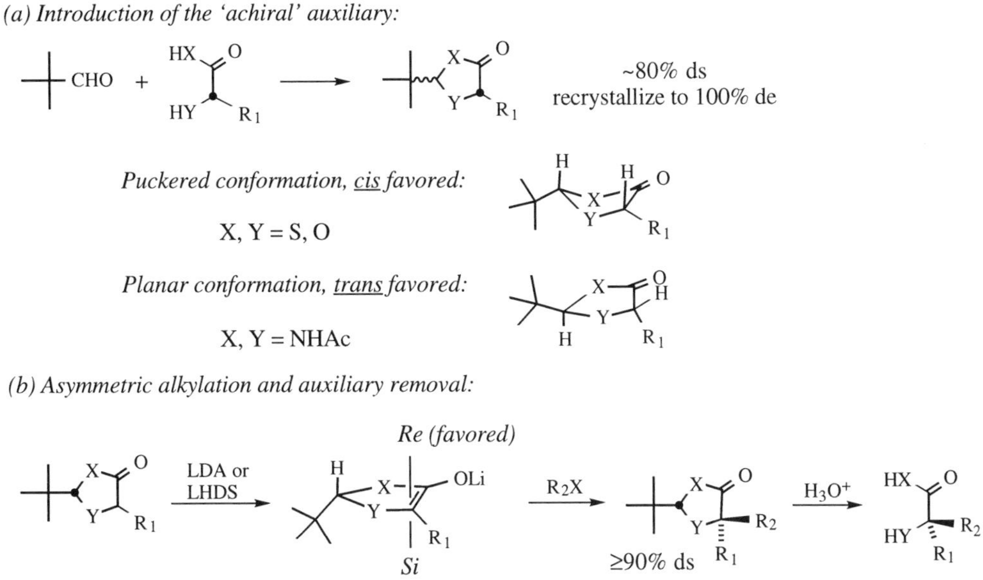
(a) Introduction of the 'achiral' auxiliary:
CHO
HX
O
HY
R1
X
Y
~80% ds
recrystallize to 100% de
Puckered conformation, cis favored:
X, Y = S, O
Planar conformation, trans favored:
X, Y = NHAc
H
(b) Asymmetric alkylation and auxiliary removal:
Re (favored)
LDA or
LHDS
OLi
Si
R2X
≥90% ds
R2
H3O+

Scheme 3.9. Self-regeneration of chirality centers [62-66].

Table 3.1. Selected examples of Seebach's "self-regeneration" of chirality centers (Scheme 3.9).

X/Y	R_1/R_2	dr[1]	% Yield	Reference
O/O	Me/Et	94:6	82	[62]
O/O	Ph/*n*-Pr	90:10	84	[62]
O/S	Me/allyl	>96:4	92	[62]
O/NCOPh	Me/Bn	>96:4	93	[64]
O/NCOPh	Bn/Me	>96:4	88	[64]
O/NCOPh	*i*-Pr/Me	100:0	53	[64]
MeN/NCOPh	Me/Et	>90:10	90	[65]
MeN/NCOPh	Me/Bn	>90:10	73	[65]
MeN/NBOC[2]	H/Bn	100:0	64	[63]
MeN/NBOC[2]	H/allyl	100:0	85	[63]
MeN/Cbz[2]	H/*i*-Pr	100:0	59	[63]

1 Diastereomer ratio after purification.
2 Obtained enantiomerically pure by resolution.

The concept of self-regeneration has been employed by other groups. For example, Vedejs has shown that condensation of a formamidino amino acid sodium salt with PhBF$_2$ affords a mixture of oxazaborolidine diastereomers that are enriched in one isomer by an asymmetric transformation of the second kind.[11] Recrystallization gives the pure diastereomer shown in Scheme 3.10a. Deprotonation and alkylation is highly diastereoselective, with alkylation occuring preferentially on the *Si* face, trans to the *B*-phenyl group. The major product can be separated from its diastereomer by crystallization and/or chromatography. Removal of the boron and formamidino groups then give the enantiopure quaternary amino acid in good yields [67]. A second example is the "dispoke" acetals developed by Ley and coworkers and illustrated in Scheme 3.10b [68,69]. Condensation of lactic acid with the dihydropyran gives an 85% yield of the spiro tricyclic acetal shown with 92% diastereoselectivity. Purification by recrystallization and alkylation gives dialkyl-lactates in excellent diastereoselectivity. The more reactive electrophiles (allyl, benzyl) afford higher diastereoselectivities. The enolate is preferentially alkylated on the *Si* face, trans to the 1,3-diaxial acetal oxygen [69].

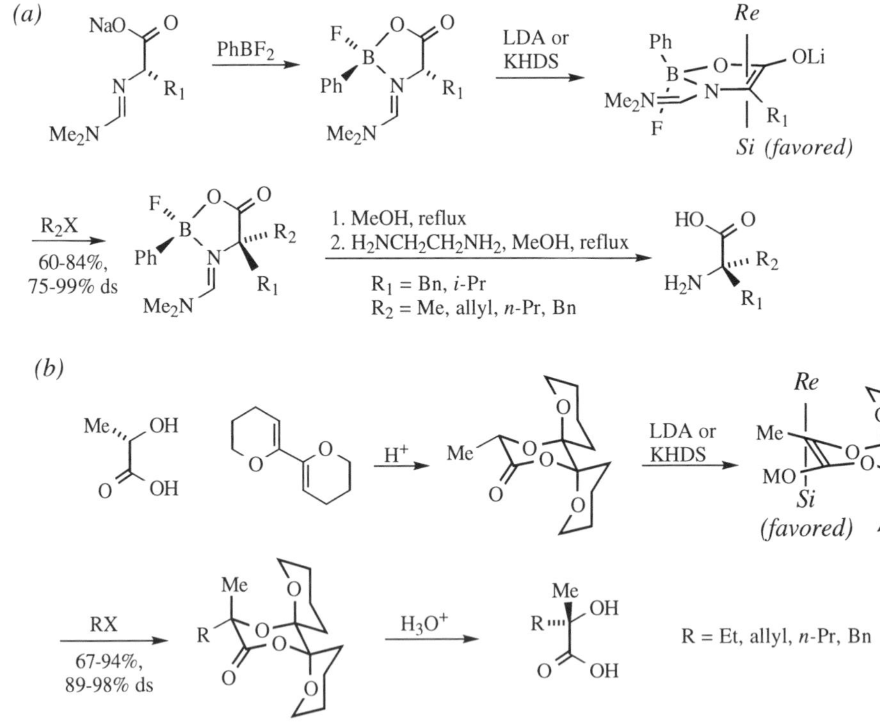

Scheme 3.10. Extensions of the self-regeneration concept: *(a)* Vedejs oxazaborolidines [67]. *(b)* Ley's 'dispoke' acetals [68,69].

Two groups have developed methods for the asymmetric alkylation of "glycine" enolates using intraannular asymmetric induction. The first, developed by Schöllkopf [70], involves condensation of two amino acid esters to a diketo-

---

[11] See glossary, section 1.6, for definition of the two types of asymmetric transformation.

piperazine, one of which serves as the chiral auxiliary for α-alkylation of the other. *O*-Alkylation gives the bis-lactim ether shown at the top right of Scheme 3.11. After deprotonation, alkylation occurs stereoselectively such that the electrophile approaches the anion anti to the isopropyl. Typical selectivities for this process are listed in Table 3.2. Advantages of this process are that selectivities are high and that it makes chiral quaternary carbons. Disadvantages are that the electrophiles must often be activated (*i.e.,* allylic, benzylic), and that the alkylated amino ester and the amino ester chiral auxiliary must be separated at the end.

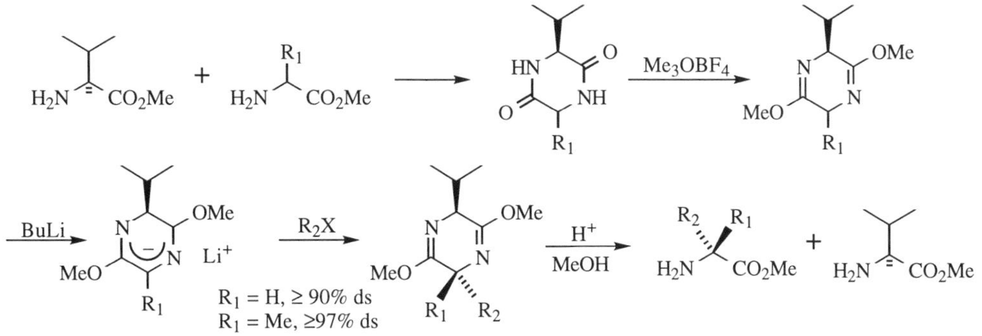

Scheme 3.11. Shöllkopf's bis-lactim ether amino acid synthesis [70].

Table 3.2. Examples of Shöllkopf's amino acid synthesis (Scheme 3.11 [70]).

R₁/R₂	% ds	% Yield
H/Bn	96	81
H/PhCH=CHCH₂	97.5	90
H/n-C₇H₁₅	87	62
Me/Bn	98	68
Me/PhCH=CHCH₂	98	89
Me/n-C₇H₁₅	98	43

A second method for amino acid synthesis was developed by Williams [71]. As shown in Scheme 3.12, the chiral diphenyloxazinone may be alkylated using LHDS or NHDS with excellent diastereoselectivity provided the alkylating agent is activated, such as a benzyl, allyl, or methyl halide. The stereoselectivity is ≥98% and the conformation shown in Scheme 3.12 was postulated to explain the selectivity. After the first alkylation, a second alkylation may be executed. After purification of the crystalline oxazinones, reductive cleavage of the benzylic–heteroatom bonds liberates the amino acid. This destruction of the "auxiliary" is a drawback to this strategy because of the high cost of the amino alcohol (>$15/g). Selected examples of this process are listed in Table 3.3. As with the Schöllkopf method, the electrophiles must be activated. However, in this self-immolative method the separation of the amino acid from the remainder of the auxiliary is not a complicating factor.

Another method for asymmetric alkylation of a masked glycine was reported by Yamada, and is shown in Scheme 3.13 [72]. In this example of a chiral glycine enolate, the Schiff base of *tert*-butyl glycine and an α-pinene-derived ketone is dilithiated with two equivalents of LDA. Presumably, the lithium alkoxide is chelat-

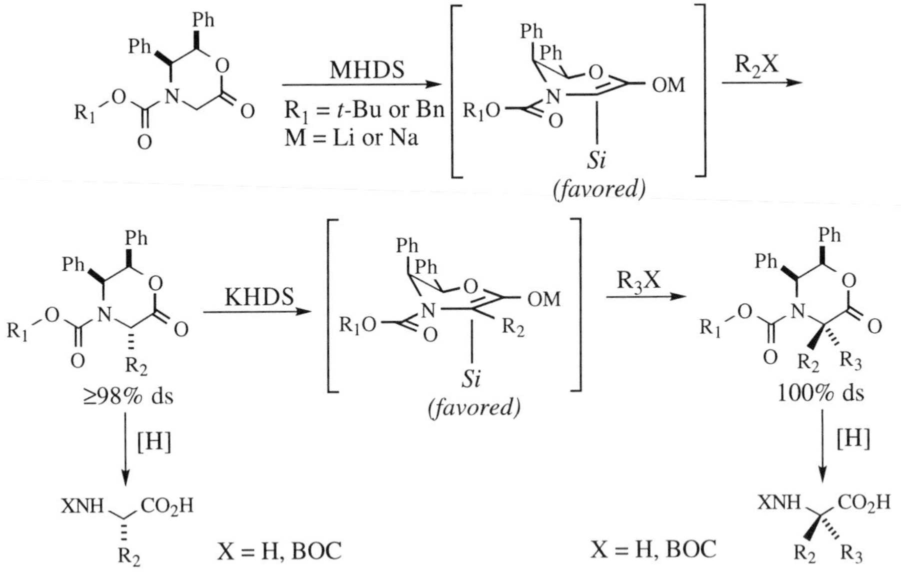

Scheme 3.12. William's oxazinone enolate amino acid synthesis [71]. The conformation shown in the two bracketed structures has the C-5 phenyl in the axial position to avoid $A^{1,3}$ interactions with the adjacent *N*-acyl group.

Table 3.3. Examples of Williams's amino acid synthesis (Scheme 3.12 [71]. In all cases, the diastereoselectivity was ≥98%.

R_1	R_2	R_3	Base	% Yield (alkylation)	% Yield (amino acid)	% ee
t-Bu	allyl	-	LHDS	86	50-70	98
t-Bu	Me	-	NHDS	91	54	97
t-Bu	Bn	-	NHDS	70	76	98
Bn	Bn	-	NHDS	77	93	>99
t-Bu	Me	allyl	KHDS[1]	87[1]	70	100
t-Bu	Me	Bn	KHDS[1]	84[1]	93	100
t-Bu	*n*-Pr	allyl	KHDS[1]	90[1]	60	100
Bn	Me	Bn	KHDS[1]	84[1]	93	100

[1]   Second alkylation.

ed to the nitrogen as shown. This tricyclic chelate is rigid and the dienolate must adopt the *s*-trans conformation in order to avoid severe nonbonding interactions with the α-pinene moiety. Nonbonding interactions that restrict conformational motion are necessary for high selectivity in examples of extraannular asymmetric induction (Figure 3.6) such as these. Approach of the electrophile from the *Re* face gives the configuration shown. Although the authors did not determine the configuration of the enolate, if we assume that the enolate is *E(O)* as shown, then the Agami approach trajectory (Figure 3.4) would be slanted towards the back of the figure, trans to the alkoxide. Approach from the *Si* face would not only be on the concave face of the structure but would also be slanted back towards the α-pinene moiety. Table 3.4 summarizes the selectivities reported for this asymmetric amino acid synthesis.

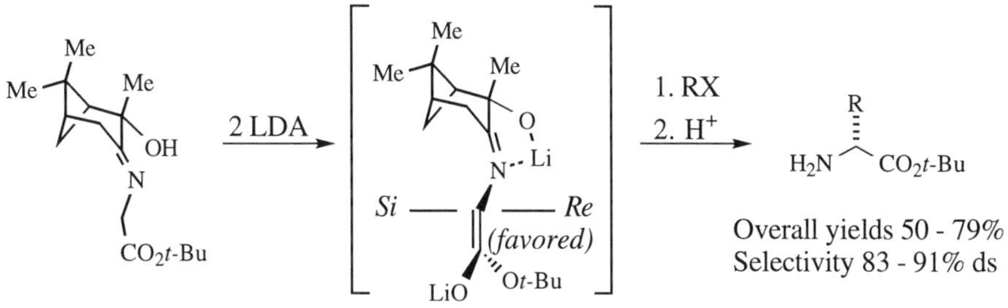

Scheme 3.13. Yamada's chiral glycine enolate [72].

Table 3.4. Stereoselective alkylations of Yamada's glycine enolate (Scheme 3.13) [72].

R	(presumed) % ds[1]	Yield[2]
Me	91	52
i-Bu	91	50
Bn	86	79
3,4–(MeO)$_2$–C$_6$H$_3$CH$_2$	83	62

[1] Calculated from %ee of product.
[2] Overall yield of amino acid ester.

During the 1980s, one of the major thrusts of asymmetric synthesis was the development of chiral auxiliaries for the alkylation and aldol addition of propionates. The following paragraphs describe some of the methods that evolved, beginning with two examples of propionate ester alkylations developed by Helmchen using chiral alcohols as the auxiliary and ending with propionic imide auxiliaries developed by Evans and Oppolzer.

The two ester enolate alkylation methods developed by Helmchen are illustrated in Schemes 3.14 and 3.15 [73-76]. These esters are designed so that one face of the enolate will be completely shielded by a second ligand appended to the camphor nucleus. As shown in Scheme 3.14, deprotonation by LICA affords the *E(O)*-enolate illustrated. Nonbonding steric interactions are thought to hold this enolate in the illustrated geometry.[12] Specifically, the OLi is thought to be syn to the illustrated endo carbinol hydrogen, since other rotamers would engender strain between various parts of the enolate and the camphor nucleus. Similarly, the most stable conformation of the carbamate shielding group has the endo C–H syn to the C=O. In this conformation, approach of the electrophile is only possible from the front (*Re*) face. After purification, LAH reduction affords enantiomerically pure alcohols as shown [73,74]. Representative examples of this procedure are listed in Table 3.5.

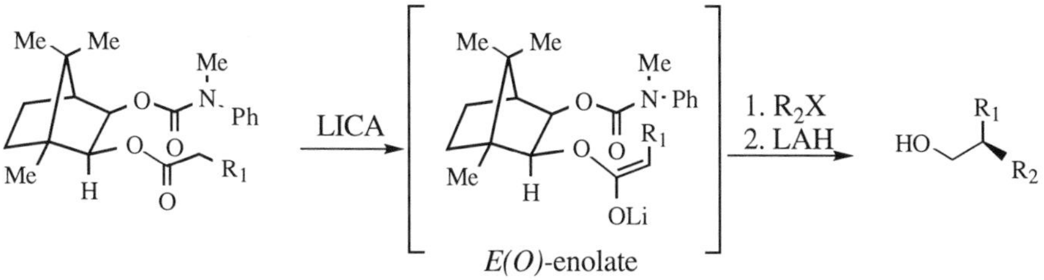

Scheme 3.14. Helmchen's asymmetric ester enolate alkylation [73,74].

---

[12] Note the similarity to the enolate conformations in Figure 3.6.

Table 3.5. Stereoselective alkylation of camphor ester enolates (Scheme 3.14).

R_1	R_2	Yield	%ds	Reference
Me	$n-C_{16}H_{33}$	83%	93	[73]
$n-C_{16}H_{33}$	Me	80%	90	[73]
Me	Bn	96%	94	[74]
Bn	Me	95%	95	[74]

Scheme 3.15 illustrates a different auxiliary derived from camphor, and which has similar design features, but which affords higher diastereoselectivity [75]. Additionally, Scheme 3.15 illustrates the selective formation of either an *E(O)*- or *Z(O)*-enolate based on the presence or absence of HMPA in the reaction mixture. Thus, deprotonation of the ester with LICA is 98% selective for the *E(O)*-enolate and deprotonation in the presence of HMPA is 96% selective for the *Z(O)*-enolate. Alkylation with benzyl bromide is more selective for the *E(O)*-enolate than for the *Z(O)*, but after diastereomer separation, reduction gives enantiomerically pure *R*- or *S*-2-methyl-3-phenylpropanol, opposite enantiomers from the same auxiliary [75].

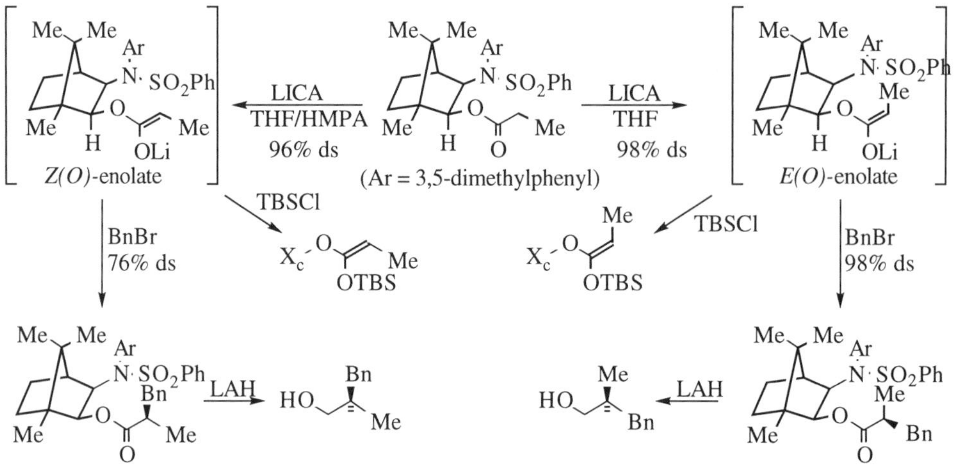

Scheme 3.15. Controlled stereoselective enolate formation and asymmetric alkylation of a "second generation" camphor ester enolate chiral auxiliary [75].

The mechanistic rationale for the selectivity of these ester enolate alkylations may be summarized as follows (Scheme 3.14 and 3.13 [74]):

1. Deprotonation in THF solvent gives *E(O)*-enolates while enolization in THF/HMPA solvent gives *Z(O)*-enolates with a high degree of selectivity.
2. After deprotonation, the enolate is oriented as illustrated in Schemes 3.14 and 3.15 such that the H–C–O–C–OLi moiety is coplanar and the OLi is syn to the endo carbinol hydrogen of the camphor.
3. Similarly, the H–C–O–C=O of the adjacent urethane or the H–C–N–S=O sulfonamide moiety is also in the conformation illustrated.
4. With the rear face of the enolate thus shielded, alkylation occurs from the front face (direction of the viewer - *Re* face of the *E(O)*-enolate and *Si* face of the *Z(O)*-enolate.

Although accounting for the gross data, this rationale is not completely satisfactory. Subsequent studies [75] showed that addition of HMPA *after enolate formation but before electrophile addition* also had an effect on the selectivity of the alkylation, leading Helmchen to speculate that the sulfonamide in Scheme 3.15 (or presumably the urethane in Scheme 3.14) may be chelated to the lithium. Another possibility may be that the enolates are aggregated, and the effect of HMPA is to disrupt the aggregation. Additionally, the difference in selectivity between the *E(O)*- and *Z(O)*-enolate alkylations (Scheme 3.15) remains unexplained.

Helmchen has used this methodology in asymmetric synthesis of the three stereoisomers of the tsetse fly pheromone [73] and the side chain of α-tocopherol [76], illustrated in Figure 3.7.

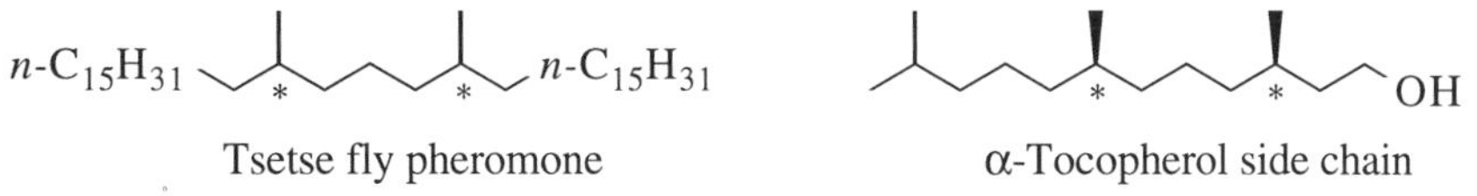
n-C15H31
n-C15H31
OH
Tsetse fly pheromone
α-Tocopherol side chain

Figure 3.7. Natural products synthesized using Helmchen's ester enolates: the tsetse fly pheromone [73] and the side chain of α-tocopherol [76].

It is generally true that restrictions on conformational mobility minimize the number of competing transition states and simplify analysis of the factors that affect selectivity. Chelation of a metal by a heteroatom often provides such restriction and also often places the stereocenter of a chiral auxiliary in close proximity to the α-carbon of an enolate. This proximity often results in very high levels of asymmetric induction. A number of auxiliaries have been developed for the asymmetric alkylation of carboxylic acid derivatives using chelate-enforced intraannular asymmetric induction. The first practical method for asymmetric alkylation of carboxylic acid derivitives utilized oxazolines and was developed by the Meyers group in the 1970's (Scheme 3.16a), whose efforts established the importance and potential for chelation-induced rigidity in asymmetric induction (reviews: [77-79]). In 1980, Sonnet [80] and Evans [81,82] independently reported that the dianions of prolinol amides afford more highly selective asymmetric alkylations (Scheme 3.16b).

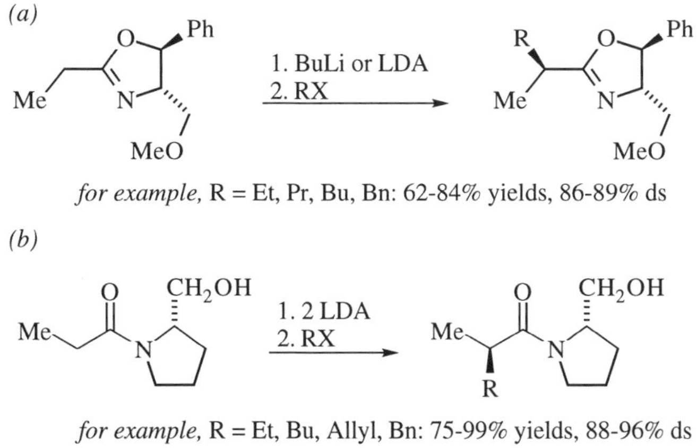
(a)
Ph
O
Me
N
MeO
1. BuLi or LDA
2. RX
R
O
Ph
Me
N
MeO
for example, R = Et, Pr, Bu, Bn: 62-84% yields, 86-89% ds
(b)
O
CH2OH
Me
N
1. 2 LDA
2. RX
O
CH2OH
Me
N
R
for example, R = Et, Bu, Allyl, Bn: 75-99% yields, 88-96% ds

Scheme 3.16. Early examples of asymmetric enolate alkylations: *(a)* Meyers's oxazolines [77-79]; *(b)* Evans's [81,82] and Sonnet's [80] proline amide alkylations .

In 1982, Evans reported that the alkylation of oxazolidinone imides appeared to be superior to either oxazolines or prolinol amides from a practical standpoint, since they are significantly easier to cleave [83]. As shown in Scheme 3.17, enolate formation is at least 99% stereoselective for the Z(O)-enolate, which is chelated to the oxazolidinone carbonyl oxygen as shown. From this intermediate, approach of the electrophile is favored from the *Si* face to give the monoalkylated acyl oxazolidinone as shown. Table 3.6 lists several examples of this process. As can be seen from the last entry in the table, alkylation with unactivated alkyl halides is less efficient, and this low nucleophilicity is the primary weakness of this method. Following alkylation, the chiral auxiliary may be removed by lithium hydroxide or hydroperoxide hydrolysis [84], lithium benzyloxide transesterification, or LAH reduction [85]. Evans has used this methology in several total syntheses. One of the earliest was the Prelog-Djerassi lactone [86] and one of the more recent is ionomycin [87] (Figure 3.8).

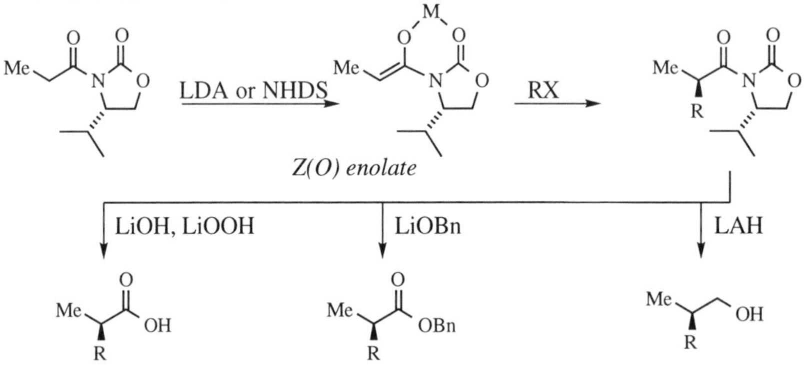

Scheme 3.17. Evans's asymmetric alkylation of oxazolidinone imides [83].

Table 3.6. Alkylations of Evans's oxazolidinone imides (Scheme 3.17 [83]. In all cases, the alkylation products were >99% pure after chromatography.

R	%ds	Yield[1]
Bn	>99	92%
methallyl	98	62%
allyl	98	71%
Et	94	36%

[1] Isolated yields after chromatography.

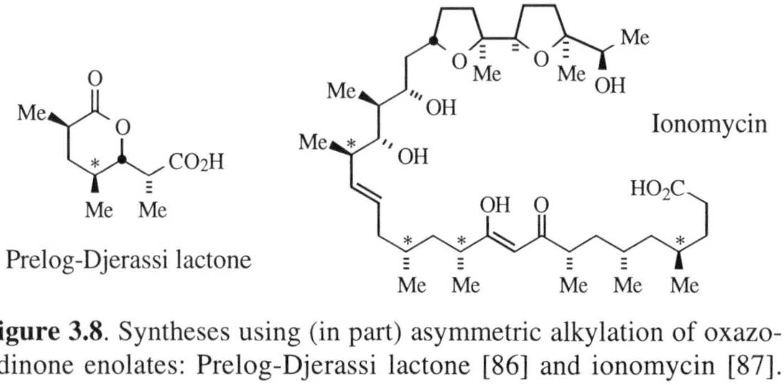

Prelog-Djerassi lactone

Ionomycin

Figure 3.8. Syntheses using (in part) asymmetric alkylation of oxazolidinone enolates: Prelog-Djerassi lactone [86] and ionomycin [87]. Stereocenters created by alkylation are indicated (*).

In addition to these examples of alkylations that employ chelate-enforced intraannular asymmetric induction, Evans's imides are useful in asymmetric aldol (Section 5.2.2 and 5.2.3), and Michael additions (Section 5.3.2), Diels-Alder reactions (Section 6.2.2), and enolate oxidations (Section 8.4), making this one of the most versatile auxiliaries ever invented.

In 1989 Oppolzer reported that the enolates of *N*-acyl sultams derived from camphor afford highly diastereoselective alkylation products with a variety of electrophiles *including those which are not allylically activated* [88]. The sultam is deprotonated using either butyllithium with a catalytic amount of cyclohexyl isopropyl amine, or butyllithium alone, or sodium hexamethyldisilyl amide.[13] As illustrated in Scheme 3.18, alkylation occurs selectively from the *Re* face of the *Z(O)*-enolate to give monoalkylated sultams which can be cleaved by LAH reduction or lithium hydroperoxide catalyzed hydrolysis. Representative examples are listed in Table 3.7.

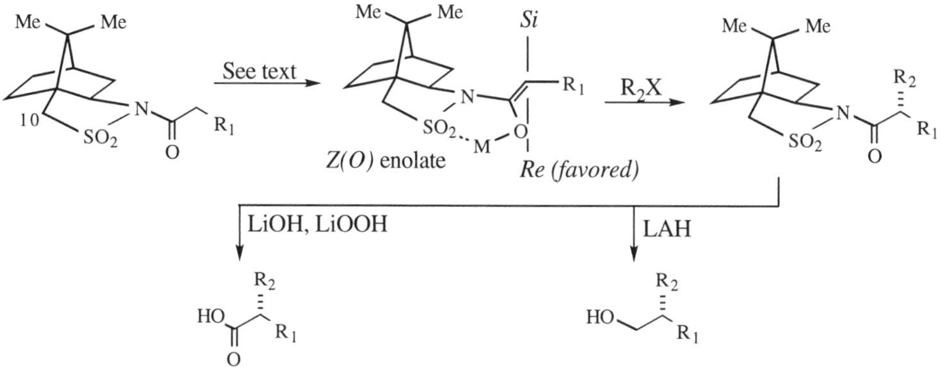

Scheme 3.18. Oppolzer's asymmetric sultam alkylation [88].

Table 3.7. Asymmetric alkylation of Oppolzer's sultams (Scheme 3.18) [88].

R_1	R_2	%ds	dr[1]	Yield[1]
Me	Bn	98.5	>99:1	89%
Bn	Me	97.4	>99:1	88%
Me	allyl	98.3	>98:2	74%
allyl	Me	97.7	>99:1	2
Me	methallyl	89.6	>99:1	70%
Me	$n\text{-}C_5H_{11}$	98.9	98:2	81%
$n\text{-}C_5H_{11}$	Me	98.1	98:2	2

[1] Diastereomer ratio after recrystallization.
[2] Not reported.

---

[13] These conditions are necessary to avoid competitive deprotonation at C_{10}.

The selectivity of this reaction is based on the following mechanistic rationale of chelate-enforced intraannular asymmetric induction[14] (Scheme 3.18):

1. Following the Ireland model (Scheme 3.4), deprotonation gives the *Z(O)*-enolate.
2. The lithium of the enolate is chelated to the sultam which also has a pyramidal nitrogen.
3. The *Si* face is shielded by the bridging methyls, and approach is therefore from the *Re* face, opposite the nitrogen lone pair.

A versatile method for the synthesis of compounds containing quaternary centers (using an intraannular asymmetric induction strategy) was developed by Meyers and uses the bicyclic lactams illustrated in Scheme 3.19 [90-96]. The bicyclic lactams may be synthesized by condensation of an amino alcohol with a keto acid as illustrated [90,95], or by condensation of an amino alcohol with an anhydride followed by reductive cyclization [91]. Sequential alkylations proceed with differing degrees of stereoselectivity. The first alkylation is not very selective, but the second is highly so, as shown by the examples listed in Table 3.8. Note that a different auxiliary is used for the two ring systems. Specifically, for the 5,5-bicyclic system (n=0), an auxiliary derived from valinol is used ($R_1 = i\text{-Pr}$, $R_2 = H$). For the 5,6-bicyclic system, an auxiliary containing a free hydroxyl group is required in order to enable reduction of the carbonyl by intramolecular delivery of hydride at a later stage [93]. In all of the cases reported to date, the two diastereomeric dialkylated bicyclic lactams have been separable by chromatography, insuring enantiomerically pure products at the end. Scheme 3.19 details one protocol for the elaboration of these bicyclic lactams into cyclohexenones, and Scheme 3.20 summarizes different ways that have been developed for the elaboration of the alkylated bicyclic lactams into enantiomerically pure cylopentenones and cyclohexenones containing chiral quaternary centers of differing substitution patterns [94] (see also ref. [97,98]).

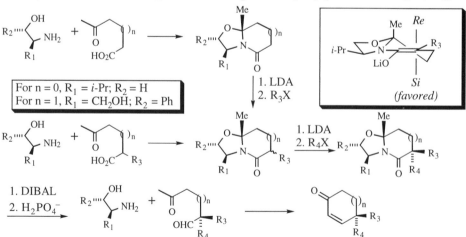

Scheme 3.19. Meyers's asymmetric alkylations of bicyclic lactams [90-96]. When n = 0, $R_1 = i\text{-Pr}$ and $R_2 = H$; when n = 1, $R_1 = CH_2OH$ and $R_2 = Ph$.

---

[14] An alternative explanation, based on analogy of the sultam to a *trans*-2,5-disubstituted pyrrolidine, has also been offered [89].

Table 3.8. Selected examples of asymmetric alkylation of Meyers's bicyclic lactams (Scheme 3.19 [90,92,93]).

n	R$_3$	R$_4$	%ds	% yield	Reference
0	allyl	Bn	97	77	[92]
0	Bn	allyl	95	63	[92]
0	Et	Bn	95	74	[92]
1	Me	Bn	97	48	[90,93]
1	Bn	Me	75	"50-80"	[93]
1	Bn	allyl	82	"50-80"	[93]

In all of the examples reported to date, the second alkylation occurs on the *Si* face as illustrated in Scheme 3.19. The origin of this stereoselectivity is not clear, and there are a probably a number of subtle factors which combine to produce the observed chirality sense. Among the factors that favor approach from the *Si* face are that this is the axial direction, and it is anti to the angular methyl group. On the other hand, the Agami trajectory (Figure 3.4a) for the approach of an electrophile would be slanted toward the β-carbon, trans to the enolate oxygen (Figure 3.4b), which would seem to favor the *Re* face (at least in the 6-membered ring), due to the axial hydrogen on the γ-carbon. Meyers has suggested that the selectivity may be due to stereoelectronic factors related to perturbation of the enolate π bond by the nitrogen lone pair [99].

Figure 3.9 illustrates a number of targets that have been synthesized using this methodology. In each case, the stereocenter formed in the asymmetric alkylation is indicated with an asterisk. In all cases, the configuration at this center controlled the selective formation of the rest.

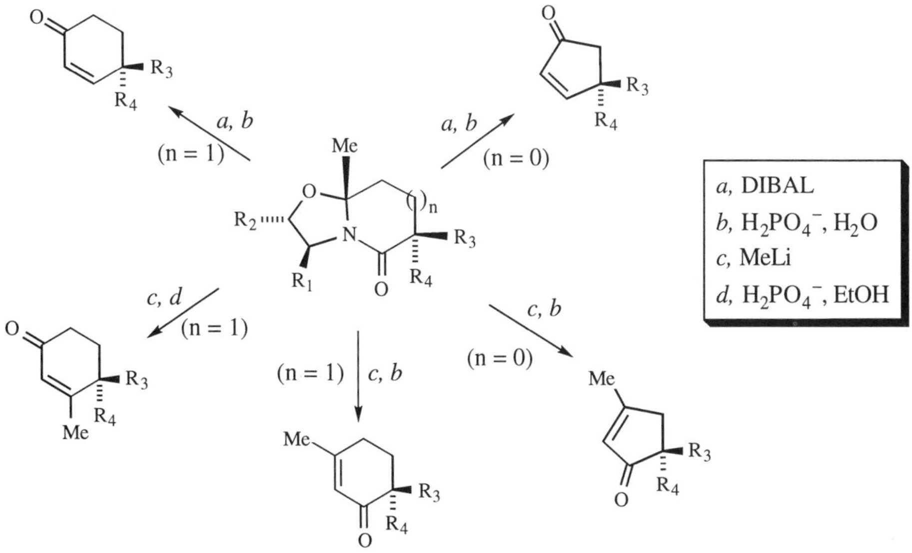

Scheme 3.20. Protocols for the elaboration of bicyclic lactams into cyclo-alkenones having different substitution patterns [94] (see also [97,98]).

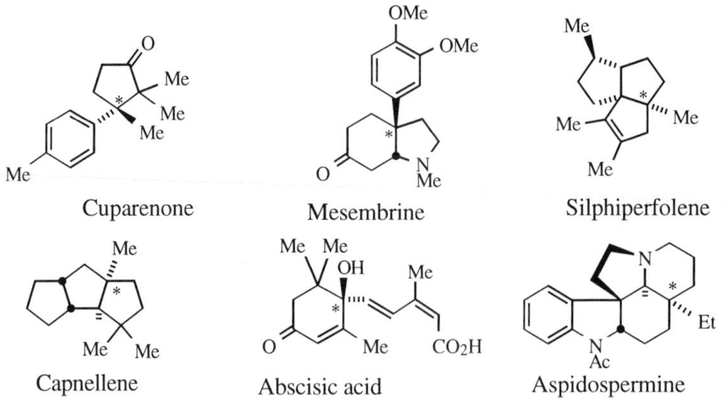

Figure 3.9. Natural products synthesized using Meyers's bicyclic lactam methodology: cuparenone [100], mesembrine [101], abscisic acid [102], capnellene [102], silphiperfolene [94], and aspidospermine [103].

For the asymmetric alkylation of ketones and aldehydes, a highly practical method was developed by the Enders group, and uses SAMP-RAMP hydrazones (reviews: [104-107]). SAMP and RAMP are acronyms for $\underline{S}$- or $\underline{R}$-1-amino-2-methoxymethylpyrrolidine. This chiral hydrazine is used in an asymmetric version of the dimethylhydrazone methodology originally developed by Corey and Enders [108,109]. These auxiliaries are available from either proline or pyroglutamic acid [104,110]. As shown in Scheme 3.21, SAMP hydrazones of aldehydes [111] and ketones [111,112] may be deprotonated by LDA and alkylated. The diastereoselectivity of the reaction may often be determined by integration of the methoxy singlet after treatment with a shift reagent.[15] After alkylation, cleavage may be effected with a number of reagents [105,106]. Among these are oxidative cleavage by ozonolysis [105], sodium perborate [113], or magnesium peroxyphthalate [114], acidic hydrolysis using methyl iodide and dilute HCl [111,112], or BF_3 and water [115,116]. Table 3.9 lists a few examples of SAMP asymmetric alkylations.

Scheme 3.22 illustrates the mechanistic rationale for this asymmetric alkylation. Deprotonation by the Ireland model (*cf.* Scheme 3.2) gives the E_{CC},Z_{CN} enolate as shown [117].[16] Cryoscopic and spectroscopic measurements indicate that the lithiated hydrazones are monomeric in THF solution [118], and a crystal structure shows the lithium σ-bonded to the azomethine nitrogen and chelated by the methoxy of the auxiliary [119]. The azomethine nitrogen is largely sp^2-hybridized, and the nitrogen is pulled 'downward' 17.5° below the azaallyl plane by the chelating methoxyl. If this structural feature is preserved in the transition state, the approach of the electrophile toward the *Si* face would be hindered by the C-5 methylene of the pyrrolidine ring and approach toward the *Re* face would be favored, as is observed [119].[17,18] Note that this substructure is equally accessible from both

---

[15] See Section 2.3.3 for a discussion of the use of lanthanide shift reagents.
[16] The deprotonation of hydrazones is not regioselective, so the Z_{CN} geometry results from equilibration *after* deprotonation.
[17] The Agami trajectory (Figure 3.5) would seem to suggest an approach trajectory that is slanted *away* from the pyrrolidine (*i.e.*, towards the viewer), decreasing the effect of the auxiliary in

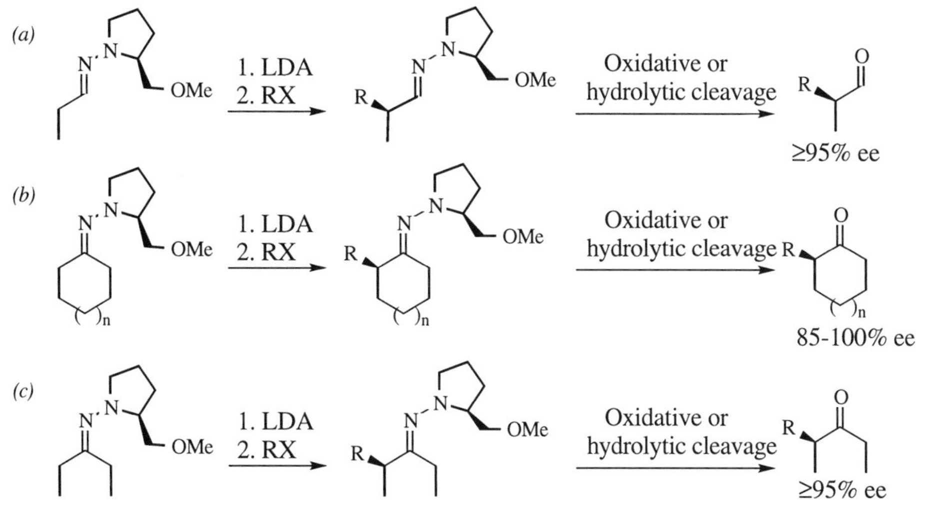

Scheme 3.21. Asymmetric alkylations of aldehydes and ketones with SAMP hydrazones.

Table 3.9. Asymmetric alkylations of SAMP-RAMP hydrazones (Scheme 3.21 [105]).

Product	Electrophile	% Yield	% ee	Reference
(aldehyde with R and Me)	EtI	71	95	[104,111]
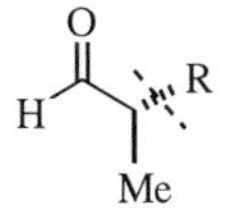	$C_6H_{13}I$	52	≥95	
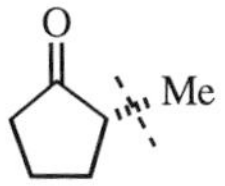 (cyclopentanone with Me)	Me_2SO_4	66	86	[111]
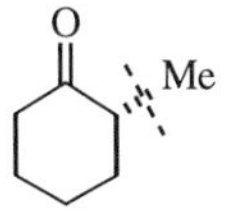 (cyclohexanone with Me)	Me_2SO_4	70	>99	[111]
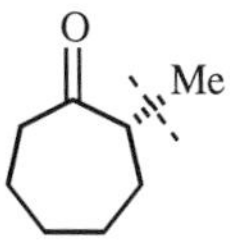 (cycloheptanone with Me)	MeI	59	94	[111]
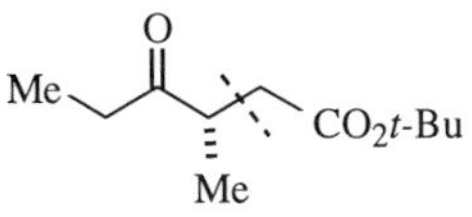 (ketone with Me, Me, CO_2t-Bu)	$BrCH_2CO_2t$-Bu	53	>95	[104]
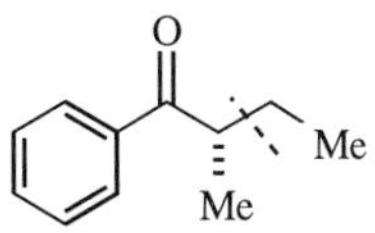 (phenyl ketone with Me, Me)	EtI	44	>97	[112]

directing the approach. On the other hand, *gem*-dimethyls at the 5-position of the auxiliary enhance the selectivity [119].

18 Another rationale, which postulates a chelated lithium that is situated on top of the π-cloud of the azaallyl anion, has also been proposed [9,106].

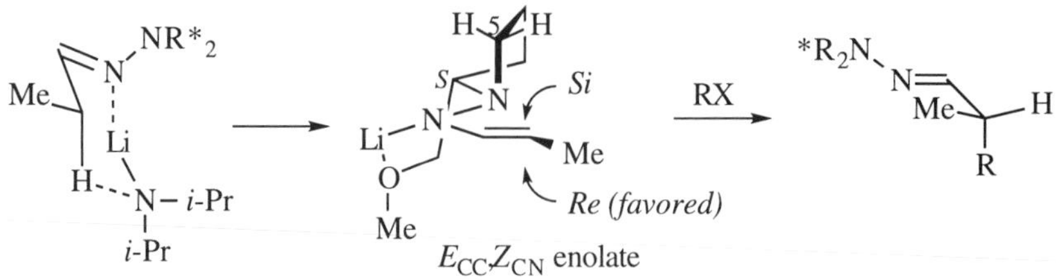

Scheme 3.22. Mechanistic rationale for face-selectivity of SAMP hydrazone alkylation [119].

cyclic and acyclic ketones as well as aldehydes. Approach from the *Re* face gives the configuration shown, which is uniformly predictable independent of the ketone or aldehyde educt.

Figure 3.10 illustrates several natural products which have been synthesized using this methodology. These include a number of insect pheromones as well as the sesquiterpene eremophilenolide and the antibiotic X-14547A. The latter two compounds have multiple stereocenters but the asymmetric alkylation using the SAMP-RAMP hydrazone method produces one stereocenter which is then used to direct the selective formation of the others.

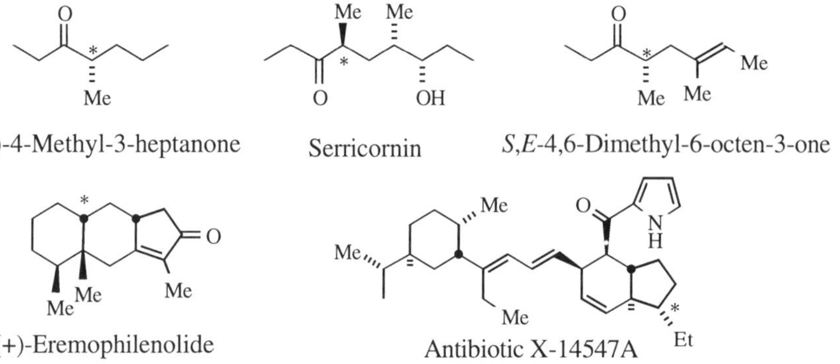

S -(+)-4-Methyl-3-heptanone   Serricornin   S,E-4,6-Dimethyl-6-octen-3-one

(+)-Eremophilenolide   Antibiotic X-14547A

Figure 3.10 Natural product synthesis imploying SAMP-RAMP hydrazones: S-(+)-4-methyl-3-heptanone, the leaf cutting ant alarm pheromone [105]; serricornin, the sex pheromone of the cigarette beetle [120]; S,E-4,6-dimethyl-6-octen-3-one, the defense substance of "daddy longlegs" [120]; (+)-eremophilenolide [120] and antibiotic X-14547A [121]. Stereocenters formed by asymmetric alkylation are indicated by *.

Interligand asymmetric induction. Group-selective reactions are ones in which heterotopic ligands (as opposed to heterotopic faces) are distinguished. Recall from the discussion at the beginning of this chapter that secondary amines form complexes with lithium enolates (pp 76-77) and that lithium amides form complexes with carbonyl compounds (Section 3.1.1). So if the ligands on a carbonyl are enantiotopic, they become diastereotopic on complexation with chiral lithium amides. Thus, deprotonation of certain ketones can be rendered enantioselective by using a chiral lithium amide base [122], as shown in Scheme 3.23 for the deprotonation of cyclohexanones [123-128]. 2,6-Dimethyl cyclohexanone (Scheme 3.23a) is meso, whereas 4-*tert*butylcyclohexanone (Scheme 3.23b) has no stereocenters. Nevertheless, the enolates of these ketones are chiral. Alkylation of the enolates affords nonracemic products and O-silylation affords a chiral enol ether which can be further manipulated by a number of means. Although crystallographic and spectroscopic characterization of chiral lithium amides have been carried out [125], a rationale explaining the relative topicity of these deprotonations has not been offered. Note that any heterotopic protons may – in principle – be distinguished by this concept. An early contribution to this area was the group-selective deprotonation of cyclohexene oxide, reported by the Whitesell group in 1980 [129], but the selectivities were not high, probably because of minimal prior complexation of the lithium base with the carbon acid.

This concept has been extended to the kinetic resolution (selective reaction of protons that are enantiotopic by external comparison) [30,130] and to selective reaction at proton pairs that are diastereotopic (double asymmetric induction) [131].

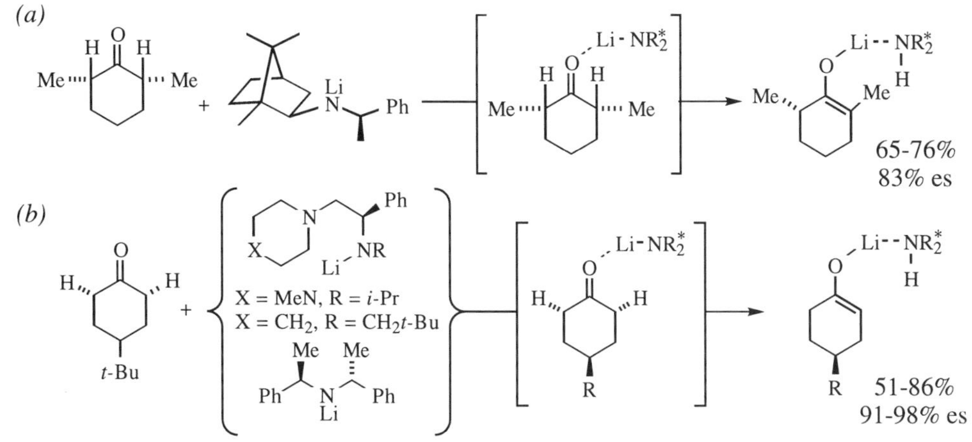

Scheme 3.23. Enantioselective deprotonation of achiral ketones with chiral lithium amide bases: *(a)* [123]. *(b)* [124-126,128].

A related concept is the selective protonation of enantiotopic faces of an enolate, which is possible because of a combination of two factors:

1. When an enolate·secondary amine complex is quenched with water, the enolate is protonated by internal return from the amine (Scheme 3.2).
2. Complexation with a chiral amine renders the enolate faces diastereotopic.

Thus, use of a chiral lithium amide base followed by protonation by internal return may be enantioselective because of interligand asymmetric induction. Scheme 3.24a shows an example reported in 1982 by Hogeveen [30,122]. In competition with protonation of the enolate by proton transfer from the amine is direct protonation by water, which has the effect of lowering the enantioselectivity of the process. A recent contribution by Vedejs [32,132] notes that the intermolecular route can be avoided by quenching the enolate·amine complex with an aprotic acid such as boron trifluoride, and excellent selectivities were obtained in certain instances (*e.g.*, Scheme 3.24b). The aggregate proposed to account for these selectivities is illustrated in Scheme 3.24b. The argument is that the amide nitrogen (NR_2) is rotated out of plane and is hydrogen bonded to the amine NH. Complexation of the amine nitrogen by boron increases the acidity of the N–H ('ammonium-like'), and the proton is then transferred to the nearest (*Si*) face of the enolate. At this point in time, no method offers a *substrate-independent* asymmetric

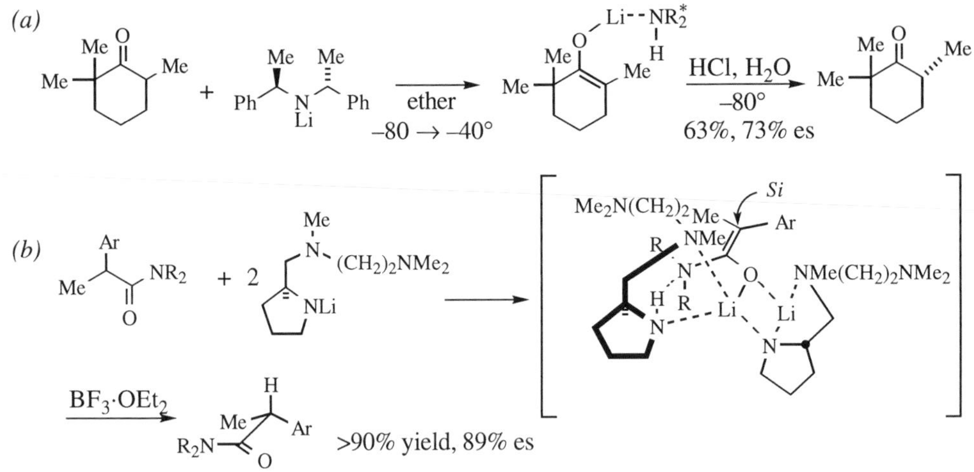

Scheme 3.24. Intrasupramolecular enantioselective protonation of an enolate. The lithium amides are illustrated as monomers for simplicity; the aggregation states are unknown. *(a)* [30,122]. *(b)* [32,132].

protonation protocol, but progress is being made (review: [133]; see also ref. [24,32,134-140]).

A further extension of these concepts is the alkylation of enolate / secondary amine complexes. Following several early observations [141-143],[19] systematic investigations were undertaken by the Koga group [24,25,147-149]. These efforts have resulted in a very selective asymmetric alkylation of cyclohexanone and α-tetralone with activated alkyl halides (Scheme 3.25). As listed in Table 3.10, alkylation of these ketones affords up to 96% enantioselectivity. During the optimization studies, Koga observed an increase in enantioselectivity and chemical yield as the reaction time increased, and ascribed the phenomenon to the formation of a mixed aggregate that includes the lithium bromide formed as the reaction proceeds. Further experiments revealed that addition of one equivalent of lithium

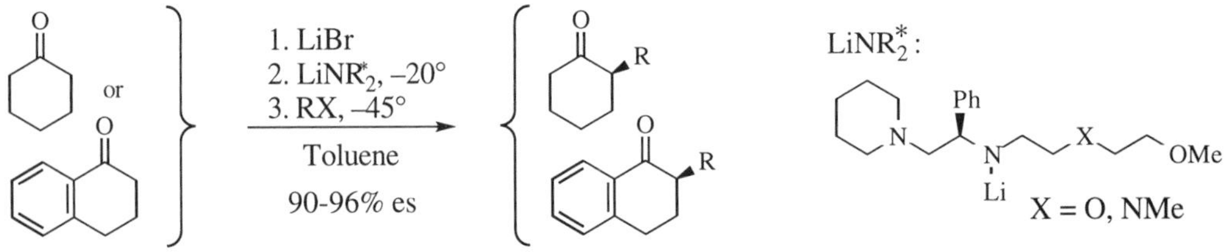

Scheme 3.25. Enantioselective alkylation of lithium enolate/secondary amine/lithium bromide complexes by interligand asymmetric induction [148,149].

---

[19] It has even been noted that deprotonation of some chiral, nonracemic carbonyl compounds by an achiral base affords an enolate that is chiral (having a chirality axis or a chelating atom that becomes stereogenic upon coordination to the lithium, for example) and nonracemic, and which affords nonracemic products upon alkylation [144-146], but a mechanistic rationale has not been established.

Table 3.10. Koga's asymmetric alkylation of ketones (Scheme 3.25 [148])

Ketone	Electrophile	Yield	% es
cyclohexanone	$PhCH_2Br$	63%	96
cyclohexanone	$PhCH=CHCH_2Br$	60%	94
cyclohexanone	$CH_2=CHCH_2Br$	41%	90
α-tetralone	$PhCH_2Br$	89%	96
α-tetralone	$PhCH=CHCH_2Br$	93%	94
α-tetralone	MeI	71%	94

bromide at the beginning of the reaction optimizes the stereoselectivity. The reactive species is thought to be a lithium enolate / secondary amine / lithium bromide mixed aggregate [148]. A rationale for the stereoselectivity of this process has yet to emerge, and the generality of it is limited. It does, however, foretell of more general successes to come.

A conceptually different approach to interligand asymmetric induction uses chiral phase transfer catalysts. Scheme 3.26 illustrates two examples of such a process using an *N*-benzylcinchonium halide catalyst. The first is an indanone methylation [150] and the second is a glycine alkylation [151]. Hughes *et al.* reported a detailed kinetic study of the indanone methylation which revealed a mechanism significantly more complicated than a simple phase-transfer process: the reaction is 0.55 order in catalyst and 0.7 order in methyl chloride, deprotonation of the indanone occurs at the interface, and methylation of the enolate (not deprotonation) is rate-determining [150]. Nevertheless, the rationale for the

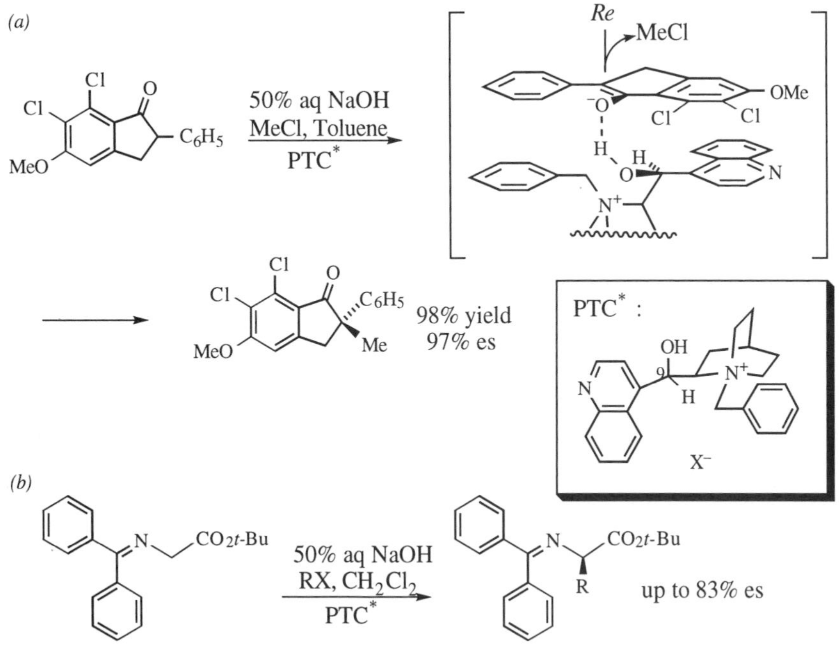

Scheme 3.26. Enantioselective alkylations using chiral phase-transfer catalysts. *(a)* [150], *(b)* [151].

enantioselectivity involves a 1:1 complex of catalyst and enolate, as illustrated in Scheme 3.26a. Molecular modeling studies and an X-ray crystal structure suggest that the most stable conformation of the catalyst has the quinoline ring, the C_9–O bond, and the *N*-benzyl group nearly coplanar [150]. Hydrogen bonding with the enolate, dipole alignment, and π-stacking of the aromatic moieties result in the assembly shown. Methylation then occurs from the *Re* face, opposite the catalyst.

O'Donnell reported the asymmetric alkylation of the Schiff base of *tert*-butyl glycinate using *N*-benzylcinchonium chloride (Scheme 3.26b, [151]). This process, which works for methyl, primary alkyl, allyl, and benzyl halides (Table 3.11), is noteworthy because the substrate is acyclic and because monoalkylation is achieved without racemization under the reaction conditions. The observed chirality sense may be rationalized by assuming an *E(O)*-enolate and π-stacking of the benzophenone rings of the enolate above the quinoline ring on the catalyst, and approach of the electrophile as before.

Table 3.11. O'Donnell's asymmetric glycine alkylations
by chiral phase transfer catalysis (Scheme 3.26b [151]).

RX	Yield	% es
MeBr	60%	71
n-BuBr	61%	76
CH$_2$=CHCH$_2$Br	75%	83
PhCH$_2$Br	75%	83

The lack of racemization and dialkylation in this process deserves comment. Apparently, the rate of deprotonation of the product is significantly slower than deprotonation of the starting material. The reasons for the reduced acidity of the product become apparent upon examination of models (Figure 3.11).[20] A^{1,3} strain considerations dictate that the α-carbon-hydrogen bond (nearly) eclipses the nitrogen-carbon double bond and also forces the syn phenyl group out of planarity. The three lowest energy conformers[21] are illustrated looking down the α-carbon–nitrogen bond. The global minimum (conformation *a*) has the α-proton near the nodal plane of the carbonyl π-system, and therefore nonacidic. The other two have the α-proton in better alignment with the carbonyl, but shielded from the approach of the base by the phenyl group at the top. Note that a proton in the position of the α-methyl in conformation *a* (as in the starting material) would be quite acidic due to overlap with *both* π-systems. Additionally, to the extent that the proximal phenyl lies in the plane of the C=N–C=C π system, the substituted enolate is significantly destabilized by A^{1,3} strain as shown in *(d)*.

As the factors that influence and control the geometry of supramolecular complexes become better understood, progress in interligand asymmetric induction will be less empirically driven and major advances will undoubtedly ensue.

---

[20] Molecular modeling calculations were done by the author using the MM2* force field as supplied in Macromodel [152].

[21] Three other accessible conformers lie ≥1.8 kcal/mole above the global minimum.

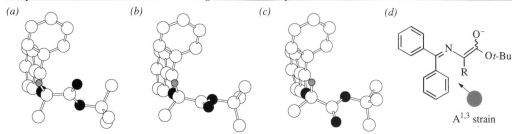

Figure 3.11. *(a)-(c)* Low energy conformations of *tert*-butyl alaninate-benzophenone Schiff base. Only the α-hydrogen is shown (hydrogen is shaded, nitrogen and oxygen are black). E_{rel} (kcal/mole): *a*, 0; *b*, +0.10; *c*, +0.14. *(d)* Substituted enolate, showing allylic strain due to the R group.

3.2 Chiral organolithiums

sec-Butyllithium is chiral, but it is usually found in racemic form.[22] Indeed, many secondary organolithiums (and Grignard reagents) are chiral, but those used in asymmetric synthesis have been mostly limited to α-heteroatom organometallics.[23] In contrast to resonance stabilized anions, α-heteroatom 'carbanions' are stabilized by inductive and dipole effects, or both, and sometimes by chelation [156]. The heteroatom may be a first row element such as nitrogen or oxygen, or main group elements such as phosphorous, sulfur, selenium, or tellurium. In most of these cases, the carbon bearing the metal is tetrahedral, and may be stereogenic. The following sections focus on organolithium species, where the heteroatom is either an oxygen or a nitrogen. Two types of species are discussed, those in which the negative charge on carbon is stabilized by a dipole, so-called dipole-stabilized anions, and those in which the inductive electron withdrawal of the heteroatom is the major contributor (Figure 3.12). There is ample evidence, both theoretical [157-159] and structural [160,161], that dipole-stabilized organometallics are chelated by the carbonyl oxygen. There is also good evidence that inductively stabilized α-heteroatom organolithiums have the metal bridged across the carbon-heteroatom bond [161,162]. Note that a distinction is made between bridging and chelation, even though the former might be called α-chelation. The simple reason is that there are distinct differences in stability and reactivity between the two types of compounds (*e.g.*, see ref. [163]).

When contemplating the use of stereogenic 'carbanions' in the synthesis of non-racemic compounds, one must consider several factors (Scheme 3.27):

1. Is the organolithium configurationally stable (Scheme 3.27a)?
2. Does the reaction with an electrophile proceed with retention or inversion of configuration at the carbanionic carbon (Scheme 3.27b)?

---

[22] Reich has shown that 2° alkyllithiums possess reasonable configurational stability, even in THF [153].

[23] An interesting recent development employs an asymmetric transformation to enantioselectively alkylate benzylic organolithiums in the presence of sparteine [154,155].

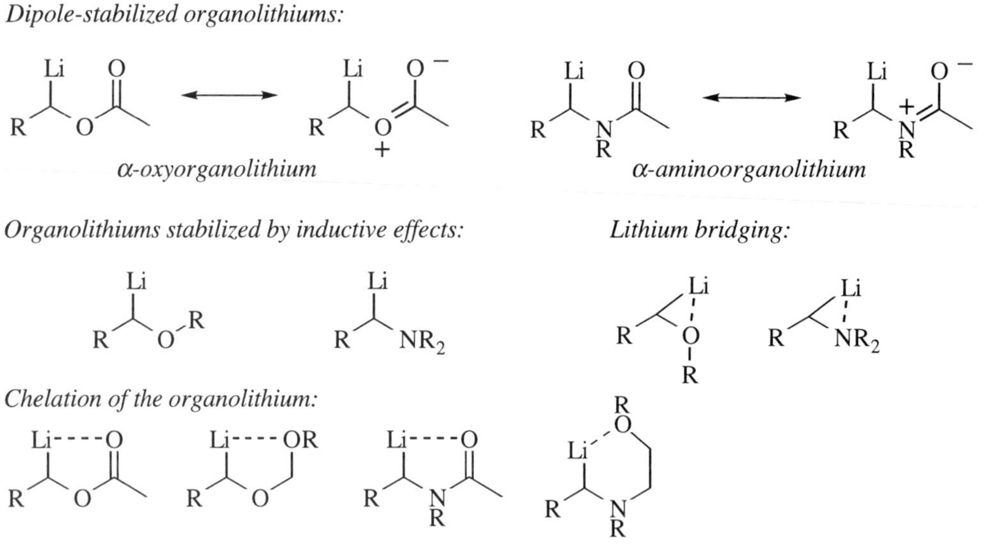

Dipole-stabilized organolithiums:

α-oxyorganolithium   α-aminoorganolithium

Organolithiums stabilized by inductive effects: *Lithium bridging:*

Chelation of the organolithium:

Figure 3.12. Classification of α-oxy- and α-aminoorganolithiums as either dipole-stabilized or inductively stabilized. Metal atom bridging and internal chelation may also play a role in both stabilization and chemical properties such as configurational stability.

3. If the electrophile is an aldehyde or an unsymmetrical ketone, does the organometallic add selectively to one of the heterotopic faces (Scheme 3.27c)?

4. What is the aggregation state of the organometallic? If there are aggregates, are they homochiral or heterochiral? If there is more than one species present in solution, which one is responsible for the observed behavior (Scheme 3.27d)?

Answering these questions is not always possible; but without answers, mechanistic interpretation is speculative, at best.

3.2.1 α-Alkoxyorganolithiums[24]

α-Alkoxy carbanions can be obtained by deprotonation or by exchange with another atom, most commonly with tin. In 1980, Still reported that the α-alkoxy-organolithium reagents derived from tin-lithium exchange of α-alkoxyorgano-stannanes are configurationally stable [165].[25] The tin-lithium exchange reaction takes place with retention of configuration [165,167],[26] so obtaining an α-alkoxy-organolithium of known configuration is predicated on having an α-alkoxyorgano-stannane of known configuration. These are made by O-alkylation of the corres-ponding α-hydroxystannanes, which are in turn formed by asymmetric reduction (Chapter 7) of an acyl stannane [169-171], kinetic resolution using a lipase enzyme [172], or oxidation of α-stannylboronates [173]. Enantiomeric purities of the α-alkoxystannanes thus obtained are often in the 95% range.

---

24 For a review of α-alkoxyorganolithiums in coupling reactions, see ref. [164].
25 For a theoretical explanation of this stability, see ref. [166].
26 The stereochemical course at tin depends on the tin ligands and the solvent [168].

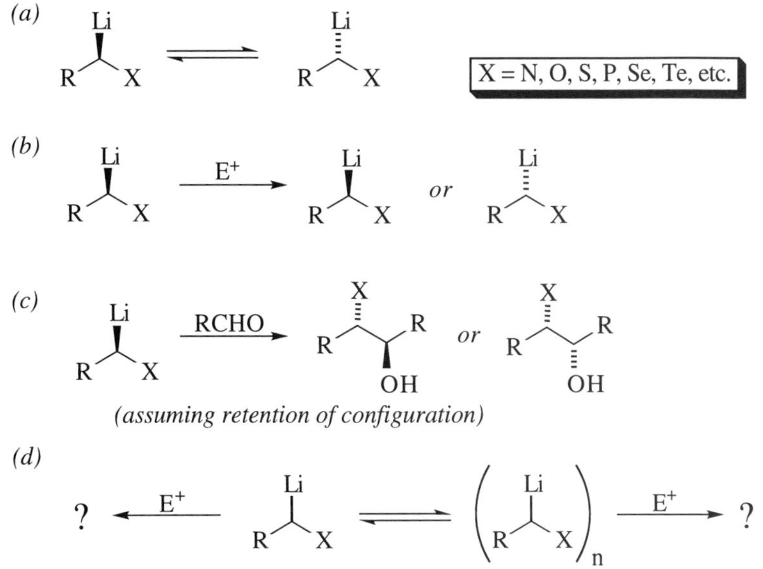

(a)

X = N, O, S, P, Se, Te, etc.

(b)

or

(c)

or

(assuming retention of configuration)

(d)

Scheme 3.27. Factors to consider in evaluating reactions of chiral organo-lithiums.

The reason tin-lithium exchange proceeds with retention may be understood by consideration of the two transition structures ($I^{\ddagger}$ and $R^{\ddagger}$) for bimolecular substitution shown in Scheme 3.28. The S_N2 reaction occurs with inversion of configuration. Note that in the $I^{\ddagger}$ transition structure for the reaction $I^- + RCl$ (Scheme 3.28a), the nucleophile and the leaving group both carry partial negative charges, which are better accomodated by $I^{\ddagger}$ than by $R^{\ddagger}$, simply because of Coulombic repulsion. In the tin-lithium exchange (S_E2 reaction), the lithium replaces the pentavalent tin of an ate-complex, so that in the transition state, the lithium carries a partial positive charge, while there is still a partial negative charge on tin. Coulombic attraction suggests that $R^{\ddagger}$ should be favored in this case (Scheme 3.28b).[27]

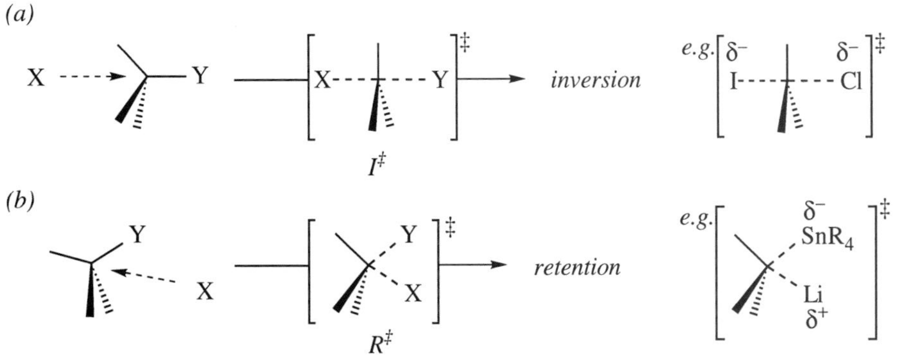

(a)

inversion

(b)

retention

Scheme 3.28. *(a)* Bimolecular inversion reaction and transition state, typified by S_N2 reaction. *(b)* Bimolecular reaction with retention, typified by tin-lithium exchange.

---

[27] In some electrophilic substitutions, reagent X initially coordinates to Y and the $R^{\ddagger}$ transition state is cyclic. For a thorough account of the many types of electrophilic substitutions, see ref. [174].

Deprotonation of ethers is another route to the α-alkoxy anions, but this pathway is often precluded by a kinetic barrier. Unless the α-carbon is benzylic [175], surmounting this barrier usually requires conditions that are not favorable to the survival of the anion [164]. Notable exceptions are the hindered aryl esters studied by Beak [176], Figure 3.13a, and the carbamates studied by Hoppe [177], shown in Figure 3.13b. In both cases, *sec*-butyllithium is required for deprotonation, and the carbonyls which direct the metalation by a complex-induced proximity effect [178] must be shielded from the base by large alkyl groups. Once formed, the organolithiums are chelated and stabilized by the heteroatom-induced dipole [179].

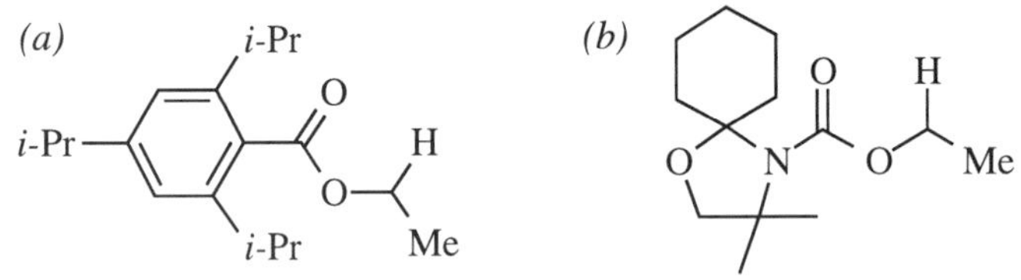

Figure 3.13. Substrates that may be deprotonated by butyllithium bases α to nitrogen. In both cases the bulk of the carbonyl moiety opposite the ethoxy group shields the carbonyl from nucleophilic attack. *(a)* Trisopropyl benzoates [176]. *(b)* Oxazolidine carbamates [177].

Reaction with carbonyl electrophiles is possible, so enantiopure stannanes are excellent precursors of enantiopure α-alkoxy tertiary alcohols [165,167], α-alkoxy acids and esters [180], and α-alkoxyketones [181], and γ-alkoxyhydrazides (precur-

Table 3.12. Stereospecific reactions of α-alkoxyorganolithiums with electrophiles.

Entry	Stannane	Electrophile	Product	% Yield	Ref.
1	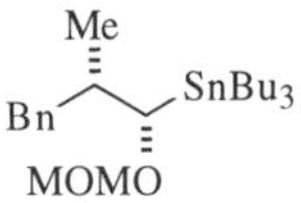	acetone	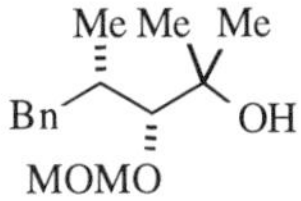	90	[165]
2	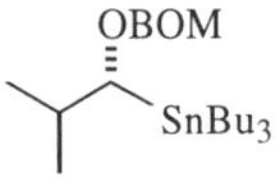	CO_2	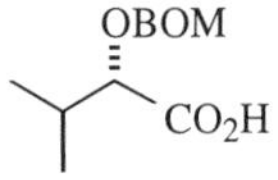	93	[180]
3	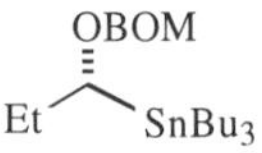	$ClCO_2Me$	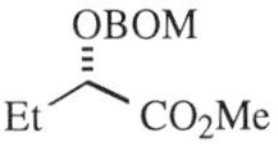	71	[180]
4	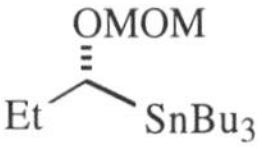	$RCONMe_2$	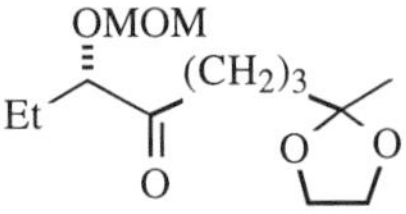	76	[181]
5	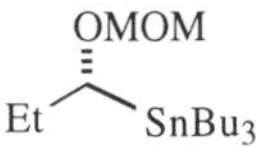	$CH_2=CHCON_2Me_3$	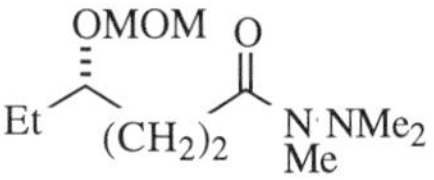	50	[182]

sors to γ-lactones) [182], as the examples in Table 3.12 illustrate. Note however, that addition of these nucleophiles to aldehydes and unsymmetric ketones is not diastereoselective. Unfortunately, the reaction of these α-alkoxyorganolithiums with alkyl halides is usually inefficient and not stereoselective due to the intervention of single electron transfer processes [167]. Methylation can be achieved with dimethyl sulfate, however [165,167], and silylation is stereospecific [165].

The carbamates in Figure 3.13b deserve special mention because Hoppe has shown that a complex of *sec*-butyllithium and sparteine (an inexpensive, chiral, tetracyclic diamine) deprotonates *O*-ethyl, *O*-butyl, *O*-isobutyl, and *O*-hexyl (*i.e.*, unactivated, nonallylic or nonbenzylic) carbamates to afford stereogenic organo-lithiums enantioselectively, as illustrated in Scheme 3.29 (review: [183]). Reaction with certain electrophiles affords high yields of product, and the oxazolidine/carbamate may be cleaved by acid hydrolysis (review: [184]). The authors suggest that the source of the enantioselectivity is the deprotonation [177]. Scheme 3.10, in the previous section, illustrated two examples of group-selective reactions where enantiotopic groups on a carbonyl were distinguished by a chiral base. In this case, the enantiotopic groups are the protons of a prochiral methylene, and the chiral base is the *sec*-butyllithium·sparteine complex.

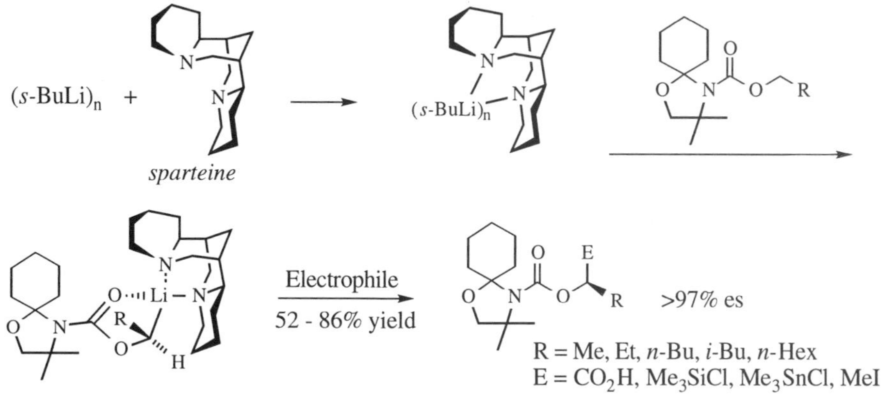

Scheme 3.29. Enantioselective deprotonation and alkylation of carbamates [183,184].

The rationale for the observed configuration (Scheme 3.29), is based on the X-ray structure of another α-carbamoyloxyorganolithium·sparteine complex [185]. After deprotonation, the chelated supramolecular complex shown in the lower left is postulated. This structure contains an adamantane-like lithium-diamine chelate, and contains new stereocenters at the lithiated carbon and at lithium itself. Note that epimerization of the lithiated carbon would produce severe van der Waals repulsion between R and the lower piperidine ring, whereas epimerization at lithium produces a similarly unfavorable interaction between the same piperidine ring and the oxazolidine substituents. Thus, the carbamate is "tailor-made" for sparteine chelation of only one enantiomer of the α-carbamoyloxyorganolithium. These effects may provide thermodynamic stability to the illustrated isomer. To the extent these effects are felt in the transition state, they are also responsible for the stereoselectivity of the deprotonation.

It may be tempting to assume that similar organolithiums would also alkylate with retention of configuration at the metal-bearing carbon. Not so. Unlike S_N2 reactions, transition states for S_E2 electrophilic substitution reactions giving retention ($R^{\ddagger}$, Scheme 3.28) and inversion ($I^{\ddagger}$) are not far apart in energy [186], and both reaction manifolds are common. For example, the carbamate shown in Scheme 3.30 affords products of either retention or inversion, depending on the electrophile: esters, anhydrides, and alkyl halides afford products of retention whereas acid chlorides, acyl cyanides, carbon dioxide, carbon disulfide, isocyanates, and tin chlorides afford products of inversion [184,187]. Interestingly, this *acyloxy*organolithium reacts well with akyl halides, unlike the *alkoxy*organo-lithiums listed in Table 3.12.

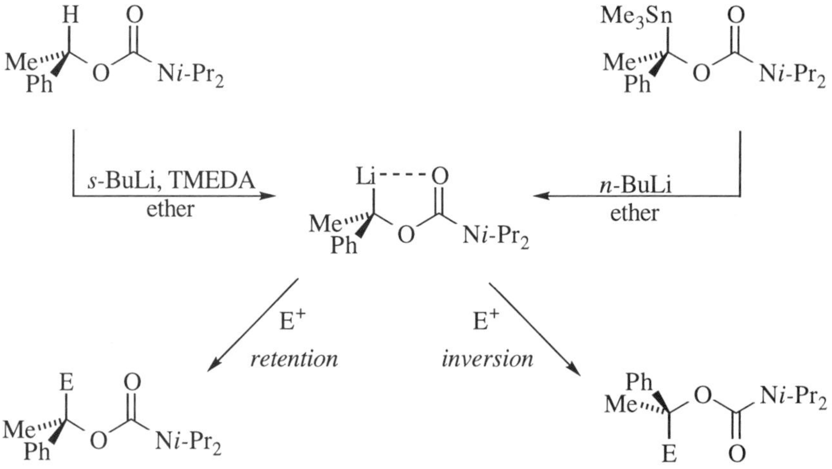

retention: E$^+$ = MeOH, RX, RCO$_2$Me, RCO$_2$COR, 53-96% yield, 65-≥95% ee

inversion: E$^+$ = HOAc, Ph$_3$CH, Me$_3$SnCl, RCOCl, ClCO$_2$Me, CO$_2$, CS$_2$,
35-95% yield, 74-≥95%ee

Scheme 3.30. The stereochemical course of the alkylation of chiral organo-lithiums may depend on the electrophile [184,187].

The authors speculate that the stereochemical divergence may be related to the ability of the electrophile to coordinate with the lithium, coupled with the presence or absence of a low-lying LUMO. Curiously, protonation by methanol proceeds with retention whereas protonation with either acetic acid or triphenyl methane proceeds with inversion. The authors speculate that, in acetic acid, protonation of the TMEDA nitrogen and internal return (*cf.* Schemes 3.2 and 3.24) may occur instead of direct protonation [184]. Presumably, direct protonation is the only mechanistic course with weak acids such as methanol and triphenylmethane and steric effects dictate inversion for the latter. Hoppe also noted that the enantiomeric purity of the products also depended on the solvent. In THF, the products were nearly racemic, and the enantiomeric purity of several of the other alkylation products was variable in solvents such as ether and pentane. This variability is due, at least in part, to the degree of covalency of the C–Li bond. In donor solvents such as THF, racemization is more facile.

3.2.2 α-Aminoorganolithiums[28]

Because nitrogen is trivalent, it is possible to attach a third substituent, often an activating group or a chiral auxiliary that facilitates either deprotonation or stereoselective alkylation, or both. Most commonly, dipole-stabilized [179,189] α-aminoorganolithiums have been used. As with the α-oxyorganolithiums discussed in the previous section, α-lithiated amines feature pyramidal carbanionic carbons, and can be formed either by deprotonation or by tin-lithium exchange, although deprotonation of an unactivated (nonallylic or nonbenzylic) position has a fairly high kinetic barrier.

The barrier to pyramidal inversion of acyclic dipole-stabilized α-aminoorgano-lithiums is considerably lower than the inversion barrier for α-alkoxyorgano-lithiums, so temperatures near –100° C are necessary to maintain configurational integrity [190,191]. For allylic or benzylic dipole-stabilized α-aminoorgano-lithiums, it appears that pyramidal inversion cannot be prevented even at such low temperature [192,193]. Recall that an enantioselective deprotonation was the source of the enantioselectivity in the alkylation of dipole-stabilized α-oxyorganolithiums (Scheme 3.29), and that benzylic dipole-stabilized α-oxyorganolithiums were found to be configurationally stable (Scheme 3.30). Detailed mechanistic studies of the lithiation of tetrahydroisoquinolines having formamidine or oxazoline chiral auxiliaries have shown that the deprotonation is stereoselective [192,194], but in the oxazolines (Scheme 3.31, H_{Re} removal favored *via* the postulated coordination complex shown), the selectivity of the bond-forming step is determined later [192]. Specifically, stereoselective deuteration at C-1 and analysis of the alkylation products for both deuterium content and diastereomer ratio showed that – even when the stereoselectivity of the deprotonation is reduced to zero by an isotope effect – the diastereomer ratio in the product is unchanged. There are two limiting

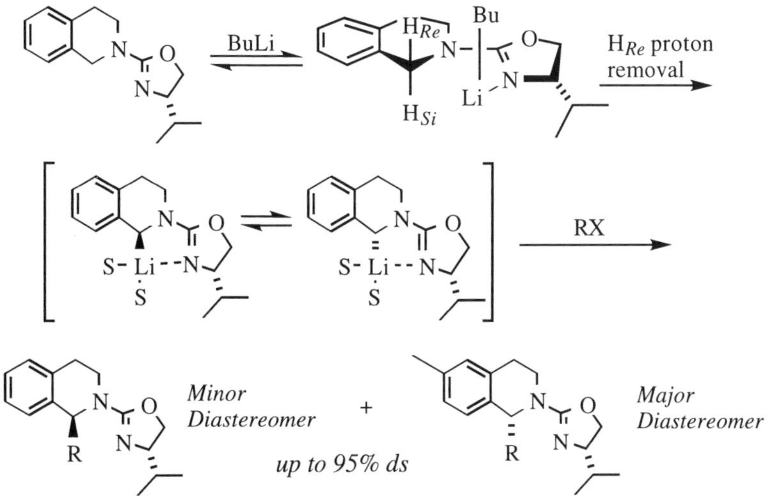

Scheme 3.31. Mechanism for asymmetric alkylation of a tetrahydro-isoquinoline using an oxazoline auxiliary [192].

---

[28] For a review of alkylations of nitrogen-stabilized carbanions, see ref. [188].

possibilities for the source of the observed selectivity: an unbalanced equilibrium of organolithium diastereomers and a fast alkylation compared to inversion (thermodynamic effect), or a fast equilibrium coupled to energetically nonequivalent transition states (Curtin-Hammett kinetics [195,196]). Because of uncertainties in the position of the organolithium equilibrium, the aggregation state of the reacting species, and the kinetics of the reaction, further insight is not possible (*cf.* Scheme 3.27 and accompanying discussion, pp. 103-4).

For asymmetric synthesis, the formamidine auxiliary developed in the Meyers laboratory has been the most useful, and has been applied to the asymmetric synthesis of a number of isoquinoline and indole alkaloids (reviews: [197,198]). The general process is illustrated in Scheme 3.32, along with several examples of tetrahydroisoquinoline [199] and β-carboline alkylations [200]. Note that the alkyllithium base selectively removes the H_{Si} proton from the illustrated isoquinoline formamidine [194,201]. Meyers has speculated that, because of the chelation illustated, that the organolithium is more stable in the configuration shown, and that alkylation occurs by inversion of configuration [194,201]. Figure 3.14 illustrates several natural products synthesized using the asymmetric alkylation strategy, with the stereocenter formed in the asymmetric alkylation indicated.

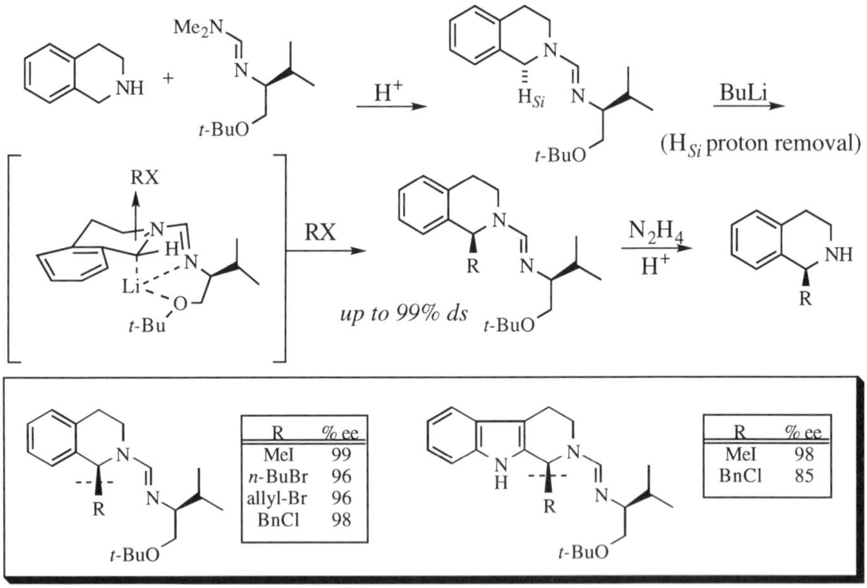

Scheme 3.32. Formamidine approach to asymmetric alkylation of tetrahydroisoquino-lines [199] and β-carbolines [200]. The indicated enantiomeric excesses were determined after auxiliary removal, derivatization and CSP-HPLC analysis (Chapter 2).

Unlike lithiated tetrahydroisoquinolines, α-lithio derivatives of saturated heterocycles are configurationally stable [202-204] (review: [163]), and they have a considerably higher kinetic barrier to deprotonation. Nevertheless, there have been a number of activating groups developed for the alkylation of α-lithio amines. In 1991, Beak showed that the complex of sparteine and *sec*-butyllithium enantioselectively deprotonates BOC-pyrrolidine, and that the derived organolithium is a good nucleophile for the reaction with several electrophiles, as shown in

reticuline, 97% ee   norcoralydine, 99% ee   yohimbone, 98% ee

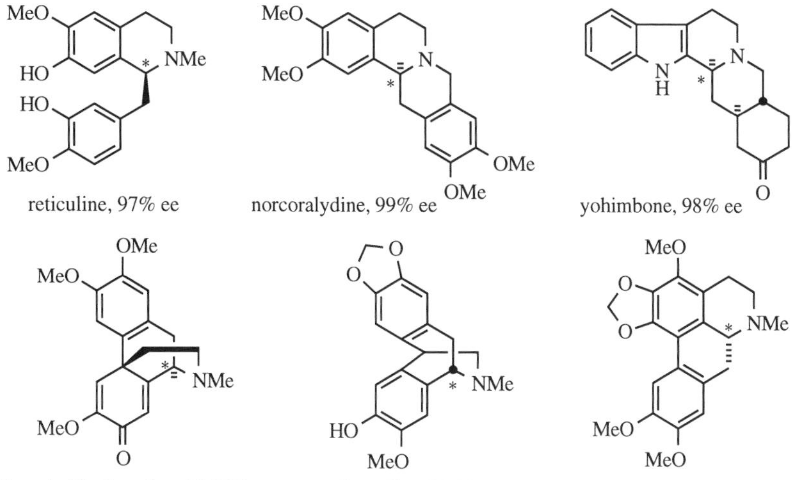

O-methylflavinantine, 94-96% ee   reframoline, 98-99% ee   ocoteine, 93% ee

Figure 3.14. Natural products synthesized using the asymmetric alkylation strategy: reticuline [205], norcoralydine [206,207], yohimbone [208], *O*-methyl-flavinantine [209], reframoline [207], and ocoteine [206,207]. The stereocenter formed by asymmetric alkylation is indicated with *.

Scheme 3.33 [204,210].[29] A limitation of this method is the failure of the organolithium to react efficiently with alkyl halide electrophiles.

The mechanism of this reaction has been studied by Beak and his group. Pertinent aspects are illustrated in Scheme 3.34. NMR studies indicate that sparteine and isopropyllithium form an unsymmetrical complex wherein one of the lithiums of the isopropyllithium dimer is chelated by sparteine while the other is not [211]. Kinetic studies indicate that when BOC-pyrrolidine is added to this complex, an equilibrium is established with a ternary complex of isopropyllithium, sparteine, and BOC-pyrrolidine (favoring the ternary complex with an equilibrium constant ≥ 300). Although the structure of this complex is not known, it is difficult to imagine that coordination of the BOC-pyrrolidine to the distal (unchelated) lithium would afford a species that is likely to react enantioselectively, so Beak suggests that the most likely possibility is the complex shown in the lower left of Scheme 3.34 [210,212]. The kinetic data further indicate that the deprotonation step is rate determining [212]. Beak suggests that a conformation such as the one illustrated presents the H$_{Si}$ proton to the alkyllithium [210].

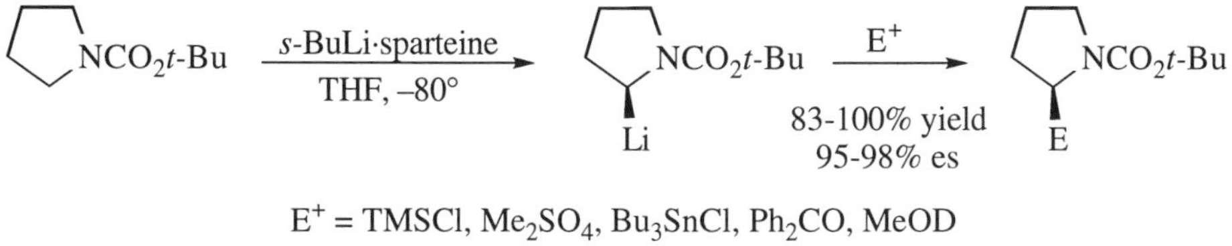

E^+ = TMSCl, Me$_2$SO$_4$, Bu$_3$SnCl, Ph$_2$CO, MeOD

Scheme 3.33. Asymmetric deprotonation and electrophilic substitution of BOC-pyrrolidine [204,210].

---

[29] Attempts to enantioselectively deprotonate BOC-piperidine with *s*-BuLi·sparteine failed [D. Hoppe, private communication].

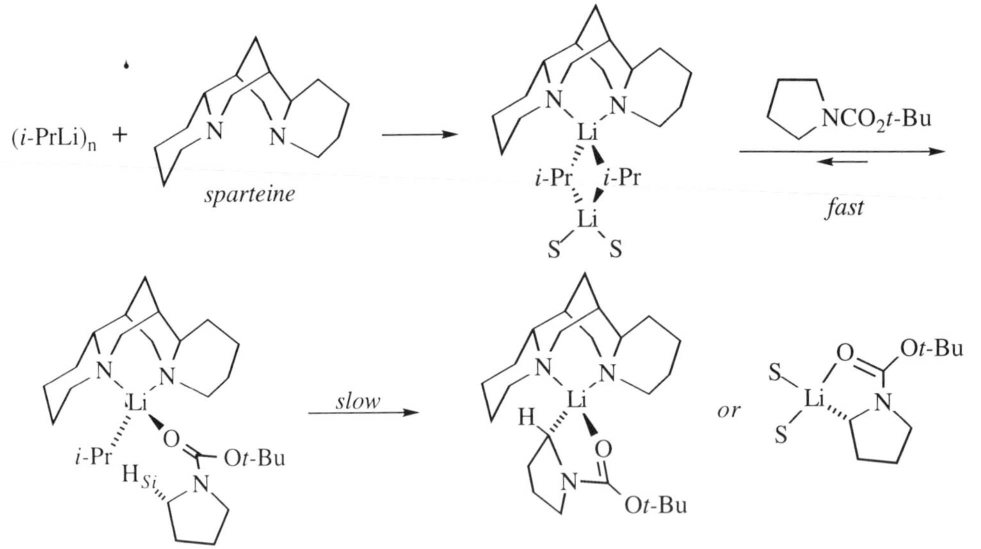

Scheme 3.34. Postulated mechanism for the asymmetric deprotonation of BOC-pyrrolidine [212].

The chemistry of lithiated *N*-methylpiperidines and *N*-methylpyrrolidines, α-aminoorganolithiums that are not dipole-stabilized, exhibits features that are quite distinct from those found for lithiated dipole-stabilized heterocycles. First of all, 2-lithio-*N*-methylpiperidine and 2-lithio-*N*-methylpyrrolidine possess the greatest configurational stability of any α-aminoorganolithium known: in the presence of TMEDA, they are configurationally stable at temperatures as high as –40° C, and are more prone to chemical decomposition than racemization [163,213]. Second, they react smoothly with alkyl halides (Scheme 3.35) more efficiently than either lithiated formamidines [214] or BOC heterocycles [215,216]. Third, the mechanistic (and stereochemical) course of their electrophilic substitution reactions depend on the electrophile in a unique way [217]. These organolithium compounds are obtained by tin-lithium exchange from the corresponding stannane; examples of their reactivity are shown in Scheme 3.35 [217]. With most carbonyl electrophiles retention of configuration is observed, whereas with alkyl halides, inversion is observed. When the electrophile is easily reduced, as with benzophenone or *tert*-butyl bromoacetate, the products are racemic. It is thought that the reactions affording racemic products proceed by a single electron transfer (radical) mechanism, while the others go by $R^{\ddagger}$ or $I^{\ddagger}$ (recall Scheme 3.28) mechanisms, as shown in the inset in Scheme 3.35 [217]. Note, however, that the dichotomy observed in these reactions bears no resemblance to the dichotomy observed by Hoppe in α-oxyorganolithium reactions, which occured with different carbonyl electrophiles and which was attributed to a low-lying LUMO (Scheme 3.30 [184]). For both types of organolithium compounds, a firm mechanistic basis for this dichotomy has yet to be established. Moreover, comparison of the varied reactivities of dipole-stabilized and inductively stabilized α-aminoorganolithiums reveal a clear difference in reactivity pattern.

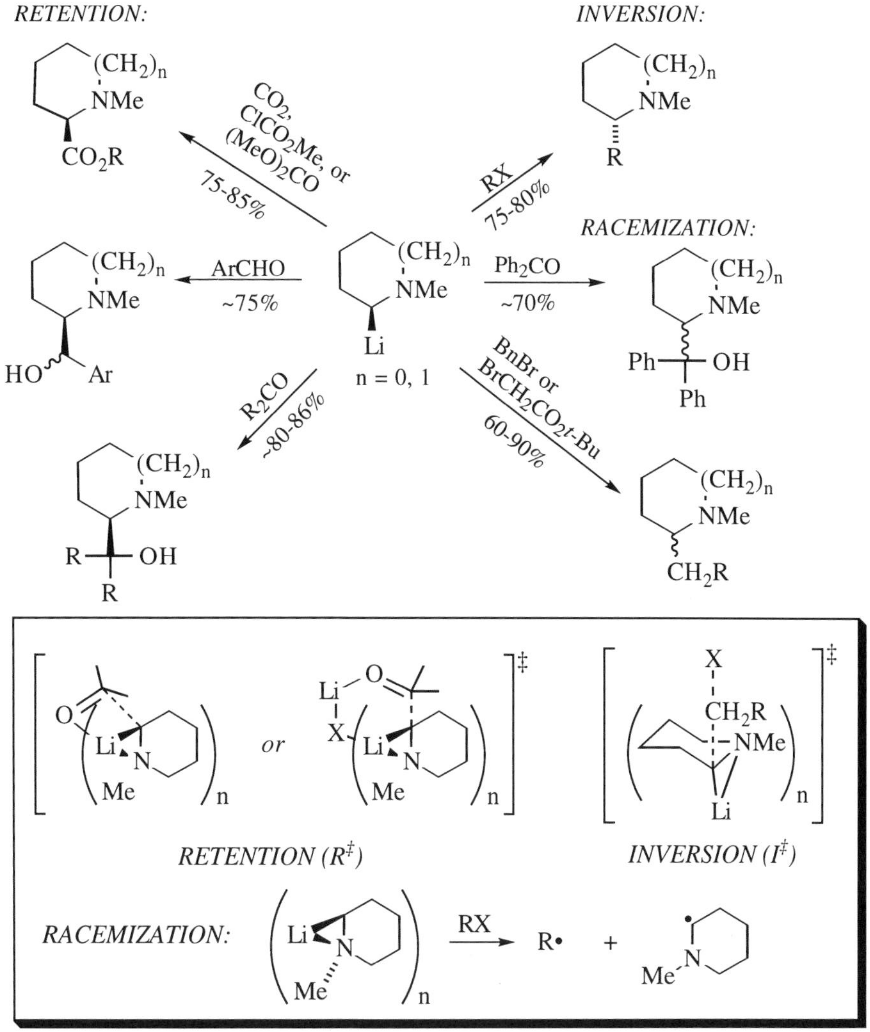

Scheme 3.35. 2-Lithio *N*-methylpiperidines and pyrrolidines are versatile reagents in electrophilic substitutions. The stereochemical course of the reaction depends on the electrophile. *Inset:* proposed transition structures for the $R^{\ddagger}$ and $I^{\ddagger}$ reactions, and SET mechanistic proposal for the electrophiles that afford racemic products [217].

3.3 References

1. D. J. Cram *Fundamentals of Carbanion Chemistry*; Academic: New York, 1965.
2. J. C. Stowell *Carbanions in Organic Sythesis*; Wiley: New York, 1979.
3. R. B. Bates *Carbanions*; Springer-Verlag: Berlin, 1983.
4. *Comprehensive Carbanion Chemistry*; E. Buncel; T. Durst, Eds.; Elsevier: Amsterdam, 1980.
5. D. Seebach *Angew. Chem. Int. Ed. Engl.* **1988**, *27*, 1624-1654.
6. G. Boche *Angew. Chem. Int. Ed. Engl.* **1989**, *28*, 277-297.
7. E. S. Gould *Mechanism and Structure in Organic Chemistry*; Holt, Rinehart and Winston: New York, 1959.

8.   D. Caine In *Carbon-Carbon Bond Formation*; R. L. Augustine, Ed.; Marcel Dekker: New York, 1979, p 85-352.

9.   D. Caine In *Comprehensive Organic Synthesis. Selectivity, Strategy, and Efficiency in Modern Organic Chemistry*; B. M. Trost, I. Fleming, Eds.; Pergamon: Oxford, 1991; Vol. 3, p 1-63.

10. J.-M. Lehn *Pure Appl. Chem.* **1978**, *50*, 871-892.

11. J.-M. Lehn *Angew. Chem. Int. Ed. Engl.* **1988**, *27*, 89-112.

12. J. d'Angelo *Tetrahedron* **1976**, *32*, 2979-2990.

13. L. M. Jackman; B. C. Lange *Tetrahedron* **1977**, *33*, 2737-2769.

14. D. A. Evans In *Asymmetric Synthesis*; J. D. Morrison, Ed.; Academic: Orlando, 1984; Vol. 3, p 1-110.

15. P. G. W. M. J. Hintze *J. Am. Chem. Soc.* **1990**, *112*, 8602-8604.

16. E. M. Arnett; F. G. Fischer; M. A. Nichols; A. A. Ribiero *J. Am. Chem. Soc.* **1990**, *112*, 801-808.

17. K. Sakuma; J. H. Gilchrist; F. E. Romesberg; C. E. Cajthami; D. B. Collum *Tetrahedron Lett.* **1993**, *34*, 5213-5216.

18. A. J. Edwards; S. Hockey; F. S. Mair; P. R. Raithby; R. Snaith; N. S. Simpkins *J. Org. Chem.* **1993**, *58*, 6942-6943.

19. M. T. Reetz; S. Hütte; R. Goddard *J. Am. Chem. Soc.* **1993**, *115*, 9339-9340.

20. L. M. Jackman; T. S. Dunne *J. Am. Chem. Soc.* **1985**, *107*, 2805-2806.

21. H. Estermann; D. Seebach *Helv. Chim. Acta* **1988**, *71*, 1824-1839.

22. K. Narasaka; Y. Ukaji; K. Watanabe *Chem. Lett.* **1986**, 1755-1758.

23. B. J. Bunn; N. S. Simpkins *J. Org. Chem.* **1993**, *58*, 533-534.

24. T. Yasukata; K. Koga *Tetrahedron Asymmetry* **1993**, *4*, 35-38.

25. Y. Hasegawa; H. Kawasaki; K. Koga *Tetrahedron Lett.* **1993**, *34*, 1963-1966.

26. P. L. Creger *J. Am. Chem. Soc.* **1970**, *92*, 1396-1397.

27. D. Seebach; M. Boes; R. Naef; W. B. Schweizer *J. Am. Chem. Soc.* **1983**, *105*, 5390-5398.

28. J. D. Aebi; D. Seebach *Helv. Chim. Acta* **1985**, *68*, 1507-1518.

29. T. Laube; J. D. Dunitz; D. Seebach *Helv. Chim. Acta* **1985**, *68*, 1373-1393.

30. M. B. Eleveld; H. Hogeveen *Tetrahedron Lett.* **1986**, *27*, 631-634.

31. E. Juaristi; A. K. Beck; J. Hansen; T. Matt; T. Mukhopadhyay; M. Simson; D. Seebach *Synthesis* **1993**, 1271-1290.

32. E. Vedejs; N. Lee *J. Am. Chem. Soc.* **1995**, *117*, 891-900.

33. D. B. Collum *Acc. Chem. Res.* **1992**, *25*, 448-454.

34. H. B. Mekelburger; C. S. Wilcox In *Comprehensive Organic Synthesis. Selectivity, Strategy, and Efficiency in Modern Organic Chemistry*; B. M. Trost, I. Fleming, Eds.; Pergamon: Oxford, 1991; Vol. 2, p 99-131.

35. I. E. Kopka; Z. A. Fataftah; M. W. Rathke *J. Org. Chem.* **1987**, *52*, 448-450.

36. R. E. Ireland; P. Wipf; J. D. Armstrong, III *J. Org. Chem.* **1991**, *56*, 650-657.

37. L. Xie; W. H. Saunders, Jr. *J. Am. Chem. Soc.* **1991**, *113*, 3123-3130.

38. P. L. Hall; J. H. Gilchrist; D. B. Collum *J. Am. Chem. Soc.* **1991**, *113*, 9571-9574.

39. R. E. Ireland; R. H. Mueller; A. K. Willard *J. Am. Chem. Soc.* **1976**, *98*, 2868-2877.

40. D. W. Moreland; W. G. Dauben *J. Am. Chem. Soc.* **1985**, *107*, 2264-2273.

41. Z. A. Fataftah; I. I. Kopka; M. W. Rathke *J. Am. Chem. Soc.* **1980**, *102*, 3959-3960.

42. E. J. Corey; A. W. Gross *Tetrahedron Lett.* **1984**, *25*, 495-498.

43. K. Hattori; H. Yamamoto *J. Org. Chem.* **1993**, *58*, 5301-5303.

44. P. L. Hall; J. H. Gilchrist; A. T. Harrison; D. J. Fuller; D. B. Collum *J. Am. Chem. Soc.* **1991**, *113*, 9575-9585.

45. M. J. Munchhof; C. H. Heathcock *Tetrahedron Lett.* **1992**, *33*, 8005-8006.

46. F. E. Romesberg; D. B. Collum *J. Am. Chem. Soc.* **1995**, *117*, 2166-2178.

47. E. J. Corey *J. Am. Chem. Soc.* **1954**, *76*, 175-179.

48. E. J. Corey; R. A. Sneen *J. Am. Chem. Soc.* **1956**, *78*, 6269-6278.

49. I. Fleming *Frontier Orbitals and Organic Chemical Reactions*; Wiley-Interscience: New York, 1976.
50. C. Agami *Tetrahedron Lett.* **1977**, 2801-2804.
51. C. Agami; M. Chauvin; J. Levisalles *Tetrahedron Lett.* **1979**, 1855-1858.
52. C. Agami; J. Levisalles; B. L. Cicero *Tetrahedron* **1979**, *35*, 961-967.
53. K. N. Houk; M. N. Paddon-Row *J. Am. Chem. Soc.* **1986**, *108*, 2659-2962.
54. N. G. Rondan; M. N. Paddon-Row; P. Caramella; K. N. Houk *J. Am. Chem. Soc.* **1981**, *103*, 2436-2438.
55. K. N. Houk *Pure Appl. Chem.* **1983**, *55*, 277-282.
56. D. Seebach; J. Zimmerman; U. Gysel; R. Ziegler; T.-K. Ha *J. Am. Chem. Soc.* **1988**, *110*, 4763-4772.
57. H. O. House; B. A. Tefertiller; H. D. Olmstead *J. Org. Chem.* **1968**, *33*, 935-942.
58. I. Kuwajima; E. Nakamura; M. Shimizu *J. Am. Chem. Soc.* **1982**, *104*, 1025-1030.
59. R. W. Hoffmann *Chem. Rev.* **1989**, *89*, 1841-1860.
60. P. Caramella; N. G. Rondan; M. N. Paddon-Row; K. N. Houk *J. Am. Chem. Soc.* **1981**, *103*, 2438-2440.
61. G. J. McGarvey; J. M. Williams *J. Am. Chem. Soc.* **1985**, *107*, 1435-1437.
62. D. Seebach; R. Naef; G. Calderari *Tetrahedron* **1984**, *40*, 1313-1324.
63. R. Fitzi; D. Seebach *Tetrahedron* **1988**, *44*, 5277-5292.
64. D. Seebach; A. Fadel *Helv. Chim. Acta* **1985**, *68*, 1243-1250.
65. D. Seebach; J. D. Aebi; R. Naef; T. Weber *Helv. Chim. Acta* **1985**, *68*, 144-154.
66. D. Seebach; B. Lamatsch; R. Amstutz; A. K. Beck; M. Dobler; M. Egli; R. Fitzi; M. Gautschi; B. Herradón; P. C. Hidber; J. J. Irwin; R. Locher; M. Maestro; T. Maetzke; A. Mouriño; E. Pfammatter; D. A. Plattner; C. Schickli; W. B. Schweizer; P. Seiler; G. Stucky; W. Petter; J. Escalante; E. Juaristi; D. Quintana; C. Miravitlles; E. Molins *Helv. Chim. Acta* **1992**, *75*, 913-934.
67. E. Vedejs; S. C. Fields; M. R. Schrimpf *J. Am. Chem. Soc.* **1993**, *115*, 11612-11613.
68. R. Downham; K. S. Kim; S. V. Ley; M. Woods *Tetrahedron Lett.* **1994**, *35*, 769-772.
69. G.-J. Boons; R. Downham; K. S. Kim; S. V. Ley; M. Woods *Tetrahedron* **1994**, *50*.
70. U. Schöllkopf *Tetrahedron* **1983**, *39*, 2085-2091.
71. R. M. Williams; M.-N. Im *J. Am. Chem. Soc.* **1991**, *113*, 9276-9286.
72. S. Yamada; T. Oguri; T. Shioiri *J. Chem. Soc.* **1976**, 136-137.
73. E. Ade; G. Helmchen; G. Heiligenmann *Tetrahedron Lett.* **1980**, *21*, 1137-1140.
74. R. Schmierer; G. Grotemeier; G. Helmchen; A. Selim *Angew. Chem. Int. Ed. Engl.* **1981**, *20*, 207-208.
75. G. Helmchen; A. Selim; D. Dorsch; I. Taufer *Tetrahedron Lett.* **1983**, *24*, 3213-3216.
76. G. Helmchen; R. Schmierer *Tetrahedron Lett.* **1983**, *24*, 1235-1238.
77. A. I. Meyers; E. D. Mihelich *Angew. Chem. Int. Ed. Engl.* **1976**, *15*, 270-281.
78. A. I. Meyers *Acc. Chem. Res.* **1978**, *11*, 375-381.
79. K. A. Lutomski; A. I. Meyers In *Asymmetric Synthesis*; J. D. Morrison, Ed.; Academic: Orlando, 1984; Vol. 3, p 213-273.
80. P. E. Sonnet; R. R. Heath *J. Org. Chem.* **1980**, *45*, 3137-3139.
81. D. A. Evans; J. M. Takacs *Tetrahedron Lett.* **1980**, *21*, 4233-4236.
82. D. A. Evans; J. M. Takacs; L. R. McGee; M. D. Ennis; D. J. Mathre; J. Bartroli *Pure Appl. Chem.* **1981**, *53*, 1109-1127.
83. D. A. Evans; M. D. Ennis; D. J. Mathre *J. Am. Chem. Soc.* **1982**, *104*, 1737-1739.
84. J. R. Gage; D. A. Evans *Organic Syntheses* **1993**, *Coll. Vol. VIII*, 339-343.
85. D. A. Evans; T. C. Britton; J. A. Ellman *Tetrahedron Lett.* **1987**, *28*, 6141-6144.
86. D. A. Evans; J. Bartroli *Tetrahedron Lett.* **1982**, *23*, 807-810.
87. D. A. Evans; R. L. Dow; T. L. Shih; J. M. Takacs; R. Zahler *J. Am. Chem. Soc.* **1990**, *112*, 5290-5313.
88. W. Oppolzer; R. Moretti; S. Thomi *Tetrahedron Lett.* **1989**, *30*, 5603-5606.

89.   B. H. Kim; D. P. Curran *Tetrahedron* **1993**, *49*, 293-318.

90.   A. I. Meyers; M. Harre; R. Garland *J. Am. Chem. Soc.* **1984**, *106*, 1146-1148.

91.   A. I. Meyers; B. A. Lefker; T. J. Sowin; L. J. Westrum *J. Org. Chem.* **1989**, *54*, 4243-4246.

92.   A. I. Meyers; K. T. Wanner *Tetrahedron Lett.* **1985**, *26*, 2047-2050.

93.   A. I. Meyers; B. A. Lefker; K. T. Wanner; R. A. Aitken *J. Org. Chem.* **1986**, *51*, 1936-1938.

94.   A. I. Meyers; B. A. Lefker *Tetrahedron* **1987**, *43*, 5663-5676.

95.   A. I. Meyers; D. Berney *Organic Syntheses* **1990**, *69*, 55-65.

96.   D. Romo; A. I. Meyers *Tetrahedron* **1991**, *47*, 9503-9569.

97.   A. I. Meyers; L. J. Westrum *Tetrahedron Lett.* **1993**, *34*, 7701-7704.

98.   L. Snyder; A. I. Meyers *J. Org. Chem.* **1993**, *58*, 7507-7515.

99.   A. I. Meyers; R. H. Wallace *J. Org. Chem.* **1989**, *54*, 2509-2510.

100. A. I. Meyers; B. A. Lefker *J. Org. Chem.* **1986**, *51*, 1541-1544.

101. A. I. Meyers; R. Hanreich; K. T. Wanner *J. Am. Chem. Soc.* **1985**, *107*, 7776-7778.

102. A. I. Meyers; M. A. Stuyers *Tetrahedron Lett.* **1989**, *30*, 1741-1744.

103. A. I. Meyers; D. Berney *J. Org. Chem.* **1989**, *54*, 4673-4676.

104. D. Enders In *Asymmetric Synthesis*; J. D. Morrison, Ed.; Academic: Orlando, 1984; Vol. 3, p 275-339.

105. D. Enders; H. Kipphardt; P. Fey *Organic Syntheses* **1987**, *65*, 183-202.

106. D. E. Bergbreiter; M. Momongan In *Comprehensive Organic Synthesis. Selectivity, Strategy, and Efficiency in Modern Organic Chemistry*; B. M. Trost, I. Fleming, Eds.; Pergamon: Oxford, 1991; Vol. 2, p 503-526.

107. D. Enders In *Stereoselective Synthesis: Lectures Honouring Prof. Dr. h. c. Rudolf Wiechert*; E. Ottow, K. Schöllkopf, B. G. Schulz, Eds.; Springer: Berlin, 1994, p 63-90.

108. E. J. Corey; D. Enders *Tetrahedron Lett.* **1976**, 3-6.

109. E. J. Corey; D. Enders *Chem. Ber.* **1978**, *111*, 1337-1361.

110. D. Enders; P. Fey; H. Kipphardt *Organic Syntheses* **1987**, *65*, 173-182.

111. D. Enders; H. Eichenauer *Chem. Ber.* **1979**, *112*, 2933-2960.

112. D. Enders; H. Eichenauer; U. Baus; H. Schubert; K. A. M. Kremer *Tetrahedron* **1984**, *40*, 1345-1359.

113. D. Enders; V. Bhushan *Z. Naturforsch.* **1987**, *42*, 1595-1596.

114. D. Enders; A. Plant *Synlett* **1990**, 725-726.

115. R. E. Gawley; E. J. Termine *Synth. Commun.* **1982**, *12*, 15-18.

116. D. Enders; H. Dyker; G. Raabe *Angew. Chem. Int. Ed. Engl.* **1992**, *31*, 618-620.

117. K. G. Davenport; H. Eichenauer; D. Enders; M. Newcomb; D. E. Bergbreiter *J. Am. Chem. Soc.* **1979**, *101*, 5654-5659.

118. D. Enders *Chem. Scripta* **1985**, *25*, 139-147.

119. D. Enders; G. Bachstädter; K. A. M. Kremer; M. Marsch; K. Harms; G. Boche *Angew. Chem. Int. Ed. Engl.* **1988**, *27*, 1522-1524.

120. K. Mori; H. Nomi; T. Chuman; M. Kohno; K. Kato; M. Noguchi *Tetrahedron* **1982**, *38*, 3705-3711.

121. K. C. Nicolaou; D. P. Papahatjis; D. A. Claremon; R. E. Dolle, III *J. Am. Chem. Soc.* **1981**, *103*, 6967-6969.

122. H. Hogeveen; L. Zwart *Tetrahedron Lett.* **1982**, *23*, 105-108.

123. C. M. Cain; N. S. Simpkins *Tetrahedron Lett.* **1987**, *28*, 3723-3724.

124. R. Shirai; M. Tanaka; K. Koga *J. Am. Chem. Soc.* **1986**, *108*, 543-545.

125. D. Sato; H. Kawasaki; I. Shimada; Y. Arata; K. Okamura; T. Date; K. Koga *J. Am. Chem. Soc.* **1992**, *114*, 761-763.

126. R. P. C. Cousins; N. S. Simpkins *Tetrahedron Lett.* **1989**, *30*, 7241-7244.

127. K. Aoki; H. Noguchi; K. Tomioka; K. Koga *Tetrahedron Lett.* **1993**, *34*, 5105-5108.

128. K. Koga *Pure Appl. Chem.* **1994**, *66*, 1487-1492.

129. J. K. Whitesell; S. W. Felman *J. Org. Chem.* **1980**, *45*.

130. H. Kim; H. Kawasaki; M. Nakajima; K. Koga *Tetrahedron Lett.* **1989**, *30*, 6537-6540.

131. M. Sobukawa; K. Koga *Tetrahedron Lett.* **1993**, *34*, 5101-5104.

132. E. Vedejs; N. Lee *J. Am. Chem. Soc.* **1991**, *113*, 5483-5485.

133. S. Hünig In *Stereoselective Synthesis*; G. Helmchen, R. W. Hoffmann, J. Mulzer, E. Schaumann, Eds.; Georg Thieme: Stuttgart, 1995; Vol. E21d, p Chapter 2.1.

134. I. T. Barnish; M. Corless; P. J. Dunn; D. Ellis; P. W. Finn; J. D. Hardstone; K. James *Tetrahedron Lett.* **1993**, *34*, 1323-1326.

135. K. Fuji; K. Tanaka; H. Miyamoto *Tetrahedron Asymmetry* **1993**, *4*, 247-249.

136. C. Fehr; O. Guntern *Helv. Chim. Acta* **1992**, *75*, 1023-1028.

137. C. Fehr; I. Stempf; J. Galindo *Angew. Chem. Int. Ed. Engl.* **1993**, *32*, 1042-1044.

138. C. Fehr; I. Stempf; J. Galindo *Angew. Chem. Int. Ed. Engl.* **1993**, *32*, 1044-1046.

139. A. Yanagisawa; T. Kurabayashi; T. Kikuchi; H. Yamamoto *Angew. Chem. Int. Ed., Engl.* **1994**, *33*, 107-109.

140. E. Vedejs; N. Lee; S. T. Sakata *J. Am. Chem. Soc.* **1994**, *116*, 2175-2176.

141. T. Yamashita; H. Mitsui; W. H; N. Nakamura *Bull. Chem. Soc. Jpn.* **1982**, *55*, 961-962.

142. H. Hogeveen; W. M. P. B. Menge *Tetrahedron Lett.* **1986**, *27*, 2767-2770.

143. A. Ando; T. Shiori *J. Chem. Soc., Chem. Commun.* **1987**, 656-658.

144. D. Seebach; D. Wasmuth *Angew. Chem. Int. Ed. Engl.* **1981**, *20*, 971.

145. T. Kawabata; K. Yahiro; K. Fuji *J. Am. Chem. Soc.* **1991**, *113*, 9694-9696.

146. T. Kawabata; T. Wirth; K. Yahiro; H. Suzuki; K. Fuji *J. Am. Chem. Soc.* **1994**, *116*, 10809-10810.

147. K. Tomioka; M. Shindo; K. Koga *Chem. Pharm. Bull.* **1989**, *37*, 1120-1122.

148. M. Murakata; M. Nakajima; K. Koga *J. Chem. Soc., Chem. Commun.* **1990**, 1657-1658.

149. M. Imai; A. Hagihara; H. Kawasaki; K. Manabe; K. Koga *J. Am. Chem. Soc.* **1994**, *116*, 8829-8830.

150. D. L. Hughes; U.-H. Dolling; K. M. Ryan; E. F. Schoenewaldt; E. J. J. Grabowski *J. Org. Chem.* **1987**, *52*, 4745-4752.

151. M. J. O'Donnell; W. D. Bennett; S. Wu *J. Am. Chem. Soc.* **1989**, *111*, 2353-2355.

152. F. Mohamadi; N. G. J. Richards; W. C. Guida; R. Liskamp; M. Lipton; C. Caufield; G. Chang; T. Hendrickson; W. C. Still *J. Comput. Chem.* **1990**, *11*, 440-467.

153. H. J. Reich; M. A. Medina; M. D. Bowe *J. Am. Chem. Soc.* **1992**, *114*, 11003-11004.

154. P. Beak; H. Du *J. Am. Chem. Soc.* **1993**, *115*, 2516-2518.

155. S. Thayumanavan; S. Lee; C. Liu; P. Beak *J. Am. Chem. Soc.* **1994**, *116*, 9755-9756.

156. A. Krief *Tetrahedron* **1980**, *36*, 2531-2640.

157. N. G. Rondan; K. N. Houk; P. Beak; W. J. Zajdel; J. Chandrasekhar; P. v. R. Schleyer *J. Org. Chem.* **1981**, *46*, 4108-4110.

158. R. D. Bach; M. L. Braden; G. J. Wolber *J. Org. Chem.* **1983**, *48*, 1509-1514.

159. L. J. Bartolotti; R. E. Gawley *J. Org. Chem.* **1989**, *54*, 2980-2982.

160. D. Seebach; J. Hansen; P. Seiler; J. M. Gromek *J. Organomet. Chem.* **1985**, *285*, 1-13.

161. G. Boche; M. Marsch; J. Harbach; K. Harms; B. Ledig; F. Schubert; J. C. W. Lohrenz; H. Albrecht *Chem. Ber.* **1993**, *126*, 1887-1894.

162. P. v. R. Schleyer; T. Clark; A. J. Kos; G. W. Spitznagel; C. Rohde; D. Arad; K. N. Houk; N. G. Rondan *J. Am. Chem. Soc.* **1984**, *106*, 6467-6475.

163. R. E. Gawley; Q. Zhang *Tetrahedron* **1994**, *50*, 6077-6088.

164. D. R. Cheshire In *Comprehensive Organic Synthesis. Selectivity, Strategy, and Efficiency in Modern Organic Chemistry*; B. M. Trost, I. Fleming, Eds.; Pergamon: Oxford, 1991; Vol. 3, p 193-205.

165. W. C. Still; C. Sreekumar *J. Am. Chem. Soc.* **1980**, *102*, 1201-1202.

166. J.-M. Lehn; G. Wipff *J. Am. Chem. Soc.* **1976**, *98*, 7498-7505.

167. J. S. Sawyer; A. Kucerovy; T. L. Macdonald; G. J. McGarvey *J. Am. Chem. Soc.* **1988**, *110*, 842-853.

168. H. J. Reich; J. P. Borst; M. B. Coplein; N. H. Phillips *J. Am. Chem. Soc.* **1992**, *114*, 6577-6579.

169. J. A. Marshall; W. Y. Gung _Tetrahedron Lett._ **1988**, _29_, 1657-1660.

170. P. C.-M. Chan; J. M. Chong _J. Org. Chem._ **1988**, _53_, 5584-5586.

171. J. A. Marshall; G. S. Welmaker; B. W. Gung _J. Am. Chem. Soc._ **1991**, _113_, 647-656.

172. J. M. Chong; E. K. Mar _Tetrahedron Lett._ **1991**, _32_, 5683-5686.

173. D. S. Matteson; P. B. Tripathy; A. Sarkur; K. N. Sadhu _J. Am. Chem. Soc._ **1989**, _111_, 4399-4402.

174. M. H. Abraham In _Comprehensive Chemical Kinetics_; C. H. Bamford, C. F. H. Tipper, Eds.; Elsevier: Amsterdam, 1973; Vol. 12, p 1-256.

175. B. A. Barner; R. S. Mani _Tetrahedron Lett._ **1989**, _30_, 5413-5416.

176. P. Beak; L. G. Carter _J. Org. Chem._ **1981**, _46_, 2363-2373.

177. D. Hoppe; F. Hintze; P. Tebben _Angew. Chem. Int. Ed. Engl._ **1990**, _29_, 1422-1423.

178. P. Beak; A. I. Meyers _Acc. Chem. Res._ **1986**, _19_, 356-363.

179. P. Beak; D. B. Reitz _Chem. Rev._ **1978**, _78_, 275-316.

180. P. C.-M. Chan; J. M. Chong _Tetrahedron Lett._ **1990**, _31_, 1985-1988.

181. J. M. Chong; E. K. Mar _Tetrahedron_ **1989**, _45_, 7709-7716.

182. J. M. Chong; E. K. Mar _Tetrahedron Lett._ **1990**, _31_, 1981-1984.

183. D. Hoppe; F. Hintze; P. Tebben; M. Paetow; H. Ahrens; J. Schwerdtfeger; P. Sommerfeld; J. Haller; W. Guarnieri; S. Kolczewski; T. Hense; I. Hoppe _Pure Appl. Chem._ **1994**, _66_, 1479-1486.

184. A. Carstens; D. Hoppe _Tetrahedron_ **1994**, _50_, 6097-6108.

185. M. Marsch; K. Harms; O. Zschage; D. Hoppe; G. Boche _Angew. Chem. Int. Ed. Engl._ **1991**, _30_, 321-323.

186. E. D. Jemmis; J. Chandrasekhar; P. v. R. Schleyer _J. Am. Chem. Soc._ **1979**, _101_, 527-533.

187. D. Hoppe; A. Carstens; T. Krämer _Angew. Chem. Int. Ed. Engl._ **1990**, _29_, 1424-1425.

188. R. E. Gawley; K. Rein In _Comprehensive Organic Synthesis. Selectivity, Strategy, and Efficiency in Modern Organic Chemistry_; B. M. Trost, I. Fleming, Eds.; Pergamon: Oxford, 1991; Vol. 3, p 65-83.

189. P. Beak; W. J. Zajdel; D. B. Reitz _Chem. Rev._ **1984**, _84_, 471-523.

190. W. H. Pearson; A. C. Lindbeck _J. Am. Chem. Soc._ **1991**, _113_, 8546-8548.

191. J. M. Chong; S. B. Park _J. Org. Chem._ **1992**, _57_, 2220-2222.

192. K. Rein; M. Goicoechea-Pappas; T. V. Anklekar; G. C. Hart; G. A. Smith; R. E. Gawley _J. Am. Chem. Soc._ **1989**, _111_, 2211-2217.

193. A. I. Meyers; J. Guiles; J. S. Warmus; M. A. Gonzalez _Tetrahedron Lett._ **1991**, _32_, 5505-5508.

194. A. I. Meyers; D. A. Dickman _J. Am. Chem. Soc._ **1987**, _109_, 1263-1265.

195. D. Y. Curtin _Rec. Chem. Progr._ **1954**, _15_, 111-128.

196. J. I. Seeman _Chem. Rev._ **1983**, _83_, 83-134.

197. T. K. Highsmith; A. I. Meyers In _Advances in Heterocyclic Natural Product Synthesis_; W. H. Pearson, Ed.; JAI: Greenwich, CT, 1991; Vol. 1, p 95-135.

198. A. I. Meyers _Tetrahedron_ **1992**, _48_, 2589-2612.

199. A. I. Meyers; M. Boes; D. A. Dickman _Angew. Chem. Int. Ed. Engl._ **1984**, _23_, 458-459.

200. M. F. Loewe; A. I. Meyers _Tetrahedron Lett._ **1985**, _26_, 3291-3294.

201. A. I. Meyers; J. S. Warmus; M. A. Gonzalez; J. Guiles; A. Akahane _Tetrahedron Lett._ **1991**, _32_, 5509-5512.

202. R. E. Gawley; G. Hart; M. Goicoechea-Pappas; A. L. Smith _J. Org. Chem._ **1986**, _51_, 3076-3078.

203. R. E. Gawley; G. C. Hart; L. J. Bartolotti _J. Org. Chem._ **1989**, _54_, 175-181 and 4726.

204. S. T. Kerrick; P. Beak _J. Am. Chem. Soc._ **1991**, _113_, 9708-9710.

205. A. I. Meyers; J. Guiles _Heterocycles_ **1989**, _28_, 295-301.

206. D. A. Dickman; A. I. Meyers _Tetrahedron Lett._ **1986**, _27_, 1465-1468.

207. A. I. Meyers; D. A. Dickman; M. Boes _Tetrahedron_ **1987**, _43_, 5095-5108.

208. A. I. Meyers; D. B. Miller; F. H. White _J. Am. Chem. Soc._ **1988**, _110_, 4778-4787.

209. R. E. Gawley; G. A. Smith _Tetrahedron Lett._ **1988**, _29_, 301-302.
210. P. Beak; S. T. Kerrick; S. Wu; J. Chu _J. Am. Chem. Soc._ **1994**, _116_, 3231-3239.
211. D. J. Gallagher; S. T. Kerrick; P. Beak _J. Am. Chem. Soc._ **1992**, _114_, 5872-5873.
212. D. J. Gallagher; P. Beak _J. Org. Chem._ **1995**, _60_, 7092-7093.
213. R. E. Gawley; Q. Zhang _J. Am. Chem. Soc._ **1993**, _115_, 7515-7516.
214. A. I. Meyers; P. D. Edwards; W. F. Reiker; T. R. Bailey _J. Am. Chem. Soc._ **1984**, _106_, 3270-3276.
215. P. Beak; W.-K. Lee _Tetrahedron Lett._ **1989**, _30_, 1197-1200.
216. P. Beak; W. K. Lee _J. Org. Chem._ **1993**, _58_, 1109-1117.
217. R. E. Gawley; Q. Zhang _J. Org. Chem._ **1995**, _60_, 5763-5769.

Chapter 4

1,2 and 1,4 Additions to Carbonyls

Some of the earliest attempts to understand stereoselectivity in organic reactions were the rationalizations and predictive models made in the early 1950s by Curtin [1], Cram [2] and Prelog [3] to explain the addition of achiral nucleophiles such as Grignard reagents to the diastereotopic faces of ketones and aldehydes having a proximal stereocenter.[1] In the decades since, there has been a steady stream of additional contributions to the understanding of these phenomena.

In this book, a distinction is made between additions that involve allylic nucleophiles and those that do not. For the purposes of this discussion, the addition of enolates and allylic nucleophiles will be labeled π-transfers, and nonallylic nucleophiles will be labeled σ-transfers, as illustrated in Figure 4.1. Note that for σ-transfers aggregation is possible, so that the addition may proceed through a transition state featuring either a four-membered ring or a six-membered ring. This chapter covers 1,2- and 1,4 additions to carbonyls by σ-transfer; the addition of enolates and allyls (π-transfer) is detailed in Chapter 5.

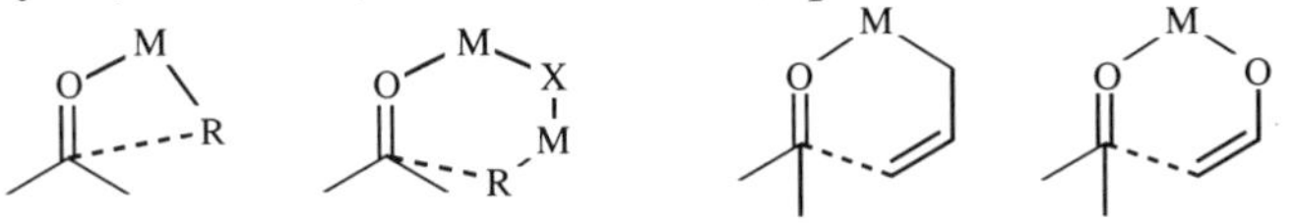

"σ-transfer" transition structures   *"π-transfer" transition structures*

Figure 4.1. Classification of nucleophilic additions to carbonyls.

This chapter begins with a detailed examination of the evolution of the theory of nucleophilic attack on a chiral aldehyde or ketone, from Cram's original "rule of steric control of asymmetric induction" to the Felkin-Anh-Heathcock formulation. Then follows a discussion of Cram's simpler "rigid model" (chelate rule), then carbonyl additions using chiral catalysts and chiral (nonenolate) nucleophiles. The chapter concludes with asymmetric 1,4-additions to conjugated carbonyls and azomethines.

4.1 Cram's rule: open-chain model

About one hundred years ago, the stereoselective addition of cyanide to a chiral carbonyl compound, the Kiliani-Fischer synthesis of carbohydrates, was proclaimed by Emil Fischer to be "the first definitive evidence that further synthesis with asymmetric systems proceeds in an asymmetric manner" [5]. By the mid-twentieth century, enough experimental data had accumulated that attempts to rationalize the selectivity of such additions could be made. The most useful of these was made by Cram in 1952 (Figure 4.2a, [2]). In this model, Cram proposed that coordination of

---

[1]   For a review of the early literature on the stereoselective reactions of chiral aldehydes, ketones, and α-keto esters, and also of the addition of Grignards and organolithiums to achiral ketones and aldehydes in the presence of a chiral complexing agent or chiral solvent, see ref. [4].

the metal of (for example) a Grignard reagent to the carbonyl oxygen rendered it the bulkiest group in the molecule. It would tend to orient itself between the two least bulky groups, as shown. In 1959 [6], the model was redrawn as in Figure 4.2b, which also implies a second, less favored conformation, Figure 4.2c.

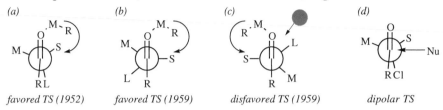

favored TS (1952)   *favored TS (1959)*   *disfavored TS (1959)*   *dipolar TS*

Figure 4.2. *(a-c)* Cram's models for predicting the major isomer of a nucleophilic addition to a carbonyl having a stereocenter in the α position [2,6]. *(d)* Cornforth's dipole model for α-chloro ketones [7]. S, M, and L refer to the small, medium, and large groups, respectively.

These models correctly predict the major diastereomer of most asymmetric additions. A notable exception is Grignard addition to α-chloro ketones, which led Cornforth to propose a model where the halogen plays the role of the large substituent so that the C=O and C–Cl dipoles are opposed (Figure 4.2d, [7]).

4.1.1 The Karabatsos model

The predictive value of Cram's rule notwithstanding, the rationale was speculative, and as spectroscopic methods developed, it was called into question. For example, Karabatsos studied the conformations of substituted aldehydes [8] and dimethylhydrazones [9] by NMR, and concluded that one of the ligands at the α position eclipses the carbonyl. It was felt that in the addition reaction, the organometallic probably *did* coordinate to the carbonyl oxygen as Cram had suggested, and Karabatsos used the conformations of the dimethylhydrazone as a model for the metal-coordinated carbonyl. He concluded that since the aldehyde and the hydrazone have similar conformations, so should the metal-complexed carbonyl [10]. He also assumed that the transition state is early, so that there is little bond breaking or bond making in the transition states (Hammond postulate [11]), and that the arrangement of the three ligands on the α carbon are therefore the same in the transition state as they are in the starting materials: eclipsed.

Thus Karabatsos concluded that the rationale for Cram's rule was incorrect [10]. In 1967, he published a new model, which took into account the approach of the nucleophile from either side of all three eclipsed conformers [10]. He noted that the enthalpy and entropy of activation for Grignard or hydride additions to carbonyls are 8 to 15 kcal/mole and −20 to −40 eu, respectively. Since the barrier to rotation around the sp^2–sp^3 carbon-carbon bond is much lower [12], the selectivity must arise from Curtin-Hammett kinetics [13,14]. Of the six possible conformers (Figure 4.3), four were considered unlikely due to steric repulsion between the nucleophile and either the medium or large α-substituents. The two most likely transition states, 4.3a and 4.3d, have the nucleophile approaching closest to the smallest group on the α carbon, and are distinguished by the repulsive interactions between the carbonyl oxygen and the α substituent (either M or L), with 4.3a favored.

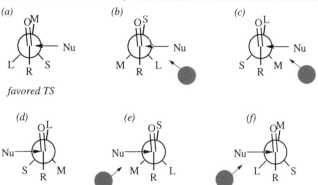

Figure 4.3. Karabatsos's transition state models [10].

4.1.2 Felkin's experiments

In 1968, Felkin noted that neither the Cram nor the Karabatsos models predict the outcome of nucleophilic addition to cyclohexanones [15], and fail to account for the effect of the size of R on the selectivity [16]. The point about cyclohexanones is particularly well-taken, since it is unlikely that the mechanisms of Grignard and hydride additions to cyclic and acyclic ketones differ significantly. The data in Table 4.1 indicate that as the size of the substituent "on the other side" increases, so does the selectivity, except for the single example where the "large" substituent is cyclohexyl and the carbonyl is flanked by a *tert*-butyl.

Table 4.1. Stereoselectivity (% ds) of reductions of $R_1MeCHC(=O)R_2$ by $LiAlH_4$ [16].

Large Subs.	$R_2 = Me$	$R_2 = Et$	$R_2 = i\text{-}Pr$	$R_2 = t\text{-}Bu$
$R_1 = c\text{-}C_6H_{11}$	62	66	80	62
$R_1 = Ph$	74	76	83	98

To explain these results, Felkin proposed a new model [16], in which the incoming nucleophile attacks the carbonyl from a direction that is antiperiplanar to the large substituent (Figure 4.4), while maintaining the notion of an early transition state. Whereas the Cram and Karabatsos models dictate that the nucleophile's approach eclipses (Cram dihedral 0°) or nearly eclipses (Karabatsos dihedral 30°) the small substituent on the α carbon, Felkin proposed that the nucleophile bisects the bond between the medium and small substituents, as in conformers 4.4a and 4.4b (60° dihedral). Felkin suggested that the factor controlling the relative energy of the transition states is the repulsive interaction between R and either the small or medium ligands on the stereocenter, and assumed that there is no energy differential resulting from the interaction between the carbonyl oxygen and either the small or medium substituents on the α carbon.[2] Thus, conformer 4.4a is

---

[2] This rationale is a major weakness of Felkin's theory [17]. First, it assumes that *intramolecular interactions in the substrate* are responsible for the selectivity of a *bimolecular* reaction. Note that the following distances are identical in both transition states: Nu–O, Nu–R, Nu–S, Nu–M. Second, it is hard to accept that R=H is *more* sterically demanding than oxygen, as would be required for aldehydes (H/S and H/M interactions more important than O/S and O/M).

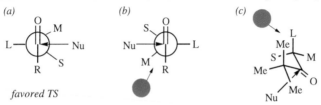

(a)   (b)   (c)

favored TS

Figure 4.4. *(a-b)* Felkin's transition state models. *(c)* De-stabilized 'favored' transition state with a flanking *tert*-butyl [16].

favored. The higher selectivities observed across the board (Table 4.1) when the "large" group is phenyl was explained by the greater electronegativity of phenyl over cyclohexyl (*i.e.,* increased differential between 4.4a and 4.4b). Felkin also postulated that when one of the substituents was a chlorine, it would assume the role of the "large" antiperiplanar substituent due to polar effects, thus obviating the need for the Cornforth model (Figure 4.2d). To explain the seemingly anomalous result with a *tert*-butyl substituent, Felkin suggested that the normally preferred conformation is destabilized by a severe 1,3-interaction between the large substituent and one of the methyls of the *tert*-butyl, as in 4.4c.[3] An accompanying paper extended these theories to the cyclohexanone problem [15] (see also ref. [17-19]).

4.1.3 The Bürgi-Dunitz trajectory: a digression.

Note that these three models vary in their assumptions about the trajectory of the incoming nucleophile, but *all are entirely speculative*. How might the approach trajectory te determined? Professor Dunitz suggested "turning on the lights."[4] Bürgi, Dunitz, and Schefter took the position that an observed set of static structures, obtained by X-ray crystallography, when arranged in the right sequence might provide a picture of the changes that occur along the reaction pathway [21]. The model system chosen was nucleophilic approach to a carbonyl by a tertiary amine. Figure 4.5 illustrates the series of compounds whose crystal structures were compared. In the structures of A - E, the nitrogen interacts with the carbonyl carbon to varying degrees, while in F it is covalently bonded, making an acetal. It was noted that in all cases the nitrogen, and the carbonyl carbon and oxygen atoms lie in an approximate local mirror plane (the "normal" plane), but that the carbonyl carbon deviates significantly from the plane defined by the oxygen and the two α substituents. This deviation increased as the N–C distance decreased, but the N–C–O and R–C–R' angles varied only slightly from their mean values.

---

[3] This is a 2,3-*P*-3,4-*M* gauche pentane conformation, which is equivalent to 1,3-diaxial substituents on a cyclohexane. Note that – because the carbonyl substituent is a *tert*-butyl – it cannot be avoided by rotation around the *tert*-butyl–carbonyl bond. For further elaboration of this effect, see Figure 5.5 and the accompanying discussion. For an explanation of the *P,M* terminology, see the glossary, Section 1.6.

[4] "The difference between a chemist and a crystallographer can be compared to two people who try to ascertain what furniture is present in a darkened room; one probes around in the dark breaking the china, while the other stays by the door and switches on the light." (J. D. Dunitz, quoted in ref. [20]).

A: methadone B: cryptopine C: protopine

D: clivorine E: retusamine F: *N*-brosylmitomycin A

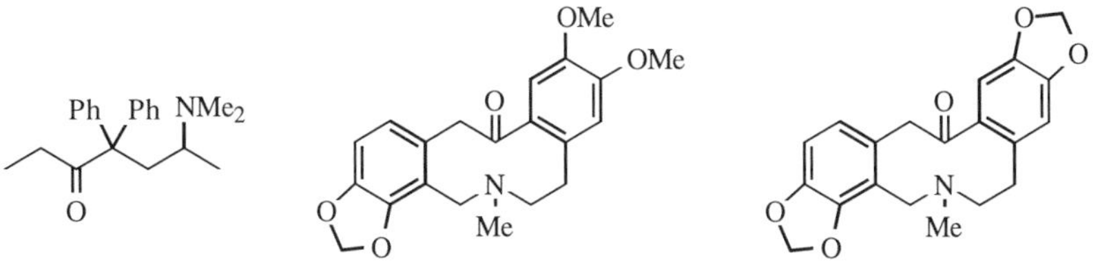

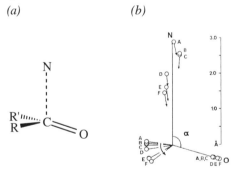

Figure 4.5. Compounds whose X-ray structures provided the basis for the "Bürgi-Dunitz" trajectory.

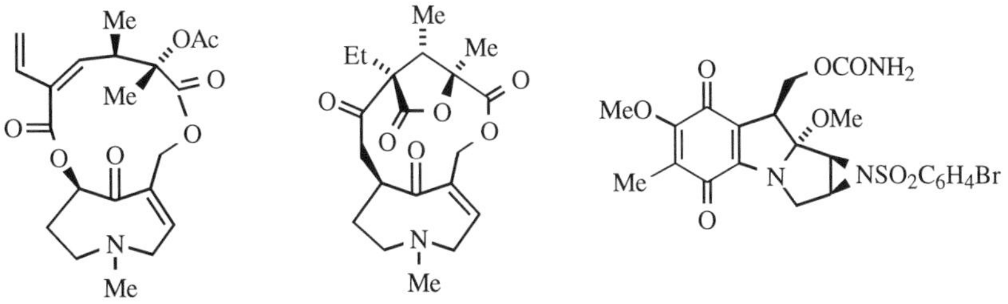

When the coordinates of the carbonyl carbon atoms and the direction of the C–O bonds are superimposed on a three dimensional graph, and the position of the nitrogen is plotted on the normal plane, the trajectory of approach is revealed: it *"is not perpendicular to the C–O bond but forms an angle of 107° with it"* (Figure 4.6) [21]. Also revealed is the variation in C–O bond length and the distortion of the RCR plane as the nitrogen nears bonding distance. The small arrows indicate the presumed direction of the nitrogen lone pair.

(a) *(b)*

Figure 4.6. *(a)* Orientation of the superimposed carbonyl and nitrogen atoms. *(b)* Superimposed plot of the N, C, and O atoms of structures A-F, and the variance of the RRC plane from the RRO plane. α is the "Bürgi-Dunitz angle," 107°. Reprinted with permission from ref. [21], copyright 1973, American Chemical Society.

The crystal structure data are appealing (as far as they go), but the extent to which substituent effects and crystal packing forces influenced the arrangement of the atoms could not be evaluated. Also, the structural data could provide no information about energy variations along (or variant from) the proposed reaction path. In 1974 Bürgi, Lehn, and Wipff studied the approach of hydride to formaldehyde using computational methods [22]. Thus, a hydride was placed at varying distances from formaldehyde and the minimum energy geometry was located. By superimposing these geometries, the theoretical approach trajectory could be deduced. The results (Figure 4.7), can be summarized as follows. At H^-–C distances of >3.0Å, the hydride approaches along the X axis. At an H⁻–C distance of 3.0Å, the H⁻ and formaldehyde hydrogens are about 2.7Å apart. At this point, the hydride leaves the HCH plane and glides over the formaldehyde hydrogens until it senses the optimal direction for its attack on the carbonyl, 105±5°.

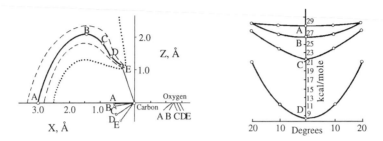

Figure 4.7. *(a)* Minimum energy path for addition of hydride to formaldehyde. Points A, B, C, D, and E correspond to H⁻–C distances of 3.0, 2.5, 2.0, 1.5, and 1.12 Å. The dashed and dotted curves show paths that are 0.6 and 6.0 kcal/mole higher than the minimum energy path. *(b)* Energy profiles for lateral displacement out of the normal (XZ) plane. Reprinted with permission from ref. [22], copyright 1974, Elsevier Science, Ltd.

Although the energy profile of the trajectory illustrated in Figure 4.7 drops continuously and never passes through a transition structure, its similarity to the X-ray structural data is striking. Taken together [22], these studies provide strong support for an approach trajectory that is at or near the Bürgi-Dunitz angle of 107°.

4.1.4 Back to the Cram's rule problem (Anh's analysis)

In 1977, Anh [23] used ab initio methods to evaluate the energies of all the postulated transition structures (Figures 4.2 - 4.4) for the reaction of 2-methylbutanal and 2-chloropropanal (the former to test the Cram, Karabatsos, and Felkin models, and the latter to test the Felkin and Cornforth models). The nucleophile was H⁻, located 1.5Å from the carbonyl carbon, at a 90° angle, on each face of the carbonyl. Rotation of the C_1–C_2 carbon-carbon bond then provided an energy trace which included structures close to all of the previously proposed conformational models. The results for both compounds clearly showed the Felkin transition states to be the lowest energy conformers for attack on either face of the carbonyl. Inclusion of a proton or lithium ion, coordinated to the oxygen, produced similar results. It therefore appeared that Felkin's notion of attack antiperiplanar to the large substituent was correct.

The Felkin *geometries* have the lowest energy, but that did not necessarily mean that the Felkin *rationale* was correct. Recall that Felkin assumed that a hydrogen is more sterically demanding than an oxygen.[2] In their calculations, Anh and Eisenstein held the geometry of the carbonyl rigid (in the Felkin conformation) and varied the angle of hydride attack on the two aldehydes coordinated to a cation. They found optimum angles of 100°, but also found that the energy difference between the two transition states was *amplified* in this geometry [23]. Thus, the Felkin model was revised to include the Bürgi-Dunitz trajectory. Nonperpendicular attack increases the eclipsing effect with either the small or medium substituents, and also increases the interaction of the nucleophile with R, while decreasing the interaction with the oxygen. With Anh's modifications, the Felkin transition states appear to be on a firm theoretical footing, as illustrated in Figure 4.8.

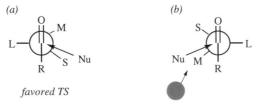

(a) *(b)*

favored TS

Figure 4.8. The Felkin-Anh transition state models for asymmetric induction [17,23].

4.1.5 Heathcock's refinement

Heathcock, in 1983 [24], proposed that the increase in selectivity seen as the size of the "other" substituent increased (Table 4.1, [16]), or when the carbonyl is complexed to a Lewis acid [24] might be explained by deviations of the attack trajectory from the normal plane. In 1987 [25], Heathcock reported the results of a semi empirical study of the angle of approach for the attack of pivaldehyde by hydride. The results, illustrated in Figure 4.9a, illustrate that the approach deviates significantly away from the normal plane, away from the *tert*-butyl group. Although not illustrated, the Bürgi-Dunitz component was variable, but was about the same as found for attack on formaldehyde (108-115°). Although the potential surface near the transition state for nucleophilic additions to unhindered carbonyls is fairly flat [22,26], and has room for some "wobble" in the approach (*cf*. Figure 4.7b), Heathcock showed [25] that constraining the hydride to the normal plane in approach to pivaldehyde is higher in energy, especially at longer bond distances. At 2.5 Å, the energy difference reached its maximum of 0.7 kcal/mole. Figure 4.9b shows Heathcock's rationale for Felkin's observations [16] listed in Table 4.1. When R is small, the "Flippin-Lodge angle", ϕ,[5] is large, and the nonbonded interactions resulting from interaction of the nucleophile with the substituents in R* are diminished. As the size of R increases, the approach trajectory is pushed back toward the normal plane, increasing the nonbonded interactions with R*, and amplifying the selectivity.

In his 1977 paper, Anh also addressed the issue of which substituent would assume the role of the "large" substituent anti to the incoming nucleophile. A simple rule was offered [23]: the substituents should be ordered according to the energies of the antibonding, σ^* orbitals. The preferred anti substituent will be that one having the lowest lying σ^* orbital, not necessarily the one that is the most demanding sterically. This rule explains the α-chloro ketone anomaly, since the σ^* orbital of the carbon-chlorine bond is lower in energy than a carbon-carbon bond. However in 1987, Heathcock tested this hypothesis [28], and concluded that the rule is only partly correct.

---

[5] Professor Heathcock named this angle after his two collaborators, Lee Flippin and Eric Lodge [27].

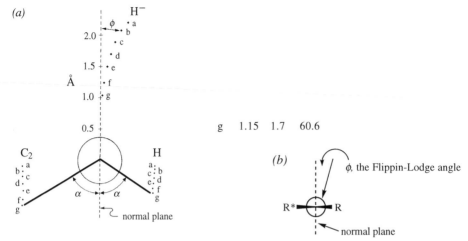

Figure 4.9. *(a)* Deviation of the attack trajectory from the normal plane in the reaction of hydride with pivaldehyde. Reprinted with permission from ref. [25], copyright 1987, American Chemical Society. *(b)* Newman projection of a ketone, with an approaching nucleophile, and the Flippin-Lodge angle of deviation from the normal plane, away from the larger substituent, R* (after ref. [27]).

Specifically, Heathcock examined a series of aldehydes designed to evaluate the relative importance of steric and orbital energy effects. Aldehydes having a substituent with a low energy σ^* orbital (methoxy and phenyl) as well as a sterically variable substituent (methyl, ethyl, isopropyl, *tert*-butyl, phenyl) were synthesized and evaluated. The data are summarized in Table 4.2.[6]

If the antiperiplanar substituents in the Felkin-Anh model (L in Figure 4.8) are those with low-lying σ^* orbitals (X in Table 4.2), one would expect a gradual increase in selectivity as the steric bulk of the remaining substituent (M in Figure 4.8) increased. The data in Table 4.2 show that this is clearly not the case. In the methoxy series, the expected trend is observed for methyl, ethyl, and isopropyl. But the *tert*-butyl and the phenyl groups are anomalous, if one considers the standard *A* values[7] as a measure of steric bulk. In the phenyl series, there is no apparent pattern, and when R = *tert*-butyl, the Anh hypothesis predicts the wrong product.

These data may be interpreted using the four-conformer model shown in Figure 4.10. Simply put, *both steric and electronic effects determine the favored anti substituent*. Thus in the methoxy series (Figure 4.10a), conformers A and B are favored when R is methyl, ethyl, or isopropyl, and attack is favored *via* conformer A. When R is *tert*-butyl, its bulk begins to compensate for the σ^* orbital effect, and conformations C and D become important, with D favored. A rationale for the observed (93% ds) selectivity for the *tert*-butyl ligand is that a very high selectivity results from the preference of A over B, but is tempered by an offsetting selectivity of D over C. When R is phenyl, the bulk of the phenyl as well as its low-lying C_{sp^3}–

---

[6]   Note that the nucleophile in this study is an enolate, not a Grignard reagent.
[7]   The free energy differences ($-\Delta G°$), *A* values, between equatorial and axial conformations of a substituted cyclohexane ring are (kcal/mole): Cl = 0.52, MeO = 0.75, Me = 1.74, Et = 1.75, *i*-Pr = 2.15, Ph = 2.7, *t*-Bu = 4.9 (taken from ref. [29]).

Table 4.2. Cram's rule stereoselectivities (% ds) for aldol additions to aldehydes (negative value indicates anti-Cram is favored), assuming X is the large substituent in the Felkin-Anh model [28]:

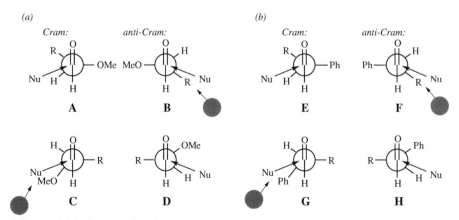

X	R = Me	R = Et	R = i-Pr	R = t-Bu	R = Ph
OMe	58	76	93	93	83
Ph	78	86	70	−63	–

C_{sp^2} σ^* orbital play a role. A prediction made on the basis of its bulk alone (*A* values[5]) would predict a selectivity greater than when R is isopropyl (still assuming an anti methoxy), but the phenyl σ^* orbital is lower in energy than a C_{sp^3}–C_{sp^3} σ^* orbital, which increases the importance of conformers C and D (anti-Cram D is favored).

In the phenyl series (Figure 4.10b), when R is methyl or ethyl, conformers E and F are dominant, with E favored. Note that the selectivity in the phenyl series for methyl and ethyl ligands is greater than in the methoxy series (Table 4.2). This is because the phenyl group is bulky *and* has a low energy σ^* orbital, so that the electronic and steric effects act in concert. For the isopropyl and *tert*-butyl ligands, the importance of the G/H conformers increases, and when R is *tert*-butyl they predominate.

Heathcock refers to conformers C, D, G, and H as "non-Anh" conformations, since they have one of the ligands with a *higher* σ^* orbital energy anti to the nucleophile. The non-Anh conformations are more important in the phenyl series because there is less difference in the σ^* orbital energies between C_{sp^3}–C_{sp^3} and C_{sp^3}–C_{sp^2} bonds than between carbon–carbon and carbon–heteroatom bonds.

Figure 4.10. Heathcock's four-conformer model for 1,2-asymmetric induction [28]. *(a)* Electronic effects favor methoxy as anti ligand (A and B) while steric effects may favor C and D. *(b)* Electronic effects favor phenyl as anti ligand (E and F) while steric effects favor G and H for very large alkyl groups.

4.1.6 The bottom line (hasn't been written yet)

Theoretical investigations into the origins of Cram's rule selectivity continue. For example, Dannenberg has shown that the energies of the frontier orbitals change as a function of the dihedral angle [19], and Frenking has concluded that "the most important factor for the π-facial diastereoselectivity in nucleophilic addition reactions to carbonyl compounds originates from simple conformational effects" [30] (see also ref. [31-33]).

To predict the major stereoisomer in a "Cram's rule situation", a thorough analysis should include consideration of the following points:

1. The nucleophile will approach along the Bürgi-Dunitz trajectory, approximately 100-110° from the carbonyl oxygen (Figures 4.6 and 4.7).
2. For ketones, the approach may be in or near the normal plane, but for aldehydes, there will be a deviation from this plane, toward the hydrogen and away from the stereocenter (Figure 4.9).
3. If there is a strong electronic or steric preference by one ligand that is not offset by another ligand, the Felkin-Anh two conformer model (Figure 4.8) may be used with the following order of preference for the anti position: MeO>*t*-Bu>Ph>*i*-Pr>Et>Me>H [28].
4. A complete evaluation of the selectivity requires (at least) a four conformer analysis (Figure 4.10) with the electronic effect dictating an anti preference of MeO>Ph>R>H, while the steric effect leads to the order *tert*-Bu>Ph>*i*-Pr>Et>Me>H [28].

4.2 Cram's rule: rigid, chelate, or cyclic model

In his 1952 paper [2] Cram also considered a cyclic model that may be invoked when chelation is possible. In 1959 [6] the model was examined in detail for α-hydroxy and α-amino ketones, since the cyclic and acyclic models predict different outcomes for these systems. The cyclic model (Figure 4.11) has stood the test of time rather well, and has recently received direct experimental confirmation, in the form of NMR observation of a chelate as an intermediate in the addition of dimethylmagnesium to α-alkoxy ketones [34]. The cyclic model is applicable to cases where there is a chelating heteroatom on the α-carbon, when that carbon is also a stereocenter (reviews: [35,36]).

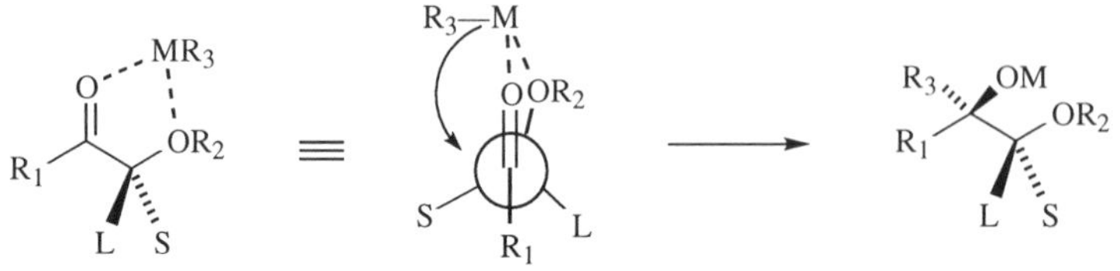

Figure 4.11. Cram's cyclic model for asymmetric induction. L and S are large and small substituents, respectively [2,6]).

Table 4.3 lists selected examples where exceptionally high stereoselection has been encountered. Solvent effects play an important role in achieving high selectivity. For example the >99% diastereoselectivities for the addition of

Grignard reagents to α-alkoxy ketones in THF (entry 1) were greatly diminished in ether, pentane, or methylene chloride [37]. Eliel demonstrated similar selectivites for additions by dimethylmagnesium in THF (entry 2). With aldehydes, there have been conflicting reports. Still reported a 90% diastereoselectivity in the reaction of methylmagnesium bromide with 2-(benzyloxymethoxy)propanal [38], but Eliel [39] and Keck [40] observed poor selectivities in THF. Eliel found good selectivities (90-94% ds) in ether (*e.g.,* entry 3) for the addition of a Grignard to the benzyl or MOM ethers of a 2-hydroxyundecanal. For a number of additions of less reactive

Table 4.3. Selected examples of nucleophilic addition to α-alkoxy carbonyls.

Entry	Educt	Conditions	Product (%ds)	Reference
1	Me, C_7H_{15}, OR [1]	BuMgBr THF,[2] $-78°$	HO, Bu, Me, OR, C_7H_{15} (>99% ds)	[37]
2	Ph, Me, OR [3]	Me_2Mg THF, $-70°$	HO, Me, Ph, OR, Me (>99% ds)	[34]
3	H, $C_{10}H_{21}$, OR [4]	$Ph(CH_2)_3MgBr$ Et_2O,[5] $-78°$	OH, $Ph(CH_2)_3$, OR, $C_{10}H_{21}$ (94% ds)	[39]
4	H, OBn, CH_2CO_2Me	$MgBr_2·OEt_2$ $CH_2=CHMgBr$ CH_2Cl_2,[5] $-78°$	OH, OBn, CH_2CO_2Me (>99% ds)	[40]
5	H, OBOM, Me	Me_2CuLi Et_2O, $-78°$	OH OBOM, Me, Me (97% ds)	[38]
6	Ph, Me, $OSi(i-Pr)_3$	Me_2Mg THF, $-70°$	HO, Me, Ph, $OSi(i-Pr)_3$, Me (58% ds)	[34]

1 R = MEM (methoxyethoxymethyl-), MOM (methoxymethyl-), MTM (methylthiomethyl-), CH_2–furyl, Bn (benzyl-), BOM (benzyloxymethyl-).
2 Pentane, ether, and methylene chloride afforded much lower selectivities.
3 R = Me, $SiMe_3$.
4 R = Bn, MOM
5 THF affords much lower selectivity.

nucleophiles, Reetz has shown that prior organization of the chelate by complexation with a Lewis acid improves results with aldehydes [41]. Along these lines, Keck has reported [40] that prior coordination of an α-alkoxy aldehyde with magnesium bromide in methylene chloride, followed by addition of a vinyl Grignard affords excellent selectivity (entry 4). In order to achieve high selectivity, the THF in which the Grignard was formed had to be distilled away and replaced by methylene chloride [40].

The cyclic model applies mainly for α-alkoxy carbonyls (*5-membered chelate*), whereas β-alkoxy carbonyls (*6-membered chelate*) are less selective in most cases. An exception is the addition of cuprates to β-alkoxy aldehydes having an α-stereo-center (entry 5).

Two features of the cyclic model are particularly important synthetically. The first is that the selectivities can be significantly higher than for the acyclic category. Compare entries 2 and 6 of Table 4.3: the methoxy and trimethylsilyloxy groups chelate the magnesium (entry 2) whereas the triisopropylsilyloxy group does not (entry 6). This poorly selective example reacts by the acyclic pathway (also compare entries 1-5 with Tables 4.1 and 4.2). The second noteworthy point is that the product predicted by the cyclic and acyclic models are sometimes different. As shown in Scheme 4.1, the predictions of the acyclic and cyclic models are different for Table 4.3, entry 1 (see also entries 2 and 6).

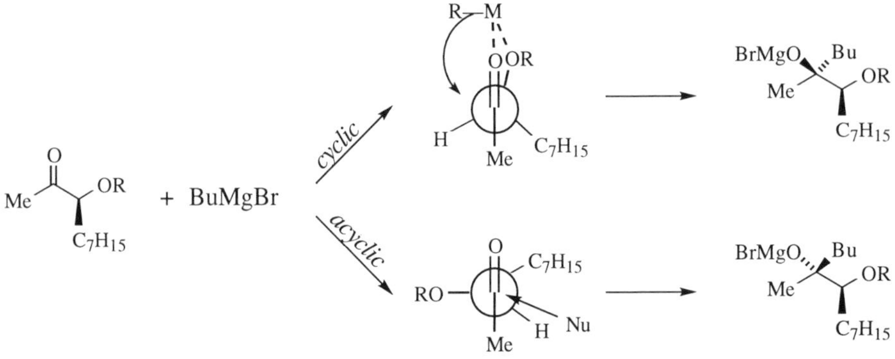

Scheme 4.1. Cyclic and acyclic models often predict opposite outcomes.

Study of the mechanism of Grignard addition (RMgX) via the chelate pathway is complicated by the presence of Schlenck equilibria, but Eliel has examined the mechanism of the addition of dimethylmagnesium (R$_2$Mg) to α-alkoxy ketones (*e.g.,* Table 4.3, entries 2 and 6) in detail, since dimethylmagnesium is a well-characterized monomer in THF solution. Scheme 4.2 summarizes the current picture of the mechanism [34]. Beginning with the educt in the middle of the scheme, there are two competing pathways for the addition reaction. One involves chelated (cyclic) intermediates (to the right of the scheme), while the other involves nonchelated (acyclic) intermediates (shown on the left). One should also recognize that there are two distinct issues that must be considered for these competing pathways: their *relative rates*, and their *stereoselectivities*.

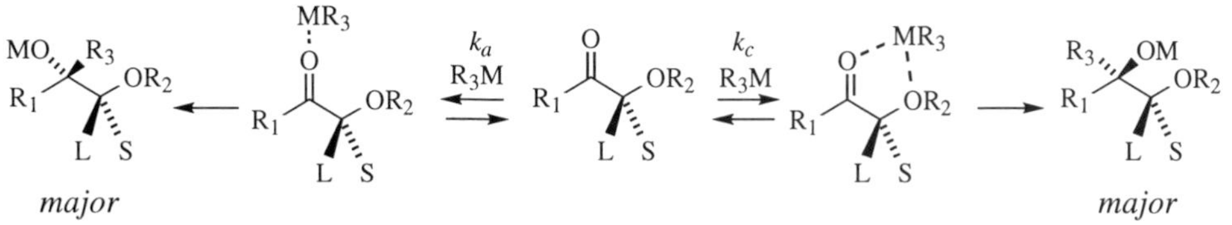

Scheme 4.2. The acyclic and cyclic mechanisms compete for the consumption of substrate.

The chelate rule will only be applicable if addition via the chelate is faster than addition by the acyclic mechanism (*i.e*, $k_c > k_a$ in Scheme 4.2). Because the chelate is rigid, it is often considerably more stereoselective as well.[8] *However, the relative rate issue is independent of the stereoselectivities of the two processes.* For example, chelation can be used to control regiochemistry: selective reduction of a diester is achieved by preferential chelation to a 5-membered ring over a 6-membered ring by magnesium bromide (Scheme 4.3, [40]).

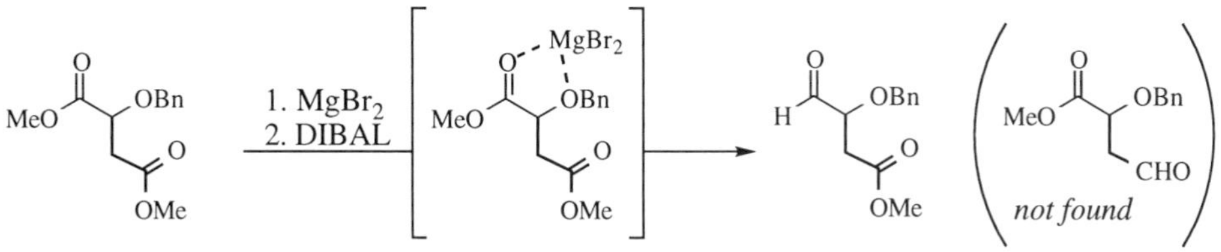

Scheme 4.3. Independent of stereochemical issues, chelation can determine reactivity [40].

For the chelate path to be faster than the acyclic path, chelation must lower the energy of activation relative to the acyclic path, as shown in Figure 4.12 [34]. The two individual steps illustrated in this diagram deserve comment. First, note that the chelated intermediate is lower in energy and has a smaller energy of activation for its formation than the monodentate intermediate on the acyclic pathway. That the chelate is more stable than the monodentate complex is no surprise. However, the increased organization of the chelated transition state ($\Delta S^{\ddagger}$ less positive) and the increased steric interactions that result ($\Delta H^{\ddagger}$ more positive) would seem to dictate a slower reaction,[9] but these effects are offset by the enthalpy gained by complexation of the alkoxy ligand to the metal and the entropy gained by liberation of an additional solvent from the metal by the bidentate ligand. Regarding the second step, whereby the chelate reacts faster than the monodentate complex, it is pertinent that the kinetics of the addition of dimethylmagnesium are first order in organometallic [34]. This requires the intramolecular transfer of an R_3 ligand *via* a four-membered ring transition state. The distance between the metal ligand (R_3) and the carbonyl carbon is greater in the (linear) acyclic transition state than in the chelated one, so the chelate is further along the reaction coordinate than the linear complex [34].

---

8   This may seem contrary to the reactivity-selectivity principle, wherein one expects a decrease in selectivity to accompany an increase in reactivity, but this principle has a number of limitations. For an extensive discussion of the reactivity-selectivity principle, see ref. [42].

9   Recall (Chapter 1) that $k = \left(e^{-\Delta H^{\ddagger}/RT}\right)\left(e^{\Delta S^{\ddagger}/R}\right)$.

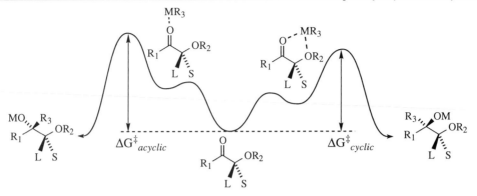

Figure 4.12. Energetics of the Cram-chelate (acyclic) model. $\Delta G^{\ddagger}acyclic > \Delta G^{\ddagger}cyclic$ (after ref. [34]).

The relative energies of the intermediates and transition structures along the reaction coordinates are subject to the influence of solvation, which may alter relative stabilities and rates. This may explain the solvent effects discussed earlier (*cf.* Table 4.3, entries 1, 3 and 4). The energetic features outlined above may also explain the lack of selectivity in the nucleophilic additions to β-alkoxy carbonyl compounds. It is possible that even though 6-membered chelates are formed, their rates of formation are slower than addition via the nonchelated path, or that they are less reactive than a 5-membered chelate. Either of these circumstances (or a combination of both) would raise the transition state energy for the chelate path and the primary addition mode could be shifted to the less selective nonchelated mechanism.[10]

Because of the high selectivities observed in chelation-controlled additions, it is often used in stereoselective total syntheses. For example, highly selective additions of Grignards were used in the synthesis of the ionophores monensin [43,44] and lasalocid [45,46], shown in Figure 4.13.

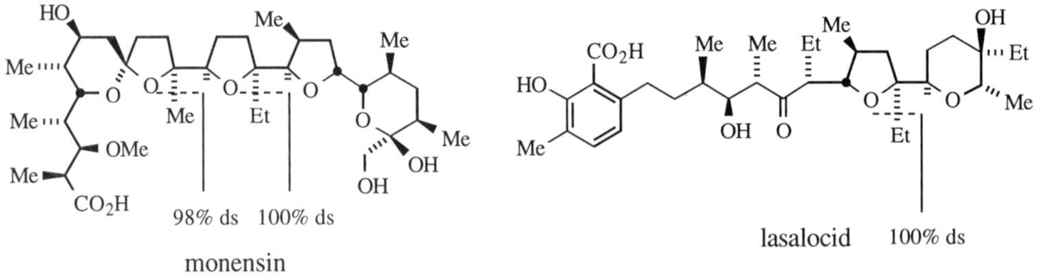

Figure 4.13. Chelation-controlled addition of Grignards to ketones figured prominently in the synthesis of monensin [43] and lasalocid [45,46]. The disconnections used and the selectivities achieved are indicated for the stereocenters formed by the Grignard addition.

4.2.1 Cram's cyclic model in asymmetric synthesis

Auxiliaries have been designed to exploit the high selectivities of chelation-controlled processes in asymmetric synthesis. Among these are the oxathiane [35,47-50]

---

[10] Another possibility is that the intrinsic selectivity of reaction *via* a 6-membered chelate is lower.

and oxazine [51,52] systems developed by Eliel. As shown in Scheme 4.4, the heterocyclic system is held rigid by its *trans*-decalin-like geometry. In both heterocyclic systems, the metal is chelated by the carbonyl oxygen and the ether oxygen (the latter in preference to either the sulfur or the nitrogen). Approach of the electrophile from the less hindered *Re* face is favored.

Both auxiliaries are synthesized from (+)-pulegone, with the sulfur version available as an *Organic Syntheses* prep [47]. Hydrolysis of the acetal after the addition removes the chiral auxiliary (recovered in good yield) and liberates an α-hydroxy aldehyde, which may be reduced to a glycol or oxidized to an α-hydroxy acid. Table 4.4 lists several examples of the addition. Entries 2/3 and 7/10 illustrate the selective formation of either possible stereoisomer by reversal of the "R" and "Nu" groups. Entries 4 and 5 illustrate a case of matched and mismatched double asymmetric induction (Chapter 1), where the distal stereocenter of the chiral nucleophile affects the selectivity of the addition. Comparison of entries 1-6 and 7-12 indicate that both the sulfur and the nitrogen auxiliaries are useful, so that the conditions necessary for cleavage may dictate the choice of auxiliary. Figure 4.14 shows several natural products that have been synthesized using this methodology.

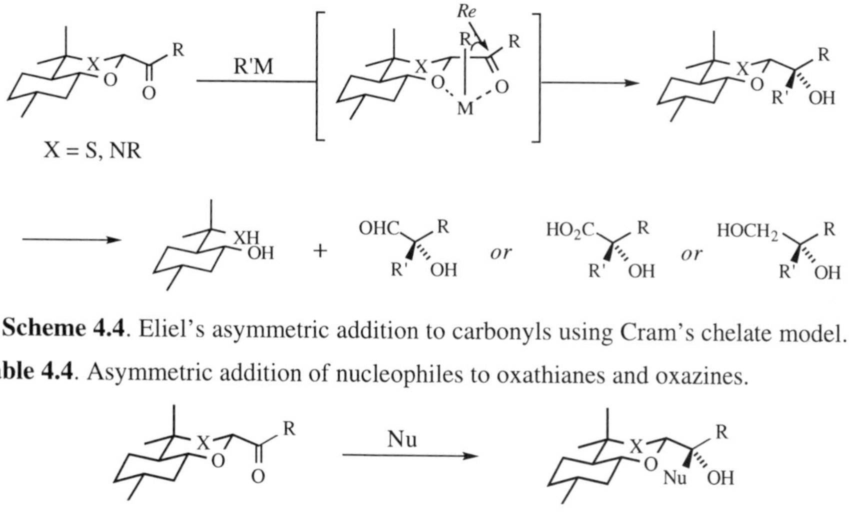

Scheme 4.4. Eliel's asymmetric addition to carbonyls using Cram's chelate model.

Table 4.4. Asymmetric addition of nucleophiles to oxathianes and oxazines.

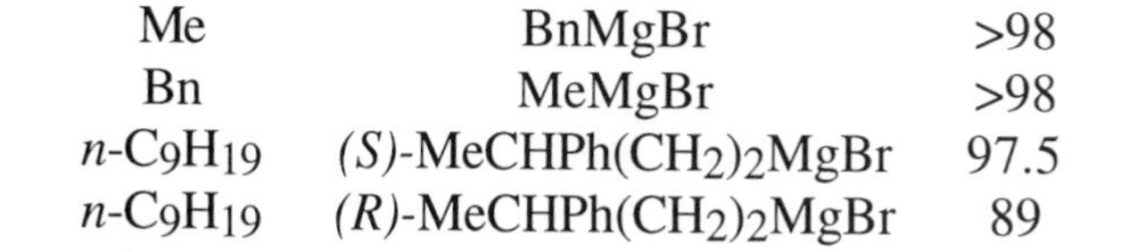

Entry	X	R	Nu	% ds	Reference
1	S	Me	CH$_2$=CHMgBr	92	[53]
2	S	Me	BnMgBr	>98	[54]
3	S	Bn	MeMgBr	>98	[54]
4	S	n-C$_9$H$_{19}$	(S)-MeCHPh(CH$_2$)$_2$MgBr	97.5	[55]
5	S	n-C$_9$H$_{19}$	(R)-MeCHPh(CH$_2$)$_2$MgBr	89	[55]
6	S	n-C$_{10}$H$_{21}$	LiBH(s-Bu)$_3$	91	[39]
7	NBn	Me	PhMgBr	95.5	[52]
8	NBn	Me	EtMgBr	92	[52]
9	NBn	Me	NaBH$_4$	95.5	[52]
10	NBn	Ph	MeMgBr	>98	[51]
11	NBn	Ph	EtMgBr	>98	[51]
12	NMe	Ph	MeMgBr	96	[52]

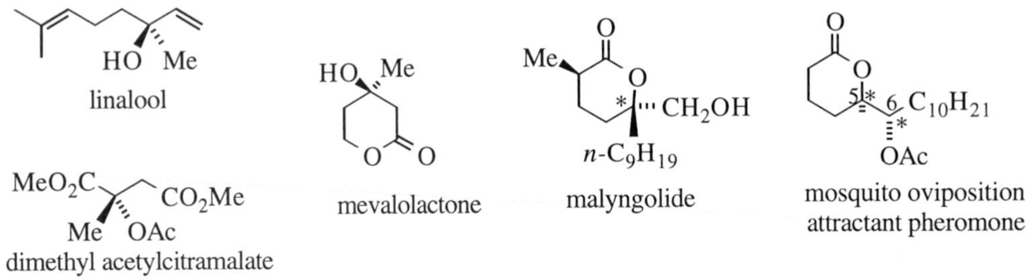

linalool

mevalolactone

malyngolide

mosquito oviposition attractant pheromone

dimethyl acetylcitramalate

Figure 4.14. Applications of oxathianes: linalool [53], dimethyl acetylcitramalate [54], mevalolactone [56], malyngolide [55], and the mosquito oviposition attractant [39]. For the latter, the C-5 stereocenter was formed by a chelate-controlled reduction while the C-6 position could be produced as either epimer by a chelate or acyclic mechanism, depending on the reducing agent.

4.3 Additions using chiral catalysts or chiral nucleophiles

The preceding discussion summarizes a great deal of work done over the last forty years on the stereoselective additions of achiral carbanionic nucleophiles to carbonyls having a neighboring stereocenter. The knowledge gained during these studies has aided in the development of two different approaches to stereoselective additions to heterotopic carbonyl faces: (i) those using chiral nucleophiles with achiral carbonyl compounds [57]; and (ii) a potentially more useful process, one in which neither partner is chiral, but a chiral catalyst is used to induce stereoselectivity (reviews: [58-60] and chapter 5 in ref. [61]).

All of the reactions discussed in this chapter require coordination of a carbonyl to a metal. This coordination activates the carbonyl toward attack by a nucleophile, and may occur by two intrinsically different bonding schemes: σ or π (Figure 4.15). The best evidence to date indicates that σ coordination predominates for Lewis acids such as boron or tin [62,63], and (more importantly) σ-bonding produces a more reactive species [64]. In the following discussions, it will be assumed that σ bonding to the carbonyl oxygen is operative.

Figure 4.15. Geometries and relative reactivities of coordinated carbonyls [64].

The potential utility of an asymmetric addition to a prochiral carbonyl can be seen by considering how one might prepare 4-octanol (to take a structurally simple example) by asymmetric synthesis. Figure 4.16 illustrates four possible retrosynthetic disconnections. Note that of these four, two present significant challenges: asymmetric hydride reduction requires discrimination between the enantiotopic faces of a nearly symmetrical ketone *(a)*, and asymmetric hydroboration-oxidation requires a perplexing array of olefin stereochemistry and regiochemical issues *(b)*. In contrast, the addition of a metal alkyl to an aldehyde offers a much more realistic prospect *(c)* or *(d)*.

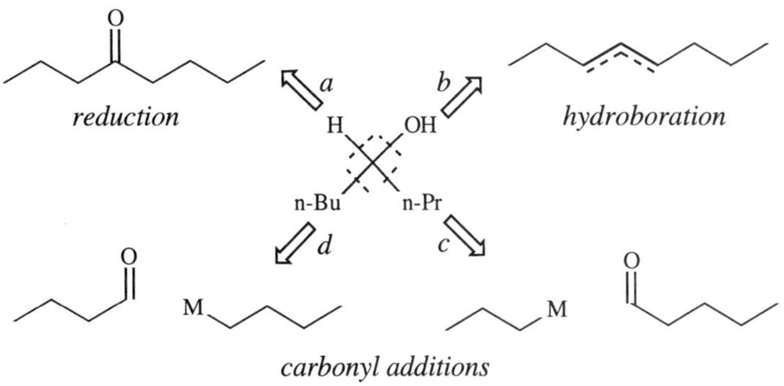

Figure 4.16. Simple retrosynthetic strategies for synthesis of 4-octanol.

4.3.1 Catalyzed Addition of organometallics

A number of organometals have been evaluated for this type of reaction, but because of fewer side reactions (such as deprotonation of the aldehyde), the substrate studied most often is benzaldehyde. Perhaps the best understood of these reactions is the addition of organozincs, especially dimethyl- and diethylzinc (reviews: [58-60,65-68]). The reactivity of alkylzincs is low, and at or below room temperature the rate of addition of, for example, diethyl zinc to benzaldehyde is negligible. Addition of a Lewis acid, however, causes rapid addition. Replacement of one of the alkyl substituents with an alkoxide produces a more reactive species as well, and amino alcohols have been found to be very useful catalysts for the addition reaction [69,70]. At least part of the reason for the increased reactivity is a rehybridization of the zinc from linear to bent upon complexation to an alkoxide, and to tetrahedral upon bidentate coordination. Additionally, donor ligands such as oxygen and nitrogen render the alkyl group more nucleophilic. Figure 4.17 illustrates some of the catalysts that afford good yields and high enantioselectivities in the diethylzinc reaction with benzaldehyde.

The mechanisms that have been proposed for the amino alcohol-catalyzed reaction all involve two zinc atoms, one amino alcohol and three alkyl groups on the active catalyst [65,71-74]. A composite mechanism is illustrated in Scheme 4.5 for a "generic" β-amino alcohol.[11] NMR evidence [71] indicates dynamic exchange of the alkyl groups on zinc as shown in the brackets (a bridged species has also been proposed [71]). In experiments done with a polymer-bound amino alcohol catalyst, Frechet has noted that the alkoxide product is not bound to the catalyst and that the alkyl transfer must have therefore occured from diethylzinc in solution.

It might be expected that use of an amino alcohol of less than 100% enantiomeric purity would place an upper limit on the enantiomeric purity of the product. However, Noyori reported that when a catalyst (Figure 4.17b) of 15% ee was used in the diethylzinc reaction, 1-phenyl-1-propanol of 95% ee was isolated in 92% yield [71]. As it turns out, the zinc alkoxide produced after the reaction of one equivalent of diethylzinc dimerizes (Scheme 4.6). When both enantiomers of the amino alcohol are present, both homochiral and heterochiral dimers may be formed.

---

[11]   For a discussion of the various mechanistic models and a detailed analysis, see ref. [58,75].

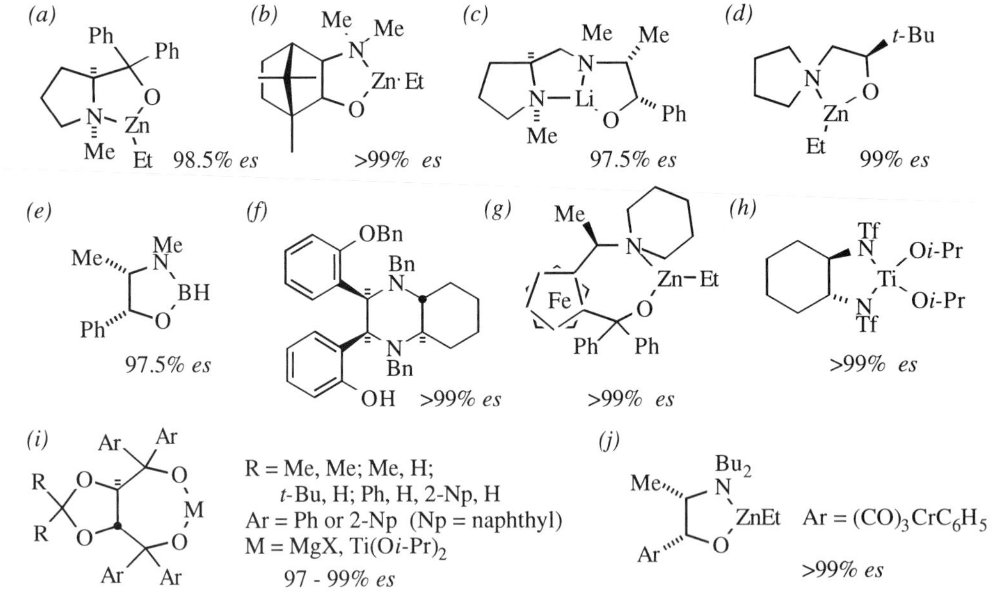

Figure 4.17. Catalysts for the diethylzinc reaction with benzaldehyde: *(a)*, [76]; *(b)*, [71]; *(c)*, [73]; *(d)*, [77]; *(e)*, [78]; *(f)*, [79]; *(g)*, [80]; *(h)*, [81,82]; *(i)*, [83,84]; *(j)*, [85].

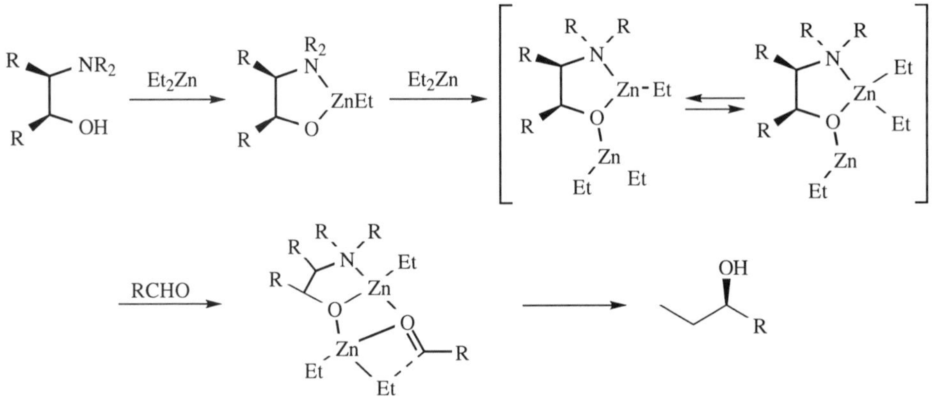

Scheme 4.5. Proposed mechanistic scheme for amino alcohol catalyzed diethylzinc reaction (after ref. [60])

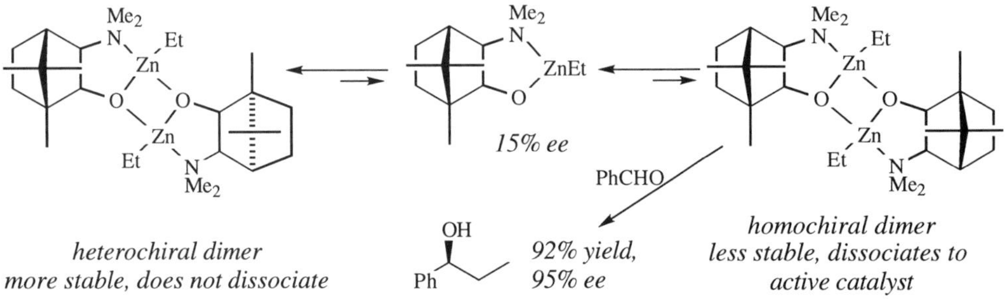

Scheme 4.6. Amplification of enantiomer excess by the Noyori catalyst [71].

Table 4.5. Catalyzed additions of organometallics (RM) to aldehydes and ketones. Numbers in the catalyst column refer to Figure 4.17.

Entry	Carbonyl	RM	Catalyst	% Yield	%es	Ref
1	n-C$_6$H$_{13}$CHO	Et$_2$Zn	4.17a	96	95.5	[76]
2	i-BuCHO	Et$_2$Zn		92	96.5	[72]
3	n-C$_6$H$_{13}$CHO	Me$_2$Zn	"	70	95	[72]
4	2-NpCHO	Ph$_2$Zn	"	83	90	[86]
5	c-C$_6$H$_{11}$CHO	Et$_2$Zn	4.17g	92	99	[80]
6	t-BuCHO	Et$_2$Zn	4.17g	93	99	[80]
7	n-C$_6$H$_{13}$CHO	Et$_2$Zn	4.17h	78	>99	[81]
8	PhCHO	Et$_2$Zn		91	96	[73]
9	PhCHO	Vinyl$_2$Zn		96	93.5	[87]
10	n-C$_5$H$_{11}$CHO	Vinyl$_2$Zn	"	90	98	[87]
11	c-C$_6$H$_{11}$CHO	Vinyl$_2$Zn	"	83	91	[87]
12	c-C$_6$H$_{11}$CHO	Bu$_2$Zn	4.17i, M = Ti(Oi-Pr)$_2$	35	95	[88]
13	PhCHO	(MOMO-(CH$_2$)$_6$)$_2$Zn	4.17i, M = Ti(Oi-Pr)$_2$	68	92	[88]
14	PhCHO	(C$_2$H$_3$-(CH$_2$)$_2$)$_2$-Zn	4.17i, M = Ti(Oi-Pr)$_2$	83	95	[88]
15	1- or 2-Np	Et$_2$Zn	4.17j	98	>99	[85]
16	PhCHO	n-BuLi		77	97.5	[70]
17	PhCHO	Et$_2$Mg	"	74	96	[70]
18	PhCOCH$_3$	EtMgBr	4.17i, M = MgX	62	99	[89]

With the Noyori catalyst, the heterochiral dimer is considerably more stable than the homochiral dimer. The latter decomposes to the active, monomeric catalyst immediately upon exposure to a dialkylzinc or an aldehyde, whereas the heterochiral dimer does not. Thus, the minor enantiomer of the catalyst is "tied up" by the major enantiomer.[12]

To provide an overview of the scope of such processes, Table 4.5 lists some of the more selective examples of this type of addition for a variety of substrates and organometallics. It would be premature to say that the process of asymmetric additions of achiral nucleophiles is a general procedure at this time (*i.e.*, that any organometallic and carbonyl can be made to couple enantioselectively), but the current rate of progress suggests that the realization of this goal will not be long in coming. Particularly noteworthy are the isolated examples of organolithium and Grignard additions (entries 16-18).

4.3.2 Hydrocyanations

The addition of cyanide to an aldehyde or ketone (hydrocyanation) is an old reaction, but it has been the subject of renewed interest since Reetz's discovery that a chiral Lewis acid could be used to catalyze the asymmetric addition of trimethyl-silylcyanide to isobutyraldehyde ([91]; reviews: [59,92]). The general process, illustrated in Scheme 4.7, usually employs trimethylsilylcyanide because hydrogen cyanide itself catalyzes the addition as well (nonselectively). Most of the catalysts are chiral titanium complexes; some of the more selective examples are shown in Table 4.6. A clear mechanistic picture of the titanium catalyzed additions has not yet emerged.[13]

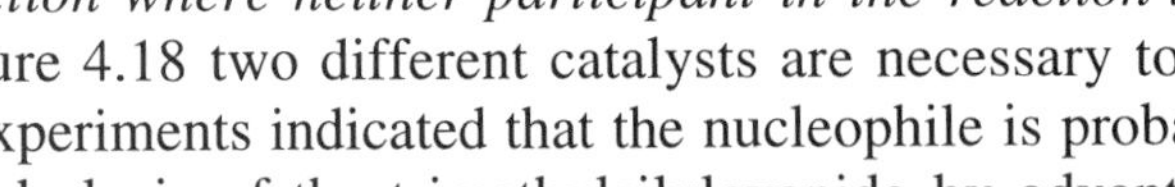

$$\underset{R}{\overset{O}{\underset{H}{\|}}} + Me_3SiCN \xrightarrow{\quad catalyst \quad} \underset{R}{\overset{OSiMe_3}{\underset{CN}{|}}}$$

Scheme 4.7. General asymmetric addition of tri-methylsilylcyanide to an aldehyde.

Experiments described by Corey constitute a noteworthy example of *double asymmetric induction where neither participant in the reaction is chiral* [95]! As illustrated in Figure 4.18 two different catalysts are necessary to achieve the best results. Control experiments indicated that the nucleophile is probably free cyanide, introduced by hydrolysis of the trimethylsilylcyanide by adventitious water, and continuously regenerated by silylation of the alkoxide product. Note that the 82.5% enantioselectivity in the presence of the magnesium complex shown in Figure 4.18a is improved to 97% upon addition of the bisoxazoline illustrated Figure 4.18b as a cocatalyst. Note also that the bisoxazoline 4.18b alone affords almost no enantio-selectivity, and that the enantioselectivity is much less when the *enantiomer* of the bisoxazoline (Figure 4.18b) when used as the cocatalyst. Thus 4.18a and 4.18b constitute a "matched pair" of co-catalysts and 4.18a and *ent*-4.18b are a "mis-matched pair" (see Chapter 1 for definitions). The proposed transition structure

---

[12] The phenomenon of nonlinear optical yields is sometimes called asymmetric amplification. For detailed analyses, see ref. [58,75,90].

[13] For mechanistic hypotheses, see ref. [93,94].

Table 4.6. Catalytic asymmetric hydrocyanation of aldehydes. Numbers in the catalyst column refer to Figure 4.18 (p. 142).

Entry	Carbonyl	Catalyst	% Yield	% es	Ref
1	i-BuCHO		85	94	[96]
2	PhCHO		67	92	[94]
3	2-NpCHO	"	76	86	[94]
4	E-CH$_3$CH=CHCHO	"	70	94	[94]
5	PhCHO		83	95	[93]
	(Trp = tryptophan)				
6	2-NpCHO	"	55	95	[93]
7	n-C$_8$H$_{17}$CHO		·85	96	[97]
8	Ph(CH$_2$)$_2$CHO	"	88	95	[97]
9	n-C$_6$H$_{13}$CHO	4.18a & 4.18b	88	97	[95]
10	Et$_2$CHCHO	"	86	95	[95]
11	c-C$_6$H$_{11}$CHO	"	94	97	[95]
12	t-BuCHO	"	57	95	[95]
13	E-n-PrCH=CHCHO	"	59	93	[95]

for the matched pair has the hydrogen cyanide complexed to 4.18a and the aldehyde complexed to the magnesium atom of 4.18b.

4.3.3 Additions to the C=N bond

The stereoselective addition of organometallics to azomethines (C=N bond) has not been as fully developed as additions to carbonyls for several reasons (review: [98]). First, imines are not as electrophilic as carbonyls, and so are less susceptible

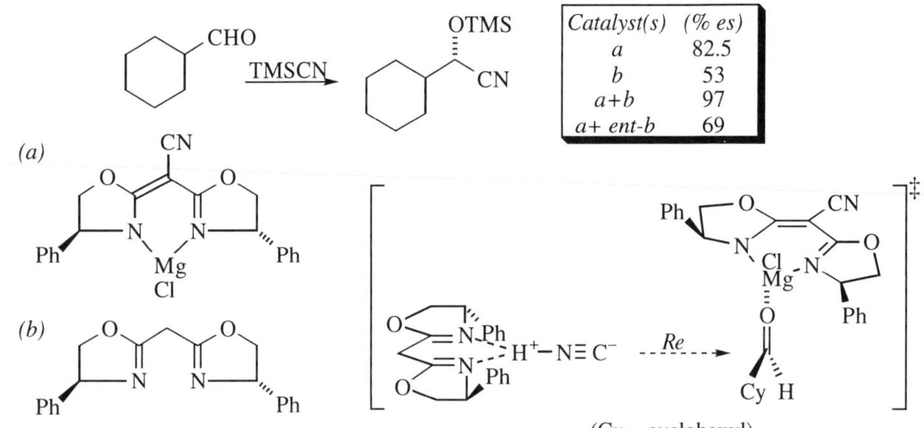

Figure 4.18. Corey's dual catalyst system for asymmetric hydrocyanation of aldehydes [95].

to nucleophilic attack. Second, many organometallic reagents are sufficiently basic that the preferential mode of reaction is abstraction of an α proton. Third, imines are susceptible to *E/Z* isomerization (often catalyzed by the Lewis acids that are a prerequisite to nucleophilic attack), which complicates the issue of stereochemical predictability. Nevertheless, the importance of amines in chemistry and medicine has furnished ample motive to pursue this method of synthesis. In fact, since the nitrogen is substituted (C=NR instead of C=O), azomethines provide an opportunity for auxiliary-based stereochemical control that is not available to carbonyls. The following examples are arranged according to the charge on the nitrogen: addition to imines and hydrazones (neutral nitrogen) is followed by addition to iminium ions.

An asymmetric synthesis of amino alcohols by asymmetric addition of Grignard reagents to chiral α-bromoglycine esters provides a convenient synthesis of α-amino esters (Scheme 4.8, [99]). Hydrolysis of the product ester produces racemized amino acids, but reduction affords amino alcohols that can be subsequently oxidized to the amino acids with no loss of enantiomeric purity. Note that in the proposed transition structure, the phenyl effectively shields the *Re* face (toward the viewer) of the imine, which is chelated to the carbonyl by magnesium halide formed in the dehydrohalogenation.

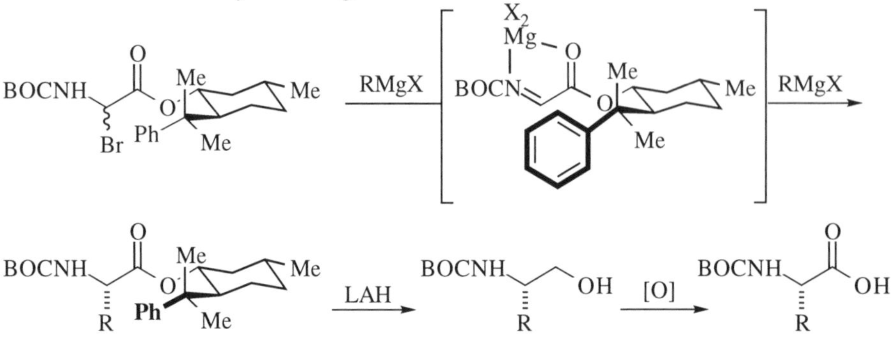

R = Me (71%, 96% ds); *i*-Pr (54%, 94% ds); *i*-Bu (65%, 97% ds); Ph (78%, 91% ds)

Scheme 4.8. Synthesis of amino alcohols and amino acids by nucleophilic additions [99].

A strategy similar to that shown in Scheme 4.8 employs a Grignard addition to a cyclic α-bromoglycine derivative. As shown in Scheme 4.9, elimination of bromide affords an iminium ion that is selectively attacked on the *Si*-face, opposite the two phenyl groups [100]. Reductive cleavage of the benzylic C–N and C–O bonds provides ready access to amino acids.

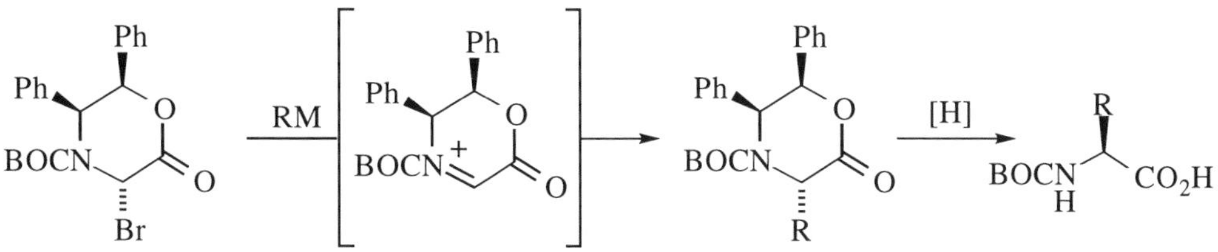

RM = MeZnCl (46%, 98% ds); Bu$_2$Cu(CN)Li (48%, >99% ds)

Scheme 4.9. Oxazinones as chiral electrophilic glycine equivalents [100].

The addition of organometallics to *S*AMP and *R*AMP hydrazones has been studied by the Enders [101-106] and Denmark groups [107-109]. The best selectivities result from addition of organolanthanide reagents; table 4.7 illustrates several of the more highly selective examples. In conjunction with reductive cleavage of the hydrazone by hydrogenolysis [101,102] or dissolving metal reduction [110], the addition provides a convenient synthesis of α-branched primary amines (*c.f.*, Figure 4.16, p. 137). The intermediate hydrazines are somewhat unstable, but *N*-acylation makes for easier handling [105,110]. A mechanistic model has not been proposed to account for the observed configuration.

Table 4.7. Asymmetric addition of organoceriums to hydrazones.

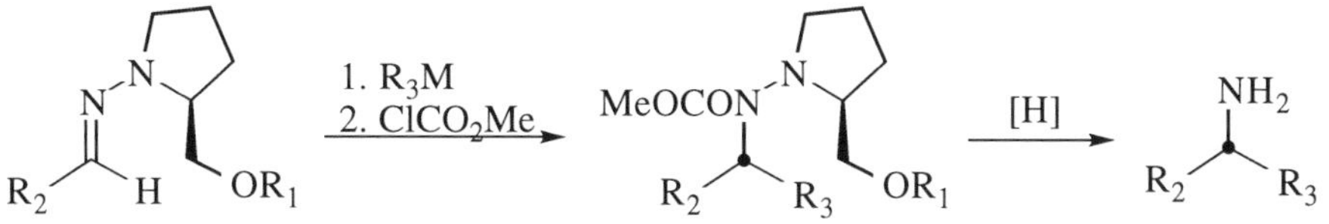

Entry	R$_1$	R$_2$	R$_3$	% Yield	% ds	Ref
1	Me	(EtO)$_2$CH	EtLi/CeCl$_3$	91	96	[102]
2	Me	"	*n*-BuLi/CeCl$_3$	92	97	[102]
3	Me	Ph(CH$_2$)$_2$	MeLi/CeCl$_3$	81	98	[107]
4	Me	"	PhLi/CeCl$_3$	72	96	[107]
5	Me	PhCH$_2$	MeLi/CeCl$_3$	66	96	[107]
6	Me	*E*-CH$_3$CH=CH	"	82	96	[107]
7	Me	TBSO(CH$_2$)$_4$	*n*-PrLi/YbCl$_3$	83	>99	[105]
8	Me	*n*-Pr	TBSO(CH$_2$)$_4$Li/ YbCl$_3$			[105]
9	(CH$_2$)$_2$OMe	Ph(CH$_2$)$_2$	*n*-BuLi/CeCl$_3$	72	97	[107]
10	"	Me	Ph(CH$_2$)$_2$Li/CeCl$_3$	53	97	[108]
11	"	*t*-Bu	"	60	98	[108]
12	"	Ph	"	80	97	[108]

Stereoselective addition of Grignards to chiral pyridinium ions has been used to gain access to an important class of chiral heterocycles: substituted piperidines. Marazano uses *N*-α-methylbenzyl pyridiniums obtained by exchange of α-methylbenzyl amine with an *N*-2,4-dinitrophenylpyridinium [111], while Comins uses an *N*-acylpyridinium obtained by acylation with 8-phenylmenthyl chloroformate or a similar derivative (Table 4.8, [112-115]). Note that these processes are complicated by the symmetry of the ring system: *Si*-face attack at C-2 and *Si*-face attack at C-6 are equivalent (*i.e.*, the *Si*-faces of C-2 and C-6 are homotopic, Figure 4.19a). As a result of this equivalence, face selectivity at C-2 is topologically equivalent to regioselectivity (C-2 *vs.* C-6) from a single face. Thus, in a transition structure where (for example) attack of a nucleophile comes exclusively from the direction of the viewer, addition to C-2 and C-6 produce the same set of isomers that would result from attack at the front and back of only C-2 (Figure 4.19b). To circumvent this complication, Comins puts a large (removable) blocking group at C-3, which also blocks addition at C-4 (Figure 4.19c). Figure 4.20 illustrates several alkaloids synthesized using this approach.

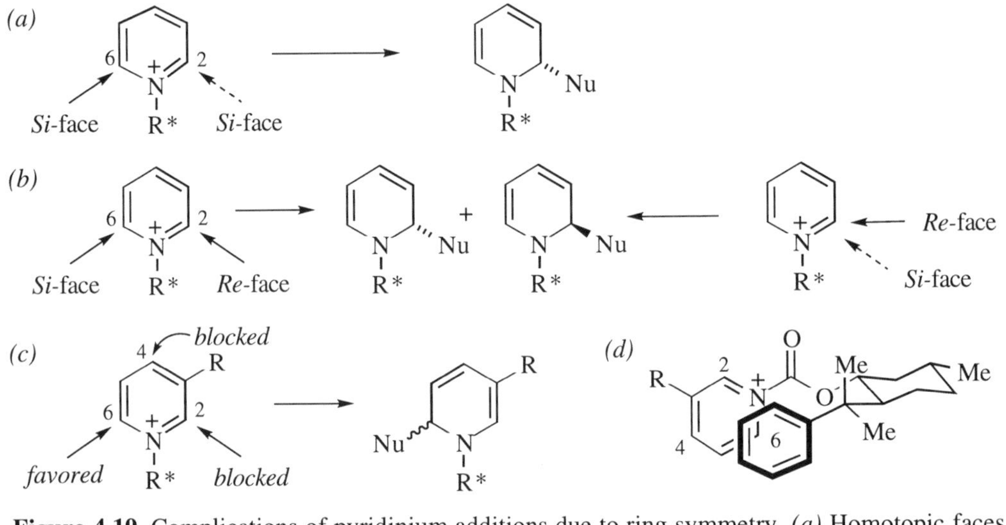

Figure 4.19. Complications of pyridinium additions due to ring symmetry. *(a)* Homotopic faces of C-2 and C-6; *(b)* Equivalence of 100% selective addition to only the front face with no regioselectivity and 100% regioselectivity with no face selectivity; *(c)* A bulky group at C-3 simplifies the situation by blocking attack at C-2 (and coincidentally C-4); *(d)* Comins's conformational model favoring *Re*-face (back side) attack at C-6 of an acylpyridinium ion [112].

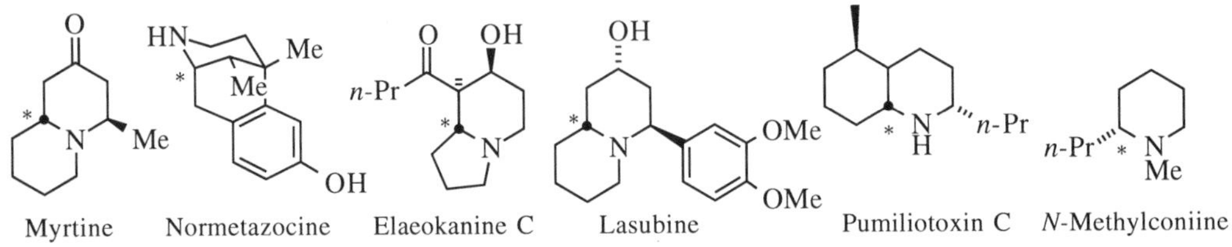

Myrtine   Normetazocine   Elaeokanine C   Lasubine   Pumiliotoxin C   *N*-Methylconiine

Figure 4.20. Alkaloids synthesized by asymmetric addition to chiral pyridiniums: myrtine ([116], normetazocine [100], elaeokanine C [117], lasubine [116], pumiliotoxin C [118], and *N*-methylconiine [113]. The stereocenter created in the addition reaction is indicated (∗).

Table 4.8. Asymmetric additions to chiral acylpyridinium esters. For the structure of R*, see Figure 4.19d.

$$\text{RMgX}$$

Entry	Educt	R	% Yield	% ds	Ref
1	Sn(*i*-Pr)$_3$ (CO$_2$R*)	*n*-Pr	72	91	[113]
2	"	*c*-C$_6$H$_{11}$	81	95	[113]
3	"	PhCH$_2$	58	88	[113]
4	"	CH$_2$=CH	71	95	[113]
5	"	Ph	85	94	[113]
6	OMe Si(*i*-Pr)$_3$ (CO$_2$R*)	Me	92	95	[112]
7	"	*i*-Bu	95	96	[112]
8	"	*c*-C$_6$H$_{11}$	90	90	[112]
9	"	Ph	88	96	[112]

4.4 Conjugate additions[14]

Two strategies have been used for asymmetric 1,4-additions: those that are based on a chiral auxiliary that is covalently attached to one of the reactants, and those that rely on chiral ligands on the metal (reviews: [120-122]). As yet the former afford the higher selectivities, but progress is being made in the development of the latter, which has the most potential for cost effectiveness via chiral catalysis. The following discussion is organized by electrophile.

4.4.1 Esters

Since esters exhibit a strong preference for a conformation in which an alkoxy C–H is synperiplanar to the carbonyl, the job of the auxiliary is to then project an appendage back over the enoate π-system, leaving only one face open to attack by a nucleophile. Figure 4.21 illustrates three of the more selective auxiliaries for this purpose. These auxiliaries are illustrated with the esters in their most stable conformations, with the alkoxy C–H and the carbonyl synperiplanar, and the enoate

---

[14] For a monograph on conjugate additions, see ref. [119].

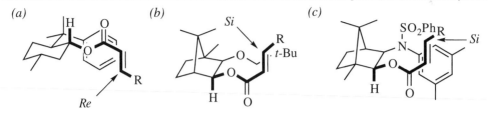

Figure 4.21. Chiral auxiliaries for asymmetric 1,4-addition to (the illustrated front face) of esters. Note the C–H/C=O coplanarity and the *s*-trans enone in the illustrated ground state conformations. *(a)* [123,124]; *(b)* [124]; *(c)* [125].

in the *s*-trans conformation. Presumably the ground state preference for this conformation is also felt in the transition state, which has the rear face shielded. Table 4.9 lists several examples of asymmetric additions to *E*-enoates. Less success has been realized in asymmetric additions to *Z*-enoates and to di- and trisubstituted double bonds.

Interligand asymmetric induction is observed in the 1,4-addition of certain organolithiums to hindered aryl esters in the presence of a chiral ligand. For example, Tomioka has shown that a chiral diether ligand affords affords good to excellent enantioselectivities in the conjugate additions of aryllithiums to the BHA esters shown in Scheme 4.10 [126]. Addition of butyllithium is much less selective, but similar selectivities can be achieved in aryllithium additions to BHA esters of 2-naphthoic acid. The additions are about 10-20% less selective when the ligand is used in catalytic quantities (10-20 mol%), but control experiments showed that the ligand accelerates the addition when the reaction is conducted in toluene.

Table 4.9. Asymmetric 1,4-addition to unsaturated esters of chiral alcohols. Numbers in the OR* column refer to Figure 4.21.

Entry	R	OR*	Nucleophile	% Yield	% es	Ref
1	Me	4.21a	PhCuBF$_3$	76	>99	[123]
2	"	4.21c	"	97	>99	[125]
3	"	4.21c	VinylCuBF$_3$	94	>99	[125]
4	"	4.21c	EtCuBF$_3$	90	>99	[125]
5	"	4.21a	*n*-BuCuBF$_3$	75	>99	[123]
6	"	4.21b	Me$_2$C=C(CH$_2$)$_2$-CuP(*n*-Bu)$_3$BF$_3$	81	99	[124]
7	"	4.21c	*i*-PrCuBF$_3$	92	>99	[125]
8	Et	4.21c	MeCuBF$_3$	86	99	[125]
9	*n*-Bu	4.21a	MeCuP(*n*-Bu)$_3$BF$_3$	96	93	[124]
10	"	4.21b	"	82	97	[124]
11	*n*-C$_8$H$_{17}$	4.21b	"	90	99	[124]
12	*i*-Pr	4.21c	MeCuBF$_3$	92	>99	[125]

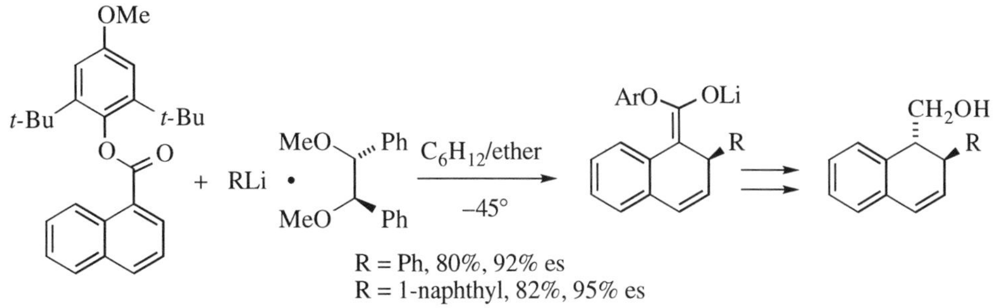

Scheme 4.10. Ligand induced asymmetric addition to naphthoic acid BHA (butylated hydroxy anisole esters) [126].

Figure 4.22 illustrates several natural products synthesized using auxiliary-modified esters. Particularly noteworthy is the ability of the method to produce the correct relative and absolute configuration of the alkyl branches on these acyclic frameworks. The illustrated structure for norpectinatone is the one originally postulated [127], but was proven incorrect by asymmetric synthesis [128].

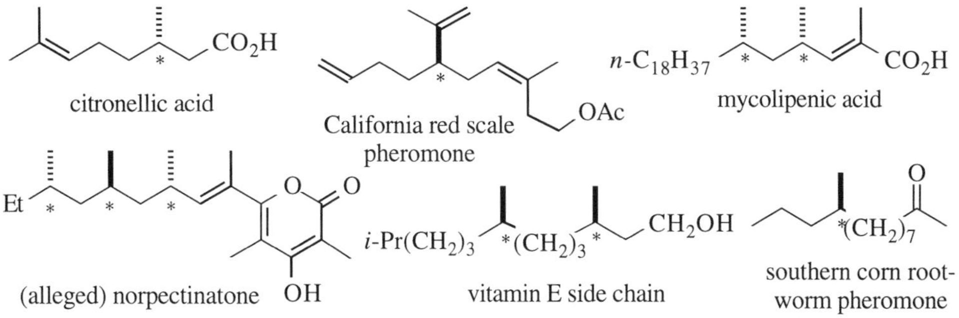

Figure 4.22. Natural products synthesized by asymmetric 1,4-addition of cuprates to esters: citronellic acid [124]; California red scale pheromone [129]; mycolipenic acid (W. Oppolzer; T. Godel, unpublished, quoted in [130]); the alleged norpectinatone [128]; vitamin E side chain (W. Oppolzer; R. Moretti, unpublished, quoted in [130]); southern corn rootworm pheromone [131]. Stereocenters created in the asymmetric conjugate addition are marked (*).

4.4.2 Amides and imides

A number of amides have been screened for their selectivity in conjugate additions of organometallics to acyclic enamides [120]. Two of the more useful auxiliaries are illustrated in Scheme 4.11. Both systems add Grignard reagents with considerable selectivity. Mukaiyama's ephedrine amides (Scheme 4.11a) require excess Grignard, and work best with organomagnesium *bromides* [132]. Oppolzer's sultam imide (Scheme 4.11b) offers several useful features [133]: in addition to the usual crystallinity of camphor derivatives (helpful for purification and diastereomer enrichment), the enolate may be alkylated (recall Scheme 3.18 and Table 3.7) with 87-88% selectivity for one of the four possible α,β-disubstituted stereoisomers. Additionally, 2-methacryloyl sultams can be protonated with a high degree of selectivity, giving 2-methyl-3,3-dialkyl amides of >97% purity [133].

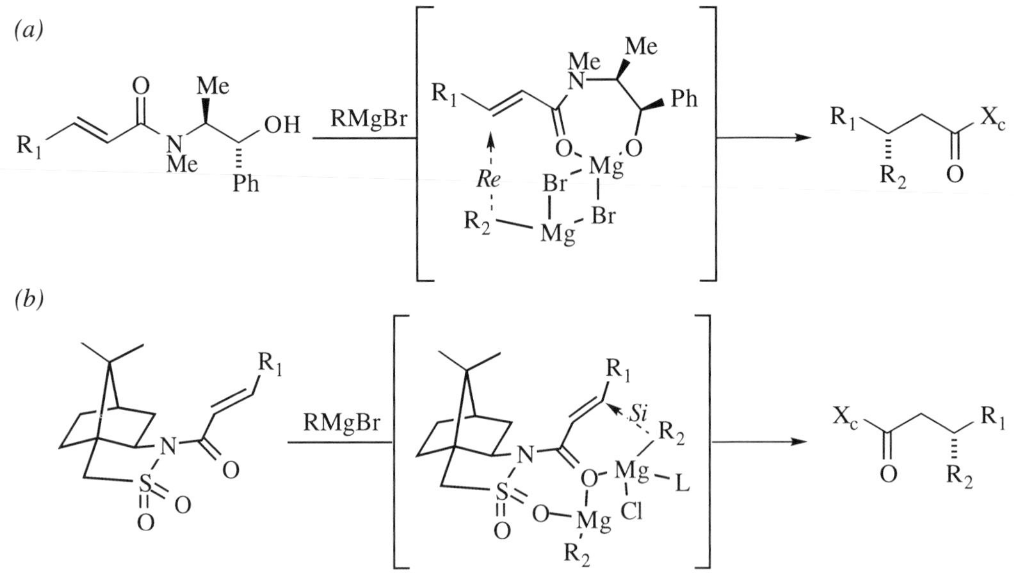

(a)

(b)

Scheme 4.11. Auxiliaries for the asymmetric 1,4-addition of Grignards to acyclic amides [132,133].

The transition structures illustrated in Scheme 4.11 have been proposed by the authors to account for the absolute configuration of the major product. Note that both groups invoke aggregation of the nucleophile with a magnesium species chelated by the enone carbonyl and a heteroatom on the auxiliary. This chelation reduces conformational motion in the ground state as well as the transition state, and reduces the possible number of competing nucleophile approach trajectories. For the ephedrine amides, the stereocenters on the auxiliary are quite remote from the site of attack. Although attack on the face opposite the methyl and phenyl groups in this chelate (as drawn) accounts for the configuration of the product, it is not clear how this steric effect is transmitted across the metallocycle chelate to the external double bond. It may be that the methyl and phenyl substituents induce a

Table 4.10. Asymmetric 1,4-additions to enamides (auxiliaries illustrated in Scheme 4.11).

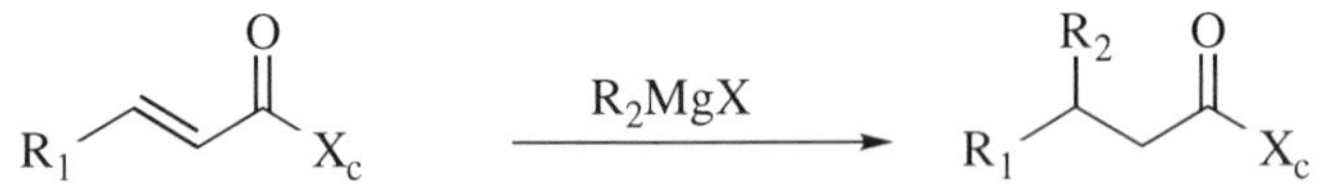

Entry	R_1	R_2MgX	Auxiliary	% Yield	% ds	Ref
1	Me	PhMgBr	a	63	95	[132]
2	Me	EtMgBr	a	79	98	[132]
3	Ph	"	a	48	98	[132]
4	Et	PhMgBr	a	76	93	[132]
5	Et	$n\text{-}C_4H_9MgBr$	a	59	79	[132]
6	$n\text{-}C_4H_9$	EtMgBr	a	69	99	[132]
7	Me	EtMgCl	b	80	94	[133]
8	Me	$i\text{-}Pr$	b	92	86	[133]
9	Et	$n\text{-}C_4H_9MgCl$	b	89	95	[133]

curved shape to the chelate ring that favors approach from the convex face, or perhaps the substrate is an aggregate of unknown structure. For the sultam (Scheme 4.11b), the situation is more clear: the bridge methyls of the camphor hinder approach from the *Re* face, similar to the situation with enolate alkylation of the same auxiliary (Scheme 3.18). Table 4.10 lists several examples of additions to these auxiliaries.

4.4.3 Dioxinones.

Incorporation of an auxiliary into a cyclic system has been used for the diastereoselective addition of cuprates to unsaturated 6-membered ring dioxinones, which are perhaps less important for their synthetic potential than for the mechanistic insight they provide. The dioxinones shown in Scheme 4.12a were obtained from *R*-3-hydroxybutanoic acid using the "self-regeneration of chirality centers" concept discussed in Chapter 3 (*cf.,* Scheme 3.9 and 3.10). After the addition, hydrolytic removal of the "achiral auxiliary" (pivaldehyde) liberates a 3-alkyl-3-hydroxybutyrate that is essentially enantiomerically pure [134].

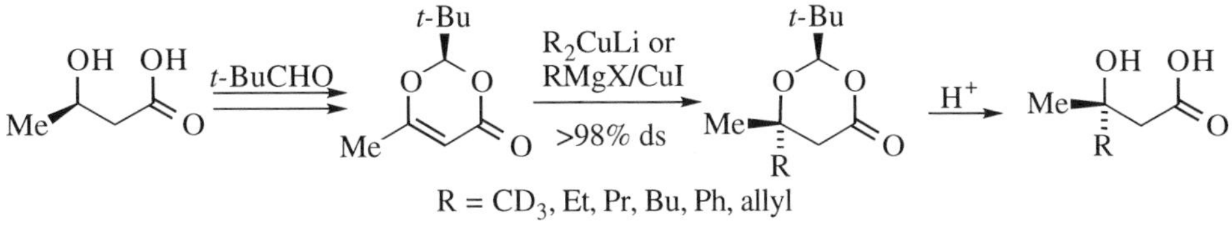

R = CD$_3$, Et, Pr, Bu, Ph, allyl

Scheme 4.12. Asymmetric conjugate addition of cuprates to dioxinones [134].

The additions are all >98% diastereoselective (the limit of detection), which is surprising since the dioxinone ring is in a sofa conformation, with only the acetal carbon significantly out of plane, leaving approach from either face essentially unhindered (recall the low selectivities for alkylation of *t*-butylcyclohexanone enolates, Scheme 3.7). Interestingly, examination of a number of X-ray crystal structures revealed that dioxinone acetals such as these have the common feature of pyramidalized carbonyl and β-carbon atoms [134]. Empirically, additions occur from the direction of the β-carbon's pyramidalization (see also ref. [135]). The reason for the pyramidalization in the substrate is the relief of torsional strain (however, calculations indicated that the energy required to flatten the pyramidal atoms is very small, ~0.1 kcal/mole). Seebach suggests [134] that approach of the nucleophile from the direction of pyramidalization should minimize the strain even more (see also ref. [15]). Since the reaction is kinetically controlled, and the selectivity is therefore determined in the transition state ($\Delta\Delta G^{\ddagger}$), this hypothesis (which is based on ground state arguments) may seem a risky infringement of the Curtin-Hammet principle [13,14]. Nevertheless, the strain that produces the pyramidalization (ΔG for the flat and pyramidal geometries) in the ground state and the energy differences in the transition state ($\Delta\Delta G^{\ddagger}$) have the same origin, and approach from the direction of pyramidalization relieves the strain while approach from the opposite direction increases it (Figure 4.23a). Thus, the energy difference between the two pyramidal ground states is *amplified* in the transition state (see

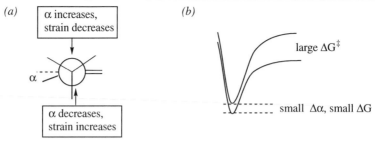

Figure 4.23. *(a)* Schematic showing how torsional strain is affected by the direction of attack on a pyramidalized trigonal center. *(b)* Linear perturbation of a Morse function that produces small distortions in the ground state can lead to large energy differences in the transition structure. (After ref. [134]).

also the two Morse curves in Figure 4.23b). Seebach also noted that the pyramidalization was evident in a computed model structure, which (since X-ray structural information is not always available) makes the following hypothesis all the more valuable: *"The steric course of attack on a trigonal center can be predicted from the direction of its pyramidalization"* [134].

4.4.4 Azomethines

The conjugate addition of organometallics to unsaturated azomethines (in the form of oxazolines, Scheme 4.13a) was one of the first carbon-carbon bond forming reactions that proceeded with >95% enantioselectivity [136-138] (review: [139]). The proposed mechanistic rationale [137,140] has the alkyllithium coordinating to the lone pair of the oxazoline nitrogen, and chelated by the methoxy group at the 4-position.[15] The alkyl group of the organometallic is oriented away from the side of the 4-substituent, and transfer occurs from the *Si* face. This alkyl transfer is reminiscent of a symmetry allowed [141] suprafacial 1,5-sigmatropic rearrangement [142]. Early on, the Meyers group showed that the 5-phenyl substituent had little effect on the selectivity [137] (see also ref. [142]); more recently [140,143], they have shown that a chelating group at the 4-position is not necessary either (Scheme 4.13b). The conditions necessary to hydrolyze the robust oxazoline nucleus initially limited the usefulness of this method, but subsequent work [142] has shown that the oxazoline may be alkylated with methyl triflate and reduced to an oxazolidine (in one pot), which is then easily hydrolyzed to an aldehyde.

The early (1975) contributions from the Meyers laboratory (Scheme 4.13a) paved the way for a number of related methods in subsequent years. Figure 4.24 illustrates a number of conceptually related conjugate additions. In several of these examples, there is a crucial difference from the examples in Scheme 4.13: in all except Figure 4.24e the α-carbon is prochiral, and two stereocenters are formed in the reaction. Fortunately, it is possible to either alkylate or protonate the azaenolate stereoselectively, such that two new stereocenters are produced in a single

---

[15] An alternative transition structure, placing the lithium on the π-cloud of the oxazoline, has also been proposed [140].

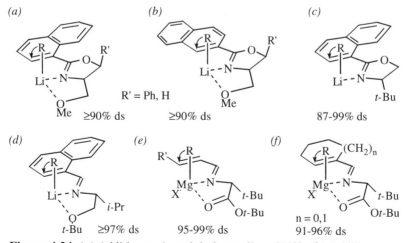

Scheme 4.13. Asymmetric addition of organolithiums to oxazolines: (*a*) [136,137]; (*b*) [140]; (*c*) Tandem asymmetric addition and alkylation of naphthalenes.

operation. Depending on the method, the two alkyl groups may be introduced in either a cis or a trans fashion. For example, the naphthalene oxazolines (Figure 4.24a-c) alkylate trans to the first alkyl group, whereas the cycloalkenyl imines (Figure 4.24f) may alkylate either cis or trans selectively, depending on the method used. A generalized example (for 1-naphthalenes) is shown in Scheme 4.13c.

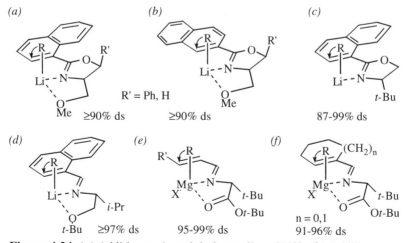

Figure 4.24. (*a*) Addition to 1-naphthyloxazolines [142]; (*b*) Addition to 2-naphthyloxazolines [142]; (*c*) addition to 1-naphthyloxazolines lacking a chelating group [144]; (*d*) addition to 1-naphthaldehyde imines [145]; (*e*) addition to crotyl amino acid imines [146,147]; (*f*) addition to cyclohexene and cyclopentene aldehyde amino acids imines [148].

When the site of nucleophilic attack has an alkoxy substituent, the azaenolate adduct undergoes spontaneous elimination of alkoxide. Since aryllithiums add efficiently to 2-alkoxyaryloxazolines, the process may be used in an asymmetric synthesis of chiral biaryls. Two strategies for auxiliary-based asymmetric induction have been evaluated: using an oxazoline as a chiral auxiliary [149-152], and using a chiral alkoxide (leaving group) as an auxiliary [153]. Scheme 4.14 illustrates several examples. Note that here again, the early reports used an oxazoline that contains a chelating substituent (Scheme 4.14a,b), but later reports indicated that a bulky substituent at the oxazoline 4-position will suffice (Scheme 4.14c). Figure 4.25 illustrates several natural products that have been made using asymmetric addition to unsaturated azomethines.

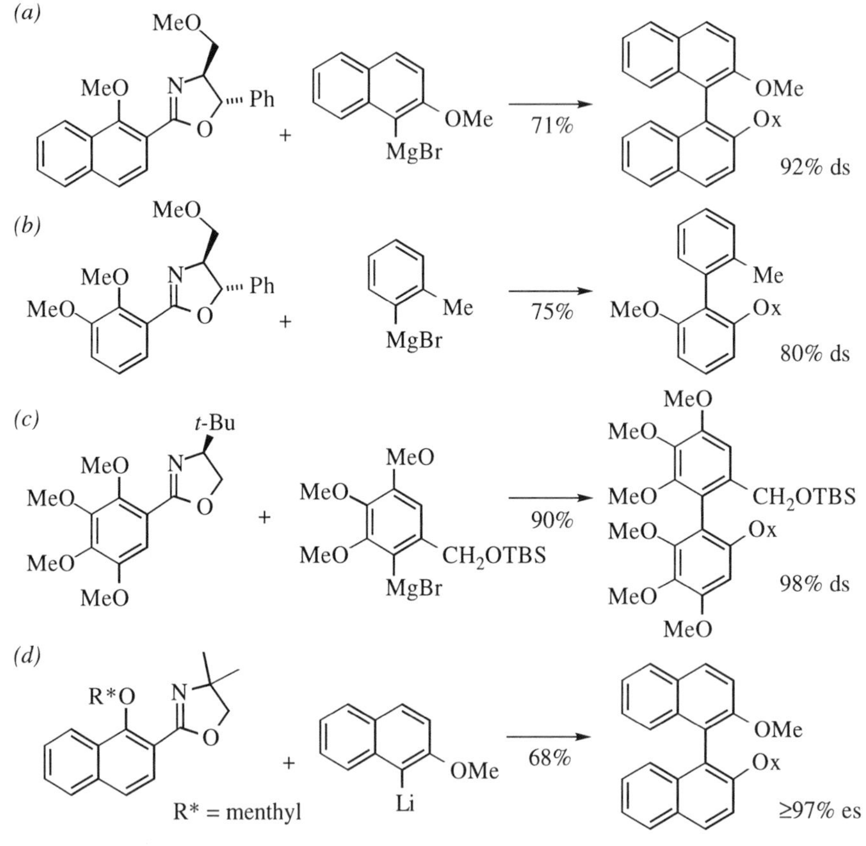

Scheme 4.14. Asymmetric synthesis of biaryls: *(a)* binaphthyls using a chelating oxazoline [149]; *(b)* biphenyls using a chelating oxazoline [150]; *(c)* biphenyls using a nonchelating oxazoline [151]; *(d)* binaphthyls using a chiral leaving group [153].

An interesting development in asymmetric additions to azomethines employs chiral ligands (chelating agents) on the organometallic. Tomioka has shown that the same ligand used for addition of aryllithiums to unsaturated esters (*cf.* Scheme 4.10) also works for unsaturated imines, as illustrated in Scheme 4.15 [154].[16] The

---

[16] For a related study of the stereoselectivity of addition/alkylation to cyclohexylimines in the racemic series, see ref. [155].

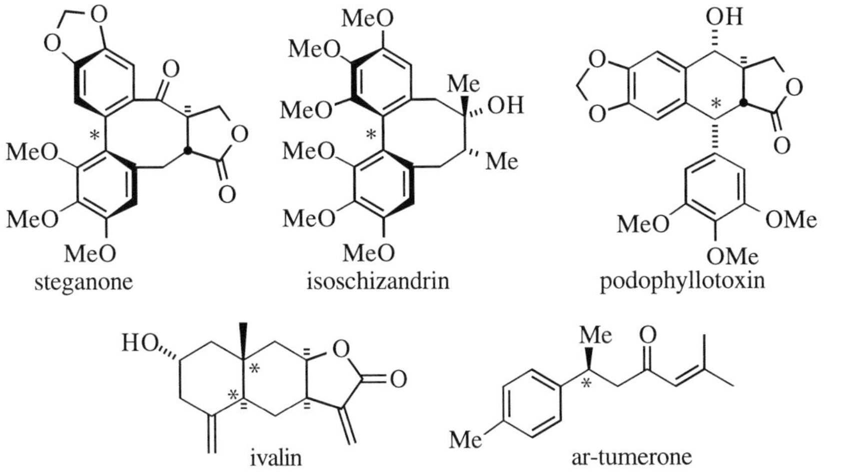

steganone   isoschizandrin   podophyllotoxin

ivalin   ar-tumerone

Figure 4.25. Natural products synthesized using asymmetric addition to chiral azomethines: steganone [156], isoschizandrin [157], podophyllotoxin [158], ivalin [159], ar-tumerone [160]. The stereogenic units formed in the conjugate addition step are marked (*).

C_2-symmetric ether is thought to chelate the alkyllithium (and thereby break up the alkyllithium aggregate), which then coordinates to the azomethine nitrogen. Note that the phenyls force the methyls into a conformation that places each of them trans to the neighboring phenyl in the chelate (see inset). Upon complexation of the azomethine to the vacant site on the lithium, the large cyclohexyl is oriented into the vacant quadrant of the chelate, as shown.[17] Suprafacial transfer of the alkyl group then gives the observed product with enantioselectivities above 90%, and usually in the 97-99% range. Examples include crotyl, cycloalkenyl, and 1-naphthyl imines [154]. Using the same ether and the 2,6-diisopropylphenyl imine of 1-fluoro-2-naphthaldehyde, a chiral binaphthyl is formed by asymmetric addition of 1-naphthyllithium in >99% yield and 95% es [161].

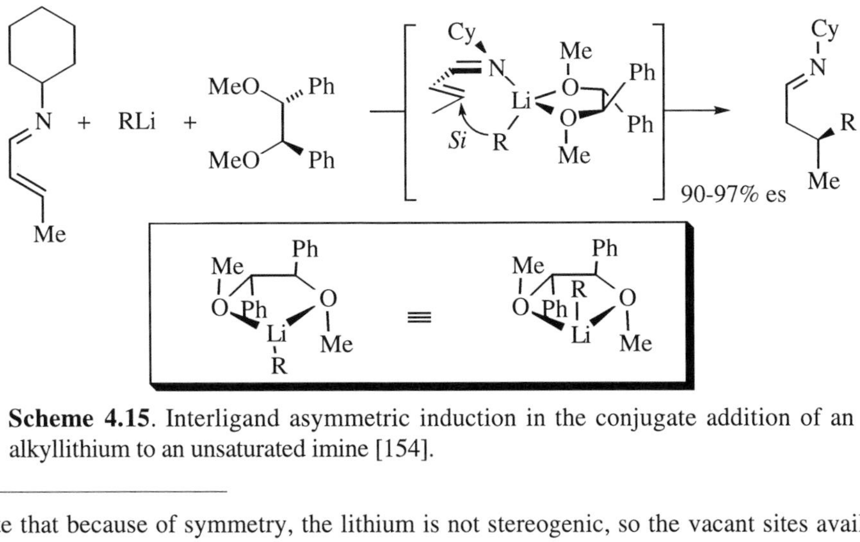

Scheme 4.15. Interligand asymmetric induction in the conjugate addition of an alkyllithium to an unsaturated imine [154].

---

[17] Note that because of symmetry, the lithium is not stereogenic, so the vacant sites available by inversion of the tetrahedral lithium are equivalent.

4.4.5 Ketones and lactones

In addition to the "self-regeneration of chirality" principle discussed in Section 4.4.3, strategies for the asymmetric 1,4-addition to enones have included both auxiliary-based methods and interligand asymmetric induction. The most fully developed auxiliary method is Posner's use of vinyl sulfoxides, illustrated in Scheme 4.16 (reviews: [162-164]). The sulfur atom is pyramidal, and therefore stereogenic. The method is most useful with 5-membered α,β-unsaturated ketones and lactones (butenolides), and may employ strategies in which chelates are involved (Scheme 4.16a) or not (Scheme 4.16b), as illustrated. Zinc bromide is the most effective for chelating the sulfoxide and carbonyl oxygens. In the 'non-chelate' strategy, dialkylmagnesiums are used as the organometallic, sometimes in the presence of crown ethers. The authors' mechanistic rationale has the organometallic adding from the side that is opposite the aryl group of the sulfoxide in either the chelate or opposing-dipole (non-chelate) conformations; improved selectivities resulted when anisyl sulfoxides were used in place of tosyl [165], and when the poorly coordinating dimethyltetrahydrofuran was used as solvent. Figure 4.26 illustrates several natural products synthesized using this method.

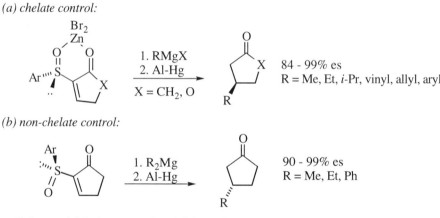

(a) chelate control:

1. RMgX
2. Al-Hg

X = CH₂, O

84 - 99% es
R = Me, Et, *i*-Pr, vinyl, allyl, aryl

(b) non-chelate control:

1. R₂Mg
2. Al-Hg

90 - 99% es
R = Me, Et, Ph

Scheme 4.16. Asymmetric addition of organomagnesiums to vinylic sulfoxides [162-164].

Schultz has reported a conceptually related method, which affords higher selectivities for cyclohexenones than is possible with the sulfoxide method, as shown in Scheme 4.17 [166]. The 2-carboxamidecyclohexenones are prepared by Birch reduction and hydrolysis of the 2-methoxybenzamide. Conjugate addition of Grignard reagents in the presence of Lewis acids, affords good yields of addition products with high selectivities for most nucleophiles (allyl is the notable exception). Presumably, the stereochemical rationale is similar, with the Lewis acid chelating the two carbonyls and the nucleophile approaching from the face opposite the methoxymethyl. Hydrolysis of the auxiliary and decarboxylation affords the 3-substituted cyclohexanones.

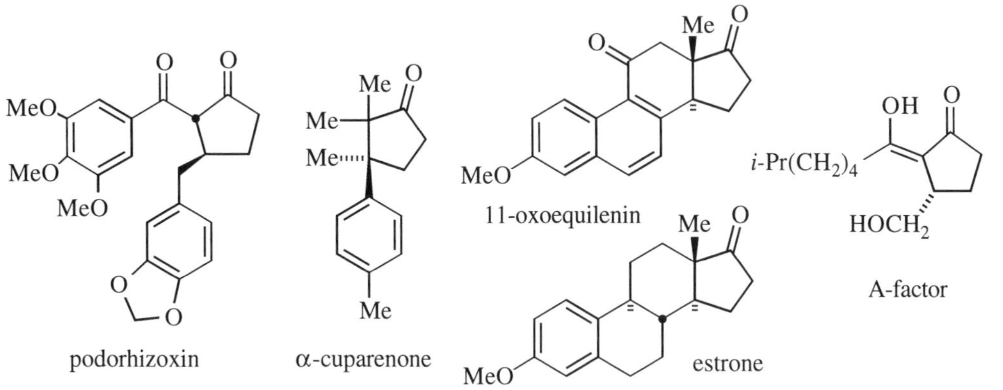

Figure 4.26. Natural products synthesized by 1,4-addition to unsaturated cyclopentenones and butenolides: podorhizoxin [167], α-cuparenone [168], 11-oxoequilenin [169], estrone [169,170], A-factor [163].

The asymmetric addition of cuprates to achiral cycloalkenones using a chiral ligand on the metal (interligand asymmetric induction) has been studied extensively,[18] but obtention of uniformly high yields with a variety of substrates and nucleophiles has not been achieved because the selectivity is dependent on a number of factors, including substrate and cuprate structure, solvent, concentration, temperature, and the presence of added salts. Two of the more highly selective ligands are illustrated in Scheme 4.18, and a mechanistic rationale for the first is also shown. Unfortunately, these processes are reported to be hypersensitive to the presence of impurities in the reaction mixture. In the first example (Scheme 4.18a), the presence of alkoxides in the alkyllithium diminishes the selectivity, and methyl iodide must be added to the recipe as an alkoxide scavenger [171]. A related approach uses a phosphine ligand, but the selectivity of these additions are highly dependent on the source of the copper [172]. The second example (Scheme 4.18b) is an optimized procedure for the asymmetric synthesis of muscone [173].

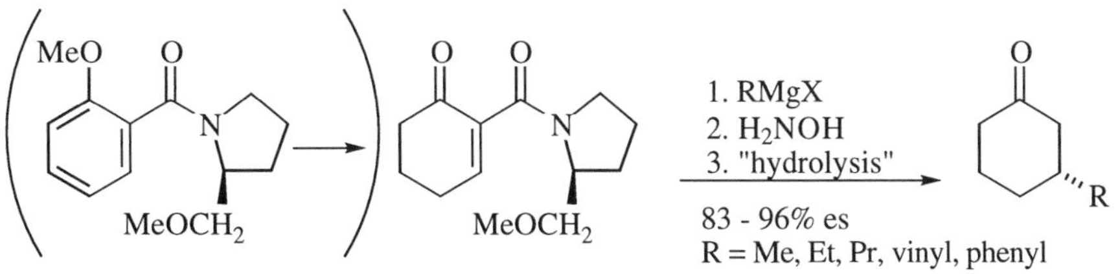

Scheme 4.17. Asymmetric addition of Grignards to cyclohexenones [166].

---

[18] For a survey of the ligands tested, see ref. [120].

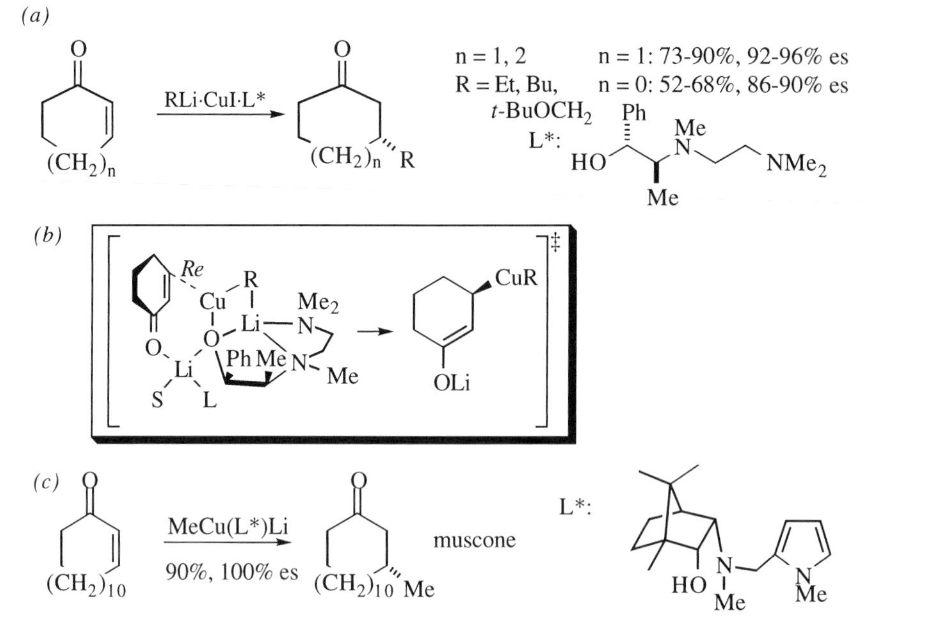

(a)

n = 1, 2   n = 1: 73-90%, 92-96% es
R = Et, Bu,   n = 0: 52-68%, 86-90% es
t-BuOCH$_2$ Ph
L*:

(b)

(c)

Scheme 4.18. (a) Asymmetric addition of cuprates to cycloalkenones [171]. (b) Mechanistic rationale for *a* [171]. (c) Asymmetric synthesis of muscone [173].

4.5 References

1.   D. Y. Curtin; E. E. Harris; E. K. Meislich *J. Am. Chem. Soc.* **1952**, *74*, 2901-2904.
2.   D. J. Cram; F. A. A. Elhafez *J. Am. Chem. Soc.* **1952**, *74*, 5828-5835.
3.   V. Prelog *Helv. Chim. Acta* **1953**, *36*, 308-319.
4.   J. D. Morrison; H. S. Mosher *Asymmetric Organic Reactions*; Prentice-Hall: Englewood Cliffs, NJ, 1971.
5.   E. Fischer *Chem. Ber.* **1894**, *27*, 3189-3232, see p. 3210.
6.   D. J. Cram; K. R. Kopecky *J. Am. Chem. Soc.* **1959**, *81*, 2748-2755.
7.   J. W. Cornforth; R. H. Cornforth; K. K. Matthew *J. Chem. Soc.* **1959**, 112-127.
8.   G. J. Karabatsos; N. Hsi *J. Am. Chem. Soc.* **1965**, *87*, 2864-2870.
9.   G. J. Karabatsos; R. A. Taller *Tetrahedron* **1968**, *24*, 3923-3937.
10. G. J. Karabatsos *J. Am. Chem. Soc.* **1967**, *89*, 1367-1371.
11. G. S. Hammond *J. Am. Chem. Soc.* **1955**, *77*, 334-338.
12. G. J. Karabatsos; D. J. Fenoglio In *Topics in Stereochemistry*; E. L. Eliel, N. L. Allinger, Eds.; Wiley-Interscience: New York, 1970; Vol. 5, p 167-203.
13. D. Y. Curtin *Rec. Chem. Progr.* **1954**, *15*, 111-128.
14. J. I. Seeman *Chem. Rev.* **1983**, *83*, 83-134.
15. M. Chérest; H. Felkin *Tetrahedron Lett.* **1968**, 2205-2208.
16. M. Chérest; H. Felkin; N. Prudent *Tetrahedron Lett.* **1968**, 2199-2204.
17. N. T. Anh *Topics in Current Chemistry* **1980**, *88*, 145-162.
18. G. Frenking; K. F. Köhler; M. T. Reetz *Angew. Chem. Int. Ed. Engl.* **1991**, *30*, 1146-1149.
19. X. L. Huang; J. J. Dannenberg *J. Am. Chem. Soc.* **1993**, *115*, 6017-6024.
20. D. Seebach *Angew. Chem. Int. Ed. Engl.* **1990**, *29*, 1320-1367.
21. H. B. Bürgi; J. D. Dunitz; E. Schefter *J. Am. Chem. Soc.* **1973**, *95*, 5065-5067.

22. H. B. Bürgi; D. Dunitz; J. M. Lehn; G. Wipff *Tetrahedron* **1974**, *30*, 1563-1572.

23. N. T. Anh; O. Eisenstein *Nouv. J. Chimie* **1977**, *1*, 61-70.

24. C. H. Heathcock; L. A. Flippin *J. Am. Chem. Soc.* **1983**, *105*, 1667-1668.

25. E. P. Lodge; C. H. Heathcock *J. Am. Chem. Soc.* **1987**, *109*, 2819-2820.

26. S. Scheiner; W. N. Lipscomb; D. A. Kleier *J. Am. Chem. Soc.* **1976**, *98*, 4770-4777.

27. C. H. Heathcock *Aldrichimica Acta* **1990**, *23*, 99-111.

28. E. P. Lodge; C. H. Heathcock *J. Am. Chem. Soc.* **1987**, *109*, 3353-3361.

29. J. March In *Advanced Organic Chemistry, 4th ed.*; Wiley-Interscience: New York, 1992, p 145.

30. G. Frenking; K. F. Köhler; M. T. Reetz *Tetrahedron* **1991**, *47*, 9005-9018.

31. G. Frenking; K. F. Köhler; M. T. Reetz *Tetrahedron* **1991**, *47*, 8991-9004.

32. G. Frenking; K. F. Köhler; M. T. Reetz *Tetrahedron* **1993**, *49*, 3971-3982.

33. G. Frenking; K. F. Köhler; M. T. Reetz *Tetrahedron* **1993**, *49*, 3983-3994.

34. X. Chen; E. R. Hortelano; E. L. Eliel; S. V. Frye *J. Am. Chem. Soc.* **1992**, *114*, 1778-1784.

35. E. L. Eliel In *Asymmetric Synthesis*; J. D. Morrison, Ed.; Academic: Orlando, 1983; Vol. 2, p 125-155.

36. M. T. Reetz *Angew. Chem. Int. Ed. Engl.* **1984**, *23*, 556-569.

37. W. C. Still; J. H. McDonald, III *Tetrahedron Lett.* **1980**, *21*, 1031-1034.

38. W. C. Still; J. A. Schneider *Tetrahedron Lett.* **1980**, *21*, 1035-1038.

39. K.-Y. Ko; E. L. Eliel *J. Org. Chem.* **1986**, *51*, 5353-5362.

40. G. E. Keck; M. B. Andrus; D. R. Romer *J. Org. Chem.* **1991**, *56*, 417-420.

41. M. T. Reetz *Acc. Chem. Res.* **1993**, *26*, 462-468.

42. B. Giese *Angew. Chem. Int. Ed. Engl.* **1977**, *16*, 125-136.

43. D. B. Collum; J. H. McDonald; W. C. Still *J. Am. Chem. Soc.* **1980**, *102*, 2118-2120.

44. D. B. Collum; J. H. McDonald; W. C. Still *J. Am. Chem. Soc.* **1980**, *102*, 2120-2121.

45. T. Nakata; Y. Kishi *Tetrahedron Lett.* **1978**, 2745-2748.

46. T. Nakata; G. Schmid; B. Vranesic; M. Okigawa; T. Smith-Palmer *J. Am. Chem. Soc.* **1978**, *100*, 2933-2935.

47. S. V. Frye; E. L. Eliel; J. E. Lynch; F. Kume *Organic Syntheses* **1985**, *65*, 215-223.

48. E. L. Eliel; S. Morris-NAtschke *J. Am. Chem. Soc.* **1984**, *106*, 2937-2942.

49. J. E. Lynch; E. L. Eliel *J. Am. Chem. Soc.* **1984**, *106*, 2943-2948.

50. X. Bai; E. L. Eliel *J. Org. Chem.* **1992**, *57*, 5166-5172.

51. X.-C. He; E. L. Eliel *Tetrahedron* **1987**, *43*, 4979-4987.

52. X.-C. He; E. L. Eliel *J. Org. Chem.* **1990**, *55*, 2114-2119.

53. M. Ohwa; T. Kogure; E. L. Eliel *J. Org. Chem.* **1986**, *51*, 2599-2601.

54. S. V. Frye; E. L. Eliel *Tetrahedron Lett.* **1985**, *26*, 3907-3910.

55. T. Kogure; E. L. Eliel *J. Org. Chem.* **1984**, *49*, 576-578.

56. S. V. Frye; E. L. Eliel *J. Org. Chem.* **1985**, *50*, 3402-3404.

57. G. Solladié In *Asymmetric Synthesis*; J. D. Morrison, Ed.; Academic: Orlando, 1983; Vol. 2, p 157-199.

58. R. Noyori; M. Kitamura *Angew. Chem. Int. Ed. Engl.* **1991**, *30*, 49-69.

59. R. O. Duthaler; A. Hafner *Chem. Rev.* **1992**, *92*, 807-832.

60. K. Soai; S. Niwa *Chem. Rev.* **1991**, *92*, 833-856.

61. R. Noyori *Asymmetric Catalysis in Organic Synthesis*; Wiley-Interscience: New York, 1994.

62. M. T. Reetz; M. Hüllman; W. Massa; S. Berger; P. Rademacher; P. Heymanns *J. Am. Chem.*

Soc. **1986**, *108*, 2405-2408.

63. S. E. Denmark; N. G. Almstead *J. Am. Chem. Soc.* **1993**, *115*, 3133-3139.

64. D. P. Klein; J. A. Gladysz *J. Am. Chem. Soc.* **1992**, *114*, 8710-8711.

65. D. A. Evans *Science* **1988**, *240*, 420-426.

66. S. L. Blystone *Chem. Rev.* **1989**, *89*, 1663-1679.

67. K. Tomioka *Synthesis* **1990**, 541-549.

68. E. Erdik *Tetrahedron* **1992**, *48*, 9577-9648.

69. T. Sato; K. Soai; K. Suzuki; T. Mukaiyama *Chem. Lett.* **1978**, 601-604.

70. T. Mukaiyama; K. Soai; T. Sato; H. Shimizu; K. Suzuki *J. Am. Chem. Soc.* **1979**, *101*, 1455-1460.

71. M. Kitamura; S. Okada; S. Suga; R. Noyori *J. Am. Chem. Soc.* **1989**, *111*, 4028-4036.

72. K. Soai; S. Yokoyama; T. Hayasaka *J. Org. Chem.* **1991**, *56*, 4254-4268.

73. S. Itsuno; J. M. J. Fréchet *J. Org. Chem.* **1987**, *52*, 4140-4142.

74. E. J. Corey; F. Hannon *Tetrahedron Lett.* **1987**, *28*, 5233-5236.

75. M. Kitamura; S. Suga; M. Niwa; R. Noyori *J. Am. Chem. Soc.* **1995**, *117*, 4832-4842.

76. K. Soai; A. Ookawa; T. Kaba; K. Ogawa *J. Am. Chem. Soc.* **1987**, *109*, 7111-7115.

77. R. Noyori; S. Suga; K. Kawai; S. Okada; M. Kitamura; N. Oguni; M. Hayashi; T. Kaneko; Y. Matsuda *J. Organomet. Chem.* **1990**, *382*, 19-37.

78. N. N. Joshi; M. Srebnik; H. C. Brown *Tetrahedron Lett.* **1989**, *30*, 5551-5554.

79. T. Shono; N. Kise; E. Shirakawa; H. Matsumoto; E. Okazaki *J. Org. Chem.* **1991**, *56*, 3063-3067.

80. M. Watanabe; S. Araki; Y. Butsugan; M. Uemura *J. Org. Chem.* **1991**, *56*, 2218-2224.

81. M. Yoshioka; T. Kawakita; M. Ohno *Tetrahedron Lett.* **1989**, *30*, 1657-1660.

82. P. Knochel; W. Brieden; M. J. Rozema; C. Eisenberg *Tetrahedron Lett.* **1993**, *34*, 5881-5884.

83. D. Seebach; D. A. Plattner; A. K. Beck; Y. M. Wang; D. Hunziker; W. Petter *Helv. Chim. Acta* **1992**, *75*, 2171-2209.

84. Y. N. Ito; A. K. Beck; A. Boháč; C. Ganter; R. E. Gawley; F. N. M. Kühnle; J. A. Piquer; J. Tuleja; Y. M. Wang; D. Seebach *Helv. Chim. Acta* **1994**, *77*, 2071-2110.

85. S. B. Heaton; G. B. Jones *Tetrahedron Lett.* **1992**, *33*, 1693-1696.

86. K. Soai; Y. Kawase; A. Oshio *J. Chem. Soc., Perkin Trans. 1* **1991**, 1613-1615.

87. W. Oppolzer; R. N. Radinov *Tetrahedron Lett.* **1988**, *29*, 5645-5648.

88. J. L. v. d. Bussche-Hunnefeld; D. Seebach *Tetrahedron* **1992**, *48*, 5719-5730.

89. B. Weber; D. Seebach *Angew. Chem. Int. Ed. Engl.* **1992**, *31*, 84-86.

90. D. Guillaneaux; S.-H. Zhao; O. Samuel; D. Rainford; H. B. Kagan *J. Am. Chem. Soc.* **1994**, *116*, 9430-9439.

91. M. T. Reetz; F. Kunisch; P. Heitmann *Tetrahedron Lett.* **1986**, *27*, 4721-4724.

92. K. Narasaka *Synthesis* **1991**, 1-11.

93. H. Nitta; D. Yu; M. Kudo; A. Mori; S. Inoue *J. Am. Chem. Soc.* **1992**, *114*, 7969-7965.

94. M. Hayashi; Y. Miyamoto; T. Inoue; N. Oguni *J. Org. Chem.* **1993**, *58*, 1515-1522.

95. E. J. Corey; Z. Wang *Tetrahedron Lett.* **1993**, *34*, 4001-4004.

96. M. T. Reetz; S.-H. Kyung; C. Bolm; T. Zierke *Chem. Ind. (London)* **1986**, 824.

97. H. Minamikawa; S. Hayakawa; T. Yamada; N. Iwasawa; K. Narasaka *Bull. Chem. Soc. Jpn.* **1988**, *61*, 4379-4383.

98. R. A. Volkmann In *Comprehensive Organic Synthesis. Selectivity, Strategy, and Efficiency in Modern Organic Chemistry*; B. M. Trost, I. Fleming, Eds.; Pergamon: Oxford, 1991; Vol. 5, p

355-396.

99. P. Ermert; J. Meyer; C. Stucki *Tetrahedron Lett.* **1988**, *29*, 1265-1268.

100. R. M. Williams; P. J. Sinclair; D. Zhai; D. Chen *J. Am. Chem. Soc.* **1988**, *110*, 1547-1557.

101. D. Enders; H. Schubert; C. Nübling *Angew. Chem. Int. Ed. Engl.* **1986**, *25*, 1109-1110.

102. D. Enders; R. Funk; M. Klatt; G. Raabe; E. R. Hovestreydt *Angew. Chem. Int. Ed. Engl.* **1993**, *32*, 418-420.

103. D. Enders; D. Bartzen *Liebigs Ann. Chem.* **1993**, 569-574.

104. D. Enders; J. Schankat *Helv. Chim. Acta* **1993**, *76*, 402-406.

105. D. Enders; J. Tiebes *Liebigs Ann. Chem.* **1993**, 173-177.

106. D. Enders; M. Klatt; R. Funk *Synlett* **1993**, 226-228.

107. S. E. Denmark; T. Weber; D. W. Piotrowski *J. Am. Chem. Soc.* **1987**, *109*, 2224-2225.

108. T. Weber; J. P. Edwards; S. E. Denmark *Synlett* **1989**, *1*, 20-22.

109. S. E. Denmark; J. P. Edwards; O. Nicaise *J. Org. Chem.* **1993**, *58*, 569-578.

110. S. E. Denmark; O. Nicaise; J. P. Edwards *J. Org. Chem.* **1990**, *55*, 6219-6223.

111. Y. Génnison; C. Marzano; B. C. Das *J. Org. Chem.* **1993**, *58*, 2052-2057.

112. D. L. Comins; R. R. Goehring; S. P. Joseph *J. Org. Chem.* **1990**, *55*, 2574-2576.

113. D. L. Comins; H. Hong; J. M. Salvador *J. Org. Chem.* **1991**, *56*, 7197-7199.

114. D. L. Comins; J. M. Salvador *J. Org. Chem.* **1993**, *58*, 4656-4661.

115. D. L. Comins; S. P. Joseph; R. R. Goehring *J. Am. Chem. Soc.* **1994**, *116*, 4719-4728.

116. D. L. Comins; D. H. La Munyon *J. Org. Chem.* **1992**, *57*, 5807-5809.

117. D. L. Comins; H. Hong *J. Am. Chem. Soc.* **1991**, *113*, 6672-6673.

118. D. L. Comins; A. Dehghani *Tetrahedron Lett.* **1991**, *32*, 5697-5700.

119. P. Perlmutter *Conjugate Addition Reactions in Organic Synthesis*; Pergamon: Oxford, 1992.

120. B. E. Rossiter; N. M. Swingle *Chem. Rev.* **1992**, *92*, 771-806.

121. H.-G. Schmalz In *Comprehensive Organic Synthesis. Selectivity, Strategy, and Efficiency in Modern Organic Chemistry*; B. M. Trost, I. Fleming, Eds.; Pergamon: Oxford, 1991; Vol. 4, p 199-236.

122. K. Tomioka; K. Koga In *Asymmetric Synthesis*; J. D. Morrison, Ed.; Academic: Orlando, 1983; Vol. 2, p 201-224.

123. W. Oppolzer; H. J. Löher *Helv. Chim. Acta* **1981**, *64*, 2808-2811.

124. W. Oppolzer; R. Moretti; T. Godel; A. Meunier; H. Löher *Tetrahedron Lett.* **1983**, *24*, 4971-4974.

125. G. Helmchen; G. Wegner *Tetrahedron Lett.* **1985**, *26*, 6051-6054.

126. K. Tomioka; M. Shindo; K. Koga *Tetrahedron Lett.* **1993**, *34*, 681-682.

127. R. J. Capon; D. J. Faulkner *J. Org. Chem.* **1984**, *49*, 2506-2508.

128. W. Oppolzer; R. Moretti; G. Bernardinelli *Tetrahedron Lett.* **1986**, *27*, 4713-4716.

129. W. Oppolzer; T. Stevenson *Tetrahedron Lett.* **1986**, *27*, 1139-1140.

130. W. Oppolzer *Tetrahedron* **1987**, *43*, 1969-2004; correctly printed in erratum, p. 4057.

131. W. Oppolzer; P. Dudfield; T. Stevenson; T. Godel *Helv. Chim. Acta* **1985**, *68*, 212-215.

132. T. Mukaiyama; N. Iwasawa *Chem. Lett.* **1981**, 913-916.

133. W. Oppolzer; G. Poli; A. J. Kingma; C. Starkemann; G. Bernardinelli *Helv. Chim. Acta* **1987**, *70*, 2201-2214.

134. D. Seebach; J. Zimmerman; U. Gysel; R. Ziegler; T.-K. Ha *J. Am. Chem. Soc.* **1988**, *110*, 4763-4772.

135. A. I. Meyers; W. R. Leonard, Jr.; J. L. Romine *Tetrahedron Lett.* **1991**, *32*, 597-600.

136. A. I. Meyers; C. E. Whitten *J. Am. Chem. Soc.* **1975**, *97*, 6266-6267.

137. A. I. Meyers; R. K. Smith; C. E. Whitten *J. Org. Chem.* **1979**, *44*, 2250-2256.

138. F. E. Ziegler; P. J. Gilligan *J. Org. Chem.* **1981**, *46*, 3874-3880.

139. K. A. Lutomski; A. I. Meyers In *Asymmetric Synthesis*; J. D. Morrison, Ed.; Academic: Orlando, 1984; Vol. 3, p 213-273.

140. A. I. Meyers; M. Shipman *J. Org. Chem.* **1991**, *56*, 7098-7102.

141. R. B. Woodward; R. Hoffmann *The Conservation of Orbital Symmetry*; Academic: New York, 1970.

142. A. I. Meyers; G. P. Roth; D. Hoyer; B. A. Barner; D. Laucher *J. Am. Chem. Soc.* **1988**, *110*, 4611-4624.

143. D. J. Rawson; A. I. Meyers *J. Org. Chem.* **1991**, *56*, 2292-2294.

144. D. J. Dawson; A. I. Meyers *J. Org. Chem.* **1991**, *56*, 2292-2294.

145. A. I. Meyers; J. D. Brown; D. Laucher *Tetrahedron Lett.* **1987**, *28*, 5283-5286.

146. S. Hashimoto; S. Yamada; K. Koga *J. Am. Chem. Soc.* **1976**, *98*, 7450-7452.

147. S. Hashimoto; S. Yamada; K. Koga *Chem. Pharm. Bull.* **1979**, *27*, 771-782.

148. H. Kogen; K. Tomioka; S. Hashimoto; K. Koga *Tetrahedron* **1981**, *37*, 3951-3956.

149. A. I. Meyers; K. A. Lutomski *J. Am. Chem. Soc.* **1982**, *104*, 879-881.

150. A. I. Meyers; R. J. Himmelsbach *J. Am. Chem. Soc.* **1985**, *107*, 682-685.

151. A. I. Meyers; A. Meier; D. J. Rawson *Tetrahedron Lett.* **1992**, *33*, 853-856.

152. H. Moorlag; A. I. Meyers *Tetrahedron Lett.* **1993**, *34*, 6989-6992.

153. J. M. Wilson; D. J. Cram *J. Am. Chem. Soc.* **1982**, *104*, 881-884.

154. K. Tomioka; M. Shindo; K. Koga *J. Am. Chem. Soc.* **1989**, *111*, 8266-8268.

155. A. I. Meyers; J. D. Brown; D. Laucher *Tetrahedron Lett.* **1987**, *28*, 5279-5282.

156. A. I. Meyers; J. R. Flisak; R. A. Aitken *J. Am. Chem. Soc.* **1987**, *109*, 5446-5452.

157. A. M. Washawsky; A. I. Meyers *J. Am. Chem. Soc.* **1990**, *112*, 8090-8099.

158. R. C. Andrews; S. J. Teague; A. I. Meyers *J. Am. Chem. Soc.* **1988**, *110*, 7854-7858.

159. K. Tomioka; F. Masumi; T. Yamashita; K. Koga *Tetrahedron Lett.* **1984**, *25*, 333-336.

160. A. I. Meyers; R. K. Smith *Tetrahedron Lett.* **1979**, 2749-2752.

161. M. Shindo; K. Koga; K. Tomioka *J. Am. Chem. Soc.* **1992**, *114*, 8732-8733.

162. G. Posner In *Asymmetric Synthesis*; J. D. Morrison, Ed.; Academic: Orlando, 1983; Vol. 2, p 225-241.

163. G. Posner *Acc. Chem. Res.* **1987**, *20*, 72-78.

164. G. H. Posner In *The Chemistry of Sulphones and Sulphoxides*; S. Patai, Z. Rapaport, C. Stirling, Eds.; Wiley: New York, 1988, p 823-849.

165. G. Posner; L. L. Frye; M. Hulce *Tetrahedron* **1984**, *40*, 1401-1407.

166. A. Schultz; R. E. Harrington *J. Am. Chem. Soc.* **1991**, *113*, 4926-4931.

167. G. H. Posner; T. P. Kogan; S. R. Haines; L. L. Frye *Tetrahedron Lett.* **1984**, *25*, 2627-2630.

168. G. H. Posner; T. P. Kogan; M. Hulce *Tetrahedron Lett.* **1984**, *25*, 383-386.

169. G. H. Posner; J. P. Mallamo; M. Hulce; L. L. Frye *J. Am. Chem. Soc.* **1982**, *104*, 4180-4185.

170. G. H. Posner; M. Hulce; J. P. Mallamo; S. Drexler; J. Clardy *J. Org. Chem.* **1981**, *46*, 5244-5246.

171. E. J. Corey; R. Naef; F. J. Hannon *J. Am. Chem. Soc.* **1986**, *108*, 7114-7116.

172. A. Alexakis; S. Mutti; J. F. Normant *J. Am. Chem. Soc.* **1991**, *113*, 6332-6334.

173. K. Tanaka; H. Suzuki *J. Chem. Soc., Chem. Commun.* **1991**, 101-102.

Chapter 5

Aldol and Michael Additions of Allyls and Enolates

In this chapter, the discussion of additions to carbonyls continues with the aldol addition reaction and the mechanistically similar allyl addition reactions, both examples of "π-transfer" additions illustrated in Figure 4.1. Also discussed are asymmetric Michael addition reactions.

The aldol condensation is one of the oldest reactions in organic chemistry, dating back to the first half of the 19th century, but about 1980 it underwent a renaissance after methods were developed to stop the reaction at the stage of the initial addition product, with a high degree of stereoselectivity. Much of the excitement and interest in asymmetric synthesis since that time has been due to the development of highly selective aldol addition reactions and the mechanistically similar allyl addition reactions. We begin the chapter with the latter, because the allyl addition is irreversible and because the transition state assemblies are somewhat less complex than those of the aldol additions.

Scheme 5.1 illustrates the transition structure most often invoked to explain the selectivities observed in π-transfer 1,2-carbonyl additions (cf. Figure 4.1): the so-called Zimmerman-Traxler transition structure [1]. This model, which was originally proposed to rationalize the selectivity of the Ivanov reaction, has its shortcomings (as will be seen) and suffers from an oversimplification when applied to enolates, in that it illustrates a monomeric enolate (cf. section 3.1 and ref. [2-4]). Nevertheless, it serves the very useful purpose of providing a simple means to rationalize relative and absolute configurations in almost all of the asymmetric 1,2-additions we will see. The favored transition structure has *lk* topicity (*Si/Si*

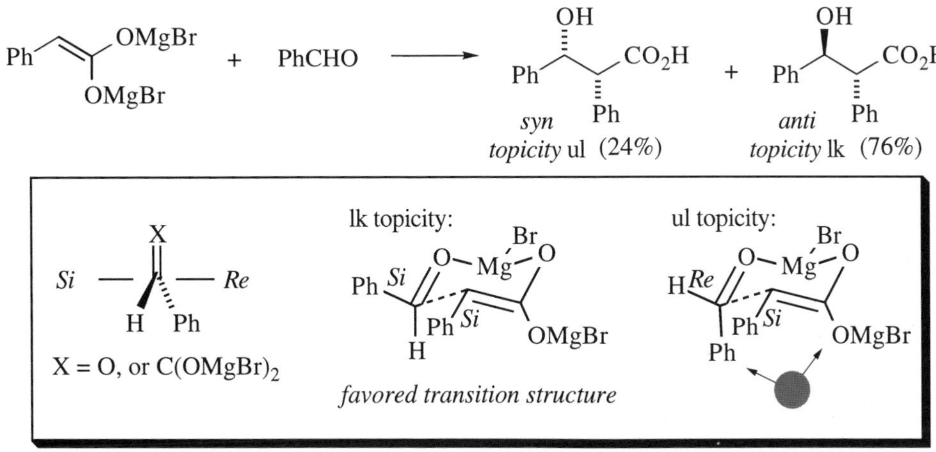

Scheme 5.1. The Zimmerman-Traxler transition state model for the Ivanov reaction [1].

161

illustrated) because the alternative has a pseudo 1,3-diaxial interaction between the aldehyde phenyl and the magnesium alkoxide. Because the magnesium alkoxide is on a trigonal carbon in the 6-membered ring, this repulsive interaction is not large, and the selectivity for the anti product is only 76% [1].[1]

5.1 1,2-Additions of allyl metals and metalloids

Most allylic organometallic or organometalloid systems are reactive enough to add to aldehyde carbonyls without the aid of additional Lewis acids, the notable exception being allyl silanes (reviews: [7-15]). Often, the allylic metal or metalloid atom itself activates the carbonyl, and a highly organized six-membered ring transition structure similar to the Zimmerman-Traxler model results. This section deals with cases where chiral ligands on the metal or on an acid catalyst induce selectivity by interligand asymmetric induction. Reactions of allyl metal compounds in which the metal-bearing carbon is stereogenic are not covered.

In order to explain the chemistry of allylic metals, the reactions of allylic boron compounds [8,12-14] are covered in detail. The boron chemistry is divided into four parts: simple enantioselectivity (addition of $CH_2=CHCH_2-$, creating one new stereocenter), simple diastereoselectivity of crotyl additions (relative configuration after $CH_3CH=CHCH_2-$ addition, where neither reagent is chiral), single asymmetric induction with chiral allyl boron compounds (one and two new stereocenters), and double asymmetric induction (both reactants chiral, one and two new stereocenters). Then follows a brief discussion of other allyl metal systems.

5.1.1 Simple enantioselectivity

Scheme 5.2 illustrates the enantiomeric chair transition structures and products for the addition of an allyl borane to acetaldehyde. Note that in assembly *a*, the *Re* face of the aldehyde is attacked, producing the *S* alcohol. Conversely, attack on the *Si* face of the aldehyde produces the *R* alcohol (assembly *b*). In the inset are shown two alternative chair transition structures, which originate by reversing the position of the aldehyde methyl and hydrogen substituents of assemblies *a* and *b* (or equivalently, by flipping the chair). These are destabilized by severe 1,3-diaxial interactions between the aldehyde methyl and one of the ligands on boron. Note that the boron ligand is fully axial (unlike the pseudoaxial magnesium alkoxide in Scheme 5.1), and the boron-oxygen bond is fairly short.[2] These two differences mean that the repulsive interaction is quite strong, and the aldehyde is preferentially oriented with its nonhydrogen substituent equatorially. Thus the simple concepts of conformational analysis of substituted cyclohexanes, applied to the Zimmerman-Traxler model, provide a basis for a "first approximation" analysis of these closed (cyclic) transition structures.

---

[1] We will use the *syn/anti* nomenclature [5] to describe the relative configuration of aldol stereoisomers, and the *lk/ul* nomenclature [6] to describe the topicity of the reaction. For definitions, see glossary, Section 1.6.

[2] A B–O bond is 1.36-1.47Å, whereas a Mg–O bond is 2.0-2.1 Å [16].

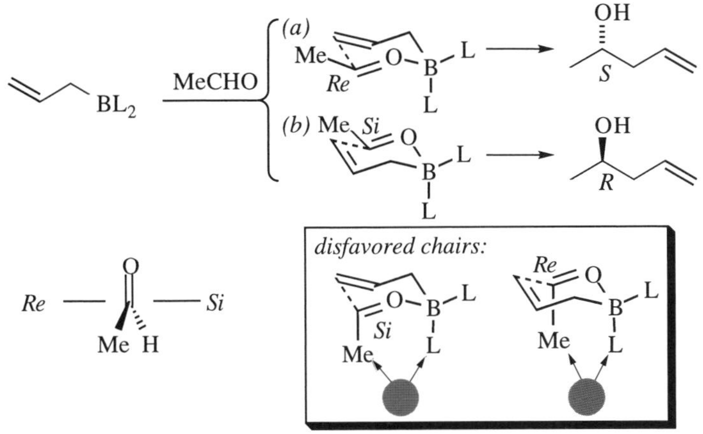

Scheme 5.2. Cyclic transition states for allyl boron additions.

Unless there is a chiral ligand on boron, assemblies *a* and *b* of Scheme 5.2 are enantiomeric and the product will be racemic. If the ligand is chiral, then the transition structures are diastereomeric and the products will be formed in unequal amounts under conditions of kinetic control (Chapter 1). Figure 5.1 illustrates several chiral boron reagents that have been tested in the allyl boration reaction, with typical enantioselectivities for each.

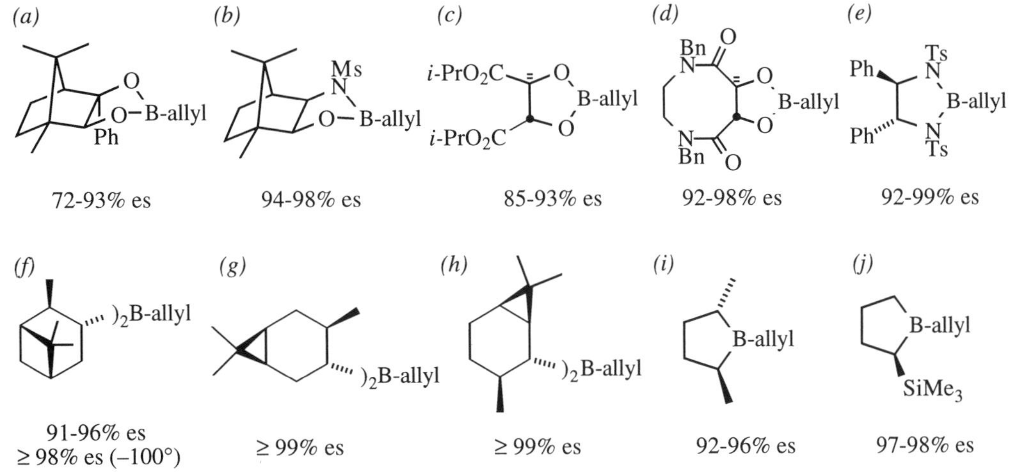

Figure 5.1. Chiral boron compounds for asymmetric allyl addition to achiral primary, secondary, and tertiary alkyl, vinyl, and aryl aldehydes, and their typical enantioselectivities (*a-e* at –78°, *g-j* at –100°). *(a)* [17]; *(b)* [18]; *(c)* [19]; *(d)* [19]; *(e)* [20]; *(f-h)* [21-24]; *(i-j)* [25].

5.1.2 Simple diastereoselectivity

When there is a substituent on the allyl double bond, geometric isomers are possible and two new stereocenters are formed. The transition structures in Scheme 5.3 illustrate how the *E*-crotyl boron compound affords racemic anti addition product and the *Z*-crotyl compound affords the syn product.[3] For the *E* isomer, the

---

[3]   This assumes that there are no isomerizations that precede the addition. For discussions of such phenomena for boranes and boronates, see ref. [26].

most stable chair presents the *Re* face of the aldehyde to the *Re* face of the double bond, or vice versa (*lk* topicity). These two transition structures are enantiomeric (and therefore isoenergetic in the absence of a chiral influence), as are the anti products. Likewise, the Z-crotyl species assembles with *ul* topicity, presenting the *Re* face of the aldehyde to the *Si* face of the double bond, or vice versa, which produces the syn addition product.

Note that reversing the face of only one component of the assembly reverses the topicity and the relative configuration of the stereocenters in the product. For example, exchanging the positions of the methyl and hydrogen in either the aldehyde or the crotyl moiety of the *lk* transition structure changes the topicity to *ul*, and the syn product would be produced. As before (Scheme 5.2) exchanging the aldehyde substituents causes severe 1,3-interactions with the axial boron ligand. Therefore, the tendency is for *lk* topicity for E-crotyl species, giving anti products and *ul* topicity for Z-crotyl compounds, giving syn products.

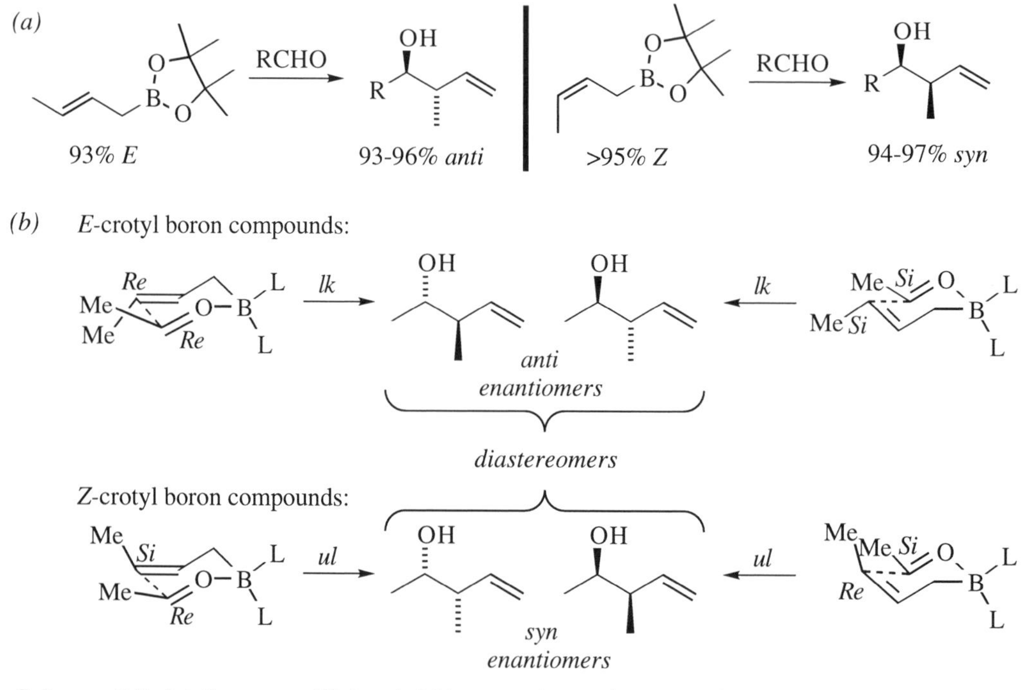

Scheme 5.3. *(a)* Stereospecificity (within experimental error) of crotyl borane additions to aldehydes, R = Me, Et, *i*-Pr, Ph [26]. *(b)* Transition structures for stereospecific addition of crotyl boron compounds to aldehydes.

5.1.3 Single asymmetric induction

Figure 5.1 lists a number of auxiliaries for asymmetric allyl addition to aldehydes. Substituted allyl boron compounds have also been used in reactions with achiral aldehydes. Table 5.1 lists several examples of 2- and 3-substituted allyl boron compounds, and the products derived from their addition. Note that for the E- and Z-crotyl compounds, the enantioselectivity indicated is for the isomer illustrated. In some cases, there was more than one of the other three possible isomers found as well.

Table 5.1. Asymmetric addition of substituted allyl boron compounds to aldehydes. Ligands are illustrated in Figure 5.1.

Entry	<u>R</u>CHO	L$_2$B<u>R</u>	Product	% Yield	% es	Ref
1	*E*-cinnamyl	-CH$_2$ (Br) L$_2$=*ent*-5.1e	Ph (OH) (Br)	79	94	[20]
2	*n*-C$_5$H$_{11}$–	"	*n*-C$_5$H$_{11}$ (OH) (Br)	77	>99	[20]
3	CH$_3$–	*Z*-crotyl L$_2$=5.1f	OH	75	95	[27]
4	CH$_3$–	*E*-crotyl L$_2$=5.1f	OH	78	95	[27]
5	CH$_3$–	-CH$_2$ (OMe) L$_2$=5.1f	R (OH) (OMe) (R = CH$_3$)	59	95	[28]
6	Ph–	"	" (R = Ph)	75	95	[28]
7	*n*-C$_6$H$_{13}$–CH$_2$ (SiMe$_2$N(*i*-Pr)$_2$) L$_2$=5.1f		R (OH) (SiMe$_2$N(*i*-Pr)$_2$) (R = *n*-C$_6$H$_{13}$)	52	>95	[29]
8	*c*-C$_6$H$_{11}$–	"	" (R = *c*-C$_6$H$_{11}$)	63	>95	[29]
9	Ph–	"	" (R = Ph)	50	>95	[29]

5.1.4 Double asymmetric induction

When the boron ligands and the aldehyde are both chiral, the inherent stereo-selectivities of each partner may be either matched or mismatched (Chapter 1). In principle, a chiral aldehyde could derive facial selectivity from either the Felkin-Anh-Heathcock model (Figures 4.8 and 4.10) or the Cram-chelate model (Figure 4.11). However, because the boron of these reagents can accept only one additional ligand, chelation is not possible. Therefore only the Felkin-Anh-Heathcock effects are operative in these reactions, and they are usually relatively weak, with diastereo-selectivities of ≤70%. The high diastereoselectivites of many of the auxiliaries illustrated in Figure 5.1 can therefore be used to control the relative and absolute configuration of both of the new stereocenters in the addition product. Table 5.2 lists selected examples of double asymmetric induction with two α-alkoxyaldehydes and several auxiliaries (the 4,5-anti isomer is favored by Cram's rule).

Table 5.2. Double asymmetric induction in addition of allyl boron compounds to aldehydes. Ligands are illustrated in Figure 5.1.

Entry	RCHO	L₂BR	Product	% Yield	% ds	Ref
1	Ph CHO / OMOM	allyl / L₂=5.1e	Ph 5 4 OH OMOM	80	96	[20]
2	Ph CHO / OMOM	allyl / L₂=5.1e	Ph 5 4 OH OMOM	–	98	[20]
3	O O CHO	allyl / L₂=5.1a	O O 5 4 OH	87	96	[30]
4	"	allyl / L₂=5.1c	"	85	93	[19,31]
5	"	allyl / L₂=5.1d	"	84	98	[19]
6	"	allyl / L₂= ent-5.1c	O O 5 4 OH	90	98	[31]
7	"	allyl / L₂= ent-5.1d	"	81	99.7	[19,31]
8	"	E-crotyl / L₂=5.1c	O O 5 4 3 OH	85	96	[8,32,33]
9	"	E-crotyl / L₂=5.1i	"	74	86	[25]

Table 5.2 (cont.). Double asymmetric induction in addition of allyl boron compounds. Ligands are illustrated in Figure 5.1.

Entry	RCHO	L₂BR	Product	% Yield	% ds	Ref
10		*E*-crotyl L₂=5.1a		85	72	[30]
11	"	*E*-crotyl L₂= *ent*-5.1c	"	87	87	[8,32,33]
12	"	*E*-crotyl L₂= *ent*-5.1i	"	71	96	[25]
13	"	*Z*-crotyl L₂=5.1a		86	>98	[30]
14	"	*Z*-crotyl L₂=5.1c	"	90	76	[8,33]
15	"	*Z*-crotyl L₂= *ent*-5.1c	"	84	>99	[8,33]
16	"	*Z*-crotyl L₂= *ent*-5.1i	"	66	92	[34]
17	"	*Z*-crotyl L₂=5.1i		65	82	[34]

Noteworthy among these examples is the ability to achieve high diastereoselectivity for both the 3,4-syn and 3,4-anti isomers, almost independent of the chirality sense of the aldehyde. Comparison of several examples show the expected trends for matched and mismatched pairs (*cf.* entry pairs 1/2, 4/6, 5/7, 9/12, 16/17). Note that either 3,4-anti diastereomer can be obtained with 96% ds (entries 8 and 12); the two 3,4-syn isomers are also available selectively (entries 13-16 and 17), although only one ligand (5.1i) is selective for the 3,4-syn-4,5-syn product (entry 17) that is a mismatched pair (*cf.* entry 16). Note that with Roush's tartrate ligand (Figure 5.1c), the *E*-crotyl mismatched pair is more selective than the matched pair (entries 8/11; for a rationale, see ref. [33]), and the matched and mismatched pair give the same major product isomer with the *Z*-crotyl compound (entries 14/15).

Several substituted allyl and crotyl derivatives have been designed to increase the usefulness of the boron-mediated allyl addition of aldehydes. For example, silanes such as those shown in Table 5.1, entries 7-9, can be stereospecifically converted to hydroxyls [29,35] or transformed into alcohols with a formal 1,3-hydroxy migration [36]. Additionally, vinyl bromides such as those shown in Table 5.1, entries 1 and 2 can be converted into a number of functional groups by standard chemical means [20]. Examples of these transformations are shown in Scheme 5.4. Note also that ozonolysis of any of these adducts give "aldol" adducts (Section 5.2).

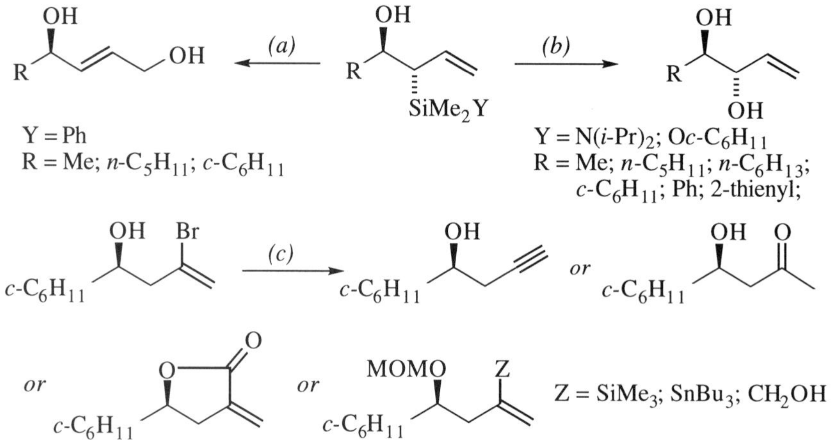

Scheme 5.4. Transformations of functionalized addition products. *(a)* [36]; *(b)* [29,35]; *(c)* [20].

5.1.5 Other allyl metals

In addition to boron, a number of other metals have been used in π-transfer addition reactions (reviews: [7-11,14,15]. Based on stereochemical tendencies and mechanistic considerations, these reagents have been classified into three groups, as illustrated in Scheme 5.5 [8,37]:

Type 1. Reagents that are stereospecific in the sense that an *E*-crotyl isomer affords the anti addition product (*lk* topicity) and a *Z*-crotyl isomer affords the syn product (*ul* topicity). The transition structure is thought to be a closed chair, analogous to the Zimmerman-Traxler transition structure (Scheme 5.1).

Type 2. Reactions that are catalyzed by Lewis acids and are stereoconvergent to syn adducts for either the *E*- or the *Z*-crotyl organometallics (*ul* topicity). The transition structure is usually considered to be open (acyclic), but the exact nature of the transition state is still a matter of discussion [8,9].

Type 3. Allyl organometallics that are (usually) generated *in situ* and which equilibrate to the more stable *E*-crotyl species, then add via a closed, Zimmerman-Traxler transition structure producing anti adducts preferentially [8].

The boron-containing compounds discussed in the previous sections are typical of Type 1 reagents. Also included in this group are reactions of allyl aluminums and uncatalyzed reactions of allyl tin reagents [8,37].

Reactions that fit Type 2 are catalyzed by Lewis acids which coordinate to the carbonyl oxygen of the aldehyde, thereby precluding coordination by the allyl metal. Such reactions proceed *via* an open transition state. As indicated previously, allyl silanes are not reactive enough to add to aldehydes without acid catalysis, so

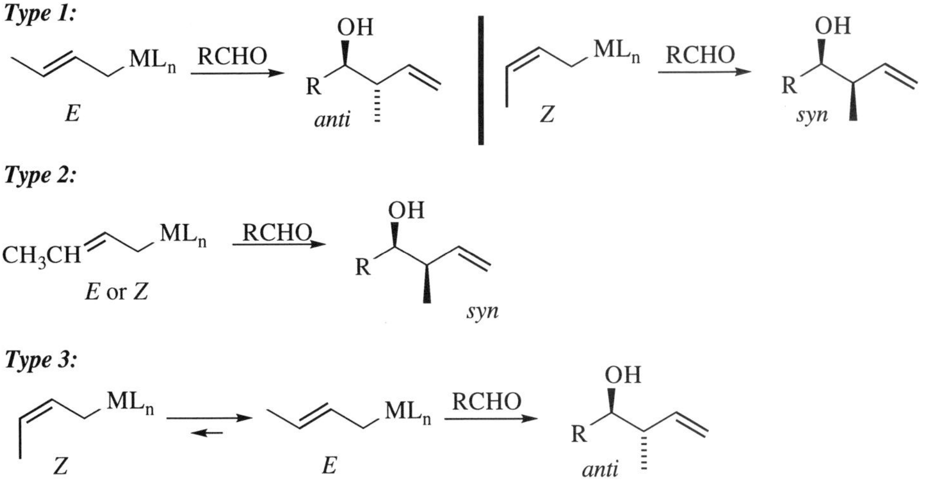

Type 1:

E → anti | Z → syn

Type 2:

E or Z → syn

Type 3:

Z → E → anti

Scheme 5.5. Mechanistic types for allyl addition to carbonyls. Types 1 and 3 proceed through transition structures similar to those in Scheme 5.3 [8,37].

they fall into this category [37]. Allyl stannane additions may be catalyzed by Lewis acids, so stannanes sometimes fall into this group [38,39], as do allyl titanium reagents [8,9,37]. Scheme 5.6 shows some enantioselective examples of allylsilane

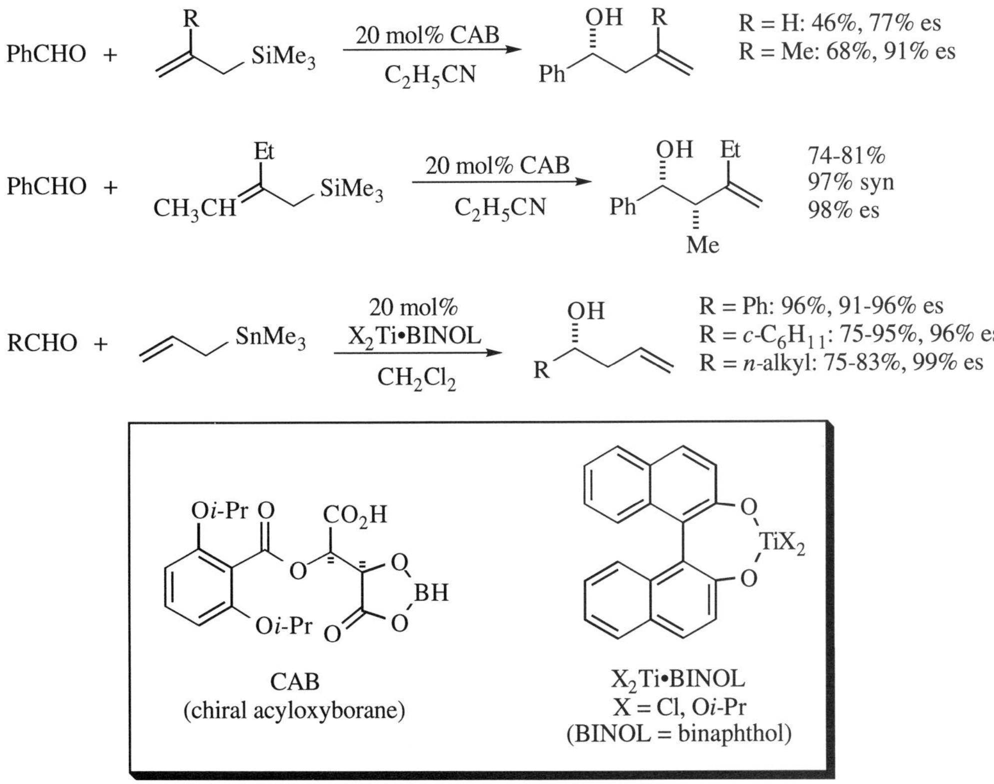

PhCHO + [R, SiMe₃] → 20 mol% CAB, C_2H_5CN → R = H: 46%, 77% es
R = Me: 68%, 91% es

PhCHO + [Et, CH₃CH, SiMe₃] → 20 mol% CAB, C_2H_5CN → 74-81%
97% syn
98% es

RCHO + [SnMe₃] → 20 mol% $X_2Ti \cdot BINOL$, CH_2Cl_2 → R = Ph: 96%, 91-96% es
R = c-C₆H₁₁: 75-95%, 96% es
R = n-alkyl: 75-83%, 99% es

CAB
(chiral acyloxyborane)

$X_2Ti \cdot BINOL$
X = Cl, Oi-Pr
(BINOL = binaphthol)

Scheme 5.6. Enantioselective additions of allyl silanes [40] and allyl stannanes [41,42], mediated by chiral catalysts.

additions [40] and allyl stannane [41,42] additions; many enantioselective additions of allylstannanes involve chirality transfer from the stannane where the allylic carbon bearing the tin is stereogenic [9,15], and are not discussed herein.

Figure 5.2 illustrates the six possible open transition structures for the Lewis acid mediated addition of allyl metals to an aldehyde. Note that for each topicity, there are two synclinal arrangements and one antiperiplanar. Several factors must be considered in explaining the observed *ul* topicity of these reactions (giving syn relative configuration in the products), and a number of rationales have been offered. If one assumes that the conformation is antiperiplanar in the transition state, then structure *a* would be favored over *d*, since this arrangement minimizes the interaction between in the aldehyde substituent, R, and the methyl of the crotyl group.

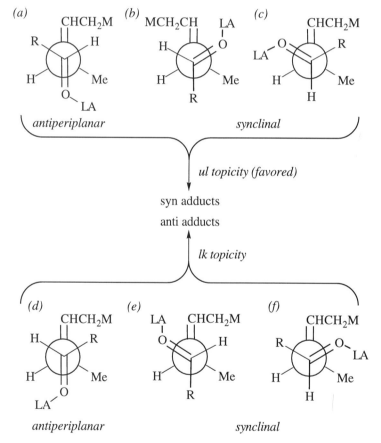

Figure 5.2. Newman projections of possible open transition structures for Lewis acid (LA) catalyzed additions to aldehydes.

On the other hand, Seebach suggested in 1981 that the topicity of a number of reactions (including these) may be explained by having the double bonds oriented in a synclinal arrangement.[4] He reasoned that steric repulsion between the R and

---

[4] For a discussion of the Seebach rule as applied to the Michael reaction, see Figure 5.9 and the accompanying discussion.

CHCH$_2$M moieties would favor *b* over *c* and *e* over *f*. Then, assuming that the nucleophile approaches the carbonyl along the Bürgi-Dunitz trajectory (Section 4.1.3), either the hydrogen (in *b*) or the methyl group (in *e*) must be squeezed in between the alkyl group and hydrogen of the aldehyde. The former would be favored. This hypothesis was offered as a "topological rule" (not a mechanism).

Later, studies of intramolecular silane [37] and stannane [39] additions offered a direct comparison between synclinal arrangement *c* and antiperiplanar arrangement *d*. The former is favored. Because of the intramolecular nature of the addition, conformations analogous to the other possibilities were not possible.

Allyl chromium, titanium and zirconium reagents fall into the Type 3 category. Enantioselective reactions in this class are relatively rare, although the diastereoselectivities can be quite high (reviews: [7,8,15]).

5.2 Aldol additions[5]

The Ivanov reaction (Scheme 5.1) is an early example of an aldol addition reaction that proceeded selectively. There has been an enormous amount of work done in this area, and only a small amount of the developmental work will be covered here. A large number of chiral auxiliaries and catalysts have been developed, but we will concentrate on only a few, which suffice to provide an understanding of the structural factors that influence selectivity. The transition structures presented in the following discussion are oversimplifications, in that the enolate and its metal are represented as monomers, when in fact they are not [2-4]. On the other hand, much of the available data may be rationalized on the basis of these structures, so the simplification is justified in the absence of detailed structural and configurational information about mixed aggregate transition structures.

5.2.1 Simple diastereoselectivity

Kinetic control. The Zimmerman-Traxler model, as applied to propionate and ethyl ketone aldol additions, is shown in Scheme 5.7 (note the similarity to the boron-mediated allyl additions in Scheme 5.3). Based on this model, we would expect a significant dependence of stereoselectivity on the enolate geometry, which is in turn dependent on the nature of X and the deprotonating agent (see section 3.1.1). In addition, the configuration and selectivity of the kinetically controlled aldol addition is dependent on the size of the substituents on the two reactants.

Figure 5.3 illustrates both enantiomers of most of the possible transition structures that have been postulated for aldol additions of R$_1$CH$_2$COX enolates. In the closed transition structures (Figure 5.3a,b), the chair conformations would normally be expected to predominate, but in certain instances a boat may be

---

5   Note the distinction between the aldol *condensation*, in which α,β-unsaturated carbonyls are formed, and the aldol *addition*, which is stopped at the β-hydroxy carbonyl stage. For reviews of the early literature, mostly focusing on the aldol condensation, see ref. [43,44]. For reviews of the aldol addition, see ref. [16,45-51] (Li and Mg enolates), [52] (B and Al enolates), and [53] (transition metal enolates).

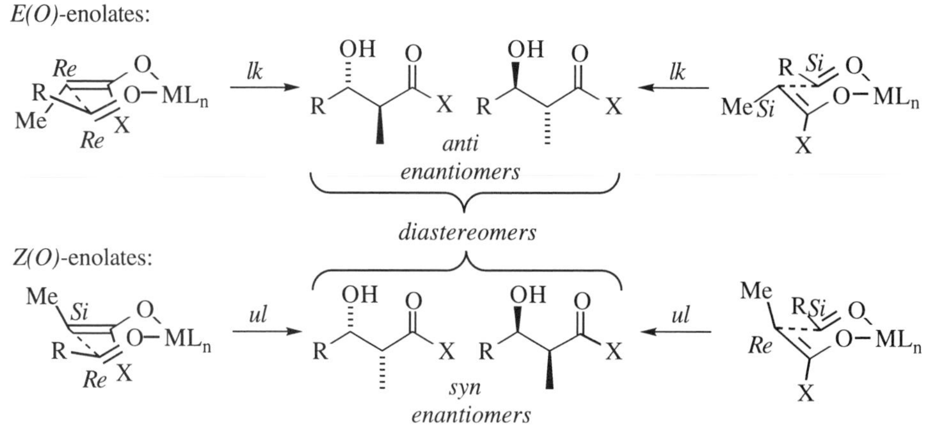

E(O)-enolates:

Z(O)-enolates:

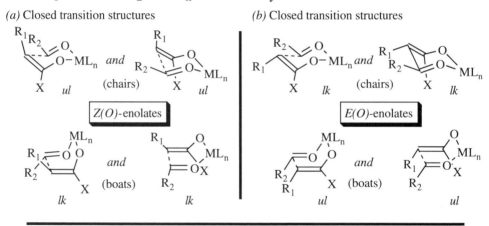

Scheme 5.7. Transition structures for stereoselective propionate additions to aldehydes.

preferable.[6] For the open transition structures, study of the intramolecular addition of silyl ethers, catalyzed by Lewis acids, showed a moderate preference for an anti conformation [59]. In intermolecular cases, the choice between open structures of *ul* or *lk* topicity will be governed by the relative magnitude of the gauche interactions between R_1 and either R_2 or ML_n on the aldehyde.

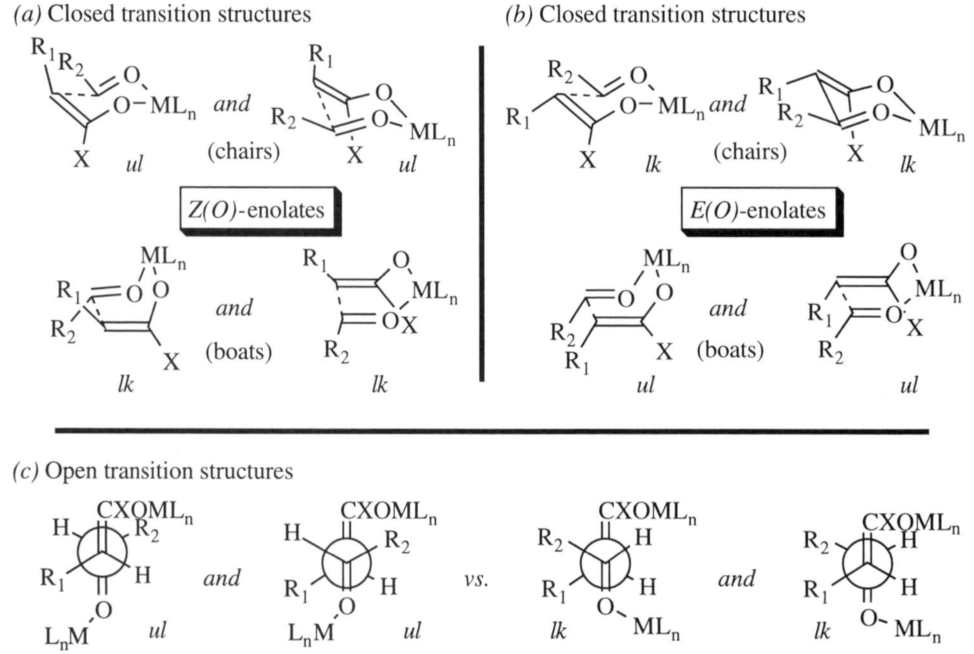

(a) Closed transition structures

(b) Closed transition structures

(c) Open transition structures

Figure 5.3. *(a)* Chair and boat transition structures for *Z(O)*-enolates. *(b)* Chair and boat transition structures for *E(O)*-enolates. *(c) ul* and *lk* open transition structures. Note that in all cases, the topicity is such that *ul* → syn; *lk* → anti.

---

6   Computational studies predict that the geometry (chair, half-chair, boat, etc) depends on the nature of R_1, R_2, and M. Theory also predicts that Z-enolates prefer a closed chair, but that *E*-enolates may prefer a boat [54-56]. For an empirical rule for predicting aldol topicity, see ref. [57]. For an investigation into the effect of metal and solvent on transition structures, see ref. [58].

The stereoselectivity of the aldol addition often depends on the selectivity of enolate formation. Ireland's rationale for the selective formation of lithum $Z(O)$-enolates of ketones, amides, and imides, and the selective formation of ester $E(O)$-enolates was outlined in section 3.1.1 [60]. The rationale for the selective formation of $E(O)$- and $Z(O)$-boron enolates by reaction with dialkylboron triflate and a tertiary amine [61] is shown in Scheme 5.8 [52,62]. The boron triflate coordinates to the carbonyl oxygen, thereby increasing the acidity of the α-proton so that it can be removed by amine bases, as shown in Scheme 5.8a. In most cases, the stereochemical situation is as shown in Scheme 5.8b. The boron is trans to the CH_2R_1, R_1 is antiperiplanar to X, and removal of the H_{Re} proton gives the $Z(O)$-enolate. Note that for amides (X = NR_2), $A^{1,3}$ strain between R_1 and NR_2 particularly destabilizes the $E(O)$-enolate. In certain instances, a repulsive van der Waals interaction between the X and BR_2OTf moieties may be particularly severe (*e.g.*, *t*-BuS– and Bu_2BOTf), such that the boron is oriented trans to X, which forces R_1 synperiplanar to X to avoid the boron ligands, as illustrated in Scheme 5.8c. Removal of the H_{Si} proton then gives the $E(O)$-enolate.

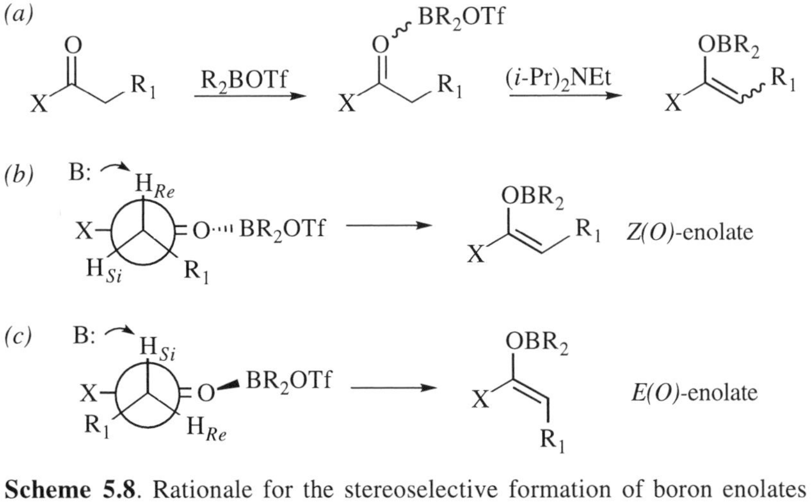

Scheme 5.8. Rationale for the stereoselective formation of boron enolates [62]. (*a*) If the boron is trans to X, $A^{1,3}$ strain considerations force R_1 syn to X, and removal of the proton from a conformation in which the C–H bond is perpendicular to the carbonyl affods the $E(O)$-enolate; (*b*) when the boron is cis to X, R_1 may orient anti to X, and the $Z(O)$-enolate ensues.

Not all aldol additions exhibit a dependence of product configuration on enolate geometry. Acid catalyzed aldols [45], some base catalyzed aldols [58], and aldols of some transition metal enolates [63,64] show no such dependency. For example, zirconium enolates afford syn adducts (*ul* topicity) independent of enolate geometry for a number of propionates [63,64]. As shown in Scheme 5.9, two explanations have been proposed to explain the behavior of zirconium enolates. One explanation (Scheme 5.9a) is that the closed transition structure changes from a chair for the $Z(O)$-enolate to a boat for the $E(O)$-enolate [16,63,65]. Another hypothesis is that these additions occur via an open transition structure. Although the original authors [64] suggested an open transition structure, they did not provide an illustration.

Recently, Heathcock proposed an open transition structure similar to the one illustrated in Scheme 5.9b for an acid-catalyzed aldol addition where the Lewis acid on the oxygen is small [66]. According to this rationale, the topicity is determined by the relative energies of the van der Waals interactions between the methyl group and either the Lewis acid or R group [66]. Heathcock postulates that when the Lewis acid is small, *ul* topicity is preferred, since it minimizes the gauche interactions between the methyl and R in the forming bond. In the case of the zirconium enolates, there is an equivalent of lithium chloride present from transmetalation of the lithium enolate with Cp_2ZrCl_2, which can act as a (small) Lewis acid. The transition structure illustrated in Scheme 5.9b is then favored because it relieves the gauche interaction between the methyl and R in the forming bond.

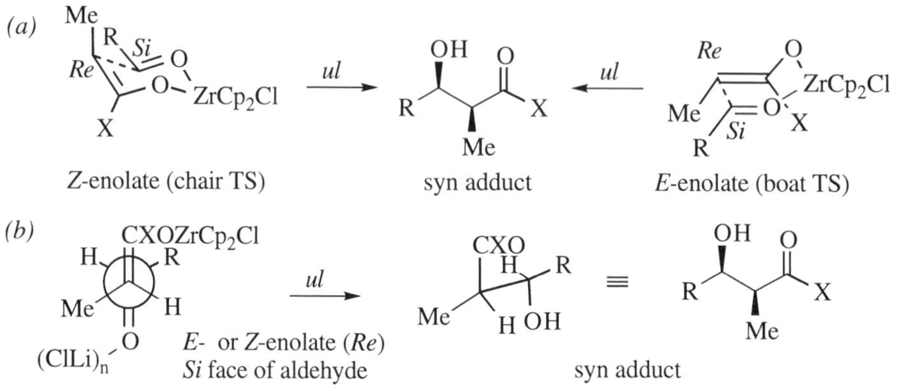

Scheme 5.9. Explanations for the *ul* selectivity of *E(O)*- and *Z(O)*-zirconium enolates: (a) *Z(O)*-enolate chair and *E(O)*-enolate boat [16,63]; (b) open structure [64] (see also ref. [66]).

Thus, two explanations rationalize the same result. The lesson is that although transition state models may serve a useful predictive value, they may or may not depict reality. The scientific method allows you to test a hypothesis, but consistency with a hypothesis does not constitute a proof: it constitutes a failure to disprove the hypothesis.

Thermodynamic control. Note that it is also possible for the aldolate adduct to revert to aldehyde and enolate, and equilibration to the thermodynamic product may afford a different diastereomer (the anti aldolate is often the more stable). The tendency for aldolates to undergo the retro aldol addition increases with the acidity of the enolate: amides < esters < ketones (the more stable enolates are more likely to fragment), and with the steric bulk of the substituents (bulky substituents tend to destabilize the aldolate and promote fragmentation). On the other hand, a highly chelating metal stabilizes the aldolate and retards fragmentation. The slowest equilibration is with boron aldolates, and increases in the series lithium < sodium < potassium, and (with alkali metal enolates) also increases in the presence of crown ethers.[7]

---

[7] For a thorough discussion of the factors affecting the equilibration of aldolates, see ref. [16]. For a procedure for thermodynamic equilibration, see ref. [67].

To achieve kinetic control in the aldol addition reaction, it does not matter if the rate for the retro aldol is fast, as long as the relative rates for syn *vs.* anti addition is large. As an example, consider the following "case study". Let us assume that the rate of *ul* addition for a Z(O)-enolate (to give syn adduct) is significantly faster than the rate of *lk* addition (giving anti adduct), such that $k_{syn}/k_{anti} = 100$. Under these conditions, a retro aldol must occur 100 times before one syn $\rightarrow$ anti isomerization can occur. The actual rates of these individual processes can be measured with experiments such as those illustrated in Scheme 5.10 [68]. In Scheme 5.10a, aldehyde exchange clearly involves a retro aldol, and has a half-life of 15 minutes. In Scheme 5.10b, isomerization to the more stable anti isomer has a half-life of 8 hours at a higher temperature. Because the retro aldol and the *ul* addition are both much faster than the unfavored *lk* addition, even the crossover is syn-selective.

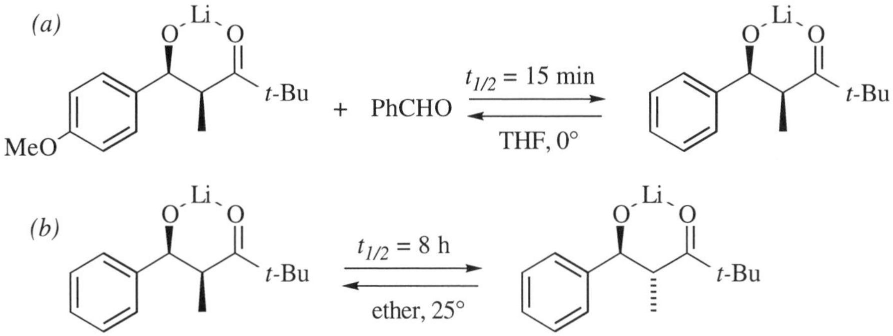

Scheme 5.10. *(a)* Aldehyde exchange and *(b)* syn-anti isomerization of aldolates [68].

In summary, the following generalizations have emerged for aldol additions under kinetic control:

1. Z(O)-enolates are highly syn-selective (*ul* topicity) when X is fairly large [51].
2. Z(O)-enolates with a large R_1 (such as an isopropyl or *tert*-butyl) give anti products (*lk* topicity) selectively [51].
3. E(O)-enolates are highly anti-selective (*lk* topicity) only with a very large X group (such as 2,6-di-*t*-butylphenol) [51].
4. For a closed transition structure, shorter M–O bond lengths amplify the van der Waals interactions between R_1, R_2, and X relative to enolates with longer bond lengths, resulting in higher stereoselectivities [16]. With boron enolates for example, Z(O)-enolates are highly syn selective [52].

5.2.2 Single asymmetric induction

For the addition of acetate and methyl ketone enolates (one new stereocenter), a number of approaches have been taken to induce enantioselectivity (review: [69]); one of these methods will be mentioned in the succeeding section, along with the propionate and ethyl ketone additions. In the open transition structures of Figure 5.3, each illustrated *lk* or *ul* pair is enantiomeric in the absence of any stereocenters in the two reactants. Introduction of a chirality element converts the paired transition structures (*i.e.,* transition structures of the same topicity) and products from enantiomers to diastereomers, and allows diastereoselection under either kinetic or thermodynamic control. There are three opportunities for introduction of chirality: a chiral auxiliary (X*), and the two sites of Lewis acid (ML*$_n$) coordination: the enolate and the aldehyde oxygens. In principle, either one, two, or three could be chiral, allowing for the possibility of single, double, and even triple asymmetric induction.

The following discussion is organized by the 'location' of the introduced chirality: X (intraligand asymmetric induction) or ML$_n$ (interligand asymmetric induction). Additionally, there is the possibility of a chirality center in the aldehyde, which will normally have an observable influence only in cases where the stereocenter is close to the carbonyl (*i.e.,* Cram's rule situations - see Chapter 4). Most of the examples that have been published to date include chirality centers in either X or ML$_n$, but not both.

Intraligand asymmetric induction. The first example of an auxiliary-based asymmetric aldol addition was reported by the Enders group, who used the enolate of a *SAMP* hydrazone in a crossed aldol [70]. This method afforded good yields, but only modest selectivities. Introduction of chirality in X (Figure 5.3) produces an enolate that affords much higher selectivities. Some of the more popular and effective auxiliaries are shown in Figure 5.4. The first of these (Figure 5.4a, R = methyl) was evaluated in racemic form by Heathcock in 1979, as its lithium *Z(O)*-enolate [71,72]. Later, a synthesis of the *S*-enantiomer from *S-tert*-leucine (*S-tert*-butylglycine) was reported [73]. A similar auxiliary was reported by the Masamune group in 1980 (Figure 5.4b, R = methyl), which afforded outstanding selectivities as its boron *Z(O)*-enolate. Initially [5] the racemate was resolved, but subsequently a chiron synthesis was reported using mandelic acid [74]. Both the Heathcock and the Masamune auxiliaries are self-immolative (*cf.* section 1.2, p. 2): 'removal' of the auxiliary by oxidative cleavage of the α-alkoxyketone to a carboxylic acid destroys the stereocenter. Figure 5.4c illustrates one of the most frequently used auxiliaries, the oxazolidinone imides developed in the Evans laboratory in 1981 [75]. These auxiliaries, which are made from amino alcohols such as valinol and phenylalaninol, can be cleaved to an acid, aldehyde, or an alcohol [76,77] *cf.* Scheme 3.16 and 3.17), and the auxiliary can be recovered in good yield. Reaction of either the boron *Z(O)*-enolates [75] or the zirconium *E(O)*- or *Z(O)*-enolates [78] are highly

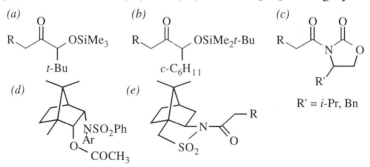

Figure 5.4. Chiral auxiliaries for asymmetric aldol additions. *(a)* racemic [71,72], from *tert*-leucine (*tert*-butyl glycine) [73]; *(b)* from mandelic acid [5,74]; *(c)* from valine or phenylalanine [75]; *(d)* from camphor [79]; *(e)* from camphor [80,81].

selective. None of the auxiliaries shown in Figure 5.4a-c are particularly selective when R is hydrogen (*i.e.*, 'acetate' enolates). The acetate shown in Figure 5.4d, reported by Helmchen in 1985 [79], is particularly good in this regard. A more recent (1988) addition to the list of effective aldol auxiliaries (Figure 5.4e) is the camphor sultam developed by Oppolzer [80-83] (*cf.* Scheme 3.18). Most of these auxiliaries (Figure 5.4a being the exception) are available as either enantiomer, making available either enantiomer of any aldol adduct. In the following discussions, only one enantiomer is illustrated, and it should be recognized that the other is also available.

The Heathcock and Masamune auxiliaries (Figure 5.4a,b) are structurally and conceptually similar, and will be discussed together. Scheme 5.11 illustrates two possibilities that can arise in these systems, depending on the metal and the substituents on silicon: a chelated or nonchelated orientation in the transition structure. Note that, for the *S*-enantiomer illustrated, the chelated enolate has the R group (*tert*-butyl or cyclohexyl) oriented to the rear, and the *front* face of the enolate is most accessible to the electrophile. Conversely, the non-chelated structure has the dipoles of the C=O and C–O bonds aligned in opposition, with the R group now projecting to the front of the structure leaving the *rear* face more accessible.

If both the *Z(O)*- and the *E(O)*-enolates can be made, and if both follow the Zimmerman-Traxler models (*i.e.*, chair transition structures), then both syn and anti adducts should be available (Scheme 5.11, path a *vs.* b or c *vs.* d). Since both enantiomers of the auxiliary are available, any desired combination of relative and absolute configurations in the products would be available.

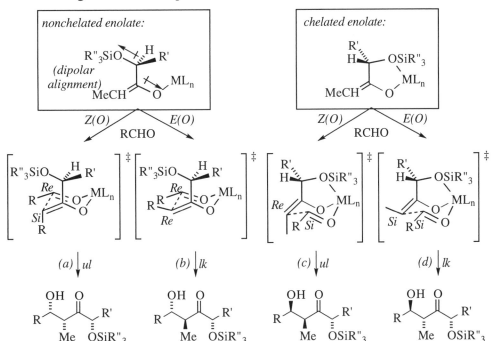

Scheme 5.11. Chelated and nonchelated pathways to aldol adduct diastereomers for the Heathcock and Masamune auxiliaries.

Note that each *E(O)*- or *Z(O)*-enolate will have a choice of two Zimmerman-Traxler transition structures. Thus (see Scheme 5.11), a *Z(O)*-enolate may add through nonchelated path a or chelated path c, both of which afford syn adducts, but of opposite absolute configuration at the two new stereocenters. Likewise, an *E(O)*-enolate may add *via* path b or d, affording diasteomeric anti adducts.

Highly selective additions of these auxiliaries have been achieved *via* all four of the postulated pathways. Table 5.3 lists several examples. For example, *Z(O)*-dibutylboron enolates (entries 1, 2) often have selectivities of >99%, and are postulated to proceed through nonchelated path a [73,74]. The reason path c cannot compete is that the boron cannot accomodate more than four ligands, and two of the ligands are non-exchangeable alkyl groups. Additionally, boron enolates are not reactive enough to add to aldehydes unless the latter are coordinated to a Lewis acid. In the absence of external acids, then, the boron of the enolate must activate the aldehyde by coordination and its two available ligand sites are occupied by the enolate and the aldehyde oxygens.

When the α-hydroxyl is silylated with a *tert*-butyldimethylsilyl group, chelation is difficult no matter what the metal. Lithium enolates of the TBS ethers are not particularly selective in their additions to aldehydes, but transmetalation to titanium affords enolates that are highly selective in their addition reactions (Table 5.3, entry 3, [84]). Acylation of the oxygen with a benzoyl group and deprotonation with LDA affords an enolate that gives the relative configuration shown in path a, although chelation by the benzoyl carbonyl oxygen is postulated (entry 4, [85]). With a smaller trimethylsilyl group, a lithium cation can simultaneously coordinate the enolate oxygen, the siloxyl oxygen, and the aldehyde oxygen. Thus, the *Z(O)*-lithium enolate affords syn adducts according to path c (entry 5, [73]).

Deprotonation of the ketone educt with *N*-(bromomagnesio)-2,2,6,6-tetramethyl-piperidide affords the *E(O)*-enolate selectively. Addition of the magnesium *E(O)*-enolate having a trimethylsiloxy group affords anti adducts (entry 6, [73]), and is postulated to occur via chelated path d (Scheme 5.11). Transmetalation of the *tert*-butyldimethylsiloxy-protected magnesium *E(O)*-enolate affords a titanium enolate that cannot chelate, and adds to aldehydes *via* path b (entry 7, [73]). In this case, only benzaldehyde afforded selectivity lower than 95%.

A highly versatile auxiliary is the Evans oxazolidinone imide (Figure 5.4c, see also Scheme 3.16), available by condensation of amino alcohols [86,87] with diethyl carbonate [86]. Deprotonation by either LDA or dibutylboron triflate and a tertiary amine affords only the *Z(O)*-enolate. Scheme 5.12 illustrates open and closed transition structures that have been postulated for these *Z(O)*-enolates under various conditions, and Table 5.4 lists typical selectivities for the various protocols. The first to be reported (and by far the most selective) was the dibutylboron enolate (Table 5.4, entry 1), which cannot activate the aldehyde and simultaneously chelate the oxazolidinone oxygen [75]. Dipolar alignment of the auxiliary and approach of the aldehyde from the *Re* face of the enolate affords syn adduct with outstanding diastereoselection, presumably *via* the closed transition structure illustrated in Scheme 5.12a [75]. The other syn isomer can be formed under two different types of conditions. In one, a titanium enolate is postulated to chelate the oxazolidinone

Table 5.3. Asymmetric additions of the Heathcock-Masamune enolates. The "Path" column indicates the product configuration and the proposed transition structure from Scheme 5.11.

Entry	Enolate	Path	RCHO	% Yield	% ds	Ref
1	Bu$_2$B, O / OTBS (cyclohexyl)	a	Et, *i*-Pr, Ph, BnO(CH$_2$)$_2$	70-85	≥97	[74]
2	Bu$_2$B, O / OTBS (*t*-Bu)	a	*i*-Pr, *t*-Bu, Ph, BnO(CH$_2$)$_2$	75-80	>95	[73]
3	(*i*-PrO)$_3$Ti, O / OTBS (cyclohexyl)	a	Et, *i*-Pr, *t*-Bu, Ph	75-88	≥98	[84]
4	Li, O, O=C–Ph (cyclohexyl)	a	Et, *i*-Pr, Ph	67-96	86-97	[85]
5	Li, O, O–SiMe$_3$ (*t*-Bu)	c	*i*-Pr, *t*-Bu, Ph	75-80	>95	[73]
6	BrMg, O, O–SiMe$_3$ (*t*-Bu)	d	*i*-Pr, *t*-Bu, Ph	75-85	92-95	[73]
7	(*i*-PrO)$_3$Ti, O / OTBS (*t*-Bu)	b	Me, *i*-Pr, *t*-Bu, Ph	85-88	80->95	[73]

oxygen [88][8] or sulfur of an oxazolidinethione [89] exposing the *Si* face of the enolate (Scheme 5.12b). Additional coordination of the aldehyde and addition via

---

[8] Recall that the titanium enolate of the Heathcock and Masamune auxiliaries (Table 5.3, entries 3 and 7) were postulated to occur by a *nonchelating* pathway. However, in those cases, the potential

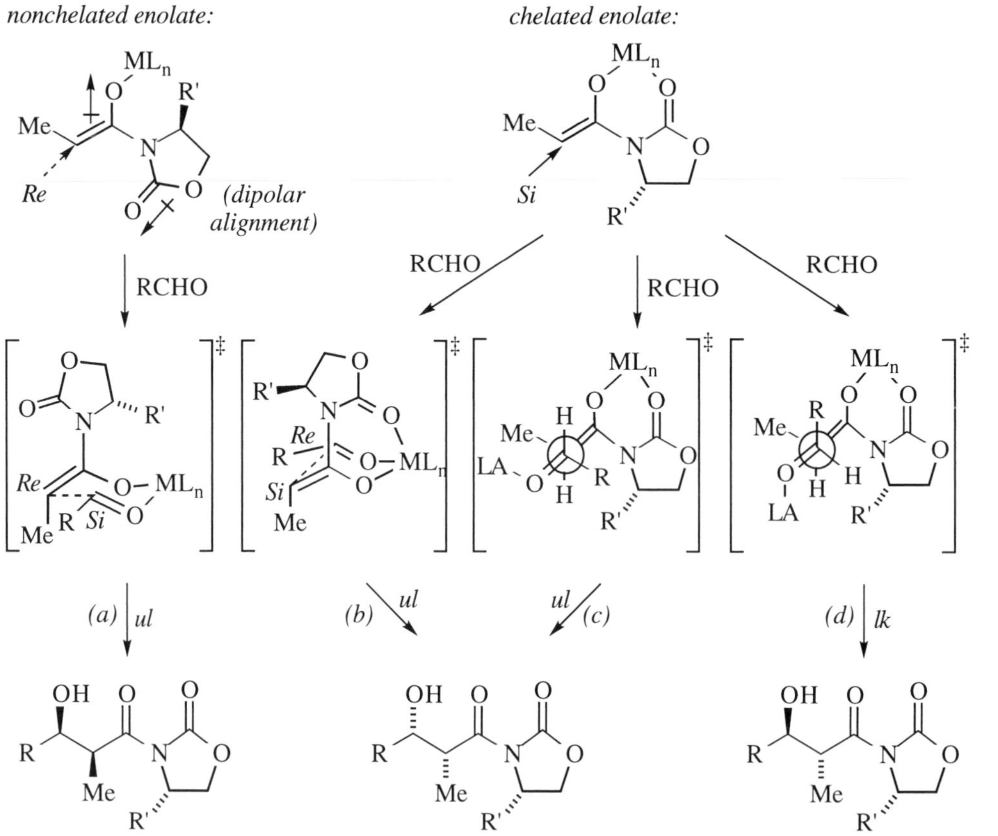

nonchelated enolate: *chelated enolate:*

Scheme 5.12. Open and closed transition structures for aldol additions of Evans's imides.

the closed transition structure shown in Scheme 5.12b affords excellent selectivity (Table 5.4, entries 2 and 3).

If the boron enolate is allowed to react with an aldehyde in the presence of another Lewis acid (LA), the addition is thought to occur *via* the open transition structures shown in Scheme 5.12c and d [66]. If the Lewis acid is small, the preferred orientation is as shown in Scheme 5.12c, which minimizes the gauche interaction between the methyl and R groups on the forming bond (Table 5.4, entry 4). Both SnCl$_4$ and TiCl$_4$ are relatively 'small' because of the long metal - oxygen bond. If the Lewis acid is large, the interaction between the Lewis acid and the methyl may outweigh the methyl/R gauche interaction. When the aldehyde is complexed to diethylaluminum chloride, the Lewis acid is effectively larger than either the tin or titanium complexes because of the shorter Al–O bond compared to either Sn–O or Ti–O, and because the ligands on the aluminum are relatively bulky. In this instance, the other face of the aldehyde will present itself to the enolate affording the anti adduct, as shown in Scheme 5.12d (Table 5.4, entry 5).

chelating atom was a severely crowded TBS ether oxygen, as opposed to the more basic and less crowded urethane carbonyl oxygen in the Evans auxiliary.

Table 5.4. Asymmetric additions of the Evans imide enolates. The "Path" column indicates the product configuration and the proposed transition structure from Scheme 5.12.

Entry	Enolate	Path	RCHO	% Yield	% ds	Ref
1	BBu$_2$	a	Bu, *i*-Pr, Ph	75-88	>99	[75]
2	Ti(O*i*-Pr)$_3$	b	Bu, *i*-Pr, Ph	56-75	85-92	[88]
3	TiCl$_3$	b	*n*-Pr, *i*-Pr, 1-propenyl, Ph	84-88	97-99	[89]
4	BBu$_2$	c	Et, *i*-Pr, *i*-Bu, *t*-Bu, 2-propenyl, Ph (·SnCl$_4$ or TiCl$_4$)	50-68	87-93	[66]
5	"	d	Et, *i*-Pr, *i*-Bu, *t*-Bu, 2-propenyl, Ph (·Et$_2$AlCl)	62-86	74-95	[66]

For syntheses requiring the syn adducts, it is more practical to use the boron enolate without additional Lewis acids [75], since the auxiliary is available as either enantiomer and is recoverable.[9] On the other hand, the anti adducts are (so far) only available by the diethylaluminim chloride/boron enolate protocol [66]. Similar principles may be used to prepare syn and anti halohydrins by aldol addition of α-halo acetate enolates of Evans imides [90,91].

A weakness of the Heathcock, Masamune, and Evans auxiliaries is their inability to selectively add methyl ketone or acetate enolates. An excellent auxiliary for this purpose is the ester developed by Helmchen, shown in Figure 5.4d and Scheme 5.13 [79]. The yields were in the 50 - 70% range. The authors proposed a closed

---

[9] Originally, Evans used the illustrated auxiliary (R' = *i*-Pr) for one product configuration and a similar norephedrine-derived auxiliary (R' = Me, plus a Ph at C-5) for the other [75]. Since that time, experience has shown [86] that a phenylalanine-derived auxiliary (R' = Bn) is usually better for practical reasons, and is available as either enantiomer.

Zimmerman-Traxler transition structure, as shown on the lower left [79], however the open structure shown in the lower right, which does not require coordination of the bulky silyloxy group to titanium, should also be considered. The aldehyde may be oriented to avoid the large *tert*-butyldimethylsilyl (TBS) group as shown, with the R group away from the TBS. Both of these models have the aldehyde approaching the enol ether from the front face, opposite the side that is shielded by the sulfonamide. Note also that the siloxy group is oriented downward, to avoid the sulfonamide. An anti-selective addition (92% ds) was also reported for the reaction of the *E(O)*-enol ether of this auxiliary with isobutyraldehyde [79].

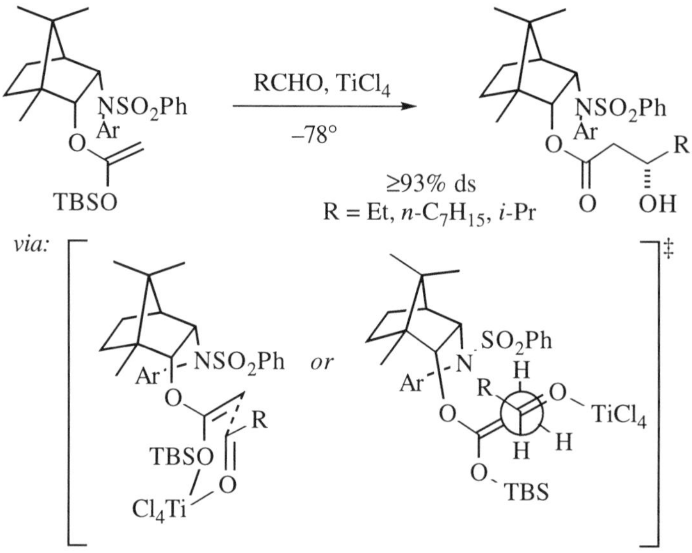

Scheme 5.13. Asymmetric addition of acetate enol silyl ethers to aldehydes [79].

Another excellent auxiliary for the asymmetric aldol addition is the camphor sultam developed in the Oppolzer laboratories (Figure 5.4e and Table 5.5). A significant feature of this auxiliary is the crystallinity of the aldol adducts, which often simplifies purification (and diastereomer enrichment). As the trimethylsilyl enol ethers (ketene acetals), acetate aldol additions afford good selectivities at –78° (Table 5.5, entry 1), and purification by recrystallization affords adducts that are >99% pure in most cases [83]. The transition structure proposed to account for the absolute configuration [83], based an an X-ray crystal structure [81], is shown in Scheme 5.14a (R_1 = H), and has the *tert*-butyldimethylsilyl group oriented toward the viewer (away from the camphor). Note also that the nitrogen is pyramidal and that there is little interaction between the nitrogen lone pair and the enolate double bond. With the silyl group occupying the front face, the *Re* face of the titanium-coordinated aldehyde (with the R_2 group oriented away from the camphor) approaches from the back. Approach of the aldehyde from the back is facilitated by the silyl group, which is antiperiplanar to the forming bond. A similar protocol affords anti aldolates from propionate-derived *tert*-butyldimethylsilyl enol ethers, as shown in Scheme 5.14a and entry 2 of Table 5.5 [81].

Table 5.5. Asymmetric additions of the Oppolzer sultam enolates. The "Path" column indicates the product configuration and the proposed transition structure from Scheme 5.14.

Entry	Enolate	Path	R$_2$CHO	% Yield	% ds	Ref
1	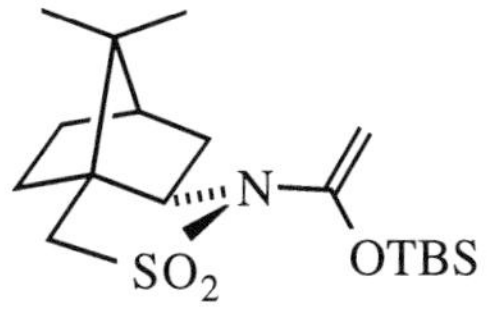	a (R$_1$=H)	Et, *n*-Bu, *i*-Pr, *i*-Bu, *c*-C$_6$H$_{11}$, Ph (& TiCl$_4$)	54-75	79-96	[83]
2	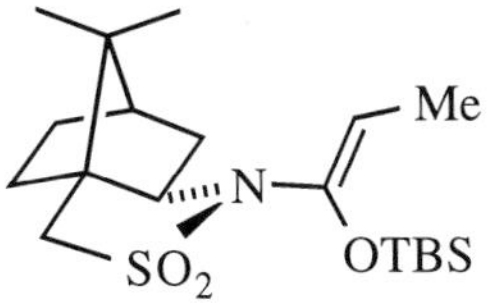	a (R$_1$=Me)	Me, Et, *i*-Pr, *i*-Bu, Ph (& TiCl$_4$ or ZnCl$_2$)	>95	≥98	[81]
3	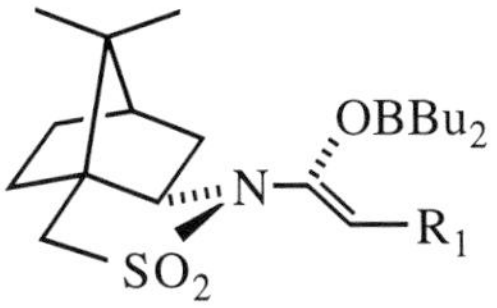	b (R$_1$=Me, Et, Bu)	Me, Et, *i*-Pr, *E*-crotyl, Ph	59-80	94->99	[80]
4	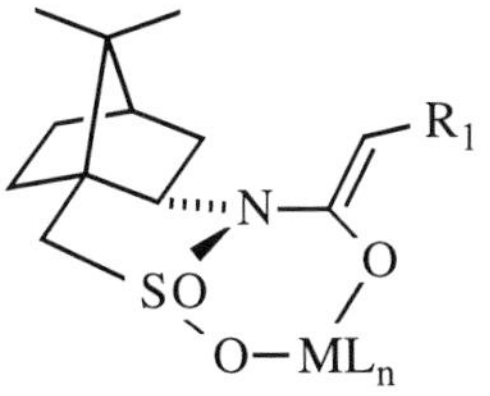 ML$_n$=SnBu$_3$ or Li	c (R$_1$=Me, Et)	*n*-Pr, *i*-Pr, *E*-orotyl, Ph	31-67	65-85	[80]

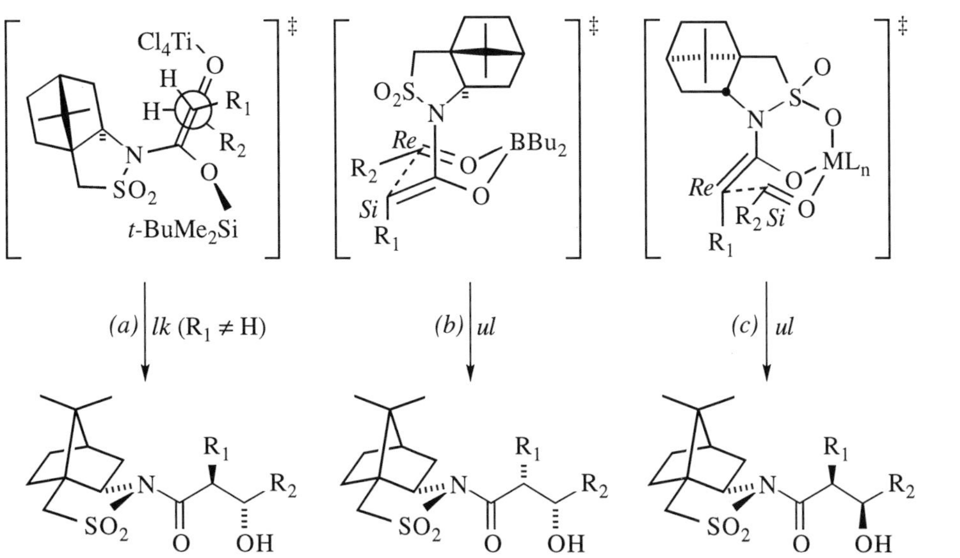

Scheme 5.14. Open and closed transition structures for the aldol addition of Oppolzer's sultams.

For syn aldolates, the boron enolate affords excellent selectivities and high yields [80], as shown in Table 5.5, entry 3. The rationale for the product configuration is shown in Scheme 5.14b, and is similar to the rationales presented above for other auxiliaries. Specifically, dipolar alignment of the C–O and S–O_2 bonds, coordination of the aldehyde to the boron, and approach of the aldehyde from the less hindered *Si* face in a Zimmerman-Traxler transition structure affords the absolute configuration shown. The lithium and tin enolates afford chelation-controlled syn adducts (Table 5.5, entry 4), as illustrated for the tin enolate in Scheme 5.14c. Both the lithium and the tin chelate the sultam oxygens while simultaneously coordinating the aldehyde oxygen. Addition again is thought to occur via a 6-membered chair Zimmerman-Traxler transition structure [80]. As with the Evans auxiliary, using the other enantiomer of the auxiliary is a more practical solution to changing the metal, if the syn isomer is desired.

Interligand asymmetric induction. Asymmetric induction by chiral ligands on the enolate metal has the advantage that the chiral moiety does not have to first be attached to one of the reactants and later removed (or destroyed). It is present only after enolate formation, and can be recovered for reuse. The introduction of chirality in the enolate metal (or metalloid) and its ligands is the intellectual stepping stone toward developing asymmetric catalysis for the aldol addition reaction, in that the stereogenic unit responsible for the asymmetric induction is not covalently bonded to either reactant. Additionally, chiral ligands on the metal allow double asymmetric induction when one of reactants is chiral, and triple asymmetric induction when both are. Most of the work that has been done in this area uses the same metals discussed in the previous section: boron, lithium, titanium, and tin.

In 1986, the groups of Masamune [92] and Paterson [93] reported (virtually simultaneously) that boron enolates containing C_2-symmetric "BR_2" moieties are effective mediators in asymmetric aldol additions. The Masamune group [92] studied the aldol addition of boron esters of *tert*-heptyl thiol acetate and propionate *E(O)*-enolates. As shown in Scheme 5.15, both types of reagents were highly selective. When R_1 is hydrogen, the selectivities are somewhat lower, because the *tert*-heptyl group can rotate away from the C_2-symmetric boracycle. When R_1 is an alkyl group, $A^{1,3}$ strain forces the *tert*-heptyl group toward the boracycle, crowding the transition structure and increasing the free energy difference ($\Delta\Delta G^{\ddagger}$) between the two illustrated transition structures. The product esters could be reduced to the corresponding primary alcohols. In spite of the high selectivities, the method has the disadvantage that the chiral boron compound is difficult to make.

Following an early lead from the Meyers group [94,95], Paterson used the readily available diisopinocampheyl (Ipc) boron triflate to make *Z(O)*-boron enolates of 3-pentanone [93] and other ketones [96], which add to aldehydes to produce syn adducts in 83 - 96% es (Scheme 5.16 and Table 5.6). Based on molecular mechanics calculations [55,56], the transition structure analysis shown in Scheme 5.16 was suggested to rationalize the enantioselectivity. The axial boron ligand rotates so that the C–H bond is over the top of the Zimmerman-Traxler six-membered ring, and the equatorial ligand orients with its C–H bond toward the

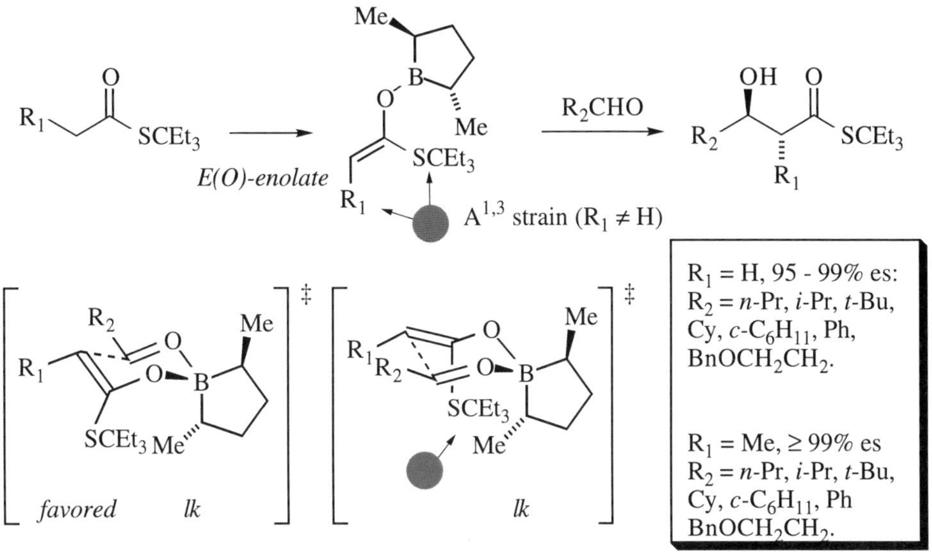

Scheme 5.15. Masamune's chiral boron enolate aldol additions [92].

axial ligand. It is interesting to note that, because of severe van der Waals interactions, the two boron-carbon bonds are conformationally locked. Note that the two methyls of the isopinocampheyl moieties are both oriented similarly, with the equatorial Ipc-methyls pointed toward the viewer. A simpler representation is to depict the carbon attached to boron as shown in the middle, with 'L' and 'S' representing the CHMe and CH₂ ligands respectively. The favored transition structure has the enolate oriented away from the 'L' ligand to avoid van der Waals repulsion between 'L' and the pseudoaxial R_2 moiety [55,56,96].

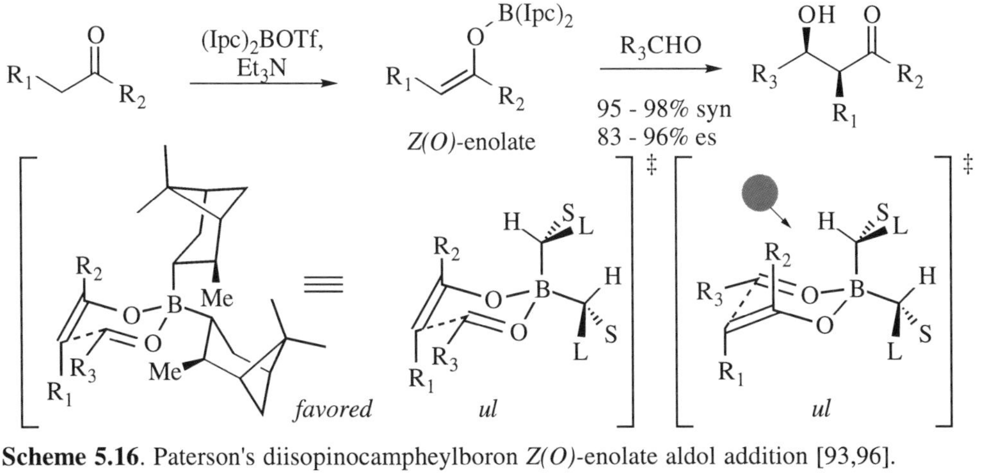

Scheme 5.16. Paterson's diisopinocampheylboron *Z(O)*-enolate aldol addition [93,96].

Use of diisopinocampheyl boron chloride in place of the triflate affords *E(O)*-enolates, but the isopinocampheyl ligands were ineffective for anti aldol reactions [48]. Encouraged by the molecular mechanics analysis of the *Z(O)*-enolate additions, Gennari and Paterson used computational methods to design a new boron ligand for use with *E(O)*-enolates [97]. The design was cued by Still's comment [98] that *cis*-2-

ethyl-1-isopropylcyclohexane has only one conformation that avoids 2,3-*P*-3,4-*M* gauche pentane interactions (Figure 5.5).[10] This conformation is analogous to a diaxial *cis*-1,3-dimethylcyclohexane interaction. They realized that replacement of the ethyl group with a CH_2B moiety would afford a molecule that is similarly conformationally constrained, and that such a molecule was available from menthone (Figure 5.5). Molecular mechanics calculations suggested that aldol additions of *E(O)*-enolates using this ligand on boron would be enantioselective [97].

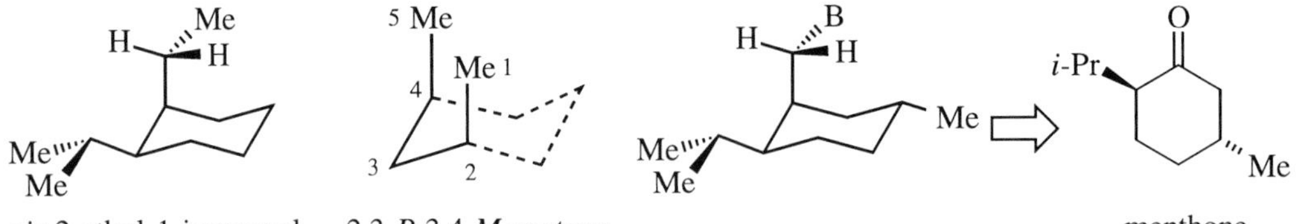

cis-2-ethyl-1-isopropyl-   2,3-*P*-3,4-*M*-pentane   menthone
cyclohexane   "gauche pentane"

Figure 5.5. The illustrated conformation of *cis*-2-ethyl-1-isopropylcyclohexane is the only one that has no destabilizing "gauche pentane" interactions [98]; similar interactions restrict the conformational motion of a boron ligand available from menthone [97].

When the "methylmenthyl" (MeMn) ligand was evaluated for selectivity in the addition of *E(O)*-enolates [97], it was found that the adducts were 86 - 100% anti, and the enantioselectivities were 78 - 94% (Scheme 5.17 and Table 5.6). The transition structure suggested to explain the chirality sense of the products again features the pseudoaxial R_2 avoiding interaction with the larger of the ligands (*i.e.*, menthyl) on the axial carbon bonded to boron. In the isopinocampheyl ligand (Scheme 5.16), the 'large' and 'small' ligands were rather similar (CH_2 *vs.* CHMe); in the present instance, the difference is huge (H *vs.* menthyl). Note that the indicated (*) bond is the one that is restricted by the "gauche pentane" interactions in the menthyl moiety.

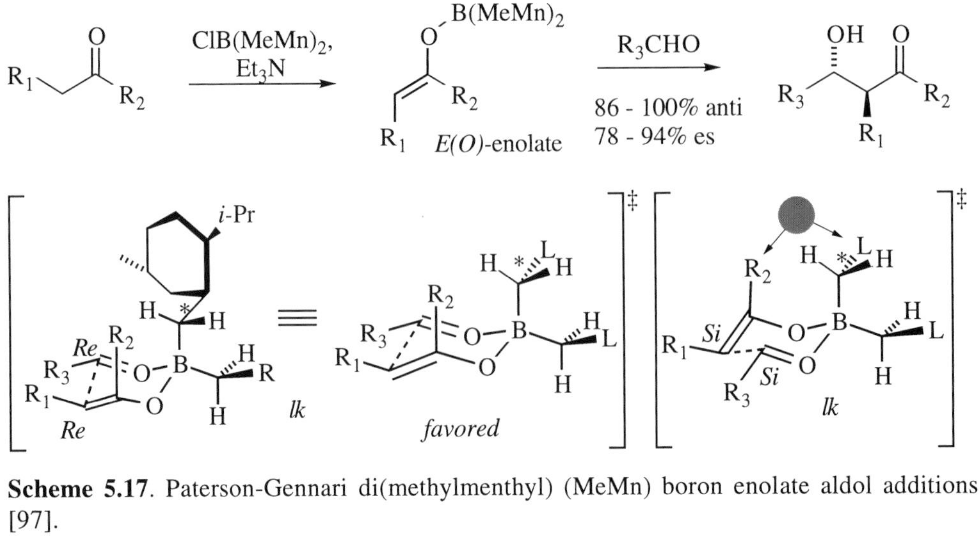

Scheme 5.17. Paterson-Gennari di(methylmenthyl) (MeMn) boron enolate aldol additions [97].

---

[10] See the glossary (Chapter 1) for an explanation of the *P*, *M* terminology.

Table 5.6. Asymmetric aldol additions of ketone enolates using chiral ligands on boron (Ipc = isopinocampheyl; MeMn = methylmenthyl). See Schemes 5.16 and 5.17.

$$R_1 \overset{O}{\underset{R_2}{\bigwedge}} \xrightarrow[\text{2. R}_3\text{CHO}]{\text{1. R}_2\text{BOTf, Et}_3\text{N}} R_3 \overset{\text{OH}}{\underset{R_1}{\bigwedge}} \overset{O}{\underset{}{\bigwedge}} R_2$$

Entry	BR$_2$	R$_1$	R$_2$	R$_3$	syn:anti	% yield	% es	Ref.
1	Ipc	Me	Et	Me	97:3	91	91	[96]
2	Ipc	Me	Et	2-propenyl	98:2	78	95	[96]
3	Ipc	Me	Et	n-Pr	97:3	92	90	[96]
4	Ipc	Me	Et	E-C$_3$H$_5$	98:2	75	93	[96]
5	Ipc	Me	Et	i-Pr	96:4	45	83	[96]
6	Ipc	Me	Et	2-furyl	96:4	84	90	[96]
7	Ipc	Me	Ph	2-propenyl	98:2	97	95	[96]
8	Ipc	Me	i-Pr	2-propenyl	95:5	99	94	[96]
9	Ipc	Me	i-Bu	2-propenyl	97:3	79	93	[96]
10	MeMn	Me	Et	2-propenyl	3:97	62	88	[97]
11	MeMn	Me	i-Pr	2-propenyl	0:100	51	94	[97]
12	MeMn	Me	Et	Et	8:92	50	90	[97]
13	MeMn	Me	i-Pr	Et	3:97	50	92	[97]
14	MeMn	Me	i-Pr	c-C$_6$H$_{11}$	0:100	54	87	[97]
15	MeMn	-(CH$_2$)$_3$-		2-propenyl	0:100	60	87	[97]
16	MeMn	-(CH$_2$)$_4$-		2-propenyl	0:100	59	78	[97]
17	MeMn	Me	Ph	2-propenyl	14:86	60	93	[97]
18	MeMn	H	i-Pr	2-propenyl	–	66	88	[97]
19	MeMn	H	i-Bu	2-propenyl	–	80	77	[97]
20	MeMn	H	t-Bu	2-propenyl	–	62	88	[97]
21	MeMn	H	Me	2-propenyl	–	65	80	[97]
22	MeMn	H	Ph	2-propenyl	–	81	85	[97]
23	MeMn	H	Me	n-Pr	–	65	87	[97]
24	MeMn	H	Et	2-propenyl	–	51	81	[97]

In addition to their usefulness for the asymmetric addition of achiral aldehydes, it will be seen in the section 5.2.3 that the Paterson strategy is particularly useful for the aldol addition of chiral fragments such as the large, polyfunctional ketone and aldehyde fragments needed for convergent macrolide synthesis.

In 1989, Corey reported that diazaborolidines are efficient reagents for asymmetric aldol additions of acetate and propionate thioesters [99]. Thioesters add to aldehydes giving syn adducts, whereas *tert*-butyl esters give anti adducts [100]. Both react via closed, Zimmerman-Traxler transition states; the difference in the topicity is due to different enolate geometries for the two ester types. Corey's rationale for the divergent enolate geometries involves competing mechanisms for deprotonation of the zwitterion shown in Scheme 5.18. Complexation of the boron reagent with the ester produces the zwitterionic complex (boxed), which may undergo either E$_1$ or E$_2$ elimination of HBr. Ionization (E$_1$) is favored when RX is thiophenyl and disfavored when RX is *tert*-butoxy; E_1 is also favored when the base is bulky, while smaller bases facilitate E_2 reaction. Note that E_1 ionization can only occur when X can easily stabilize the positive charge by resonance, which is only possible when the substituent on X (R) becomes coplanar with the rest of the molecule. Deprotonation by an E_2 mechanism is faster with the less bulky triethyl amine than with diisopropylethyl amine. Corey suggests that both E_1 and E_2 reactions occur from the illustrated conformation of the zwitterion [100].

For esters, deprotonation is effected with triethyl amine (which favors E_2), while E_1 ionization is disfavored because it requires moving the bulky *tert*-butyl group into planarity. For thioesters, E_1 reaction is facilitated by the thiophenyl group, while E_2 reaction is slowed by use of the bulky diisopropyl ethyl amine.[11]

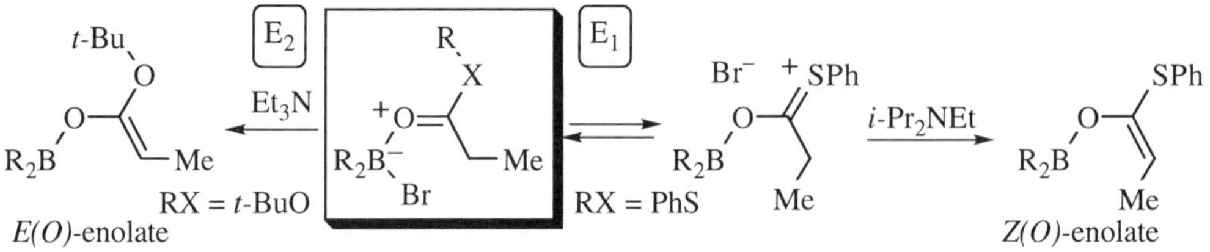

Scheme 5.18. Rationale for boron enolate stereochemistry [100].

The diazaborolidines mediate the diastereoselective and enantioselective formation of syn [99,100] anti aldols [100], as summarized in Scheme 5.19. The aryl group of the sulfonamide must be electron withdrawing, or else the boron is not a strong enough Lewis acid to mediate the process. The process has also been used for the formation of anti halohydrins [102,103], and in aldol additions to azomethines [101]. The illustrated transition structure (Scheme 5.19a) has been postulated to account for the observed enantioselectivity in the *Z(O)*-enolate addition [99]. Corey suggests that the trans phenyl substituents force the sulfonamide aryl groups into a conformation that places each aryl ring in a trans orientation to its neighbor. This *conformation* is reminiscent of the *configuration* engineered by Masamune earlier (Scheme 5.15), and may have similar control features. It is interesting to note, however, that a similar chair transition structure employing the *E(O)*-enolate predicts the wrong enantiomer (Scheme 5.19b) of the anti addition product [104]!

Duthaler and colleagues have used diacetone glucose as a ligand on titanium to induce enantioselectivity in the addition of acetate and propionate enolates (Scheme 5.20 [105,106]. The most interesting feature of the addition of the titanium enolate of *tert*-butyl acetate (Scheme 5.20a) is that the best selectivities were achieved at room temperature, making this procedure one of the most promising for scaleup [105]. Deprotonation of 2,6-dimethylphenyl propionate gives the *E(O)*-enolate, which is transmetalated slowly to the titanium enolate at −78° [106]. Addition to a

---

[11]   A rationale similar to this may be used to explain the selective formation of *E(O)*-enolates of *tert*-heptylthio- and *tert*-butylthiopropionates (Scheme 5.15 and ref. [101]): the Et3CS– replaces the Me3CO– in the Scheme 5.18 rationale) and the selective formation of ketone *Z(O)*-enolates with dialkyl boron triflates and *E(O)*-enolates with dialkylboron halides (Scheme 5.16 and 5.17: the triflates are more likely to ionize than the halides, thus favoring ionization over direct deprotonation of the zwitterion).

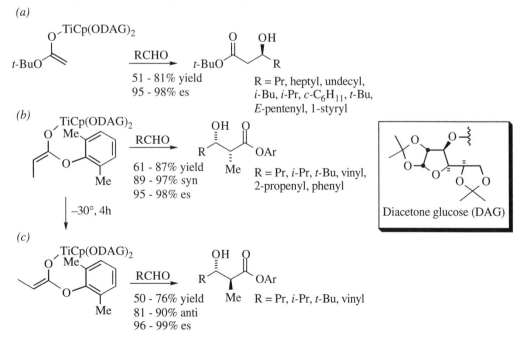

number of aldehydes affords predominantly syn adducts in excellent diastereo- and enantioselectivities. Warming the titanium enolate to −30° results in isomerization to the *Z(O)*-enolate, which adds to aldehydes with varying degrees of diastereo-selectivity. Some of the more selective examples are shown in Scheme 5.20b. Note that these examples are an exception to the generalization that *Z(O)*-enolates afford syn

Scheme 5.20. Duthaler's diacetone glucose titanium enolate aldol additions [105,106].

adducts and $E(O)$-enolates afford anti adducts. The authors suggest a boat transition structure to account for the fact (*cf.* Figure 5.3b), but do not speculate on the conformation of the diacetone glucose ligands and do not suggest a model to account for the chirality sense of the product. Finally, a single example (not illustrated here) of a $Z(O)$-enolate of an (achiral) oxazolidinone propionimide was reported to add to isobutyraldehyde in 50% yield, 88% diastereoselectivity (anti), and 97% enantio-selectivity [106]. Other groups have examined chiral diamine ligands on achiral tin [107] and lithium enolates [108,109], but the selectivities are not as high as reported for the titanium diacetone glucose aldols.

In all of the examples considered so far, the chiral element has been employed in stoichiometric quantities. Ultimately, it would be desirable to require only a small investment from the chirality pool. This is only possible if the chiral species respon-sible for enantioselectivity is catalytic. It is worth stating explicitly that, in order to achieve asymmetric induction with a chiral catalyst, the catalyzed reaction must proceed faster than the uncatalyzed reaction. One example of an asymmetric aldol addition that has been studied is variations of the Mukaiyama aldol reaction [110] whereby silyl enol ethers react with aldehydes with the aid of a chiral Lewis acid. These reactions proceed via open transition structures such as those shown in Figure 5.3c.[12]

In 1991 and 1992, several groups reported boron-based Lewis acids for catalytic Mukaiyama aldol additions (Figure 5.6). Three of these are oxazaborolidines derived from the reaction of borane with amino acid derivatives (Figure 5.6a-c), while the fourth (Figure 5.6d) is derived from tartrate. Examples of aldol additions using these catalysts are listed in Table 5.7. The turnover numbers are not large (20 - 100 mole-percent of catalyst being required), and the enol ether variability is somewhat limited. The Kiyooka catalyst (Figure 5.6a; Table 5.7, entries 1 - 3) and the Masamune catalyst (Figure 5.6b; Table 5.7, entry 4) are similar, and have been evaluated for the asymmetric addition of ketene acetals. The Kiyooka catalyst only becomes catalytic (*cf.* entries 1 and 2) in nitromethane solvent. The Corey (Table 5.7, entry 5) and Yamamoto (Table 5.7, entry 6) catalysts are effective with enol ethers of ketones, but not ketene acetals.

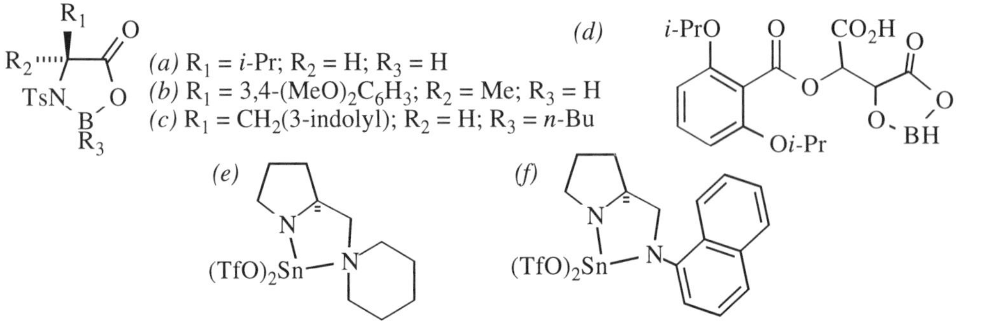

Figure 5.6. Chiral catalysts for the Mukaiyama aldol reaction: *(a)* Kiyooka catalyst [112,113]; *(b)* Masamune catalyst [114]; *(c)* Corey catalyst [115]; *(d)* Yamamoto catalyst [116,117]; *(e-f)* Kobayashi-Mukaiyama catalysts [118-120].

----

[12]   For an asymmetric Mukaiyama aldol that proceeds by an 'ene' mechanism, see ref. [111].

Table 5.7. Catalytic Mukaiyama aldol additions. The catalyst column refers to the structures in Figure 5.6

$$R_2 \overset{OTMS}{\underset{R_1 \quad X}{>=<}} + R_3CHO \xrightarrow{\text{catalyst}} \underset{R_1 \quad R_2}{\overset{TMSO \quad O}{R_3 \diagup \diagdown X}}$$

Entry	Catalyst (mole %)	R_1, R_2	X	R_3	% Yield	% es	Ref.
1	a (100)	Me, Me	OEt	Ph, E-PhCH=CH–, Ph(CH₂)₂–	80-87	91-96	[112]
2	a (20)	Me, Me	OEt	i-Pr, Ph, E-PhCH=CH–, Ph(CH₂)₂–	60-97	91-98	[113]
3	a (20)	H, H	OPh	Ph	66	90	[113]
4	b (17)	Me, Me	OEt	n-Pr, i-Bu, Ph, c-C₆H₁₁, Ph(CH₂)₂–, BnO(CH₂)₂–	68-86	92-99	[114]
5	c (20)	H, H	n-Bu, Ph	n-Pr, c-C₆H₁₁, 2-furyl, Ph	56-100	93-96	[115]
6	d (20)	H, Me (E, Z mix)	Et, n-Bu, Ph	n-Pr, n-Bu, E-CH₃CH=CH–, E-PhCH=CH–, Ph	55-99 (80 - >95% syn)	77-96	[116]
7	e (100)	H, H	SEt	i-Pr, t-Bu, Ph, Ph(CH₂)₂	77-90	91-99	[118]
8	f (100)	H, H	SEt	i-Pr, i-Bu, Ph, c-C₆H₁₁, n-C₇H₁₅, E-PhCH=CH–, E-MeCH=CH–, Ph(CH₂)₂–	70-96 (100% syn)	>99	[118]
9	f (20)	Me, H	SEt	n-C₅H₁₁, Ph, c-C₆H₁₁, E-PhCH=CH–, E-MeCH=CH–	67-80 (80-100% syn)	94-99	[121]

An interesting point is the difference between the Masamune (Figure 5.6b) and Kiyooka (Figure 5.6a) catalysts. One is catalytic and the other is not (Table 5.7, entries 1 and 4). Masamune screened a number of catalysts, including ones similar to Kiyooka's (Figure 5.6a), and suggested the catalytic cycle illustrated in Scheme

5.21, in which step two is thought to be the slow step. The critical difference is the quaternary stereocenter in the Masamune catalyst ($R_2 \neq H$). When the stereocenter in the catalyst is quaternary, the N–C–CO bond angle is compressed (Thorpe-Ingold effect – see glossary, section 1.6), thereby accelerating the silicon transfer (note the cyclic transition state) so that the catalyst can turn over more efficiently.

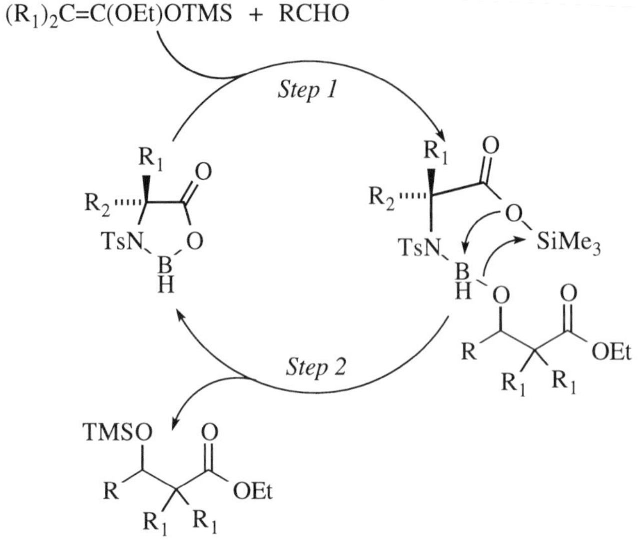

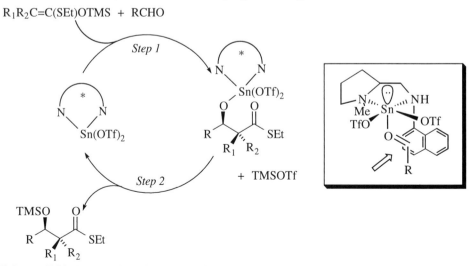

Scheme 5.21. Catalytic cycle for oxazaborolidine catalyzed Mukaiyama aldol addition (after ref. [115]).

Another catalytic system has been developed by Kobayashi and Mukaiyama. Specifically, tin triflates ligated by chiral diamines (Figure 5.6e,f) activate aldehydes toward addition by silyl enol ethers of acetate and *E(O)*-propionate thioesters (Table 5.7, entries 7-9). The catalytic version is thought to go by the two-step process shown in Scheme 5.22, with the slow step again being release of the alk-

Scheme 5.22. Proposed catalytic cycle for the Kobayashi-Mukaiyama aldol addition. Inset: proposed model for the aldehyde *Si*-face selectivity due to the catalyst [121,122].

oxide adduct by silylation [121,122]. Polar solvents such as propionitrile improve the catalytic process, presumably by increasing the rate of step two [121]. The rationale for the *Si*-face selectivity for the aldehyde is shown in the inset [122], however the authors did not postulate a transition structure to rationalize the topicity. It is clear that the mechanism does not involve silicon-tin enolate exchange [118], however unlike many acid catalyzed aldol additions, both the rate and the selectivity of the addition are dependent on enolate geometry. For the examples in Table 5.7, entry 9, an open geometry (*ul* topicity) having the methyl and the aldehyde substituent antiperiplanar (*cf.* Figure 5.3c) may be involved.

5.2.3 Double asymmetric induction and synthetic applications

Not all of the methods discussed in the preceding section have been explicitly studied with chiral aldehydes. However, most chiral aldehydes do not have a very high facial bias (see Cram's rule, section 4.1),[13] and the high selectivities obtainable by a number of the chiral aldol reagents discussed above permit "reagent-based stereocontrol" to be achieved through double asymmetric induction (see chapter 1). For example, as part of a synthesis of 6-deoxyerythronolide-B and the Prelog-Djerassi lactone (Figure 5.8, [124]), Masamune examined the selectivity of each enantiomer of his chiral enolate (Figure 5.4b) with a chiral aldehyde, as shown in Scheme 5.23. The aldehyde, which itself has a low inherent bias, and shows only 60% diastereoselectivity when allowed to react with an achiral boron enolate, may be converted selectively into either of the two possible syn adducts with 94% and 98% diastereoselectivity for the mismatched and matched cases, respectively [124].

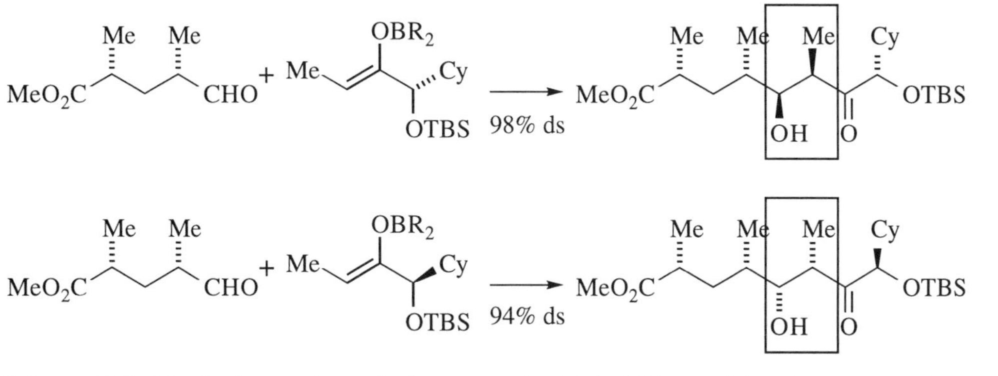

Scheme 5.23. Matched and mismatched asymmetric aldol additions of the Masamune enolate [124]. (Cy = cyclohexyl)

In his synthesis of the Prelog-Djerassi lactone, Evans tested two auxiliaries (*cf.* footnote 9) that give products with opposite absolute configurations at the two new stereocenters [125]. Here again, the facial bias inherent in the aldehyde is low. Since the two chiral enolates are not enantiomers, we cannot say which is the matched and which is the mismatched case, but it hardly matters: the selectivity is ≥99.8% for both (Scheme 5.24).

---

[13] For a thorough analysis of stereoselective aldol additions of achiral lithium and boron enolates to chiral aldehydes, see ref. [123].

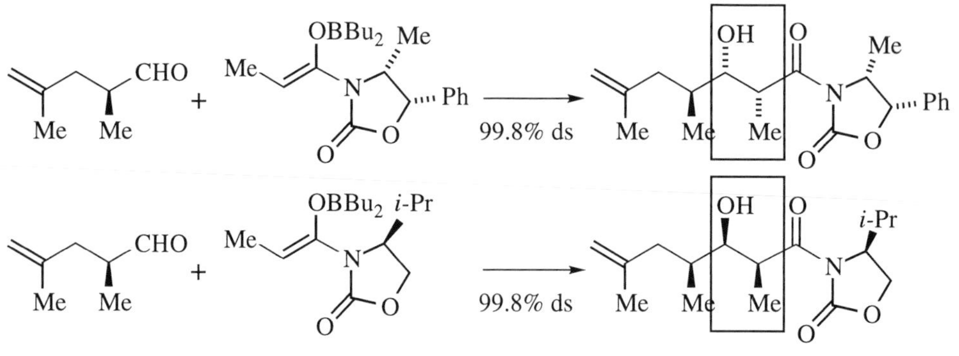

Scheme 5.24. Reagent-based stereocontrol in aldol additions using Evans imide enolates [125].

One cannot always bank on reagent-based stereocontrol, even with reagents as selective as the Evans imide enolates. For example, during the course of a synthesis of cytovaricin [126], the enolate shown in Scheme 5.25 was expected to afford the syn adduct when added to the aldehyde illustrated. Instead, an anti aldol adduct was formed as a single diastereomer. Note that the *Re* face of the enolate is preferred according to the transition state analysis presented in Scheme 5.12a, and the *Si* face of the aldehyde is preferred according to the Felkin-Anh theory (section 4.1 and Figure 4.8 or see glossary, section 1.6). Analysis of the product configuration, as shown in the inset, indicates that the preferred faces of both the enolate and aldehyde were coupled. Apparently, the *Si* facial preference of the aldehyde was sufficiently strong to disrupt the *lk* topicity preferred by the enolate.

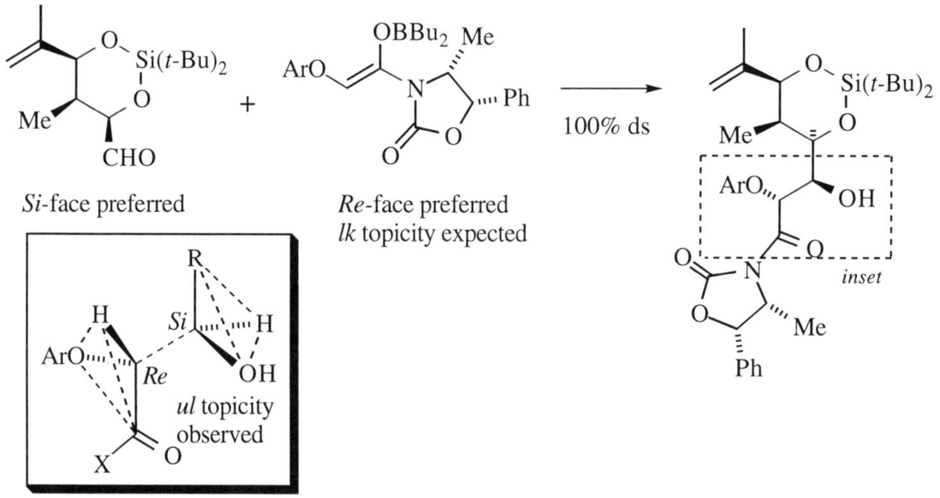

Scheme 5.25. A rare case of mismatched double asymmetric induction that is 100% diastereoselective [126].

Although the asymmetric addition of propionate enolates (outlined above) is a valuable synthetic tool, its use is restricted to targets that are amenable to a linear synthetic plan. Aldol additions may also be used to couple two large fragments in a convergent synthesis, but such reactions are not amenable to auxiliary based approaches. This weakness was recognized early in the development of the asymmetric aldol methodologies. For example, in the synthesis of 6-deoxyerythron-olide B, the Masamune group assembled two fragments that were coupled with a poor selectivity (60-71% ds) using boron enolates (Scheme 5.26). Switching to a lithium enolate increased the selectivity to 94%, but considerable background work had to be undertaken to insure success [127,128]. This result is somewhat surprising since boron enolates are typically more selective than lithium enolates.

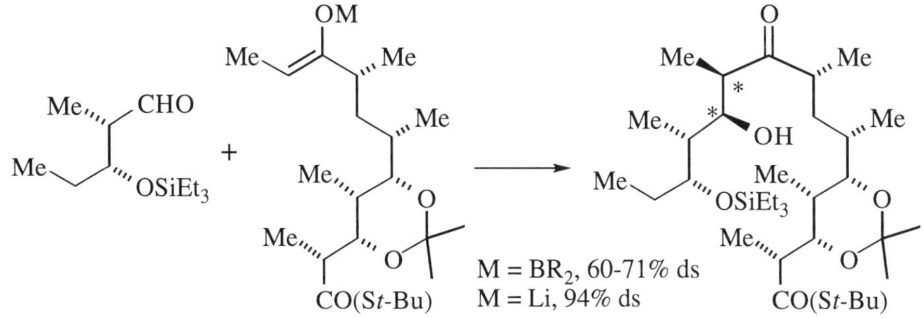

Scheme 5.26. Selective coupling of two chiral fragments (double asymmetric induction) in the asymmetric synthesis of 6-deoxyerythronolide B [124,127]. For a similar reaction in the synthesis of erythronolide B, see ref. [129].

Analysis of the major addition product of Scheme 5.26 (Figure 5.7a) indicates that the *Si* face of the enolate adds to the *Re* face of the aldehyde; the latter corresponds to anti-Cram selectivity (section 4.1). Two explanations have been offered to explain the selectivity of this aldehyde. Masamune originally suggested that the enolate adds to the aldehyde through a boat transition state that is also chelated by the silyloxy group (Cram cyclic model, section 4.2), as illustrated in Figure 5.7b. Weaknesses of this postulate are that Cram's cyclic model is more often effective when the chelate is a five-membered ring, and that the triethylsilyloxy group probably is not a good chelator [130]. Ten years after Masamune's original hypothesis was offered, Roush analyzed a considerable amount of data accumulated in the interim, and pointed out that a Zimmerman-Traxler chair, adding to the *Si* face of the aldehyde (as expected by the Felkin-Anh model), is de-

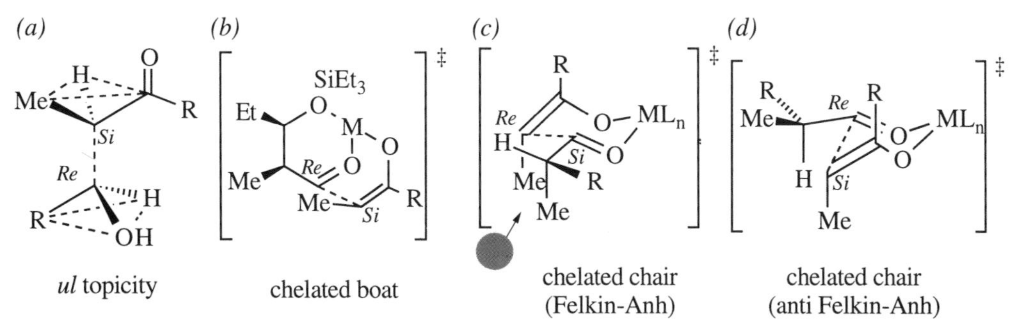

Figure 5.7. Analysis of possible transition structures for the aldol addition in Scheme 5.26: *(a)* The observed topicity; *(b)* boat transition structure postulated by Masamune [127]; *(c)* gauche pentane interaction that destabilizes the Cram (or Felkin-Anh) selectivity of the aldehyde; *(d)* anti-Cram (anti Felkin-Anh) addition via a chelated chair [123].

stabilized by a 2,3-*P*,3,4-*M* gauche pentane interaction (*cf.* Figure 5.5), as indicated in Figure 5.7c. Roush suggests that an anti Felkin-Anh (anti-Cram) chair transition structure more adequately explains the facts, as shown in Figure 5.7d [123].

Whatever the mechanism, achiral lithium enolates add to the aldehyde of Scheme 5.26 with selectivities in the 80-90% ds range; the higher selectivity observed with the chiral enolate may therefore be attributed to matched pair double asymmetric induction [127]. However, note that if the target had had the opposite absolute configuration at the indicated stereocenters, it would have been the minor isomer under any of the conditions examined.

The stereoselectivity of the aldol additions shown in Schemes 5.25 and 5.26 are obviously the result of a complex series of factors, among which are the Felkin-Anh preference dictated by the α-substituent on the aldehyde, the proximal stereocenters on the enolate, etc. Additionally, the more remote stereocenters, such as at the β-position of the aldehyde, may influence the selectivity of these types of reactions. Evans has begun an investigation into some of the more subtle effects on crossed aldol selectivity, such as protecting groups at a remote site on the enolate [131], and of β-substituents on the aldehyde component [132], and also of matched and mismatched stereocenters at the α and β positions of an aldehyde (double asymmetric induction) [133]. Further, the effect of chiral enolates adding to α,β-disubstituted aldehydes has been evaluated [134]. The latter turns out to be a case of triple asymmetric induction, with three possible outcomes: fully matched, partially matched, and one fully mismatched trio.

Another approach to the aldol problem has been investigated in the Paterson laboratory, in the hopes of using interligand asymmetric induction to control absolute configuration of the new stereocenters in the products [48,135,136]. Some examples are shown in Scheme 5.27. The chiral *E(O)*-enolate shown in Scheme

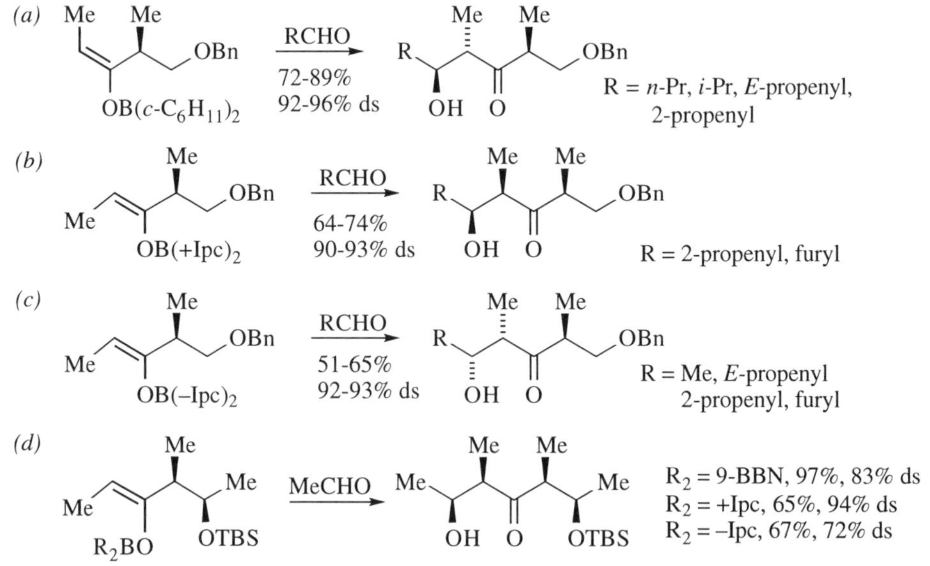

Scheme 5.27. (*a*) Anti-selective addition of ketone *E(O)*-enolate to aldehydes [137,138]; (*b, c*) Reagent controlled addition of *Z(O)*-enolate to aldehydes [126]; (*d*) Double asymmetric induction where the mismatched diastereoselectivity is decreased, not reversed [139].

5.26a adds to simple achiral aldehydes in high yield and with 95-96% diastereoselec-tivity [137]. In contrast, the *Z(O)*-enolate having achiral boron ligands, of this and similar chiral ketones, affords poor selectivity in aldol additions [126,137,138], probably because of gauche pentane interactions similar to those illustrated in Figure 5.7c. Double asymmetric induction via chiral ligands on boron can sometimes be used to control the configuration of the aldol adducts. As shown in Scheme 5.27b and c, either (+) or (–) isopinocampheyl (Ipc) ligands on boron (*cf.* Scheme 5.16) control the absolute configuration of the addition products for the enolate shown [126]. The Ipc ligands cannot always be relied upon to control inherent facial bias in the enolate, however, as shown in Scheme 5.27d [139]. In this example, the diastereoselectivity achieved with 9-BBN is enhanced with (+)-Ipc and diminished – but not reversed – with (–)-Ipc.

The aldol addition reaction, and the related crotyl metal additions (section 5.1), have figured prominently in the total synthesis of a number of complex natural products (reviews: [48,140-142]). Figure 5.8 illustrates those mentioned in the preceding discussion, along with others selected from the recent literature, with the stereocenters formed by stereoselective aldol addition indicated (∗). For the Prelog-Djerassi lactone and ionomycin, recall (Figure 3.8) that most of the other stereo-centers were formed by asymmetric enolate alkylation.

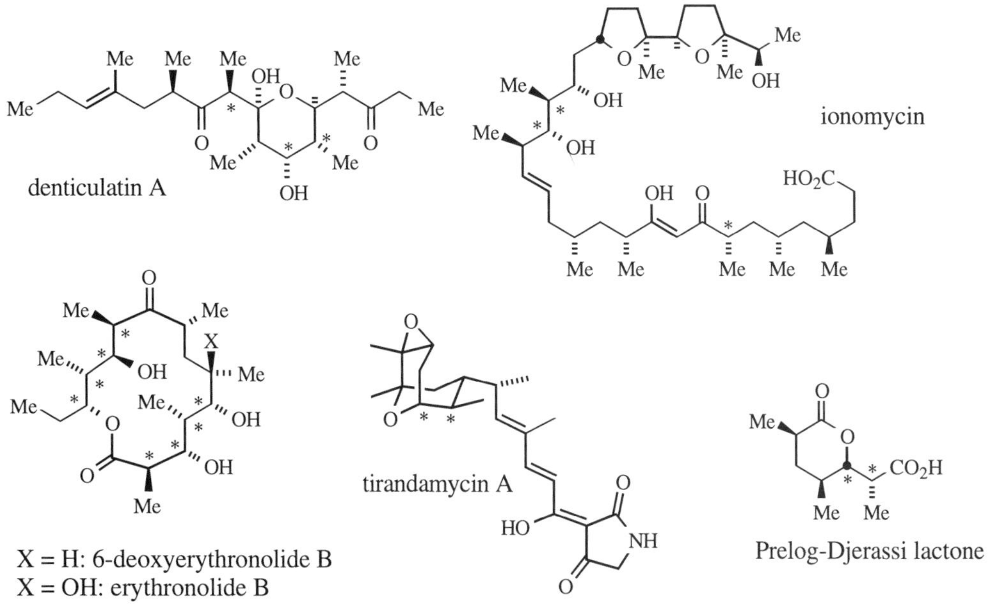

Figure 5.8. Natural products synthesized using aldol methodology: denticulatin A [143]; ionomycin [144]; 6-deoxyerythronolide B [124]; erythronolide B [129,145]; tirandimycin A [146]; Prelog-Djerassi lactone [124,125]. Stereocenters created in the aldol addition are indicated (∗).

5.3 Michael additions[14]

The term "Michael addition" has been used to describe 1,4- (conjugate) additions of a variety of nucleophiles including organometallics, heteroatom nucleophiles such as sulfides and amines, enolates, and allylic organometals to so-called "Michael acceptors" such as α,β-unsaturated aldehydes, ketones, esters, nitriles, sulfoxides, and nitro compounds. Here, the term is restricted to the classical Michael reaction, which employs resonance-stabilized anions such as enolates and azaenolates, but a few examples of enamines are also included because of the close mechanistic similarities.

5.3.1 Simple diastereoselectivity

When a prochiral acceptor ($R_1CH=A$) and a prochiral donor ($R_2CH=D$) react, the stereoisomers are labeled as either syn or anti based on the relative configurations of R_1 and R_2 when the Michael adduct is drawn in a zig-zag projection, as shown in Scheme 5.28. Using the *Re/Si* nomenclature and assuming that the CIP rank is $A>R_1>H$ and $D>R_2>H$, the syn adducts arise from *lk* topicity and anti adducts arise from *ul* topicity.

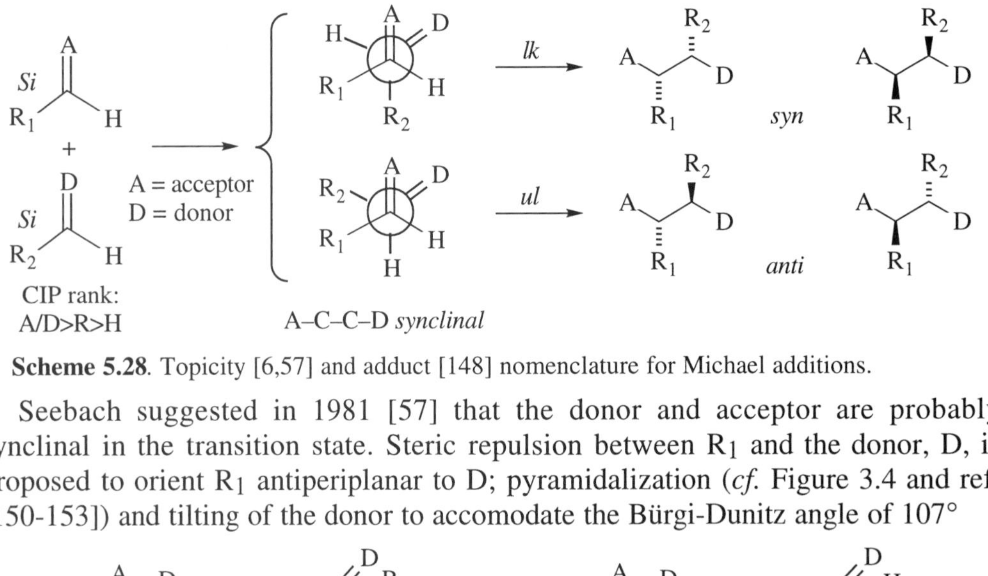

Scheme 5.28. Topicity [6,57] and adduct [148] nomenclature for Michael additions.

Seebach suggested in 1981 [57] that the donor and acceptor are probably synclinal in the transition state. Steric repulsion between R_1 and the donor, D, is proposed to orient R_1 antiperiplanar to D; pyramidalization (*cf.* Figure 3.4 and ref. [150-153]) and tilting of the donor to accomodate the Bürgi-Dunitz angle of 107°

Figure 5.9. Pyramidalization of the donor and the Bürgi-Dunitz trajectory contribute to destabilization of the *lk* topicity combination according to Seebach [57].

---

[14] For a comprehensive coverage of conjugate addition reactions, see ref. [147]. For a comprehensive review of the stereochemical aspects of base-promoted Michael reaction, see ref. [148]; for a similarly comprehensive review of acid-catalyzed Michael reactions and conjugate additions of enamines, see ref. [149].

(*cf.* Figures 4.6, 4.7, and ref. [154-156]) are proposed to disfavor the *lk* topicity, since R_2 is more sterically demanding than hydrogen (Figure 5.9).

Analysis of numerous examples [148,149] and mechanistic studies [157,158] led Heathcock to refine these hypotheses and, in consideration of the actual substituents (R_1, R_2, A, and D), place them on firmer mechanistic grounds.[15] The four transition structures in Scheme 5.29 are direct extensions of those in Scheme 5.28 and Figure 5.9. For ketone and ester enolates, there is a strong correlation between the relative configuration of the product and the enolate geometry: *Z(O)*-enolates give anti products and *E(O)*-enolates give syn adducts [157]. The rationale for this is that transition structures for paths a and c (Scheme 5.29) are favored due to repulsive interactions between Y and R_3 in paths b and d. The selectivity of *Z(O)*-enolates appears to be higher than that of *E(O)*-enolates, probably due to the destabilization of path c by the pyramidalization and trajectory considerations illustrated in Figure 5.9, which intrinsically favor paths a and d, in which a hydrogen is antiperiplanar to the enone double bond.

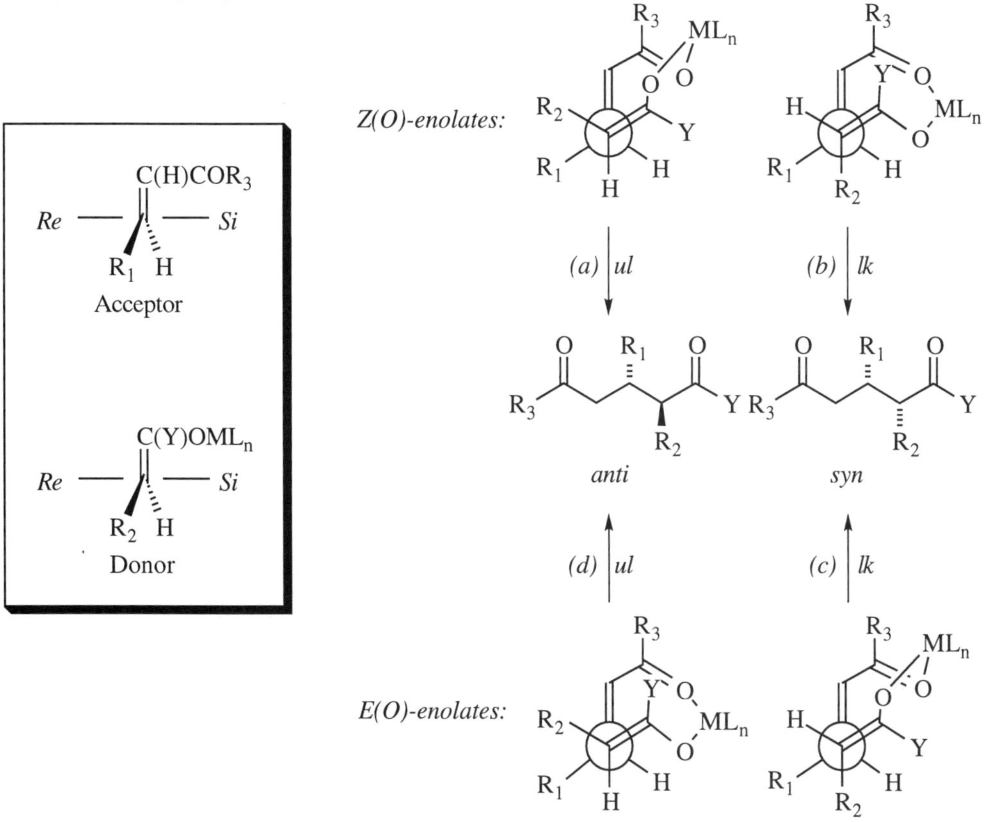

Scheme 5.29. Proposed chelated transition structures (and topicities) for Michael additions of lithium enolates of ketones, esters, and amides to enones [157,158]. Only one enantiomeric transition structure and product is shown for each topicity (*Si* face of the acceptor).

---

[15] On the other hand, a computational study [159] of the Michael addition of propionaldehyde lithium enolate adding to *E*-crotonaldehyde indicates an anticlinal conformation around the forming bond (*i.e.* A eclipsing R_2 in the *ul* topicity and A eclipsing H in the *lk* topicity of Figure 5.9).

For amide enolates, the situation is similar in that, when R$_3$ and Y are large, the transition structures of paths a and c are favored [158]. However, recall that acyclic amides invariably form Z(O)-enolates, so amide E(O)-enolates are only possible when R$_2$ and Y are joined: *i.e.,* in a lactam. In contrast to ketone and ester enolates, however, the transition structures of paths b and d appear to be intrinsically favored when Y and R$_3$ are small. This latter trend is (at least partly) contrary to what would be expected based on the simple analysis of Figure 5.9, but can be rationalized as follows. For the lactams, the R$_2$ and Y substituents present a rather flat profile, so that interaction with R$_3$ in path d is minimal. Additionally, the R$_2$–Y ring 'eclipses' the β-hydrogen of the enone in c, destabilizing this structure. For amide Z(O)-enolates and acceptors with an R$_3$ substituent such as a phenyl, there may actually be an attractive interaction between Y and R$_3$, favoring path b.

Clearly each case must be analyzed separately, but these transition structures serve as a starting point for such analyses. Note also that the structures of Scheme 5.29 all have enones in an *s*-cis conformation, which is not available to cyclic acceptors such as cyclohexenone, cyclopentenone, and unsaturated lactones.

For the purpose of asymmetric synthesis, we are interested in expanding on simple diastereoselectivity and differentiating between the two *ul* transition structures (*Re-Re* and *Si-Si*) and the two *lk* transition structures (*Re-Si* and *Si-Re*) for each enolate geometry. This is done by rendering the *Re* and *Si* faces of either component diastereotopic by the introduction of a stereogenic element. For asymmetric Michael reactions, a stereocenter in a removable substituent on the acceptor or in Y or the metal (ML$_n$) of the donor have been used to this end. Intro-duction of stereogenicity in substituents on the donor or the acceptor constitute auxiliary-based approaches, while a chiral ligand on the metal is interligand asymmetric induction. The following discussion is organized by the location of the stereogenic unit. Given the number of chiral enolate reagents developed for asymmetric alkylations and aldol additions, it should come as no surprise that many of these auxiliaries have also found use in Michael additions.

5.3.2 Chiral donors

Ester enolates. Oppolzer showed in 1983 that the Z(O)-dienolate shown in Scheme 5.30a adds to cyclopentenone with 63% diastereoselectivity [160]. Additionally, the enolate adduct can be allylated selectively, thereby affording (after purification) a single stereoisomer having three contiguous stereocenters in 48% yield. The transition structure illustrated is not analogous to any of those illustrated in Scheme 5.29 because cyclopentenone is an *s*-trans-Z-enone, whereas the enones in Scheme 5.29 are *s*-cis-E. In 1985, Corey reported the asymmetric Michael addition of the E(O)-enolate of phenylmenthone propionate to E-methyl crotonate as shown in Scheme 5.30b [161]. The product mixture was 90% syn, and the syn adducts were produced in a 95:5 ratio, for an overall selectivity of 86% for the illustrated isomer. The transition structure proposed by the authors to account for the observed selectivity is similar to that shown in Scheme 5.29c, but with the enone illustrated in an *s*-trans conformation. Intramolecular variations of these reactions were reported by Stork in 1986, as illustrated in Scheme 5.30c and 5.29d [162]. Two features of these reactions deserve comment. First, the carbonyl of the acceptor is not chelated to the enolate metal, and second, the selectivity of the camphor-derived ester is significantly higher than the similar reaction in Scheme 5.30a. The latter effect seems to be due to the position of the bridgehead methyl, which helps restrict conformational motion when it neighbors the ester enolate [162]. Note that hydrolysis of the adducts from these two reactions afford enantiomeric cyclopentanones.

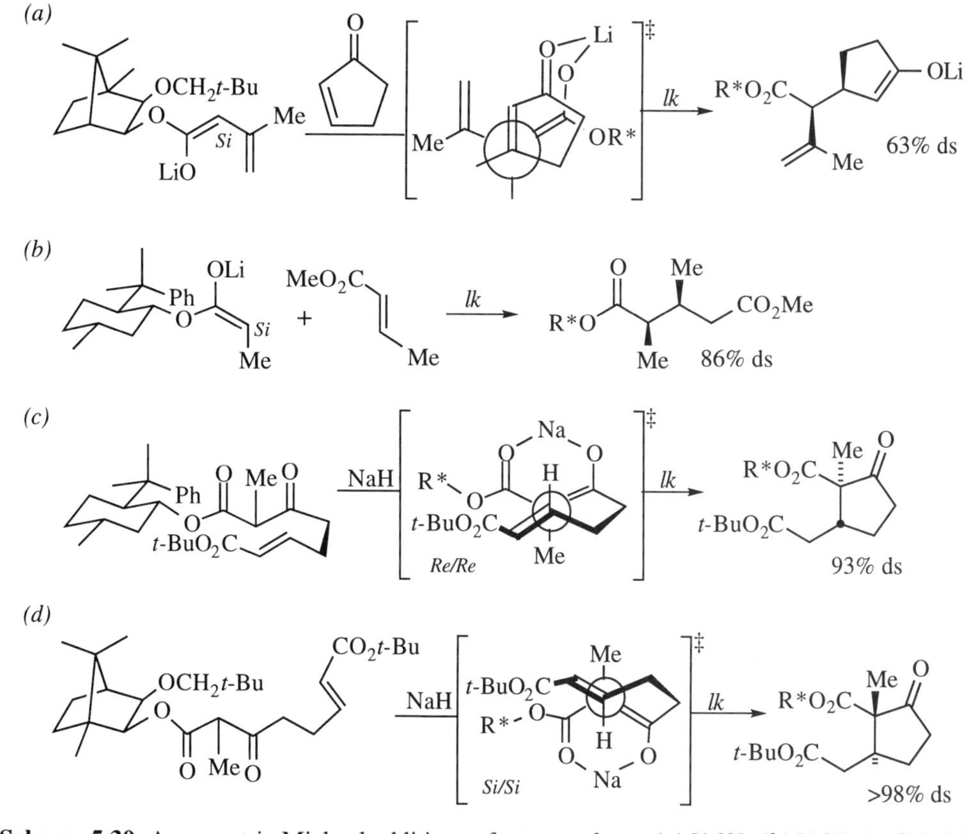

(a)

(b)

(c)

(d)

Scheme 5.30. Asymmetric Michael additions of ester enolates. *(a)* [160]. *(b)* [161]. *(c,d)* [162].

Amide and imide enolates. Scheme 5.31 illustrates several examples of asymmetric Michael additions of chiral amide and imide enolates. Yamaguchi [163] investigated the addition of amide lithium enolates to *E*-ethyl crotonate, but found no consistent topicity trend for achiral amides. The three chiral amides tested are illustrated in Scheme 5.31a-c. The highest diastereoselectivity found was with the C_2-symmetric amide shown in Scheme 5.31c.[16] Evans's imides, as their titanium enolates, afforded the results shown in Scheme 5.31d and e [164,165]. The yields and selectivities for the reaction with acrylates and vinyl ketones are excellent, but the reaction is limited to β-unsubstituted Michael acceptors: β-substituted esters and nitriles do not react, and β-substituted enones add with no selectivity [165].

---

[16] For these three examples, the syn/anti selectivity was 88-94%, and the diastereoselectivity within the major relative configuration was ≥87%.

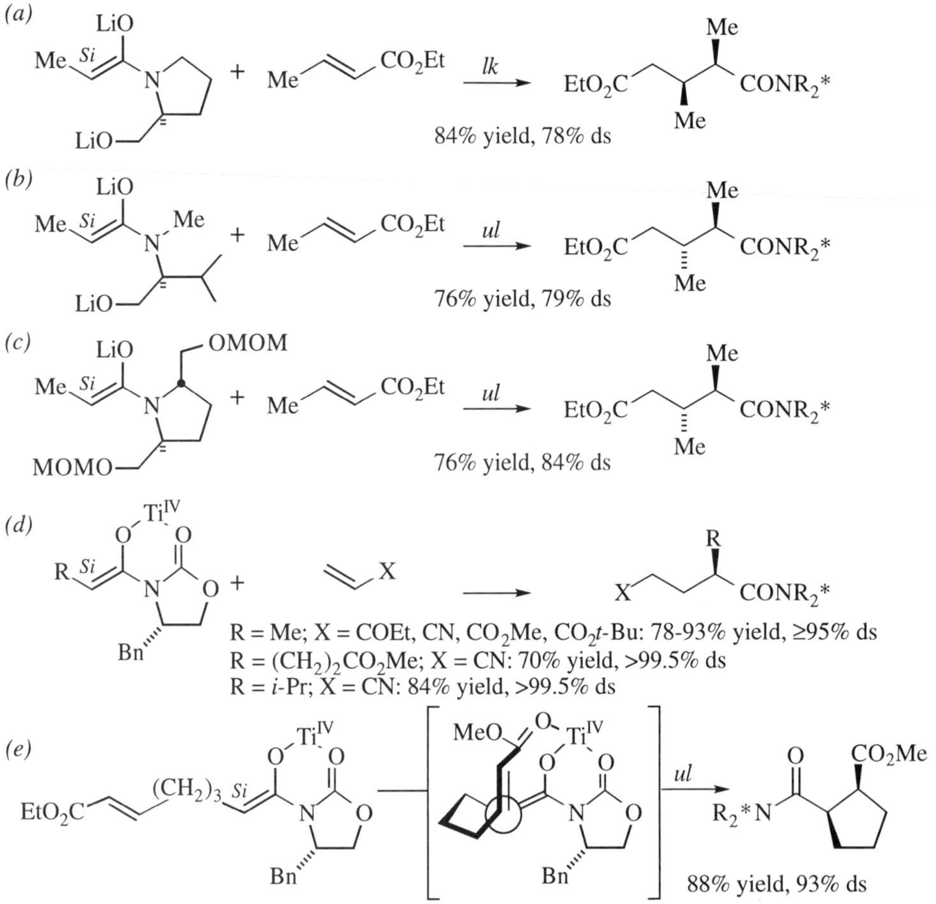

Scheme 5.31. Asymmetric Michael addition of amide and imide enolates. *(a-c)* [163]. *(d)* [164], [165]. *(e)* [165].

These four examples do not seem to comply with a consistent mechanistic model. The dilithioprolinol amide enolate in Scheme 5.31a is attacked on the enolate *Si* face, in accord with the sense of asymmetric induction observed in alkylations of this enolate [166,167]. On the other hand, the structurally similar dilithiovalinol amide enolate, while being attacked on the same face (as expected), reverses topicity. Furthermore, the *S,S*-pyrrolidine enolate in Scheme 5.31c is attacked from the *Si* face by Michael acceptors, but from the *Re* face by alkyl halides [168] and acid chlorides [169]. The titanium imide enolate in Scheme 5.31d adds Michael acceptors from the *Si* face, consistent with the precedent of aldol additions of titanium enolates (*cf.* Table 5.4, entry 2, [88]). An intramolecular addition (Scheme 5.31e) seems to follow a clear mechanistic path [165]: the *Si* face is attacked by the electrophile, and the cis geometry of the product implicates intramolecular complexation of the acceptor carbonyl. This coordination of the acceptor carbonyl is probably a function of the metal: recall the lithium ester enolates illustrated in Scheme 5.30c and d, but also metal chelation in titanium aldol additions (Table 5.4, entry 2).

Ketone and aldehyde azaenolates. Perhaps the most versatile of the auxiliaries for the asymmetric alkylation of ketones and aldehydes are the SAMP/RAMP hydrazones developed by Enders (*cf.* Schemes 3.21, 3.22, and Table 3.9). These hydrazones, as their lithium *E(O)*-enolates, also undergo highly selective Michael additions [170-173]. Several examples are illustrated in Scheme 5.32a. The rationale for the formation of the *E(O)*-enolate and for the *Re* facial selectivity of SAMP hydrazones is illustrated in Scheme 3.22. The Michael acceptors also react on the *Re* face of SAMP hydrazones, and the *ul* topicity at the new bond can be rationalized by Seebach's postulate (Figure 5.9 and Scheme 5.29d), that places the β-substituent of the Michael acceptor (R3 in Scheme 5.32a) antiperiplanar to the double bond of the donor (=D in Figure 5.9), and has the acceptor double bond (=A in Figure 5.9) bisecting the angle between the donor double bond (=D) and the donor substituent (R2 in Scheme 5.32a). Scheme 5.32b illustrates an extension for the synthesis of substituted cycloalkanes. The indicated (∗) stereocenters are formed in the Michael addition; in Scheme 5.32b, the other is formed by internal 1,2-asymmetric induction.

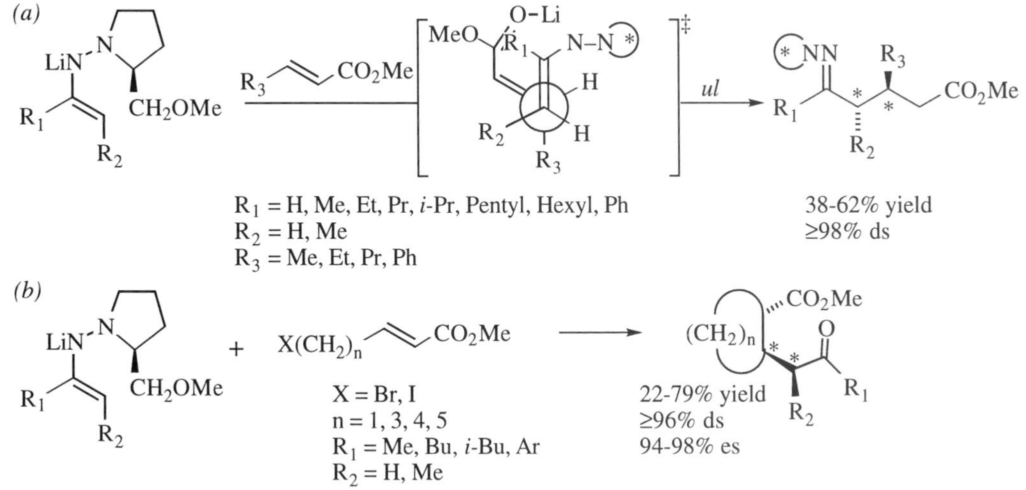

(a)

R_1 = H, Me, Et, Pr, *i*-Pr, Pentyl, Hexyl, Ph
R_2 = H, Me
R_3 = Me, Et, Pr, Ph

38-62% yield
≥98% ds

(b)

X = Br, I
n = 1, 3, 4, 5
R_1 = Me, Bu, *i*-Bu, Ar
R_2 = H, Me

22-79% yield
≥96% ds
94-98% es

Scheme 5.32. Michael additions of SAMP/RAMP hydrazones. *(a)* [170-172]. *(b)* [173]. Stereocenters formed in the Michael reaction are indicated (∗).

The Koga group has investigated the asymmetric Michael addition of β-keto esters, as their valine lithium enamides, as shown in Scheme 5.33 [174,175]. The lithium derivative adds directly to methylene malonic esters without further activation [174], but is not reactive enough to add to methyl vinyl ketone or ethyl acrylate unless trimethylsilyl chloride is also added [175]. Interestingly, the absolute configuration of the product changes when HMPA is added to the reaction mixture. The rationale for this observation is that in the absence of HMPA, the electrophile coordinates to the lithium, taking the position of L in the chelated structure shown in the inset, thus delivering the electrophile to the *Re* (rear) face. When the strongly coordinating HMPA is present, it occupies the 'L' position and blocks the *Re* face, thereby directing the electrophile to the *Si* face.

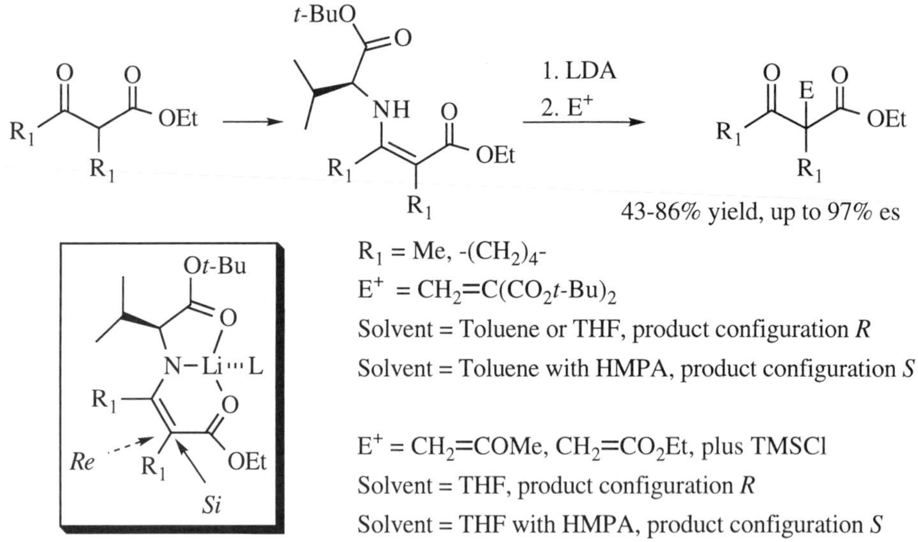

t-BuO

1. LDA
2. E⁺

43-86% yield, up to 97% es

R_1 = Me, -(CH$_2$)$_4$-

E⁺ = CH$_2$=C(CO$_2$t-Bu)$_2$

Solvent = Toluene or THF, product configuration *R*

Solvent = Toluene with HMPA, product configuration *S*

E⁺ = CH$_2$=COMe, CH$_2$=CO$_2$Et, plus TMSCl

Solvent = THF, product configuration *R*

Solvent = THF with HMPA, product configuration *S*

O*t*-Bu

N−Li···L

Re R$_1$ OEt

Si

Scheme 5.33. Koga's asymmetric Michael additions of valine enamines of β-keto esters [174,175].

Enamines.[17] The condensation of a secondary amine and a ketone to make an enamine is a well known reaction which has seen wide use in organic synthesis [176-178]. Imines of a primary amine and a ketone exist in a tautomeric equilibrium between the imine and secondary enamine forms, although in the absence of additional stabilization factors (*cf.* Scheme 5.33), the imine is usually the only detectable tautomer. Nevertheless, the enamine tautomer is very reactive toward electrophiles and Michael additions occur readily [179]. The mechanism of the Michael additions of tertiary and secondary enamines are shown in Scheme 5.34. For tertiary enamines, the Michael addition is accompanied by proton transfer from the α'-position to either the α-carbon or a heteroatom in the acceptor, affording the regioisomeric enamine as the initial adduct [180]. The proton transfer and the carbon–carbon bond forming operations may not be strictly concerted, but they are nearly so, since conducting the addition in deuterated methanol led to no deuterium incorporation [180].

    With secondary enamines, there is also transfer of a proton, but this time from the nitrogen. Again, isotope labeling studies [181] suggest that the two steps are "more or less concerted" [179], in a reaction that resembles the ene reaction (Scheme 5.34b).

    Theoretical studies indicate that these transition structures are probably influenced by frontier molecular orbitals (in addition to steric effects), as indicated in Scheme 5.34c [182]. For the reaction of aminoethylene (a primary enamine) and acrolein, the enamine HOMO and the enone LUMO have the most attractive interactions when aligned in the chair configuration shown, which has the enone in an *s*-cis conformation. Note that this orientation places the NH and the electrophile α-carbon in close proximity for proton transfer via the 'ene' transition structure.

---

[17] For a review of Michael additions of enamines, see ref. [149].

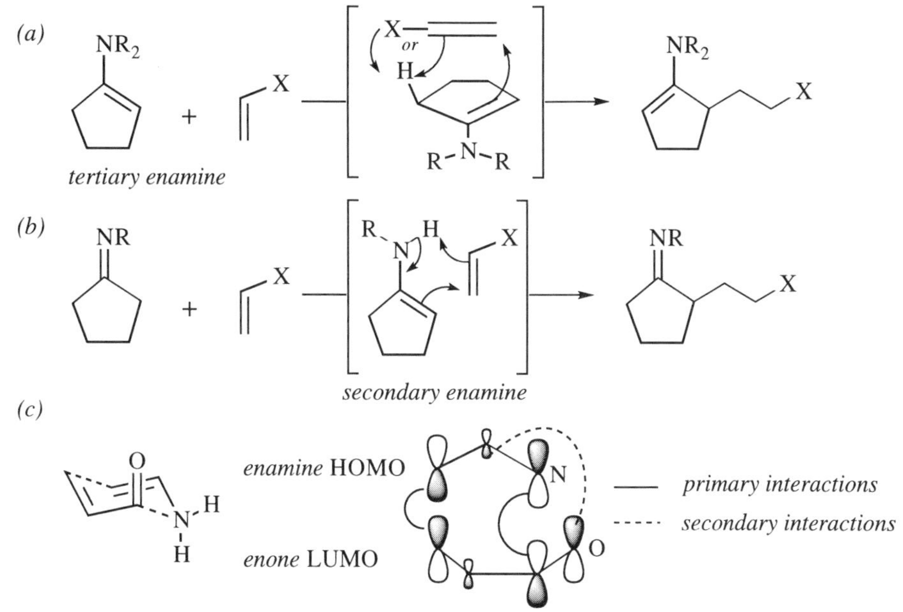

Scheme 5.34. *(a)* Suprafacial Michael addition-proton transfer of a tertiary enamine [180]. *(b)* aza-ene-like transition structure for secondary enamine Michael additions [179]. *(c)* Molecular orbital analysis of enamine and enone interactions [182].

Because the amines are removed in the subsequent hydrolytic workup, enamines are obviously amenable to an auxiliary-based asymmetric synthesis using a chiral amine. It is additionally significant from a preparative standpoint that unsymmetrical ketones alkylate at the less substituted position via tertiary enamines (*e.g.,* C_6 of 2-methylcyclohexanone) whereas the more hindered position is alkylated preferentially with secondary enamines (*e.g.,* C_2 of 2-methylcyclohexanone).

In 1969, Yamada demonstrated that the cyclohexanone enamine derived from proline methyl ester would add to acrylonitrile or methyl acrylate with 70-80% enantioselectivity (Scheme 5.35a, [183], but Ito later showed that the selectivity was much better if a prolinol ether was used instead (Scheme 5.35b, [184]. Seebach investigated the asymmetric Michael addition of enamines of prolinol methyl ether, as shown in the examples of Scheme 5.35c [185,186], and likewise found outstanding selectivities. These examples share a common sense of asymmetric induction at C_2 of the cyclohexanone, and the origin of the asymmetric induction is of interest. The example in Scheme 5.35d shows that the effect is not steric in origin, since the propyl group is isosteric and isoelectronic with the methoxymethyl group, but the selectivity is essentially lost without the oxygen (*cf.* Scheme 5.35c and d, [186]). Two possible explanations may explain these results. First, we assemble the two reactants in a synclinal orientation with the aryl group antiperiplanar to the donor double bond (Scheme 5.36, upper left; *cf.* Figure 5.9). One possibility is that the dipole of the methoxymethyl then stabilizes a zwitterionic intermediate in the nonpolar solvent, as shown in Scheme 5.36a.[18] Another is that

_____

[18] Ab initio studies suggest that a zwitterionic intermediate may normally be too high in energy to be kinetically accessible [182], but 'internal solvation' by the methoxymethyl may lower the barrier.

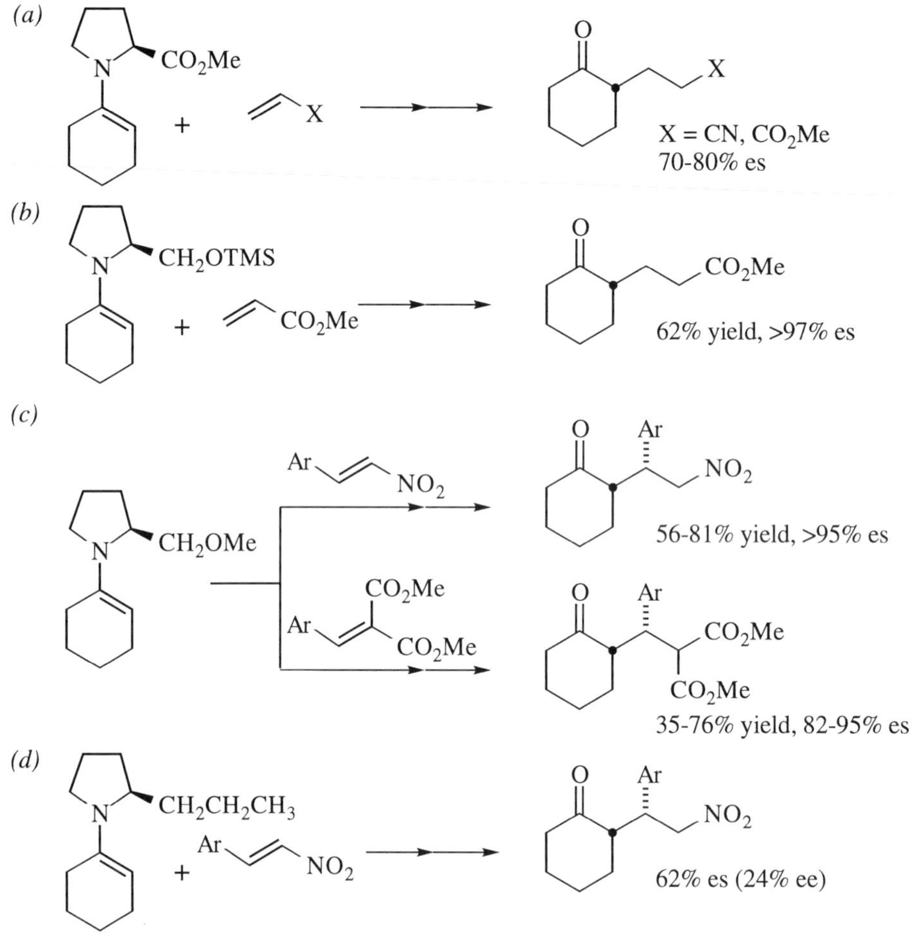

Scheme 5.35. Asymmetric Michael additions of chiral tertiary enamines. *(a)* [183]. *(b)* [184]. *(c)* [185], [186]. *(d)* [186].

the oxygen serves as a relay atom for a hydrogen transfer (another 'internal solvation' effect) such as illustrated in Scheme 5.36b (*cf.* Scheme 5.34a).

Other examples shed some light on the importance of the proton transfer in these enamine Michael additions. For example, the $\Delta^{1,2}$ enamine of β-tetralone (Scheme 5.36c) afforded high yields of *3-substituted* $\Delta^{1,2}$ enamine products, even though the $\Delta^{2,3}$ enamine isomer was not present in the reaction mixture [187] (see also ref [188]). Under the reaction conditions (toluene or ether, stirring for 3-4 days), the $\Delta^{1,2}$ isomer must isomerize to the $\Delta^{2,3}$ isomer which reacts much faster, probably due to the greater acidity of the benzylic proton of the $\Delta^{2,3}$ isomer compared to the C_3-proton of the $\Delta^{1,2}$ isomer.

The asymmetric Michael addition of secondary enamines has been reviewed by d'Angelo [179]. Some of the more selective examples of this type of reaction are listed in Table 5.8. It is significant that these Michael additions are highly regio-selective, reacting virtually exclusively at the more highly substituted carbon, which affords α,α-disubstituted (quaternary) cyclopentanones, cyclohexanones, furans,

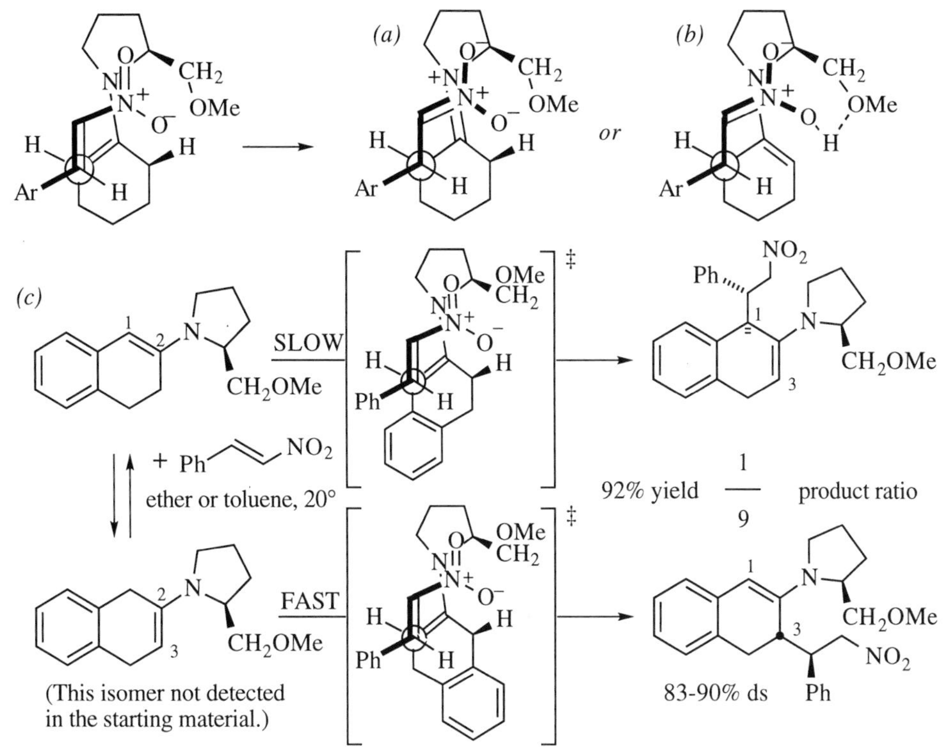

Scheme 5.36. Involvement of the methoxymethyl in the asymmetric Michael addition: *(a)* by dipolar stabilization of a zwitterion intermediate, or *(b)* by assisting in the proton transfer to the Michael acceptor. *(c)* Isolation of an addition product via enamine rearrangement [187].

and pyrans in excellent yields and selectivities. An important advantage of this process is that it is stereoconvergent: racemic 2-substituted ketones are converted into nearly enantiopure products. Limitations are that a nitrogen in place of the oxygen of entries 7 and 8 is not possible, and that a carbomethoxy group decreases the enamine reactivity such that Lewis acid catalysis is required [179]. The mild conditions of these reactions (nonpolar solvents, room temperature) and the high overall yields make this an attractive process for large scale applications. The products of these reactions will catch the eye of anyone familiar with the Robinson annelation and related reactions [189,190], as these types of compounds are used as key building blocks in numerous natural product syntheses.

What is the origin of the regioselectivity, and what determines the face-selectivity of the Michael addition? The regioselectivity results from the aza-ene-like mechanism of this reaction. As shown in Scheme 5.37, although both enamines may form, reaction of the less substituted isomer is retarded by $A^{1,3}$ strain effects. Note that in the aza-ene transition structure (Scheme 5.34b), the NH must be syn to the enamine double bond. Thus, the more highly substituted enamine isomer, in its most stable conformation, is in the proper conformation for Michael addition. In contrast, the reactive conformer of the less substituted isomer is destabilized by severe destabilizing steric interactions ($A^{1,3}$ strain) between the ring substituent and the nitrogen substituent, which increase as the carbon-nitrogen bond gains double bond character in the transition state.

Table 5.8. Michael additions to cyclic ketones and lactones via their secondary enamines. The yields listed are for the overall conversion of the ketone educt into the diketone or keto ester product.

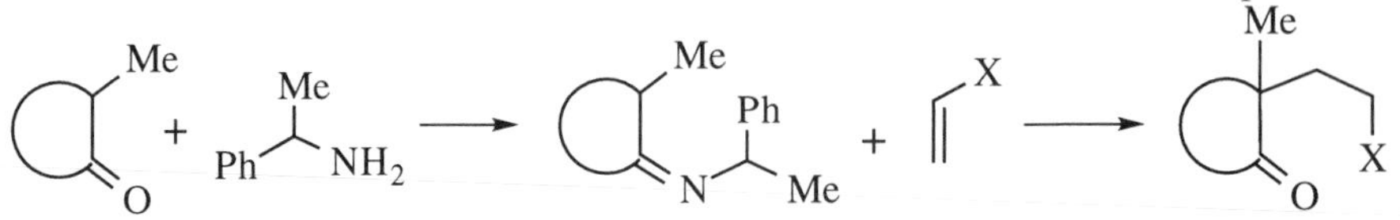

Entry	Ketone	Acceptor	Amine	Product	% Yield	% es	Ref.
1	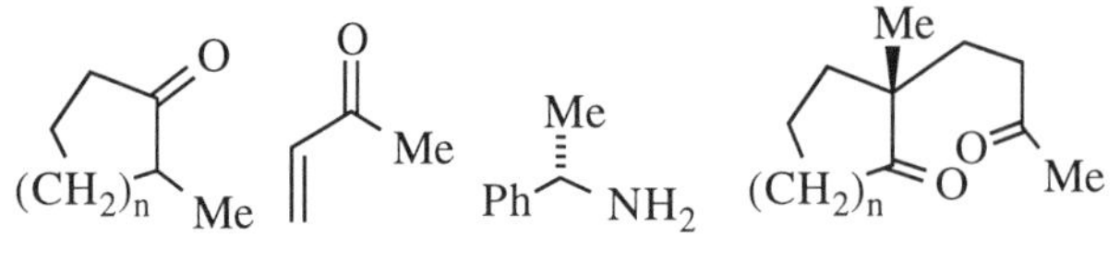				n=1, 83	94	[191]
2					n=2, 88	95	[191]
3	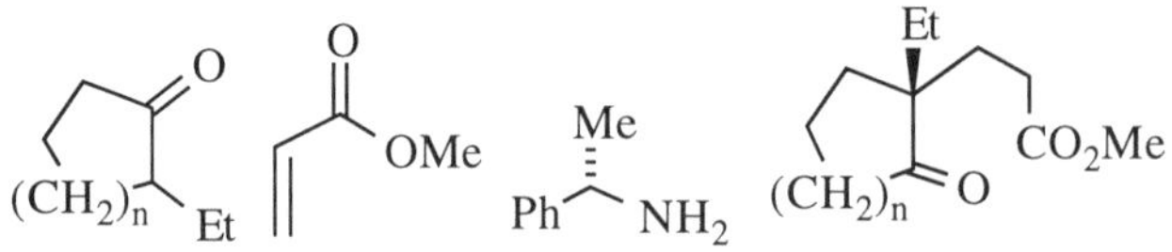				n=1, 83	95	[192]
4					n=2, 80	94	[179]
5	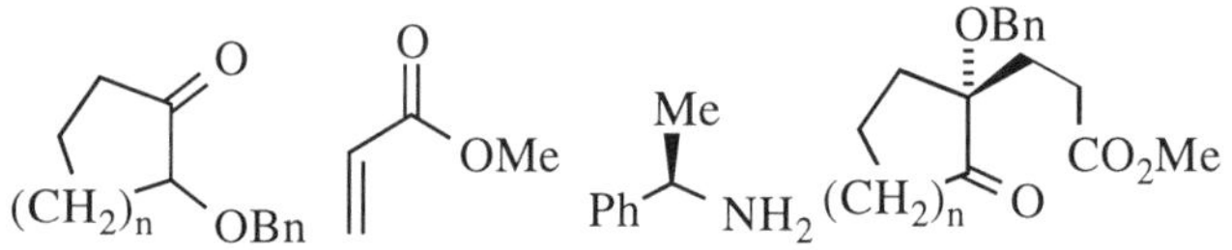				n=1, –	–	[179]
6					n=2, 75	99	[193]
7	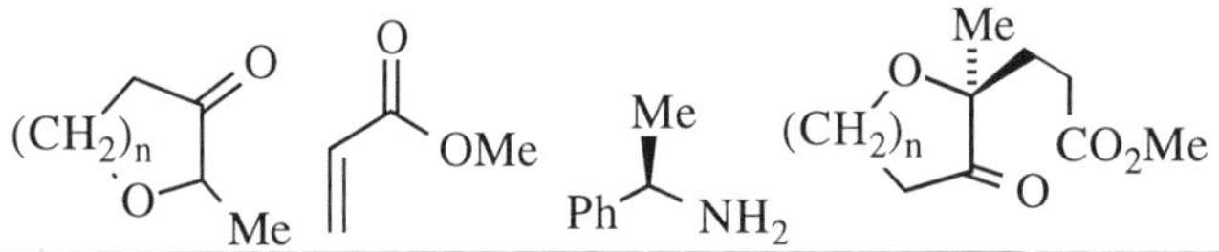				n=1, 80	98	[194]
8					n=2, 75	99	[194]

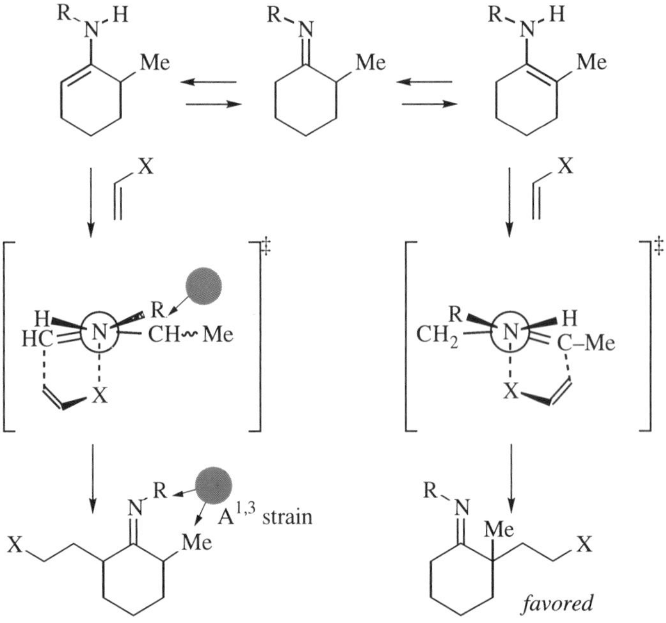

Scheme 5.37. $A^{1,3}$ strain raises the energy of the transition structure for Michael addition of the less substituted secondary enamine [179]. The dashed lines in the transition structures indicate the primary MO interactions, according to Scheme 5.34c.

The origin of the face selectivity was revealed by MNDO calculations of the chair transition structures shown in Scheme 5.38, which differ only in the face of the enamine to which the enone is attached. By constraining the two reactants into parallel planes 3Å apart and rotating around the indicated (∗) bond of each structure, the conformations shown were found to be the lowest in energy [179]. The ground state conformation probably approximates the center structure, with the benzylic carbon-hydrogen bond synperiplanar to C_1 of the cyclopentene due to repulsion of the methyl and phenyl groups by C_5. In the transition structures, the benzylic carbon-hydrogen bond rotates 60° and becomes synclinal to C_1. Comparison of these structures indicated an energy difference of about 1.1 kcal/mole, which corresponds closely to the value expected based on the observed selectivity [179].

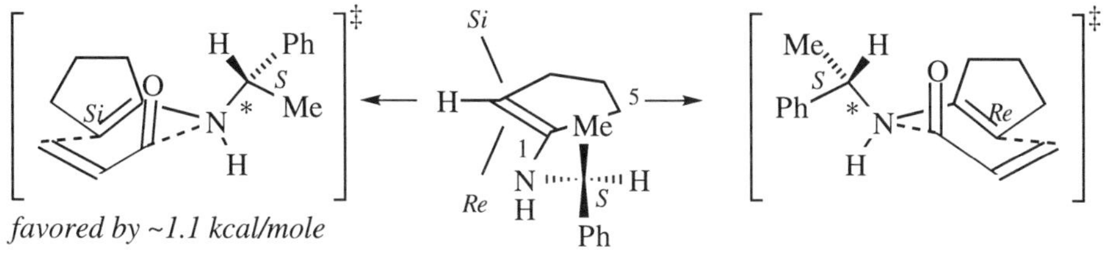

favored by ~1.1 kcal/mole

Scheme 5.38. Calculated low energy conformers for *Re* and *Si* attack of acrolein on cyclopentanone *S*-phenethyl enamine [179].

Allyl anions. The sulfur and phosphorous-stabilized allyl anions shown in Figure 5.10 have been examined by the Hua and Hanessian groups in asymmetric Michael additions to several enones. In these auxiliaries, the sulfur and the phosphorous are stereogenic, and the phosphorous additionally has chiral ligands. Some of the more selective examples of Michael additions using these ligands are listed in Table 5.9.

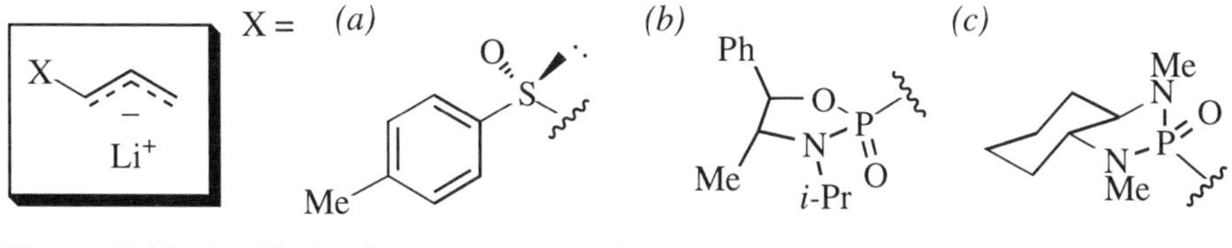

Figure 5.10. Auxiliaries for asymmetric Michael addition of allyl anions: *(a)* [195]. *(b)* [196]. *(c)* [197].

The mechanism of allylic sulfoxide addition is proposed to occur through a chelated 10-membered ring transition structure [198], as shown in Scheme 5.39a. The illustrated conformation features the favorable alignment of the molecular orbitals illustrated in the inset (*cf.* Scheme 5.34). However, it also has been suggested [148] that the reaction may proceed by sequential 1,2-addition followed by an alkoxide-accelerated Cope rearrangement,[19] as shown in Scheme 5.39b. Note that the same conformation and orbital alignment are operative in this mechanism. For the addition of the phosphorous-stabilized allyllithium of Figure 5.10b and c, 10-membered rings are postulated [196,197]. The 10-membered rings shown in Schemes 5.39c and d have conformations similar to that shown in Scheme 5.39a; conceivably the tandem 1,2-carbonyl addition/3,3-Cope rearrangement suggested

---

[19] Such a mechanism has been demonstrated in the addition of dithianyl allyl lithiums [199].

Table 5.9. Asymmetric Michael additions of sulfur and phosphorous stabilized allyllithiums. The X column refers to the auxiliaries in Figure 5.10.

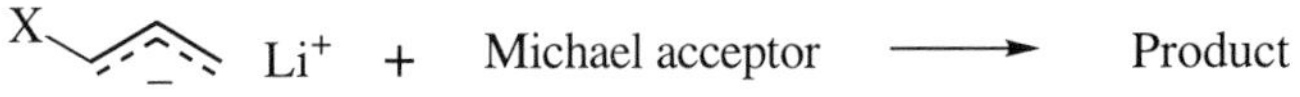

X⟋⟍⟋ Li$^+$ + Michael acceptor ⟶ Product

Entry	X	Acceptor	Product	% Yield	% es	Ref.
1	a	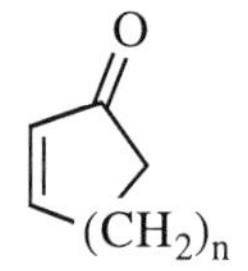	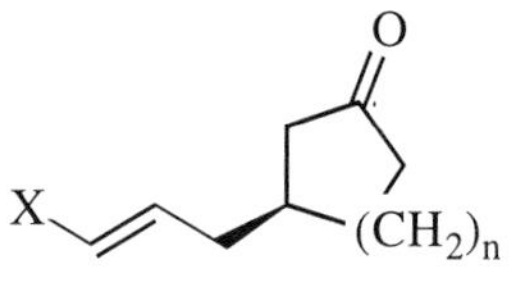	n=1, 91	98	[195]
2	b			n=1, 79	99	[196]
3	c			n=1, 88	96	[197]
4	b			n=2, 70	94	[196]
5	b			n=3, 71	97	[196]
6	a	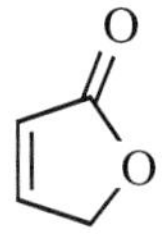	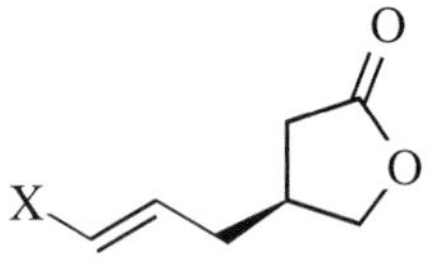	82	85	[195]
7	c			93	>99	[197]
8	a	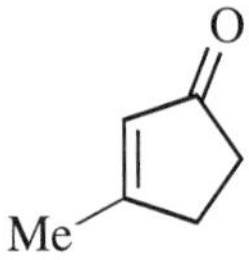	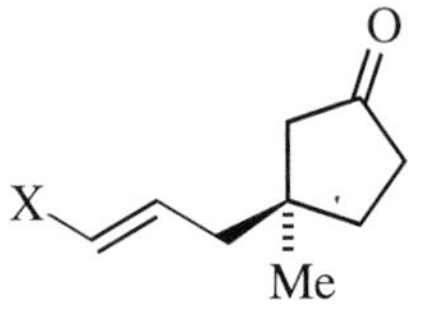	80	97	[195]
9	c			75	97	[195]
10	c	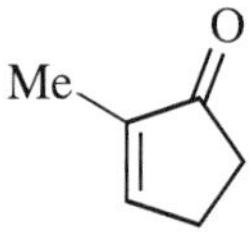	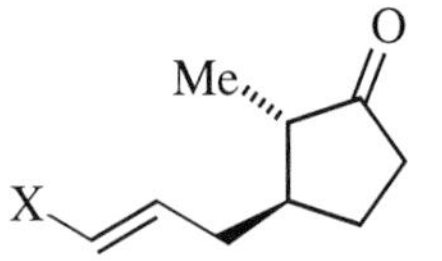	80	96	[197]
11	c		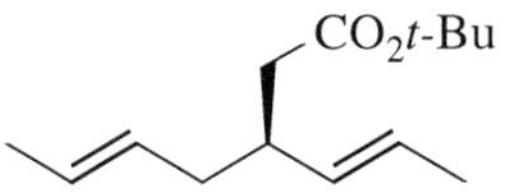	76	>99	[197]

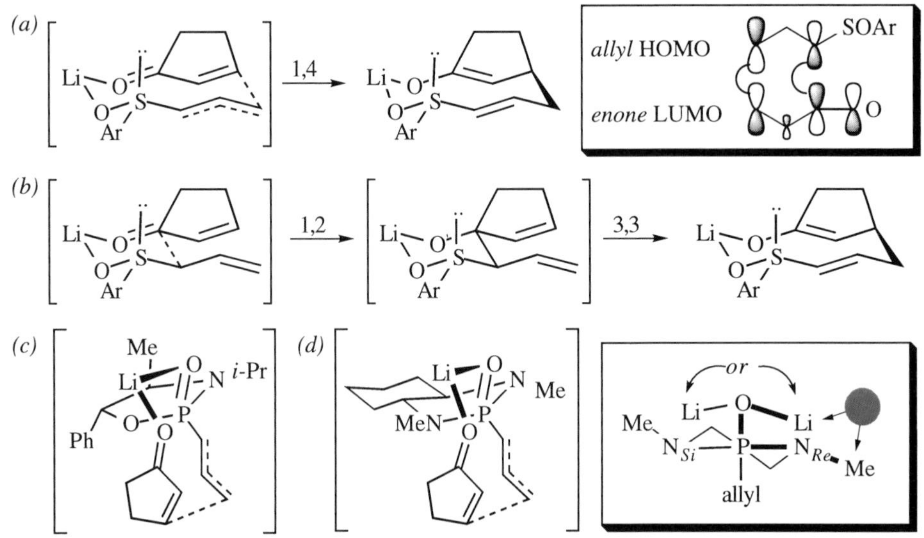

Scheme 5.39. Allyl sulfoxide additions: (a) 1,4-mechanism [198]. (b) Tandem 1,2-addition / 3,3-rearrangement mechanism [148] (see also ref. [199]). (c,d) Transition structures for allyl phosphine oxides [196,197]. *Inset:* Gauche pentane interaction between lithium and the N_{Re} methyl.

for the sulfoxides could intervene in these cases as well. For these auxiliaries, the site of lithium coordination to the phosphoryl group determines the chirality sense of the products. For the phosphaoxazolidine (Scheme 5.39c), the lithium coordinates anti to the bulky *N*-isopropyl substituent, but for the phosphaimidazolidine (Scheme 5.39d) the reason for the similar placement is not as obvious. The inset illustrates the 5-membered heterocycle in a half-chair conformation, with the *N*-methyls in pseudoequatorial configurations (the half chair is held rigid by the trans fused cyclohexane, which is deleted for clarity). The *N*-methyls are labeled according to their relative configurations on the stereogenic phosphorous. Note that coordination of the lithium syn to the N_{Re}-methyl generates a lithium/methyl interaction reminiscent of 2,3-*P*-3,4-*M* (gauche) pentane (*cf.* Figure 5.5). Coordination syn to the N_{Si}-methyl does not. Thus the latter site is preferred.

5.3.3 Interligand asymmetric induction

In considering Michael addition transition structures such as those generalized in Scheme 5.29, differentiation between two enantiomers of the same topicity can be achieved by introducing a stereogenic unit into either the donor (*vide supra*), the acceptor (*vide infra*) or the ligands on the metal. Metals can be efficiently complexed by crown ethers, and enolates form mixed aggregates with amines and lithium amides in solution. If an aggregate is chiral by virtue of a chiral ligand or a chiral crown, then interligand asymmetric induction can occur. As was true with the aldol addition (*cf.* Schemes 5.15-5.22), and enolate alkylations (*cf.* Schemes 3.23-3.26), chiral metal ligands offer the advantage of not requiring extra steps for the introduction and removal of an auxiliary, and may be amenable to catalysis. The examples illustrated below do not exhibit the outstanding selectivities that can be achieved by an auxiliary-based method, and there is little evidence upon which to base a rationale to explain the sense of asymmetric induction, but as knowledge of enolate/aggregate structures is gained, such insight will follow quickly and new systems with higher selectivities will undoubtedly emerge.

Following a 1973 lead by Långström and Bergson, who used a partially resolved amino alcohol as an asymmetric Michael catalyst [200], Wynberg used quinine as a catalyst for the asymmetric addition of 2-carbomethoxy-1-indanone to methyl vinyl ketone, obtaining 88% enantioselectivity in an optimized case (Scheme 5.40a), but the absolute configuration of the product was not determined [201]. Carbomethoxy-cyclohexanones could also be employed in this process, but the selectivities were low [201]. The Seebach group showed that cyclohexanone lithium enolates show good selectivities when complexed to chiral diamines or chiral lithium amides (Scheme 5.40b) [3]. They also noted significantly improved yields (and often improved selectivities) when an additional equivalent of lithium bromide was added to the recipe, results which clearly indicate the participation of enolate mixed aggregates in the reaction. The topicity (relative configuration of the stereocenters in the product) of this addition is consistent with a transition structure similar to that shown in Scheme 5.29c (see also Scheme 5.36). The Mukaiyama group explored the use of tin enolates complexed to chiral diamines as shown in Schemes 5.40c and d. The propionate imide enolate shown in Scheme 5.40c (when used in excess) adds to benzal acetone with excellent selectivity [202]. The topicity of this addition is consistent with a mechanism similar to that shown in Scheme 5.29a, but note that titanium enolates of chiral imides added with low selectivity to β-substituted enones (Scheme 5.31d and e, [165]). If the dithioketene acetal shown in Scheme 5.40d is added slowly to a mixture of an enone, tin triflate, and chiral diamine, good enantioselectivities are achieved with catalytic amounts of tin and diamine [203]. The slow addition is necessary to keep a low concentration of the dithioketene acetal so as to minimize a competitive nonselective addition.

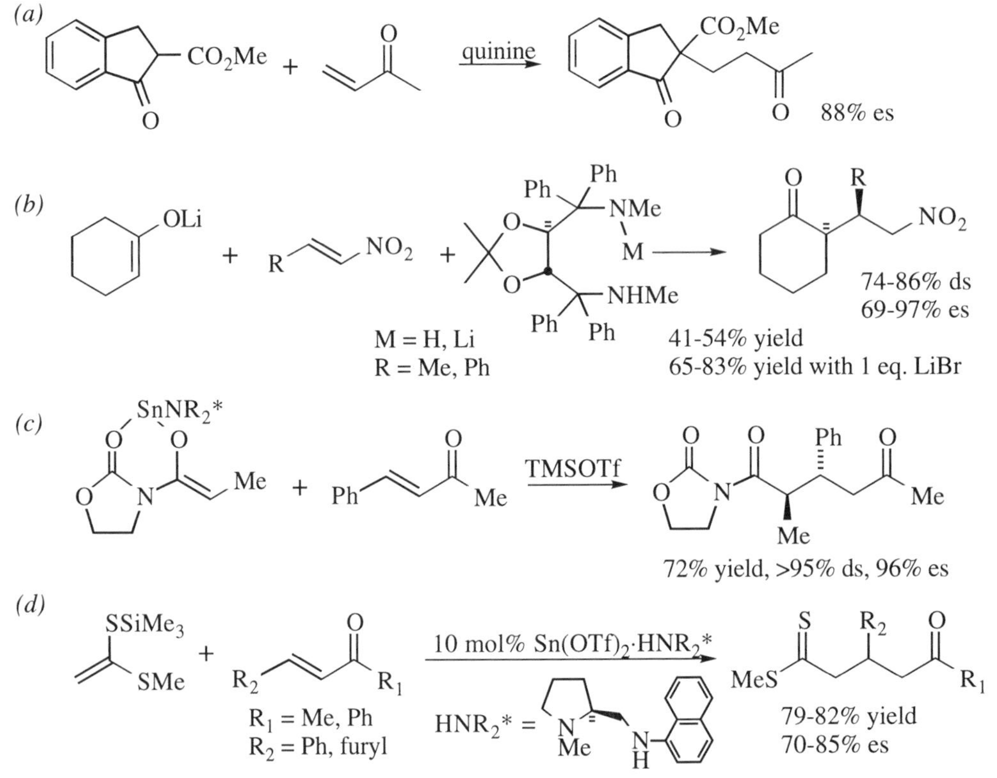

Scheme 5.40. (a) Wynberg's early example of interligand asymmetric induction in the Michael reaction [201]. (b) Seebach's investigation of cyclohexanone lithium enolate complexed to chiral diamines with extra lithium [3]. (c) Mukaiyama's imide tin enolate and chiral diamine [202]. (d) Mukaiyama's catalytic tin dithioenolate Michael addition [203].

Complexation of potassium enolates with chiral crown ethers and Michael addition of the associated enolate has been investigated by several groups, illustrated by the examples shown in Scheme 5.41. For example, Cram used C_2-symmetric crowns based on binaphthol to catalyze the addition of 2-carbomethoxyindanone to methyl vinyl ketone (Scheme 5.41a, [204]. A second example, the addition of methyl 2-phenylpropionate to methyl acrylate is shown in Scheme 5.41b [204]. Not shown are additions of methyl thiophenyl acetate enolate, which resulted in products of lower enantiomeric purity due to racemization of the product. Thus good selectivities can only be achieved when the product cannot racemize under the reaction conditions. A later example from the Penades group uses methyl thiophenyl acetate as nucleophile, and achieves reasonably high selectivities with small amounts of base and crown (Scheme 5.41c) [205]. Yamamoto added methyl thiophenyl acetate to cyclopentenone (Scheme 5.41d, [206]. The thiophenyl moiety of the addition product was reductively cleaved to afford a substituted cyclopentanone with a selectivity of 70% at the cyclopentanone β-carbon.

These four examples share the common feature of an acidic carbon stabilized by two functional groups, which permits employment of catalytic amounts of base and crown. The catalytic cycle is probably as follows [204]:

1) Crown·K+t-BuO⁻ + H–R → Crown·K+R⁻ + t-BuOH

2) Crown·K+R⁻ + C=C–C=O → R–C–C=C–O⁻K+·Crown

3) R–C–C=C–O⁻K+·Crown + H–R → R–C–CH–C=O + Crown·K+R⁻

The key step for catalyst turnover is the last one, whereby the enolate adduct deprotonates the next molecule of starting carbonyl. Clearly the initial carbon acid

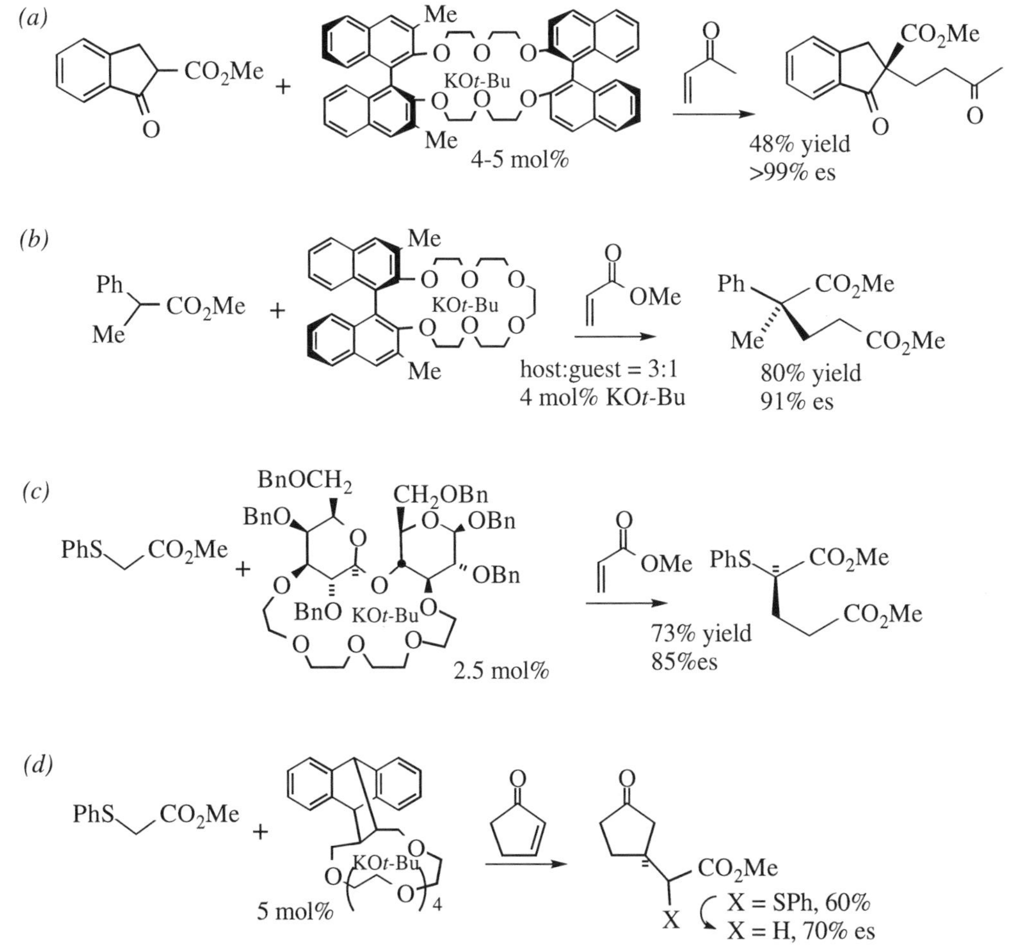

Scheme 5.41. (*a, b*) Cram's C₂-symmetric chiral crowns for asymmetric Michael addition [204]. (*c*) Penades's carbohydrate-based crown for asymmetric Michael additions [205]. (*d*) Yamamoto's chiral crown for asymmetric addition to β-substituted enones [206].

must be more acidic than the Michael product for this step to proceed. The two carbanion stabilizing groups in the above examples assure this fact, but also insure that epimerization in the product (if there are any α-protons left) can be a problem (*cf.* Scheme 5.41 c and d).

5.3.4 Chiral Michael acceptors

Posner has shown that enones having a chiral sulfoxide in the α-position are excellent receptors for conjugate addition of organometallics (Scheme 4.14, [207], and may also be used as Michael acceptors in enolate additions [208-210]. As with the addition of organometallics, the face selectivity can be rationalized based on either chelation of the metal by the enone and sulfoxide oxygens (Figure 5.11a) or by dipole alignment (Figure 5.11b) (*cf.* Scheme 4.16). In the following examples, which are chosen from others that are not as selective, the following trend emerges: enolates that are monosubstituted at the α-position follow the nonchelate (dipole) model, while α,α-disubstituted enolates follow the chelate model [211].

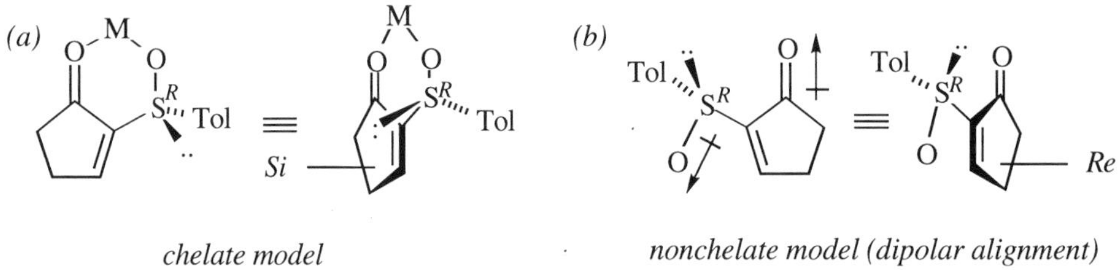

Figure 5.11. Models for face-selective addition of enolates to *R*-sulfoxides. *(a)* Chelate model predicts nucleophilic attack on *Si* face. *(b)* Nonchelate model, which has the C=O and S–O bonds antiperiplanar, predicts *Re* face attack.

The lithium enolate of methyl trimethylsilyl acetate adds to cyclopentenone and cyclohexenone sulfoxides by the nonchelate model with good to excellent selectivity, as shown in Scheme 5.42a [210]. After the Michael addition, the sulfoxide and trimethylsilyl groups are removed, and the selectivity is assessed by determining the

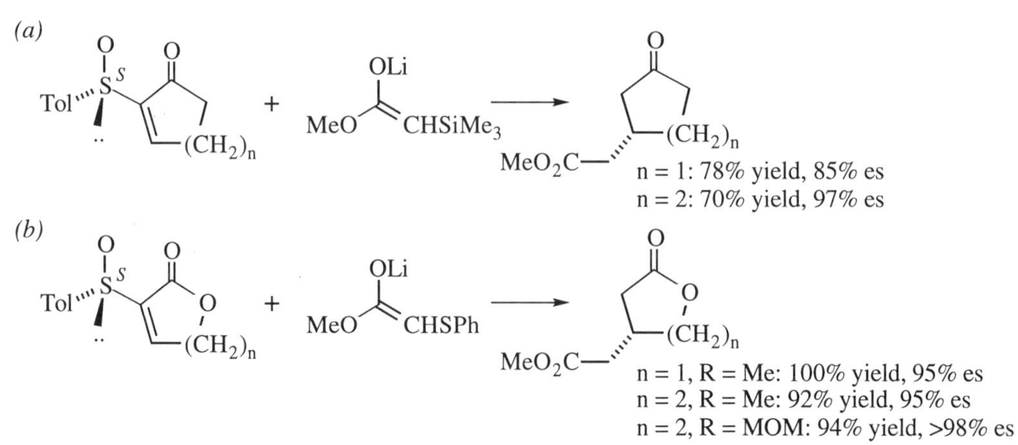

Scheme 5.42. Sulfoxide mediated asymmetric Michael additions to *(a)* cycloalkenones and *(b)* lactones. Both are postulated to proceed via the nonchelate model, Figure 5.11b [210].

enantiomeric purity of the β-substituted ketone. Similarly, lithium enolates of phenylthioacetate esters add to five and six-membered lactones as shown in Scheme 5.42b [210]. The chirality sense of these products is consistent with a nonchelate model: the nucleophile adds to the *Si* face of the *S* sulfoxide (*lk* topicity).

Scheme 5.43 illustrates three applications of this methodology to total synthesis. The first example is taken from Posner's synthesis of estrone and estradiol [211], the second from Posner's synthesis of methyl jasmonate [212], and the third from Holton's synthesis of aphidicolin [213]. The latter is particularly noteworthy in that two contiguous quaternary centers are created in the asymmetric addition with excellent selectivity. In the estrone synthesis, the chirality sense of the product is consistent with the nonchelate model, but the other two examples adhere to a chelate model. Note that the difference is the degree of substitution at the α-position of the enolate.

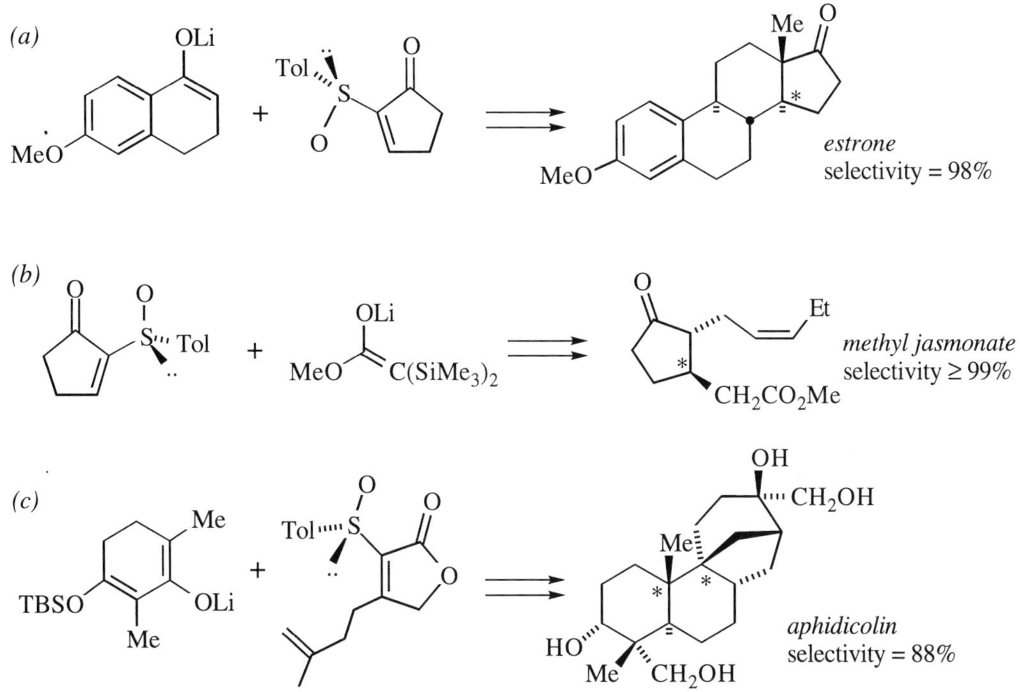

Scheme 5.43. Applications of sulfoxide Michael additions in natural product synthesis: *(a)* estrone [and estradiol] [211]. *(b)* methyl jasmonate [212]. *(c)* aphidicolin [213]. Stereocenters formed in the Michael addition are indicated (∗).

5.4 References

1.   H. E. Zimmerman; M. D. Traxler *J. Am. Chem. Soc.* **1957**, *79*, 1920-1923.
2.   D. Seebach *Angew. Chem. Int. Ed. Engl.* **1988**, *27*, 1624-1654.
3.   E. Juaristi; A. K. Beck; J. Hansen; T. Matt; T. Mukhopadhyay; M. Simson; D. Seebach *Synthesis* **1993**, 1271-1290.
4.   P. G. Williard; Q.-Y. Liu *J. Am. Chem. Soc.* **1993**, *115*, 3380-3381.

5.   S. Masamune; S. A. Ali; D. L. Snitman; D. S. Garvey *Angew. Chem. Int. Ed. Engl.* **1980**, *19*, 557-558.

6.   D. Seebach; V. Prelog *Angew. Chem. Int. Ed. Engl.* **1982**, *21*, 654-660.

7.   R. W. Hoffmann *Angew. Chem. Int. Ed. Engl.* **1982**, *21*, 555-642.

8.   W. R. Roush In *Comprehensive Organic Synthesis. Selectivity, Strategy, and Efficiency in Modern Organic Chemistry*; B. M. Trost, I. Fleming, Eds.; Pergamon: Oxford, 1991; Vol. 2, p 1-53.

9.   I. Fleming In *Comprehensive Organic Synthesis. Selectivity, Strategy, and Efficiency in Modern Organic Chemistry*; B. M. Trost, I. Fleming, Eds.; Pergamon: Oxford, 1991; Vol. 2, p 563-593.

10. Y. Yamamoto; K. Maruyama *Heterocycles* **1982**, *18*, 357-386.

11. R. Hoffmann *Angew. Chem. Int. Ed. Engl.* **1982**, *21*, 555-566.

12. R. W. Hoffmann; G. Niel; A. Schlapbach *Pure Appl. Chem.* **1990**, *62*, 1993-1998.

13. H. C. Brown; P. V. Ramachandran *Pure Appl. Chem.* **1991**, *63*, 307-316.

14. Y. Yamamoto *Acc. Chem. Res.* **1987**, *20*, 243-249.

15. Y. Yamamoto; N. Asao *Chem. Rev.* **1993**, *93*, 2207-2293.

16. D. A. Evans In *Topics in Stereochemistry*; E. L. Eliel, N. L. Allinger, Eds.; Wiley-Interscience: New York, 1982; Vol. vol. 13, p 1-115.

17. R. W. Hoffmann; T. Herold *Chem. Ber.* **1981**, *114*, 375-383.

18. M. T. Reetz; T. Zierke *Chem. Ind. (London)* **1988**, 663-664.

19. W. R. Roush; A. E. Walts; L. K. Hoong *J. Am. Chem. Soc.* **1985**, *107*, 8186-8190.

20. E. J. Corey; C.-M. Yu; S. S. Kim *J. Am. Chem. Soc.* **1989**, *111*, 5495-5496.

21. U. S. Racherla; H. C. Brown *J. Org. Chem.* **1991**, *56*, 401-404.

22. H. C. Brown; U. S. Racherla; Y. Liao; V. V. Khanna *J. Org. Chem.* **1992**, *57*, 6608-6614.

23. U. S. Racherla; Y. Liao; H. C. Brown *J. Org. Chem.* **1992**, *57*, 6614-6617.

24. H. C. Brown; S. V. Kulkarni; U. S. Racherla *J. Org. Chem.* **1994**, *59*, 365-369.

25. R. P. Short; S. Masamune *J. Am. Chem. Soc.* **1989**, *111*, 1892-1894.

26. R. W. Hoffmann; H.-J. Zeiss *J. Org. Chem.* **1981**, *46*, 1309-1314.

27. H. C. Brown; K. S. Bhat *J. Am. Chem. Soc.* **1986**, *108*, 293-294.

28. H. C. Brown; P. K. Jadhav; K. S. Bhat *J. Am. Chem. Soc.* **1988**, *110*, 1535-1538.

29. A. G. M. Barrett; J. W. Malecha *J. Org. Chem.* **1991**, *56*, 5243-5245.

30. R. W. Hoffmann; A. Endesfelder; H.-J. Zeiss *Carbohydr. Res.* **1983**, *123*, 320-325.

31. W. R. Roush; L. K. Hoong; M. A. J. Palmer; J. C. Park *J. Org. Chem.* **1990**, *55*, 4109-4117.

32. W. R. Roush; R. L. Halterman *J. Am. Chem. Soc.* **1986**, *108*, 294-296.

33. W. R. Roush; L. K. Hoong; M. A. J. Palmer; J. A. Straub; A. D. Palkowitz *J. Org. Chem.* **1990**, *55*, 4117-4126.

34. J. Garcia; B. M. Kim; S. Masamune *J. Org. Chem.* **1987**, *52*, 4831-4832.

35. W. R. Roush; P. T. Grover; X. Lin *Tetrahedron Lett.* **1990**, *31*, 7563-7566.

36. W. R. Roush; P. T. Grover *Tetrahedron Lett.* **1990**, *31*, 7567-7570.

37. S. E. Denmark; E. J. Weber *Helv. Chim. Acta* **1983**, *66*, 1655-1660.

38. Y. Yamamoto; H. Yatagai; Y. Naruta; K. Maruyama *J. Am. Chem. Soc.* **1980**, *102*, 7107-7109.

39. S. E. Denmark; E. J. Weber *J. Am. Chem. Soc.* **1984**, *106*, 7970-7971.

40. K. Ishihara; M. Mouri; Q. Gao; T. Maruyama; K. Furuta; H. Yamamoto *J. Am. Chem. Soc.* **1993**, *115*, 11490-11495.

41. A. L. Costa; M. G. Piazza; E. Tagliavini; C. Trombini; A. Umani-Ronchi *J. Am. Chem. Soc.* **1993**, *115*, 7001-7002.

42. G. E. Keck; K. H. Tarbet; L. S. Geraci *J. Am. Chem. Soc.* **1993**, *115*, 8467-8468.

43. A. T. Nielsen; W. J. Houlihan *Org. React.* **1968**, *16*, 1-438.

44. C. H. Heathcock In *Comprehensive Organic Synthesis. Selectivity, Strategy, and Efficiency in Modern Organic Chemistry*; B. M. Trost, I. Fleming, Eds.; Pergamon: Oxford, 1991; Vol. 2, p 133-179.

45. T. Mukaiyama *Org. React.* **1982**, *28*, 203-331.

46. S. Masamune; W. Choy; J. S. Petersen; L. R. Sita *Angew. Chem. Int. Ed. Engl.* **1985**, *24*, 1-76.

47. C. H. Heathcock In *Comprehensive Carbanion Chemistry, Part B*; E. Buncel, T. Durst, Eds.; Elsevier: Amsterdam, 1984, p 177-237.

48. I. Paterson *Pure Appl. Chem.* **1992**, *64*, 1821-1830.

49. C. H. Heathcock *Aldrichimica Acta* **1990**, *23*, 99-111.

50. C. H. Heathcock In *Asymmetric Synthesis*; J. D. Morrison, Ed.; Academic: Orlando, 1984; Vol. 3, p 111-212.

51. C. H. Heathcock In *Comprehensive Organic Synthesis. Selectivity, Strategy, and Efficiency in Modern Organic Chemistry*; B. M. Trost, I. Fleming, Eds.; Pergamon: Oxford, 1991; Vol. 2, p 181-238.

52. B. S. Kim; S. F. Williams; S. Masamune In *Comprehensive Organic Synthesis. Selectivity, Strategy, and Efficiency in Modern Organic Chemistry*; B. M. Trost, I. Fleming, Eds.; Pergamon: Oxford, 1991; Vol. 2, p 239-275.

53. I. Paterson In *Comprehensive Organic Synthesis. Selectivity, Strategy, and Efficiency in Modern Organic Chemistry*; B. M. Trost, I. Fleming, Eds.; Pergamon: Oxford, 1991; Vol. 2, p 301-319.

54. Y. Li; M. N. Paddon-Row; K. N. Houk *J. Org. Chem.* **1990**, *55*, 481-493.

55. A. Bernardi; A. M. Capelli; C. Gennari; J. M. Goodman; I. Paterson *J. Org. Chem.* **1990**, *55*, 3576-3581.

56. A. Bernardi; A. M. Capelli; A. Comotti; C. Gennari; M. Gardner; J. M. Goodman; I. Paterson *Tetrahedron* **1991**, *47*, 3471-3484.

57. D. Seebach; J. Golinski *Helv. Chim. Acta* **1981**, *64*, 1413-1423.

58. S. E. Denmark; B. R. Henke *J. Am. Chem. Soc.* **1989**, *111*, 8032-8034.

59. S. E. Denmark; W. Lee *J. Org. Chem.* **1994**, *59*, 707-709.

60. R. E. Ireland; R. H. Mueller; A. K. Willard *J. Am. Chem. Soc.* **1976**, *98*, 2868-2877.

61. T. Mukaiyama; T. Inoue *Chem. Lett.* **1976**, 559-562.

62. D. A. Evans; J. V. Nelson; E. Vogel; T. R. Taber *J. Am. Chem. Soc.* **1981**, *103*, 3099-3111.

63. D. A. Evans; L. R. McGee *Tetrahedron Lett.* **1980**, *21*, 3975-3978.

64. Y. Yamamoto; K. Maruyama *Tetrahedron Lett.* **1980**, *21*, 4607-4610.

65. M. T. Reetz; R. Peter *Tetrahedron Lett.* **1981**, *22*, 4691-4694.

66. M. A. Walker; C. H. Heathcock *J. Org. Chem.* **1991**, *56*, 5747-5750.

67. K. A. Swiss; W.-B. Choi; D. C. Liotta; A. F. Abdel-Magid; C. A. Maryanoff *J. Org. Chem.* **1991**, *56*, 5978-5980.

68. C. H. Heathcock; C. T. Buse; W. A. Kleschick; M. C. Pirrung; J. E. Sohn; J. Lampe *J. Org. Chem.* **1980**, *45*, 1066-1081.

69. M. Braun *Angew. Chem. Int. Ed. Engl.* **1987**, *26*, 24-37.

70. H. Eichenauer; E. Friedrich; W. Lutz; D. Enders *Angew. Chem. Int. Ed. Engl.* **1978**, *17*, 206-208.

71. C. H. Heathcock; M. C. Pirrung; C. T. Buse; J. P. Hagen; S. D. Young; J. E. Sohn *J. Am. Chem. Soc.* **1979**, *101*, 7077-7078.

72. C. H. Heathcock; M. C. Pirrung; J. Lampe; C. T. Buse; S. D. Young *J. Org. Chem.* **1981**, *46*, 2290-2300.

73. N. A. V. Draanen; S. Arseniyades; M. T. Crimmins; C. H. Heathcock *J. Org. Chem.* **1991**, *56*, 2499-2506.

74. S. Masamune; W. Choy; F. A. J. Kerdesky; B. Imperiali *J. Am. Chem. Soc.* **1981**, *103*, 1566-1568.

75. D. A. Evans; J. Bartroli; T. L. Shih *J. Am. Chem. Soc.* **1981**, *103*, 2127-2129.

76. D. A. Evans; T. C. Britton; J. A. Ellman *Tetrahedron Lett.* **1987**, *28*, 6141-6144.

77. J. R. Gage; D. A. Evans *Organic Syntheses* **1993**, *Coll. Vol. VIII*, 339-343.

78. D. A. Evans; L. R. McGee *J. Am. Chem. Soc.* **1981**, *103*, 2876-2878.

79. G. Helmchen; U. Leikauf; I. Taufer-Knöpfel *Angew. Chem. Int. Ed. Engl.* **1985**, *24*, 874-875.

80. W. Oppolzer; J. Blagg; I. Rodriguez; E. Walther *J. Am. Chem. Soc.* **1990**, *112*, 2767-2772.

81. W. Oppolzer; C. Starkemann; I. Rodriguez; G. Bernardinelli *Tetrahedron Lett.* **1991**, *32*, 61-64.

82. W. Oppolzer *Pure Appl. Chem.* **1988**, *60*, 39-48.

83. W. Oppolzer; C. Starkemann *Tetrahedron Lett.* **1992**, *33*, 2439-2442.

84. C. Siegel; E. R. Thornton *J. Am. Chem. Soc.* **1989**, *111*, 5722-5728.

85. A. Choudhury; E. R. Thornton *Tetrahedron Lett.* **1993**, *34*, 2221-2224.

86. J. R. Gage; D. A. Evans *Organic Syntheses* **1993**, *Coll. Vol. VIII*, 528-531.

87. D. A. Dickman; A. I. Meyers; G. A. Smith; R. E. Gawley *Organic Syntheses* **1990**, *Coll. Vol. VII*, 530-533.

88. M. Nerz-Stormes; E. R. Thornton *J. Org. Chem.* **1991**, *56*, 2489-2498.

89. T.-H. Yan; C.-W. Tan; H.-C. Lee; H.-C. Lo; T.-Y. Huang *J. Am. Chem. Soc.* **1993**, *115*, 2613-2621.

90. D. A. Evans; E. B. Sjogren; A. E. Weber; R. E. Conn *Tetrahedron Lett.* **1987**, *28*, 39-42.

91. L. N. Pridgen; A. F. Abdel-Magid; I. Lantos; S. Shilcrat; D. S. Eggleston *J. Org. Chem.* **1993**, *58*, 5107-5117.

92. S. Masamune; T. Sato; B. M. Kim; T. A. Wollmann *J. Am. Chem. Soc.* **1986**, *108*, 8279-8281.

93. I. Paterson; M. A. Lister; C. K. McClure *Tetrahedron Lett.* **1986**, *27*, 4787-4790.

94. A. I. Meyers; Y. Yamamoto *J. Am. Chem. Soc.* **1981**, *103*, 4278-4279.

95. A. I. Meyers; Y. Yamamoto *Tetrahedron* **1984**, *40*, 2309-2315.

96. I. Paterson; J. M. Goodman; M. A. Lister; R. C. Schumann; C. K. McClure; R. D. Norcross *Tetrahedron* **1990**, *46*, 4663-4684.

97. C. Gennari; C. T. Hewkin; F. Molinari; A. Bernardi; A. Comotti; J. M. Goodman; I. Paterson *J. Org. Chem.* **1992**, *57*, 5173-5177.

98. W. C. Still; D. Cai; D. Lee; P. Hauck; A. Bernardi; A. Romero *Lect. Heterocycl. Chem.* **1987**, *9*, 33-42.

99. E. J. Corey; R. Imwinkelreid; S. Pikul; Y. B. Xiang *J. Am. Chem. Soc.* **1989**, *111*, 5493-5495.

100. E. J. Corey; S. S. Kim *J. Am. Chem. Soc.* **1990**, *112*, 4976-4977.

101. E. J. Corey; C. P. Decicco; R. C. Newbold *Tetrahedron Lett.* **1991**, *32*, 5287-5290.

102. E. J. Corey; S. Choi *Tetrahedron Lett.* **1991**, *32*, 2857-2860.

103. E. J. Corey; D.-H. Lee; S. Choi *Tetrahedron Lett.* **1992**, *33*, 6735-6738.

104. E. J. Corey; D.-H. Lee *Tetrahedron Lett.* **1993**, *34*, 1737-1740.

105. R. O. Duthaler; P. Herold; W. Lottenbach; K. Oertle; M. Riediker *Angew. Chem. Int. Ed. Engl.* **1989**, *28*, 495-497.

106. R. O. Duthaler; P. Herold; S. Wyler-Helfer; M. Riediker *Helv. Chim. Acta* **1990**, *73*, 659-673.

107. T. Mukaiyama; T. Yura *Tetrahedron* **1989**, *45*, 1197-1207.

108. A. Ando; T. Shiori *Tetrahedron* **1989**, *45*, 4969-4988.

109. M. Muraoka; H. Kawasaki; K. Koga *Tetrahedron Lett.* **1988**, *29*, 337-338.

110. T. Mukaiyama; K. Banno; K. Narasaka *J. Am. Chem. Soc.* **1974**, *96*, 7503-7509.

111. K. Mikami; S. Matsukawa *J. Am. Chem. Soc.* **1993**, *115*, 7039-7044.

112. S.-i. Kiyooka; Y. Kaneko; M. Komura; H. Matsuo; M. Nakano *J. Org. Chem.* **1991**, *56*, 2276-2278.

113. S.-i. Kiyooka; Y. Kaneko; K.-i. Kume *Tetrahedron Lett.* **1992**, *33*, 4927-4930.

114. E. R. Parmee; O. Tempkin; S. Masamune; A. Abiko *J. Am. Chem. Soc.* **1991**, *113*, 9365-9366.

115. E. J. Corey; C. L. Cywin; T. D. Roper *Tetrahedron Lett.* **1992**, *33*, 6907-6910.

116. K. Furuta; T. Maruyama; H. Yamamoto *J. Am. Chem. Soc.* **1991**, *113*, 1041-1042.

117. K. Maruoka; H. Yamamoto In *Catalytic Asymmetric Synthesis*; I. Ojima, Ed.; VCH: New York, 1993, p 413-440.

118. S. Kobayashi; H. Uchiro; Y. Fujishita; I. Shiina; T. Mukaiyama *J. Am. Chem. Soc.* **1991**, *113*, 4247-4252.

119. S. Kobayashi; H. Uchiro; I. Shiina; T. Mukaiyama *Tetrahedron* **1993**, *49*, 1761-1772.

120. S. Kobayashi; M. Horibe *J. Am. Chem. Soc.* **1994**, *116*, 9805-9806.

121. S. Kobayashi; Y. Fujishita; T. Mukaiyama *Chem. Lett.* **1990**, 1455-1458.

122. T. Mukaiyama; S. Kobayashi; H. Uchiro; I. Shiina *Chem. Lett.* **1990**, 129-132.

123. W. R. Roush *J. Org. Chem.* **1991**, *56*, 4151-4157.

124. S. Masamune; M. Hirama; S. Mori; S. A. Ali; D. S. Garvey *J. Am. Chem. Soc.* **1981**, *103*, 1568-1571.

125. D. A. Evans; J. Bartroli *Tetrahedron Lett.* **1982**, *23*, 807-810.

126. I. Paterson; M. A. Lister *Tetrahedron Lett.* **1988**, *29*, 585-588.

127. S. Masamune In *Organic Synthesis: Today and Tomorrow*; B. M. Trost, C. R. Hutchinson, Eds.; Pergamon: Oxford, 1981, p 197-215.

128. S. F. Martin; W.-C. Lee *Tetrahedron Lett.* **1993**, *34*, 2711-2714.

129. S. F. Martin; G. J. Pacofsky; R. P. Gist; W.-C. Lee *J. Am. Chem. Soc.* **1989**, *111*, 7634-7636.

130. X. Chen; E. R. Hortelano; E. L. Eliel; S. V. Frye *J. Am. Chem. Soc.* **1992**, *114*, 1778-1784.

131. D. A. Evans; M. A. Calter *Tetrahedron Lett.* **1993**, *34*, 6871-6874.

132. D. A. Evans; J. L. Duffy; M. J. Dart *Tetrahedron Lett.* **1994**, *46*, 8537-8540.

133. D. A. Evans; M. J. Dart; J. L. Duffy; M. G. Yang; A. B. Livingston *J. Am. Chem. Soc.* **1995**, *117*, 6619-6620.

134. D. A. Evans; M. J. Dart; J. L. Duffy; D. L. Rieger *J. Am. Chem. Soc.* **1995**, *117*, 9073-9074.

135. I. Paterson; R. A. Ward; P. Romea; R. D. Norcross *J. Am. Chem. Soc.* **1994**, *116*, 3623-3624.

136. I. Paterson; R. D. N. A. Ward; P. Romea; M. A. Lister *J. Am. Chem. Soc.* **1994**, *116*, 11287-11314.

137. I. Paterson; J. M. Goodman; M. Isaka *Tetrahedron Lett.* **1989**, *30*, 7121-7124.

138. I. Paterson; R. D. Tillyer *J. Org. Chem.* **1993**, *58*, 4182-4184.

139. I. Paterson; C. K. McClure *Tetrahedron Lett.* **1987**, *28*, 1229-1232.

140. S. Masamune; P. A. McCarthy In *Macrolide Antibiotics. Chemistry, Biology, and Practice*; Academic: Orlando, 1984, p 127-198.

141. I. Paterson; M. M. Mansuri *Tetrahedron* **1985**, *41*, 3569-3642.

142. R. W. Hoffmann *Angew. Chem. Int. Ed. Engl.* **1987**, *26*, 489-594.

143. I. Paterson; M. V. Perkins *Tetrahedron Lett.* **1992**, *33*, 601-604.

144. D. A. Evans; R. L. Dow; T. L. Shih; J. M. Takacs; R. Zahler *J. Am. Chem. Soc.* **1990**, *112*, 5290-5313.

145. S. F. Martin; W.-C. Lee; G. J. Pacofsky; R. P. Gist; T. A. Mulhern *J. Am. Chem. Soc.* **1994**, *116*, 4674-4688.

146. I. Paterson; A. M. Lister; G. R. Ryan *Tetrahedron Lett.* **1991**, *32*, 1749-1752.

147. P. Perlmutter *Conjugate Addition Reactions in Organic Synthesis*; Pergamon: Oxford, 1992.

148. D. A. Oare; C. H. Heathcock In *Topics in Stereochemistry*; E. L. Eliel, N. L. Allinger, Eds.; Wiley-Interscience: New York, 1989; Vol. 19, p 87-170.

149. D. A. Oare; C. H. Heathcock In *Topics in Stereochemistry*; E. L. Eliel, N. L. Allinger, Eds.; Wiley-Interscience: New York, 1991; Vol. 20, p 227-407.

150. N. G. Rondan; M. N. Paddon-Row; P. Caramella; K. N. Houk *J. Am. Chem. Soc.* **1981**, *103*, 2436-2438.

151. K. N. Houk *Pure Appl. Chem.* **1983**, *55*, 277-282.

152. D. Seebach; J. Zimmerman; U. Gysel; R. Ziegler; T.-K. Ha *J. Am. Chem. Soc.* **1988**, *110*, 4763-4772.

153. T. Laube; J. D. Dunitz; D. Seebach *Helv. Chim. Acta* **1985**, *68*, 1373-1393.

154. H. B. Bürgi; J. D. Dunitz; E. Schefter *J. Am. Chem. Soc.* **1973**, *95*, 5065-5067.

155. N. T. Anh; O. Eisenstein *Nouv. J. Chimie* **1977**, *1*, 61-70.

156. H. B. Bürgi; D. Dunitz; J. M. Lehn; G. Wipff *Tetrahedron* **1974**, *30*, 1563-1572.

157. D. A. Oare; C. H. Heathcock *J. Org. Chem.* **1990**, *55*, 157-172.

158. D. A. Oare; M. A. Henderson; M. A. Sanner; C. H. Heathcock *J. Org. Chem.* **1990**, *55*, 132-157.

159. A. Bernardi; A. M. Capelli; A. Cassinari; A. Comotti; C. Genari; C. Scolastico *J. Org. Chem.* **1992**, *57*, 7029-7034.

160. W. Oppolzer; R. Pitteloud; G. Bernardinelli; K. Baettig *Tetrahedron Lett.* **1983**, *24*, 4975-4978.

161. E. J. Corey; R. T. Peterson *Tetrahedron Lett.* **1985**, *26*, 5025-5028.

162. G. Stork; N. A. Saccomano *Nouv. J. Chimie* **1986**, *10*, 677-679.

163. M. Yamaguchi; K. Hasebe; S. Tanaka; T. Minami *Tetrahedron Lett.* **1986**, *27*, 959-962.

164. D. A. Evans; F. Urpí; T. C. Somers; J. S. Clark; M. T. Bilodeau *J. Am. Chem. Soc.* **1990**, *112*, 8215-8216.

165. D. A. Evans; M. T. Bioldeau; T. C. Somers; J. Clardy; D. Cherry; Y. Kato *J. Org. Chem.* **1991**, *56*, 5750-5752.

166. D. A. Evans; J. M. Takacs *Tetrahedron Lett.* **1980**, *21*, 4233-4236.

167. P. E. Sonnet; R. R. Heath *J. Org. Chem.* **1980**, *45*, 3137-3139.

168. Y. Kawanami; Y. Ito; T. Kitagawa; Y. Taniguchi; T. Katsuki; M. Yamaguchi *Tetrahedron Lett.* **1984**, *25*, 857-860.

169. Y. Ito; T. Katsuki; M. Yamaguchi *Tetrahedron Lett.* **1984**, *25*, 6015-6016.

170. D. Enders; K. Papadopoulos *Tetrahedron Lett.* **1983**, *24*, 4967-4970.

171. D. Enders; K. Papadopoulos; B. E. M. Rendenbach *Tetrahedron Lett.* **1986**, *27*, 3491-3494.

172. D. Enders; B. E. M. Rendenbach *Chem. Ber.* **1987**, *120*, 1223-1227.

173. D. Enders; H. J. Scherer; J. Runsink *Chem. Ber.* **1993**, *126*, 1929-1944.

174. K. Tomioka; K. Ando; K. Yasuda; K. Koga *Tetrahedron Lett.* **1986**, *27*, 715-716.

175. K. Tomioka; W. Seo; K. Ando; K. Koga *Tetrahedron Lett.* **1987**, *28*, 6637-6640.

176. *Enamines: Synthesis, Structure, and Reactions*; 2nd ed.; A. G. Cook, Ed.; Marcel Dekker: New York, 1988.

177. *Enamines: Synthesis, Structure, and Reactions*; A. G. Cook, Ed.; Marcel Dekker: New York, 1969.

178. S. F. Dyke *The Chemistry of Enamines*; Cambridge: Cambridge, 1973.

179. J. d'Angelo; D. Desmaële; F. Dumas; A. Guingant *Tetrahedron Asymmetry* **1992**, *3*, 459-505.

180. U. K. Pandit; H. O. Huisman *Tetrahedron Lett.* **1967**, 3901-3905.

181. B. de Jeso; J.-C. Pommier *J. Chem. Soc., Chem. Commun.* **1977**, 565-566.

182. A. Sevin; J. Tortajada; M. Pfau *J. Org. Chem.* **1986**, *51*, 2671-2675.

183. S. Yamada; K. Hiroi; K. Achiwa *Tetrahedron Lett.* **1969**, 4233-4236.

184. Y. Ito; M. Sawamura; K. Kominami; T. Saegusa *Tetrahedron Lett.* **1985**, *26*, 5303-5306.

185. S. J. Blarer; W. B. Schweizer; D. Seebach *Helv. Chim. Acta* **1982**, *65*, 1637-1654.

186. S. J. Blarer; D. Seebach *Chem. Ber.* **1983**, *116*, 2250-2260.

187. S. Blarer; D. Seebach *Chem. Ber.* **1983**, *116*, 3086-3096.

188. G. Pitacco; F. P. Colonna; E. Valentin; A. Risalti *J. Chem. Soc., Perkin Trans. 1* **1974**, 1625-1627.

189. R. E. Gawley *Synthesis* **1976**, 777-794.

190. M. E. Jung *Tetrahedron* **1976**, *32*, 3-31.

191. M. Pfau; G. Revial; A. Guingant; J. d'Angelo *J. Am. Chem. Soc.* **1985**, *107*, 273-274.

192. J. d'Angelo; G. Revial; P. R. R. Costa; R. N. Castro; O. A. C. Antunes *Tetrahedron Asymmetry* **1991**, *2*, 199-202.

193. D. Desmaele; J. d'Angelo *Tetrahedron Lett.* **1989**, *30*, 345-348.

194. D. Desmaele; J. d'Angelo; C. Bois *Tetrahedron Asymmetry* **1990**, *1*, 759-762.

195. D. H. Hua; S. Venkataraman; M. J. Coulter; G. Sinai-Zingde *J. Org. Chem.* **1987**, *52*, 719-728.

196. D. H. Hua; R. Chan-Yu-King; J. A. McKie; L. Myer *J. Am. Chem. Soc.* **1987**, *109*, 5026-5029.

197. S. Hanessian; A. Gomtsyan; A. Payne; Y. Hervé; S. Beaudoin *J. Org. Chem.* **1993**, *58*, 5032-5034.

198. M. R. Binns; O. L. Chai; R. K. Haynes; A. A. Katsifis; P. A. Shober; S. C. Vonwiller *Tetrahedron Lett.* **1985**, *26*, 1569-1572.

199. F. E. Ziegler; U. R. Chakraborty; R. T. Webster *Tetrahedron Lett.* **1982**, *23*, 3237-3240.

200. B. Långström; G. Berson *Acta Chem. Scand.* **1973**, *27*, 3118-3119.

201. K. Hermann; H. Wynberg *J. Org. Chem.* **1979**, *44*, 2238-2244.

202. T. Yura; N. Iwasawa; T. Mukaiyama *Chem. Lett.* **1988**, 1021-1024.

203. T. Yura; N. Iwasawa; K. Narasaka; T. Mukaiyama *Chem. Lett.* **1988**, 1025-1026.

204. D. J. Cram; G. D. Y. Sogah *J. Chem. Soc., Chem. Commun.* **1981**, 625-628.

205. M. Alonso-Lopez; J. Jimenez-Barbero; M. Martin-Lomas; S. Penades *Tetrahedron* **1988**, *44*, 1535-1543.

206. M. Takasu; H. Wakabayashi; K. Furuta; H. Yamamoto *Tetrahedron Lett.* **1988**, *29*, 6943-6946.

207. G. Posner In *Asymmetric Synthesis*; J. D. Morrison, Ed.; Academic: Orlando, 1983; Vol. 2, p 225-241.

208. G. Posner *Acc. Chem. Res.* **1987**, *20*, 72-78.

209. G. H. Posner In *The Chemistry of Sulphones and Sulphoxides*; S. Patai, Z. Rapaport, C. Stirling, Eds.; Wiley: New York, 1988, p 823-849.

210. G. H. Posner; M. Weitzberg; T. G. Hamill; E. Asirvatham; H. Cun-heng; J. Clardy *Tetrahedron* **1986**, *42*, 2919-2929.

211. G. H. Posner; C. Switzer *J. Am. Chem. Soc.* **1986**, *108*, 1239-1244.

212. G. H. Posner; E. Asirvatham *J. Org. Chem.* **1985**, *50*, 2589-2591.

213. R. A. Holton; R. M. Kennedy; H.-B. Kim; M. E. Krafft *J. Am. Chem. Soc.* **1987**, *109*, 1597-1600.

Chapter 6

Rearrangements and Cycloadditions

This chapter examines reactions that involve molecular rearrangements and cycloadditions. The use of these terms will not be restricted to concerted, pericyclic reactions, however. Often, stepwise processes that involve a net transformation equivalent to a pericyclic reaction are catalyzed by transition metals. The incorporation of chiral ligands into these metal catalysts introduces the possibility of asymmetric induction by inter-ligand chirality transfer. The chapter is divided into two main parts (rearrangements and cycloadditions), and subdivided by the standard classifications for pericyclic reactions (*e.g.*, [1,3], [2,3], [4+2], etc.). The latter classification is for convenience only, and does not imply adherence to the pericyclic selection rules. Indeed, the first reaction to be described is a net [1,3]-suprafacial hydrogen shift, which is symmetry forbidden if concerted.

6.1 Rearrangements

Many rearrangements are highly stereoselective reactions and have found considerable application in organic synthesis. Perhaps the most common class of sigmatropic rearrangements includes such [3,3]-rearrangements as the Cope and Claisen rearrangements, the latter with its many variants (reviews: [1-8]). However, the vast majority of [3,3]-rearrangements in which stereochemistry is an important element involve enantiomerically pure starting materials, which places this class of reactions outside the purview of this book.[1] Here, we will focus on two types of rearrangements: [1,3]-hydrogen shifts and [2,3]-Wittig rearrangements. The former is a transition metal catalyzed reaction sequence that has found tremendous importance in industry. The latter is a rearrangement that (like [3,3]-rearrangements) has many applications in stereoselective reactions of enantiomerically pure compounds. But since the [2,3]-Wittig rearrangement involves anionic intermediates, a number of possibilities for asymmetric synthesis also arise. The substrates for [2,3]- (and [3,3]-) rearrangements are often derived from chiral secondary alcohols, which are in turn available by several asymmetric synthesis methods. The discussion of the Wittig rearrangement therefore includes references to methods of asymmetric synthesis of the chiral precursors, which is also relevant to many applications of [3,3]-rearrangements.

6.1.1 [1,3]-Hydrogen shifts
It has long been recognized that certain transition metal complexes can catalyze the migration of carbon-carbon double bonds.[2] When the catalyst is a transition metal hydride, the mechanism involves initial reversible addition of the metal

---

1   For an example of the Ireland-Claisen rearrangement mediated by a chiral catalyst, see ref. [9]
2   For a summary of early examples, see pp. 266-303 of ref. [10].

hydride across the double bond to produce a metal σ-alkyl. A double bond is regenerated by elimination of the metal hydride, and if a different hydrogen is eliminated, the net result to the olefin is migration (Scheme 6.1a) [10]. This mechanism is therefore not a strict 1,3-hydrogen shift, but only resembles one when starting material and product are compared. If the catalyst is not a metal hydride, the first step is π-complexation of the metal to the double bond, followed by migratory insertion of the metal, producing a π-allyl metal hydride, then reversal of the sequence at the other end of the allyl system (Scheme 6.1b) [10]. If the olefin has an allylic heteroatom, a third mechanism may intervene. With allylic amines for example (Scheme 6.1c) [11], initial coordination occurs at nitrogen, and migratory insertion yields a π-complexed iminium metal hydride. Rearrangement then yields a bidentate enamine-metal complex, and dissociation liberates the enamine.

All of these processes are under thermodynamic control, and the migration is only useful when there is an isomer that is in a thermodynamic well. For the rearrangements shown in Scheme 6.1a and b, this is the case when the rearrangement affords a more highly substituted alkene, or when the double bond moves into conjugation with a functional group such as a carbonyl. The net rearrangement can involve several individual "[1,3]-rearrangement" steps, such as migration around a ring. Such sequential shifts are blocked by a quaternary carbon. The rearrangement of an allylic amine to an enamine is also thermodynamically favored (Scheme 6.1c).

For the purposes of asymmetric synthesis, the initial alkene must be prochiral (*i.e.*, either 1,1-disubstituted or trisubstituted), so that the rearrangement produces a new stereogenic center. As shown in Figure 6.1, this is often contrathermodynamic, *but not in the case of compounds with allylic heteroatoms.*

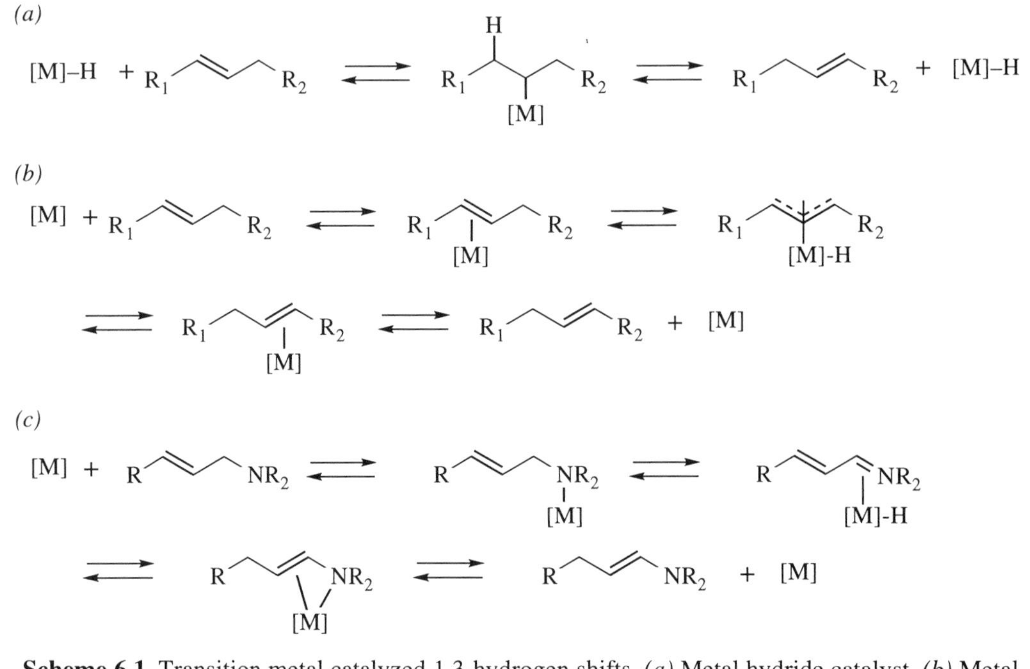

(a)

(b)

(c)

Scheme 6.1. Transition metal catalyzed 1,3-hydrogen shifts. *(a)* Metal hydride catalyst. *(b)* Metal catalyst. *(c)* Metal catalyzed rearrangement of allylic amines to enamines.

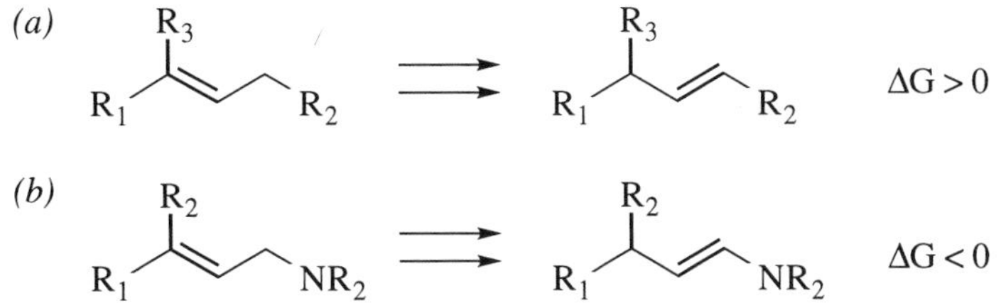

(a)

$$\Delta G > 0$$

(b)

$$\Delta G < 0$$

Figure 6.1. (a) Contrathermodynamic isomerization of a trisubstituted alkene to a disubstituted one. (b) Thermodynamically favored isomerization of an allylic amine to an enamine.

Following years of less successful attempts by other groups (≤53% enantioselectivity; reviews: [12,13] and pp. 266-303 of ref. [10]), Otsuka reported in 1978 that allylic amines could be rearranged to enamines with a chiral Co^{II} catalyst with modest (66:34) enantioselectively [14]. Further studies [11,15,16] revealed that a cationic Rh^I catalyst having arylphosphine ligands (the best is BINAP, 2,2'-bis(diphenylphosphino)-1,1'-binaphthyl) affords excellent selectivity (97-99% es) with very high catalyst turnover (300,000). This reaction has been scaled up, and is now known as the "Takasago process." It (Scheme 6.2) is used for the commercial manufacture of ~1500 tons per year (nearly 40% of the world market) of citronellal and menthol [11,16], and has been described as "the most impressive achievement to date in the area of asymmetric catalysis" [17]. It is worth mentioning that, although citronellal is available from natural sources, the enantiomer ratio of the natural product is only 90:10.

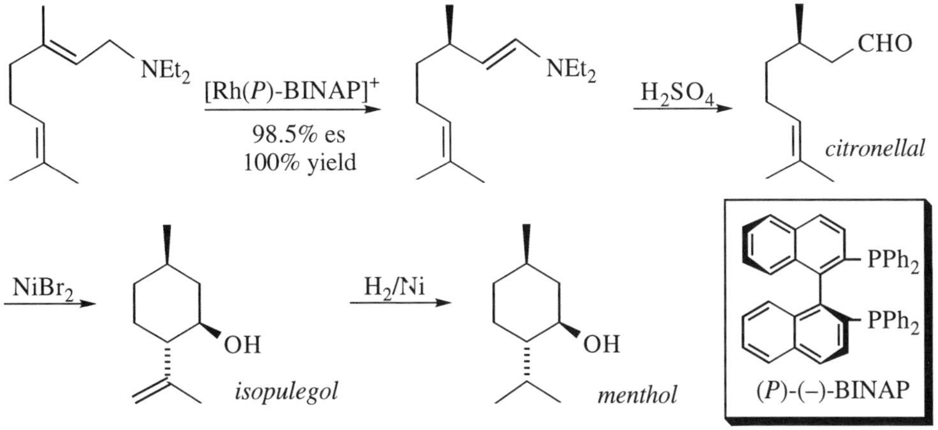

Scheme 6.2. The Takasago process for the commercial manufacture of citronellal, isopulegol, and menthol [16].[3]

Two aspects of the reaction are stereospecific. The first is that geometric isomers of the allylic amines afford enantiomeric enamines, as shown in Scheme 6.3a [19]. Note that the geometry of the enamine double bond is *not* dependent on the stereochemistry of the double bond of the allylic amine, however. The second

---

[3]   In accord with the recommendation of Prelog and Helmchen, the *P,M* nomenclature system is used to describe the configuration of molecules containing chirality axes and planes [18]. Note that *R* = *M* and *S* = *P*. See the glossary, Section 1.6, for an explanation of these terms.

stereospecific feature is revealed by the deuterium labeling studies shown in Scheme 6.3b: the R-C_1-d allylic amine, when subjected to enantiomeric rhodium catalysts, undergoes deuterium migration with M-BINAP, and hydrogen migration with R-BINAP [11]. An isotope effect was not observed, indicating that the carbon-hydrogen (or deuterium) bond breaking step is not rate determining. Furthermore, experiments (not shown) using a mixture of $-CD_2NEt_2$ and $-CH_2NEt_2$ amines revealed no crossover, indicating that the migration is intramolecular [11].

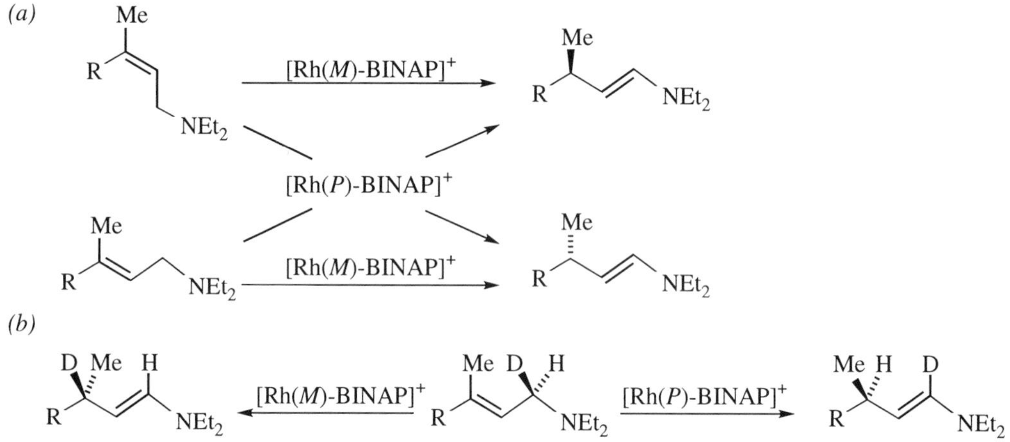

Scheme 6.3. Stereospecific aspects of rhodium catalyzed asymmetric [1,3]-hydrogen shifts.

There are two (limiting) possibilities that could explain the enantioselectivity: a group-selective metal insertion distinguishing the enantiotopic allylic protons, or a face-selective addition that distinguishes the enantiotopic double bond faces. Figure 6.2 illustrates the conformational analysis of the intermediates involved in the sequence (Scheme 6.1c). First of all, the lowest energy conformation around the

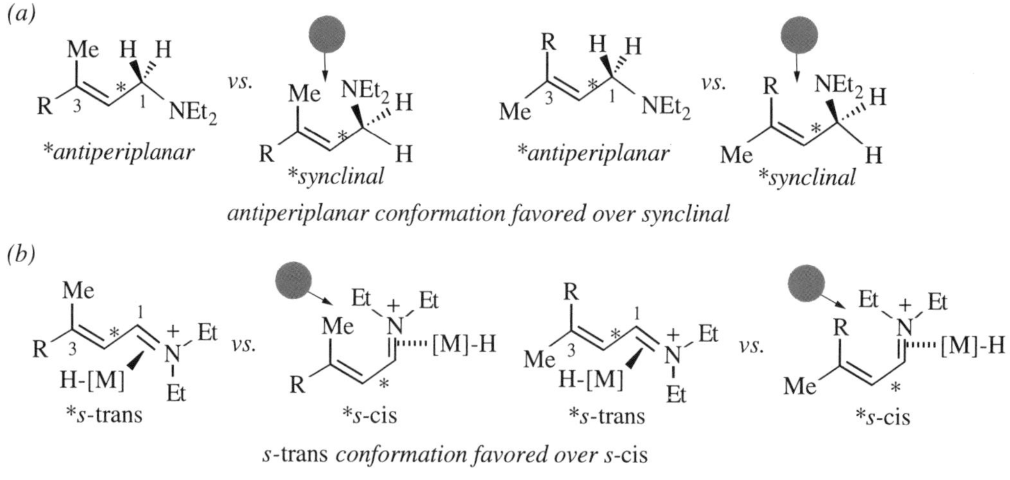

Figure 6.2. Severe conformational restrictions due to $A^{1,3}$ strain are placed on the intermediates in the asymmetric [1,3]-rearrangement of allylic amines. *(a)* The starting material (as well as the nitrogen-coordinated rhodium complex) favors the antiperiplanar conformation. *(b)* The π-bonded metal hydride intermediate is restricted to the *s*-trans conformation.

N–C_1–C_2–C_3 bond (∗) of the allylic amine is antiperiplanar due to $A^{1,3}$ strain in the synclinal conformation (Figure 6.2a). Coordination of the catalyst to the nitrogen (step 1 in Scheme 6.1c) will only increase the energetic bias in favor of the antiperiplanar conformation.[4] The second step of the reaction sequence is the migratory insertion of the metal into the C_1–H bond to give a π-bonded α,β-unsaturated iminium ion. Figure 6.2b shows that only the *s*-trans conformer of this species is accessible, because of severe $A^{1,3}$ interactions between the diethylamino substituents and the C_3 substituent in the *s*-cis conformation.

Examination of the conformers illustrated in Figure 6.2b reveals the origin of the *Re/Si* face selectivity in the transfer of hydrogen to C_3. The illustrated conformers have the metal hydride bound to the *Re* face of the iminium ion. Since the *s*-cis conformation is not accessible, and since the rearrangement corresponding to the third step of Scheme 6.1c is suprafacial, the step that determines the configuration of the π-bonded iminium metal hydride also determines the absolute configuration at C_3 in the product. This step is the migratory insertion of the metal into the C_1–H bond (*i.e.,* step 2 of Scheme 6.1c). Thus, the Takasago process is an example of a group-selective insertion of a metal into one of two enantiotopic carbon hydrogen bonds. The *M*-BINAP rhodium inserts into the C–H_{Re} bond and the *P*-BINAP rhodium inserts into the C–H_{Si} bond.

The interligand asymmetric induction from the binaphthyl moiety to C_3 of the allylic amine covers a considerable distance and deserves comment. As noted above, the enantioselectivity of the overall process is determined in the step where the metal inserts into one of the enantiotopic C_1-protons. The solid state conformation of the *P*-BINAP ligand has been established by two X-ray crystal structures (of ruthenium complexes: [20,21]), and is illustrated in Figure 6.3a, with the other ligands removed for clarity. Note that the chirality sense of the binaphthyl moiety places the four *P*-phenyl substituents in *quasi*-axial and *quasi*-equatorial orientations. It is apparent that the 'upper right' and 'lower left' quadrants (which are equivalent due to symmetry) have the most free space for accomodating additional bound ligands. Attachment of the allylic amine in the antiperiplanar C_1–C_2 conformation to the square-planar rhodium complex is illustrated in Figure 6.3b. (The two possible binding sites are equivalent due to symmetry.) The migratory insertion step must occur through a 4-membered ring transition structure, and the two possibilities are illustrated in Figures 6.3c and d. Note that insertion into the H_{Re}–C bond forces the double bond moiety into close proximity with the *quasi*-equatorial phenyl on the left (Figure 6.3c), whereas metal insertion into the H_{Si}–C bond moves the double bond into the vacant lower left quadrant. The latter is favored.

The catalytic cycle shown in Scheme 6.4 has been proposed to account for the kinetics and observable intermediates in the reaction [11]. Starting from the top, the allylic amine displaces a solvent to form the *N*-coordinated rhodium species.

---

[4] Low temperature ^{1}H and ^{31}P NMR studies indicated that only the nitrogen of allylic amines is bound to the metal. No evidence could be found for an *N*-π-chelate that might stabilize the synclinal conformation [11].

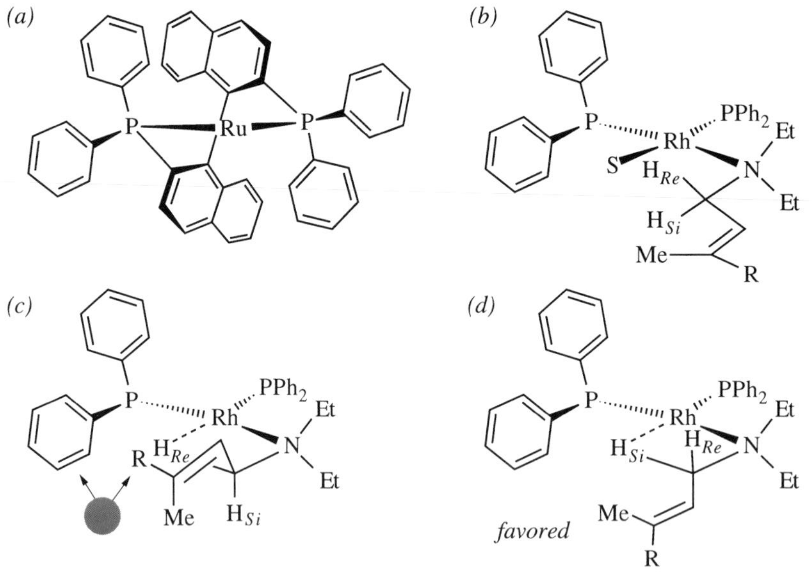

Figure 6.3. *(a)* Conformation of *P*-BINAP in two crystal structures [20,21]. *(b)* Partial structure with allylic amine bound at one of the two equivalent coordination sites. *(c)* Transition structure for insertion into C–H$_{Re}$ bond. *(d)* Transition structure for insertion into C–H$_{Si}$ bond .

Migratory insertion and hydrogen transfer then form the rhodium-enamine complex shown at the bottom, which can be isolated and characterized at low temperature. The rate determining step in the cycle is the substitution of the enamine ligand by a new allylic amine substrate, which probably proceeds via the substrate-product mixed complex shown on the left.

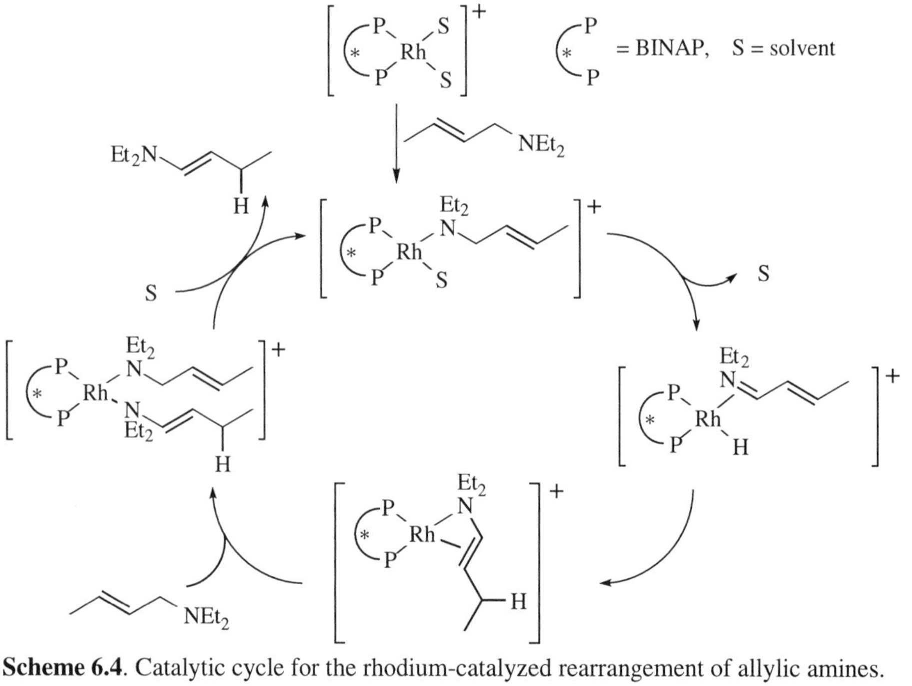

Scheme 6.4. Catalytic cycle for the rhodium-catalyzed rearrangement of allylic amines.

Investigation of the scope of the asymmetric rearrangement of allylic amines has led to the following generalizations [11]: (*i*) both C_1 and C_2 should have no alkyl substitutents (substitution at either position would erase the preference for the antiperiplanar and *s*-trans conformations, *cf.* Figure 6.2); (*ii*) C_3 may be substituted with an aryl group (or also be only monosubstituted, but the latter circumstance has no stereochemical consequence); (*iii*) the nitrogen substitutents must not be aryl (a less basic nitrogen fails to bind the rhodium and is not affected by the catalyst).

Isomerization of allylic alcohols occurs in reasonable yields but with poor enantioselectivity [22], although kinetic resolution of 4-hydroxycyclopentenone has been reported [23]. Reliable laboratory-scale procedures for the synthesis of BINAP and for the asymmetric rearrangement have been published [24,25], making this a good candidate for further applications in asymmetric synthesis.

6.1.2 [2,3]-Wittig rearrangements

What is now known as the [1,2]-Wittig rearrangement was apparently first observed in the 1920s by Schorigin [26,27], and by Schlenk and Bergmann [28], who reported that reductive metalation of benzyl alkyl ethers with lithium or sodium afforded rearranged products. In 1942, Wittig reported that benzyl ethers could be deprotonated with phenyl lithium, and similarly rearranged [29,30] (Scheme 6.5a). It is now agreed that the [1,2]-rearrangement involves successive deprotonation,

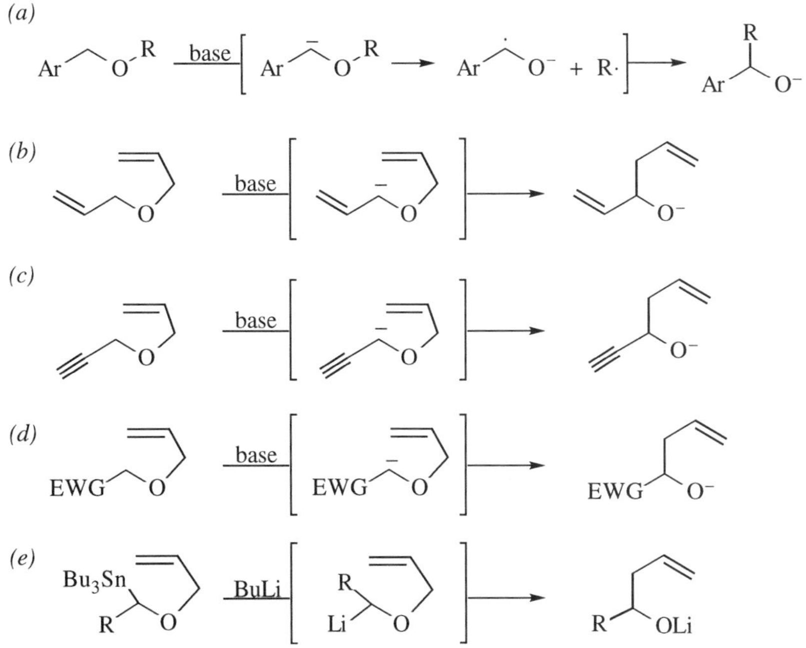

Scheme 6.5. (*a*) The [1,2]-Wittig rearrangement [29,30]. (*b*) The [2,3]-Wittig rearrangement [31]. (*c*) The [2,3]-Wittig rearrangement of propargyl allyl ethers occurs by deprotonation at the propargylic position. (*d*) Similarly, electron withdrawing groups (EWG) can be used to influence the site of deprotonation. (*e*) The Still variant of the [2,3]-Wittig, which uses a tin-lithium transmetalation to control anion formation [32].

homolysis of the opposite carbon–oxygen bond, and recombination to an alkoxide [33,34].[5] The [2,3]-variant was first observed by Wittig (although not recognized as such) in 1949 [36] and by Hauser two years later (Scheme 6.5b, [31]), and was shown in subsequent studies to proceed by a concerted S_{Ni} mechanism [37,38]. When the [1,2]- and [2,3]-rearrangements can compete, the [2,3]-Wittig rearrangement predominates at low temperatures [39-41].[6]

With unsymmetrical ethers, the problem of the regiochemistry of metalation arises. Three approaches have successfully addressed this issue. One takes advantage of the fact that propargyl allyl ethers deprotonate exclusively at the propargylic position [48,49] Scheme 6.5c). Second, an electron withdrawing group (EWG) that stabilizes the anion on one side of the ether can be used to control the site of deprotonation, although enolates may suffer competitive [3,3]-rearrangement [50,51], Scheme 6.5d). Finally, the regiochemical issue can be eliminated by using tin-lithium exchange to generate the carbanion at a specific site ([32], Scheme 6.5e).

The migration across the allyl system is suprafacial [41], as illustrated by the example shown in Scheme 6.6a [52,53]. The configuration of the carbanionic carbon[7] inverts during the rearrangement, as predicted by theory in 1990 [55], and subsequently proven by three independent studies in 1992 [56-58], the simplest of which is illustrated in Scheme 6.6b. Thus, the [2,3]-Wittig rearrangement is a $[\pi2_s + \sigma2_a + \sigma2_a]$-rearrangement, which is symmetry allowed for a concerted six electron process with two inversions [35].

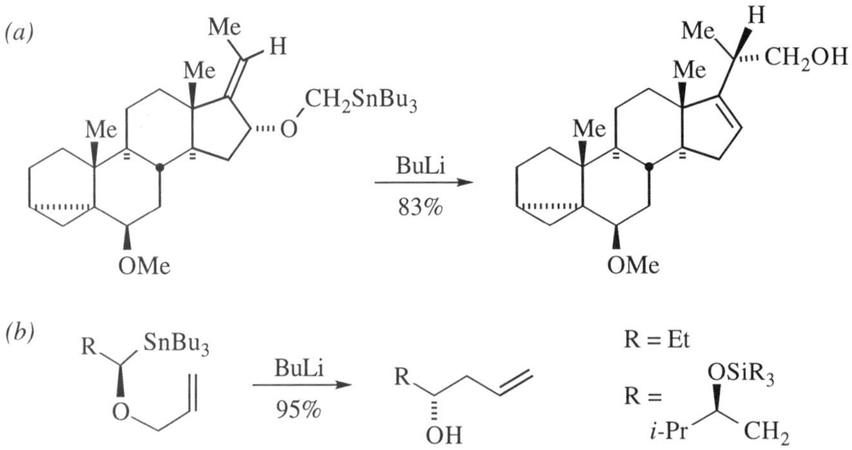

Scheme 6.6. (*a*) Example illustrating the suprafacial nature of the migration across the allyl moiety [52]. (*b*) Examples illustrating inversion of configuration at the metalated carbon [57,58].

The approximate geometries of four calculated transition structures are shown in Figure 6.4. When the lithium is included in the calculation (Figure 6.4a), all nonhydrogen atoms except the middle carbon of the allyl system are approximately coplanar [55]. When the lithium is removed, the envelope conformation is main-

---

5   Note that a concerted [1,2]-carbanion migration is symmetry forbidden [35].
6   For reviews of the Wittig rearrangements, see ref. [42-47].
7   α-Alkoxyorganolithiums are configurationally stable below about –30° (section 3.2.1, [54].

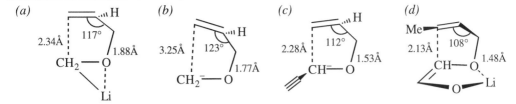

Figure 6.4. Ab initio transition structures for the [2,3]-Wittig rearrangement. *(a)* Structure including a lithium, in which the metal is antiperiplanar to both carbons of the allyl system and bridges the carbon and oxygen [55]. *(b)* Calculated transtion structure for [2,3]-rearrangement the naked $ROCH_2^-$ anion [59]. *(c)* Calculated transition structure for the [2,3]-rearrangement of a propargyl anion. Orientation of the alkynyl moiety on the convex face is favored by 2.1 kcal/mole [59]. *(d)* For the rearrangement of a lithium enolate, the endo structure is favored [59]. (The author is grateful to Professors Y. Wu and K. N. Houk, who kindly supplied the indicated bond lengths and angles in a private communication.)

tained, but the bond lengths and angles change dramatically, as shown in Figure 6.4b [59]. For the naked anion, the transition structure is extremely early, with practically no bond making or breaking having occured. When the lithium is present, the transition structure is somewhat later, which may be an artifact of the method, since the calculation requires that the lithium be unsolvated and in the gas phase. Since one cannot ignore the presence of the cation, we may assume that the real transition state geometry probably lies somewhere between these two structures. When the carbanion is stabilized by an alkynyl group (Figure 6.4c) or is an enolate (Figure 6.4d), the calculated transition structure is much more compressed [59]. Note for example, that the forming and breaking bonds are shorter than in the other two structures, and also note that the bond angle is smaller.

6.1.2.1 Simple diastereoselectivity

The aspects of diastereoselectivity in the [2,3]-Wittig rearrangement that we will be concerned with involve the geometry of the double bond and the configuration of the allylic and 'carbanionic' carbons in the allyl ether. Figure 6.5a illustrates diastereomeric transition structures for [2,3]-rearrangements of α-allyloxy organolithiums. If there is a substituent (R_1) at the allylic position, $A^{1,2}$ and $A^{1,3}$ allylic strain will play a role. If both R_1 and R_2 are not hydrogen, $A^{1,2}$ strain will disfavor the left conformer. If the alkene has the Z configuration, $A^{1,3}$ strain is particularly severe in the structure on the right. With reference to Figure 6.4, note that $A^{1,2}$ strain will be alleviated by a small allylic bond angle and that $A^{1,3}$ strain will be enhanced by a small bond angle.

Similar transition structures having stereogenic carbanionic carbons are illustrated in Figure 6.5b. For electrostatic reasons, an electron rich substituent such as an alkyl, vinyl, or alkynyl group will preferably occupy the convex face of the envelope conformation, while an electropositive substituent favors the concave side [55,59].

If the carbanionic carbon is trigonal, such as with enolates, the preference is to occupy the concave face, as shown in Figure 6.5c. This effect is reminiscent of the endo effect in Diels-Alder reactions (Section 6.6), and is also consistent with

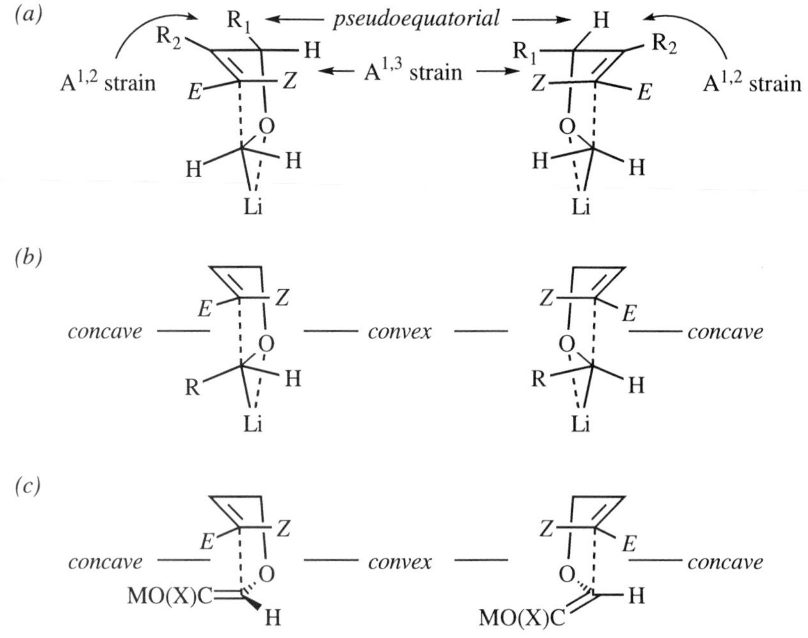

(a)

(b)

(c)

Figure 6.5. Factors influencing the relative configuration of the products in [2,3]-Wittig rearrangements: *(a)* Diastereomeric transition states illustrating the possibility of allylic strain. *(b)* The conformation having R on the convex face of the envelope is preferred for alkyl, vinyl, and alkynyl substituents. *(c)* For enolates, the concave orientation (synclinal double bonds) is preferred.

Seebach's topological rule suggesting a preference of synclinal donor/acceptor orientations in a Newman projection along the forming bond (*cf.* Figure 5.8, [60]). For Z(O)-enolates, additional stabilization can be had by metal chelation with the ether oxygen (*cf.* Figure 6.4d).

Each one of the effects illustrated in Figure 6.5 is attributable to a stereogenic element in the starting material (olefin geometry or absolute configuration at the allylic or carbanionic carbon), and is an example of single asymmetric induction. When more than one element is present, these effects can operate as matched or mismatched pairs of double asymmetric induction, and very high selectivities can be achieved when they operate in concert. Additionally, it is possible to introduce a stereogenic element elsewhere, such as a chiral auxiliary (X of Figure 6.5c). Conversely, when two elements are dissonant, lower selectivity may be expected.

The reader should recognize that these five-membered-ring transition states are considerably more flexible than, for example, a chair structure such as the Zimmerman-Traxler transition state in aldol additions (*cf.* Scheme 5.1).[8] This flexibility complicates the analysis of the various effects. A few examples serve to illustrate how these effects influence the configuration of the double bond and stereocenters in the product.

---

8   Indeed, transition state models having slightly different envelope or half-chair conformations have been proposed (*cf.* ref.[44-46,61].

Effect of allylic and double bond substitution on product configuration. Scheme 6.7 illustrates the influence of allylic strain between alkyl substituents on the double bond and allylic positions, uncomplicated by substitution at the carbanionic carbon. As shown in Scheme 6.7a, tin-lithium exchange affords an anion that rearranges (*cf.* Figure 6.5a, R_1 = *n*-Bu, R_2 = Me, E = Z = H) to give a near quantitative yield of alkene with 96-97% diastereoselectivity [32]. In this example, $A^{1,2}$ strain is relieved when the butyl group adopts the pseudoaxial orientation.

Scheme 6.7b illustrates the influence of $A^{1,3}$ strain between two alkyl groups (*cf.* Figure 6.5a, R_1 = *n*-heptyl, R_2 = H, E = H, Z = Me), this time favoring the pseudoequatorial conformation for the allylic substituent, so as to avoid the Z-methyl. [2,3]-Wittig rearrangement is 100% stereoselective for the *E*-alkene [32]. In contrast, when the alkene is unsubstituted in the "Z-position", the selectivity for a particular olefin geometry is severely diminished. Scheme 6.7c lists two such examples having E or unsubstituted alkene as educt, which are only 60-65% selective for the Z-product. It was noted (*cf.* Figure 6.4c, [59]) that propargylic anions rearrange *via* a transition structure that has significantly shorter bond lengths, and also a compressed allylic bond angle. The latter effect amplifies $A^{1,3}$ strain, and E selectivity is restored when the carbanionic carbon is propargylic (Scheme 6.7d, [48]).

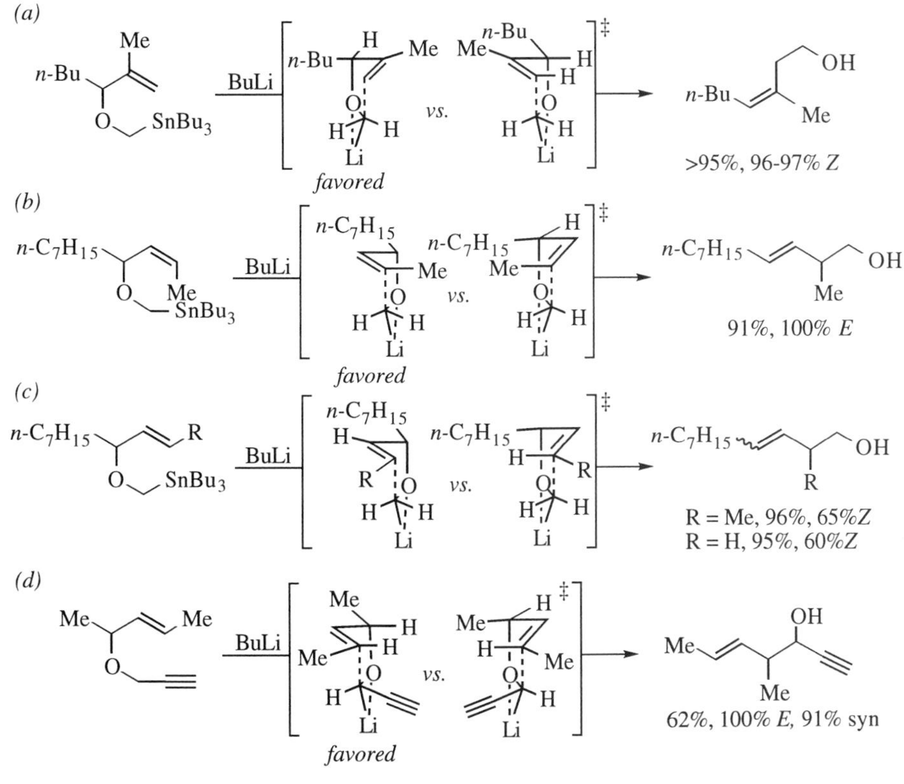

Scheme 6.7. The effects of allylic strain on the stereoselectivity of alkene formation [32]. (*a*) $A^{1,2}$ strain and the selective formation of Z-alkenes. (*b*) $A^{1,3}$ strain causes selective formation of *E*-alkenes. (*c*) If one or both of the 'partners' (*cf.* Figure 6.5a, R_1, R_2, or Z) is hydrogen, the selectivity is diminished. (*d*) $A^{1,3}$ strain produces 100% *E* selectivity when the carbanionic carbon is propargylic [48].

Effect of anion substitution on relative configuration. As seen in Scheme 6.7d, if both carbons involved in bond formation have heterotopic faces, two adjacent stereocenters are formed in the rearrangement. The topicity of these examples can be analyzed by reference to Figure 6.6, which defines the facial topicity for the components of the bond forming reaction, and also shows how these heterotopic faces are combined to form either syn or anti relative configurations in the product. Figure 6.6a and c show the topicities for Z-alkenes, while Figure 6.6b and d illustrate similar transition structures for E-alkenes. Note that the preceding discussion analyzed the combined effects of substitution on the double bond and at the allylic position. The structures in Figure 6.6 are unsubstituted at the allylic position, so that the factors affecting relative configuration can be analyzed independent of the effects of an allyl substituent.

Many examples of this type of reaction have been reported in the literature, but only with a few alkyl substituents on the metalated carbon are high selectivities consistently achieved. Table 6.1 lists several such examples, which can be rationalized by the indicated structures in Figure 6.6. Recall (Figure 6.5b and accompanying discussion) that theory predicts that electron rich alkyl substituents will prefer the convex face of the transition structures (*i.e.,* Figure 6.6a and d), for electrostatic reasons [55].

Entry 1 was the first example, reported in 1970 [40], of a highly stereoselective [2,3]-Wittig rearrangement, but comparison with entry 5 shows that only the Z isomer is selective. Entries 2-4 illustrate substituted propargyl Z allyl ethers,

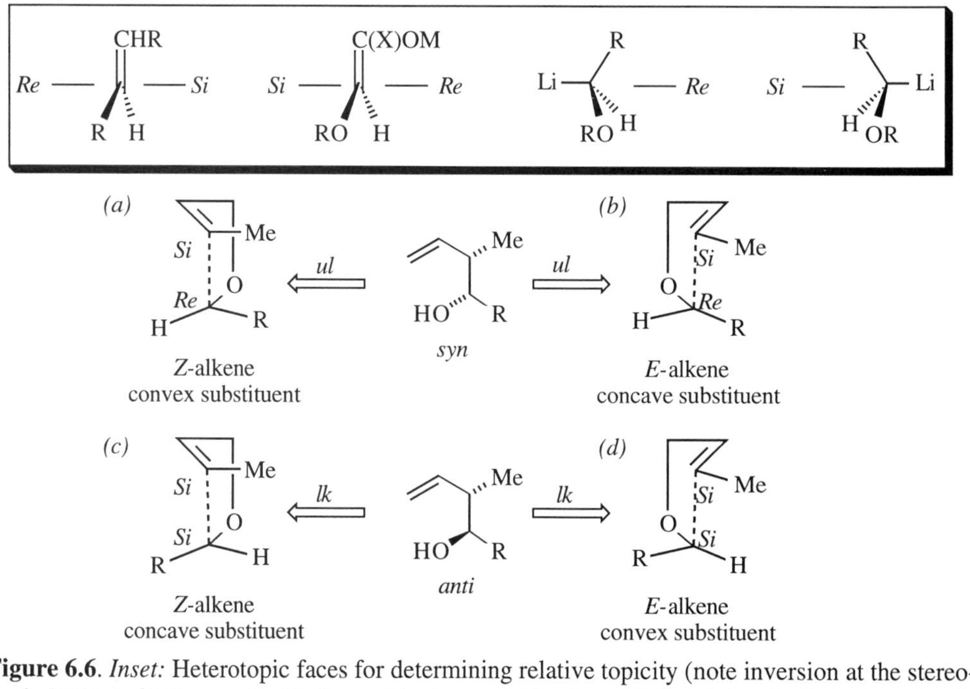

Figure 6.6. *Inset:* Heterotopic faces for determining relative topicity (note inversion at the stereogenic RLi). *(a,b)* Syn product is formed by two combinations of *ul* topicity. *(c,d)* Anti product is formed by two combinations of *lk* topicity. In transition states *a-d*, the metal is omitted. When R is an alkyl group, it would be bridged to the C–O bond, antiperiplanar to the allyl group (*cf.* Figure 6.4a, b). If R is a carbonyl, the metal will be attached to the enolate oxygen (*cf.* Figure 6.5c).

Table 6.1. Selective [2,3]-Wittig rearrangements of α-phenyl, α-propargyl, and α-alkyl organo-lithiums, showing a high $Z \rightarrow$ syn / $E \rightarrow$ anti correlation. The 'Path' column refers to the transition structures in Figure 6.6.

Entry	R	E/Z	Path	Config.	% ds	% Yield	Ref.
1	Ph	Z	a	100% syn	100	–	[40]
2	HC≡C	98%Z	a	88% syn	90	56	[48]
3	MeC≡C	98%Z	a	98% syn	100	55	[48]
4	TMSC≡C	93%Z	a (& b)	98% syn	105 (!)	74	[48]
5	Ph	E	b & d	50:50	50	–	[40]
6	HC≡C	93%E	d	93% anti	100	72	J751, [48]
7	MeC≡C	93%E	d	92%anti	99	65	J751[48]
8	TMSC≡C	93%E	b	75% syn	73	72	[48]
9	Et	E	d	99% anti	99	95	[58]

which show a consistently high $Z \rightarrow$ syn selectivity, consistent with the transition structure in Figure 6.6a being favored over Figure 6.6c [48]. Entry 4 (trimethyl-silylalkyne) is particularly striking because the product has a higher syn/anti ratio than the Z/E ratio in the starting material! This is not experimental error, as shown by Entry 8, which is also highly syn selective even though the starting material is 93% E, and anti product was expected [48]. Entries 6 and 7 show a more predict-able tendency for very high $E \rightarrow$ anti stereoselectivity (Figure 6.6d favored over Figure 6.6c), underscoring the anomalous nature of Entry 8. Entry 9 demonstrates that carbanions that are not resonance stabilized are also highly selective. In this case, the organolithium was generated by transmetalation of an organostannane, and again high $E \rightarrow$ anti stereoselectivity is observed [58].

When the alkenyl component is an *O-tert*-butyldimethylsilyl (TBDMS) enol ether, another anomaly occurs: independent of enol ether geometry, the anti product is favored (Scheme 6.8) [62]. With trimethylsilylpropargyl ethers, the anti selectivity is 95-98%, making this reaction an excellent route for the preparation of anti 1,2-diols. In these cases, transition structures similar to Figure 6.6c and d are operative, the dominant influence being mutual repulsion between the carbanion substituent, R, and the *O*-silyl group.

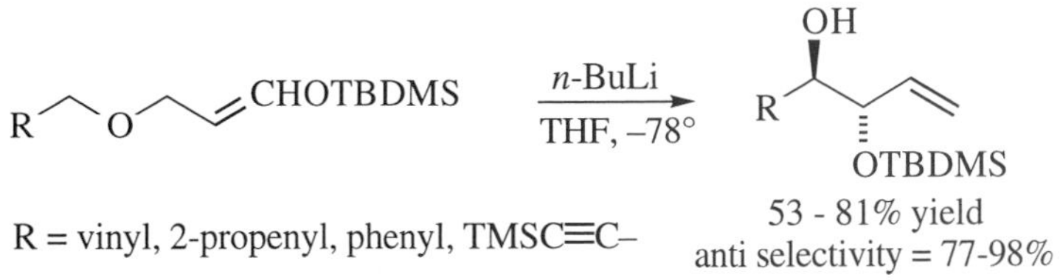

R = vinyl, 2-propenyl, phenyl, TMSC≡C–

53 - 81% yield
anti selectivity = 77-98%

Scheme 6.8. The [2,3]-Wittig rearrangement of silyl enol ethers is anti selective independent of carbanion substituent and double bond geometry [62].

For lithium enolate anions, the tendency is for the enolate to occupy the concave face of the transition structure (*cf.* Figure 6.4d and 6.5c) and therefore to prefer transition structures such as those illustrated in Figure 6.6b and c.[9] Table 6.2 lists several examples of simple acyclic diastereoselection, which show a tendency for $E \rightarrow$ syn and $Z \rightarrow$ anti selectivity, in contrast to the tendency observed for hydrocarbon substituted carbanions (Table 6.1). Entries 1 and 2 involve dianions of crotyloxy acetates, and show $E \rightarrow$ syn and $Z \rightarrow$ anti selectivity. A more complex example involving extension of a steroid side chain (similar to Scheme 6.6a), is 100% anti selective from an '*E*'-alkene, however [53].

The ester enolates illustrated in entries 3 and 4 are considerably more selective when the lithium cation is exchanged for dicyclopentadienyl zirconium [63]. It is suggested that the zirconium chelates the α-alkoxy oxygen in these examples, and that the cyclopentadienyl ligands influence the topicity in the transition state [63]. Scheme 6.9 illustrates how the *lk* topicity may be disfavored by a steric interaction between a pseudoaxial allylic hydrogen and a cyclopentadienyl ligand. The Z-alkene isomer (entry 4) is also syn-selective, although less so than the *E* isomer, and the yield is not encouraging. The rationale illustrated in Scheme 6.9 [63] implies that deprotonation of the ester affords the *Z(O)*-enolate, in contrast to the expected (Section 3.1.1) tendency of esters to afford *E(O)*-enolates. In his review of enolate formation [64], Wilcox notes that *Z(O)*-enolate formation by deprotonation of α-alkoxy esters would be expected if chelation were the dominant influence, but that the results reported in the literature show no consistent trend.[10]

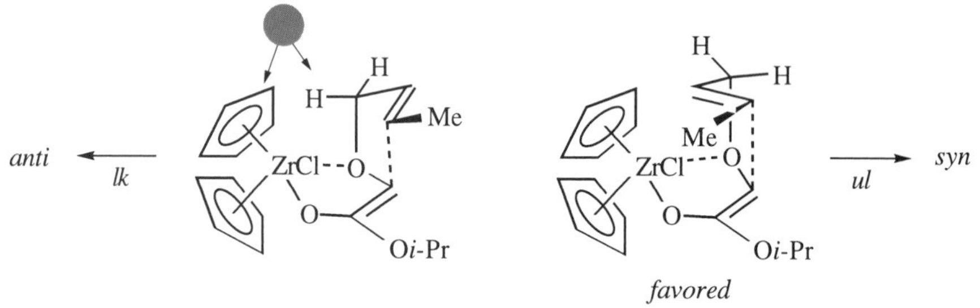

Scheme 6.9. Rationale for the *ul* selectivity of dicyclopentadienylzirconium ester enolates in [2,3]-Wittig rearrangements.

Pyrrolidinyl amides undoubtedly form *Z(O)*-enolates, and the [2,3]-Wittig rearrangement of the *E*-alkene (entry 5, [69] is highly selective. The Z-alkene was not tested, and propargylic amide enolates do not rearrange [70]. Entry 5 also shows the highest yield in the Table. As will be seen, amides of C_2-symmetric amines can be excellent chiral auxiliaries in this process.

---

[9] Note also that enolates may suffer competitive [3,3]-sigmatropic rearrangement [44,50,51].
[10] Based on the relative configuration of the products of Ireland-Claisen rearrangements, two groups have concluded that *Z(O)*-enolate formation predominates [65,66]. On the other hand, two other groups quenched α-alkoxy ester enolates with trialkylsilyl chlorides and found mixtures of enol ether (ketene acetal) isomers [67,68].

Table 6.2. [2,3]-Wittig rearrangements of α-allyloxy enolates. The 'Path' column refers to the transition structures in Figure 6.6. All examples used LDA as base; entries 3 and 4 also have Cp_2ZrCl_2 as additive (Cp = η^5-cyclopentadienyl).

Entry	X	E/Z	Path	Config.	% ds	% Yield	Ref.
1	OH	93% E	b	65% syn[11]	70	60	[71]
2	OH	95% Z	c	75% anti[9]	79	73	[71]
3	O*i*-Pr, as Cp_2ZrCl enolate	E	b	98% syn	98	47	[63]
4	O*i*-Pr, as Cp_2ZrCl enolate	Z	a	88% syn	88	15	[63]
5	pyrrolidinyl	E	b	96% syn	96	97	[69]

6.1.2.2 Chirality transfer in enantiopure educts

As seen in the previous section, substitution at the double bond, the allylic position, and the carbanionic carbon influence the configuration of the new double bond and the *relative configuration* of the stereocenter(s) in the product of a [2,3]-Wittig rearrangement. In this section, it will be seen that the *absolute configuration* of stereocenters at the allylic and carbanionic carbons determine the *absolute configuration* of the stereocenter(s) in the product (Scheme 6.10). In fact, several examples already cited involve chiral educts being transformed into chiral products (*cf.* Scheme 6.6b, Scheme 6.7b and c, Table 6.1, entry 9), although this point was not the focus of the discussion. It should come as no surprise that a transition structure that is sufficiently organized to afford good selectivity in the formation of one double bond isomer or one relative configuration, can also afford good enantioselectivity in the formation of one or two new stereocenters.

Scheme 6.10 illustrates generic schemes for the asymmetric synthesis of homoallylic alcohols using a [2,3]-Wittig reaction as a key step. In these sequences, the absolute configuration and enantiomeric purity of the starting materials are determined by their method of preparation (or commercial source), and the following examples will show that the chirality sense of the starting material controls the absolute configuration of the product via the principles of simple diastereoselectivity outlined in the preceding sections. The absolute configuration of a

---

[11] It should be noted that the original reference (J754) uses the ambiguous erythro/threo nomenclature without drawing a reference structure. Later, in a review by the same authors [44], the same nomenclature is used but apparently to indicate the opposite relative configurations. Additionally, the review [44] states a different selectivity for the *E*-alkene than is given in the original article (J754). Table 6.2 lists the selectivities from the original article with the relative configurations as drawn in the review.

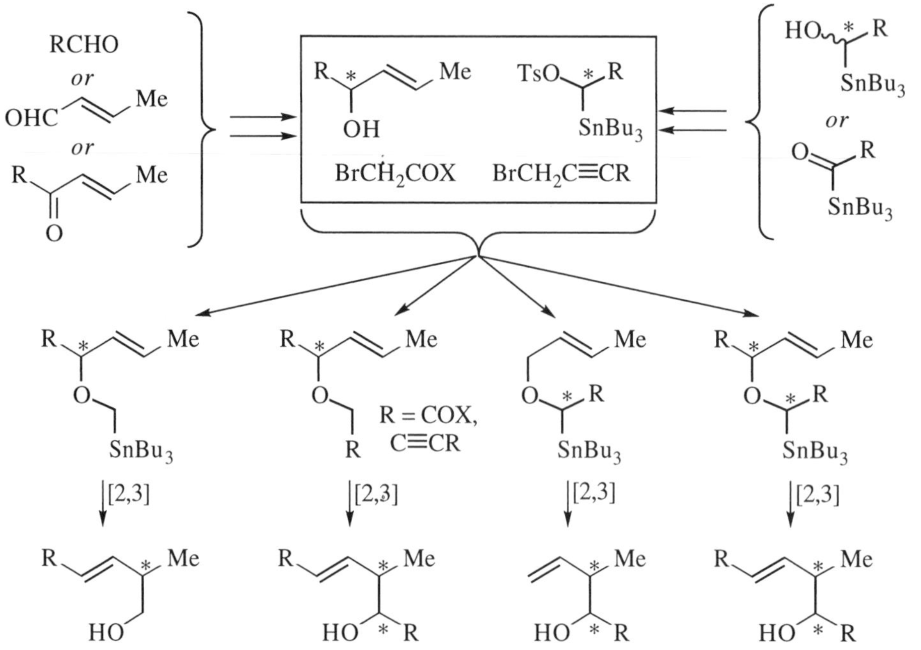

Scheme 6.10. *Top:* Some of the possible paths for the preparation of chiral building blocks for the assembly of substrates for a [2,3]-Wittig rearrangement. *Bottom:* Intramolecular asymmetric induction in [2,3]-Wittig rearrangements.

stereocenter at the allylic (or propargylic) position may be set by asymmetric reduction of an allylic or propargylic ketone (Chapter 7) or asymmetric addition to an aldehyde (Chapter 4). The absolute configuration at the tin-bearing carbon can be set by asymmetric reduction of acyl stannanes [72-74], kinetic resolution using a lipase [75], or oxidation of α-stannylborates [76]. In certain cases, the carbanion configuration can be controlled by enantioselective deprotonation.

Qualitative evidence that the [2,3]-Wittig rearrangement of nonracemic substrates might have high enantioselectivities was reported in the early 1970s (*e.g.*, see ref. [41,77], but it was some years before this aspect of the reaction was quantitated. The evidence that eventually appeared is completely consistent with the tenets of simple diastereoselectivity outlined in the preceeding section. For example in 1984, Midland [78] and Nakai [79] showed that nonracemic ethers with stereocenters at the allylic position *and* having the Z configuration at the double bond are highly selective for the product having the *E*-configured double bond and syn relative configuration at the two new stereocenters. In addition, the chirality transfer was quantitative, as illustrated in Scheme 6.11. Substituents at the allylic position and the Z-olefinic site are susceptible to severe $A^{1,3}$ strain in one of the conformers of the transition state (*cf.* Figure 6.5a, Scheme 6.7b), and this effect determines the absolute configuration at one of the two new stereocenters. Additionally, the two faces of the carbanionic carbon in these examples are heterotopic; the topicity is determined by the greater preference of the carbanionic substituent to occupy the convex face of the envelope transition structure (*cf.* Figure 6.6a). When R₁ is isopropyl, the *E*-alkene isomers show only 60-62% selectivity for the anti isomer [78].

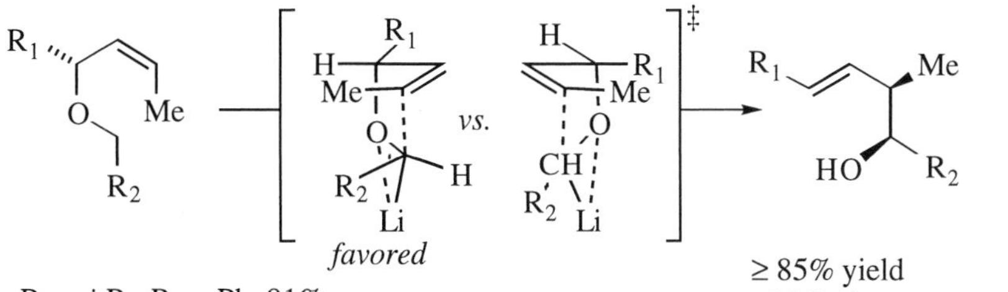

$R_1 = i$-Pr; $R_2 = $ Ph, 91% ee
$R_1 = i$-Pr; $R_2 = $ vinyl, 91% ee
$R_1 = $ Me (*ent*); $R_2 = $ C≡CSiMe$_3$, 98% ee

≥ 85% yield
≥99% *E*
92-99% syn
100% chirality transfer

Scheme 6.11. Asymmetric induction and chirality transfer in [2,3]-Wittig rearrangements of allylic benzyl [78], allyl [78], and trimethylsilylpropargyl [79] ethers.

In these cases, the isopropyl probably favors a pseudoequatorial conformation and there is only a slight preference for the phenyl or vinyl carbanion substituent to occupy the convex face of the transition structure.

Marshall reported two examples that differed only in the degree of substitution at the allylic position. In one case, with a quaternary allylic carbon, the enantiomeric purity of the product was only 59% ee (Scheme 6.12) [80]. Apparently there is less preference for the carbanionic substitutent to occupy the convex face in preference to the concave face of the transition structure. When the angular methyl is replaced by hydrogen, the chirality transfer is 100%.

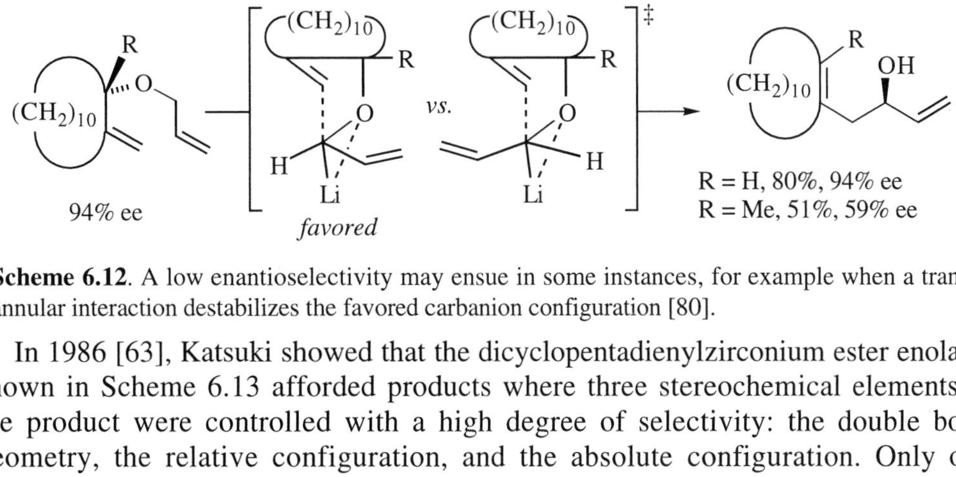

94% ee   *favored*

R = H, 80%, 94% ee
R = Me, 51%, 59% ee

Scheme 6.12. A low enantioselectivity may ensue in some instances, for example when a transannular interaction destabilizes the favored carbanion configuration [80].

In 1986 [63], Katsuki showed that the dicyclopentadienylzirconium ester enolates shown in Scheme 6.13 afforded products where three stereochemical elements in the product were controlled with a high degree of selectivity: the double bond geometry, the relative configuration, and the absolute configuration. Only one double bond isomer was observed, the syn/anti diastereoselectivity was 98-99%, and the enantioselectivity was >98%.

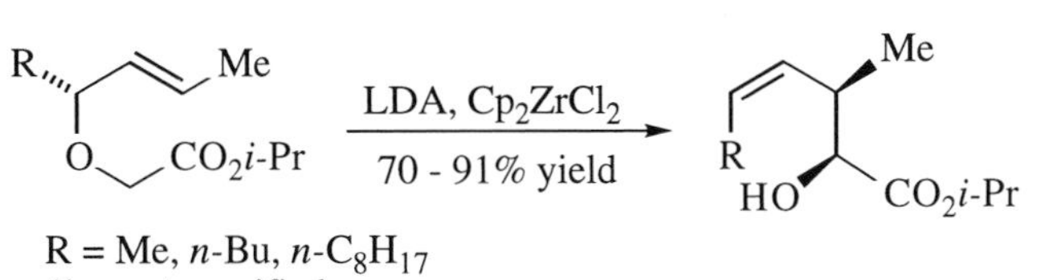

$\dfrac{\text{LDA, Cp}_2\text{ZrCl}_2}{\text{70 - 91\% yield}}$

100% *Z*
98 - 99% ds (syn/anti)
>96% ee

R = Me, *n*-Bu, *n*-C$_8$H$_{17}$
%ee not specified

Scheme 6.13. Asymmetric induction in [2,3]-Wittig rearrangements of chiral α-alkoxy esters [63].

With three stereogenic elements in the product, there are a total of eight possible stereoisomers. However, if it is assumed that the possible transition structures are similar to those shown in Scheme 6.9, there are only four possibilites for the [2,3]-rearrangement, as shown in Scheme 6.14. (Recall from Figure 6.4d that enolate transition structures have shorter bond lengths and smaller allylic bond angles than the other transition structures.) The two having *lk* topicity, Scheme 6.14a and c, are disfavored by having the pseudoaxial allylic substituent in close proximity to the cyclopentadiene ligand. Of the two possible *ul* transition structures, the R group is on the less crowded convex face of the bicyclic structure in Scheme 6.14b, but on the concave face in d, where it encounters the cyclopentadiene ligand.

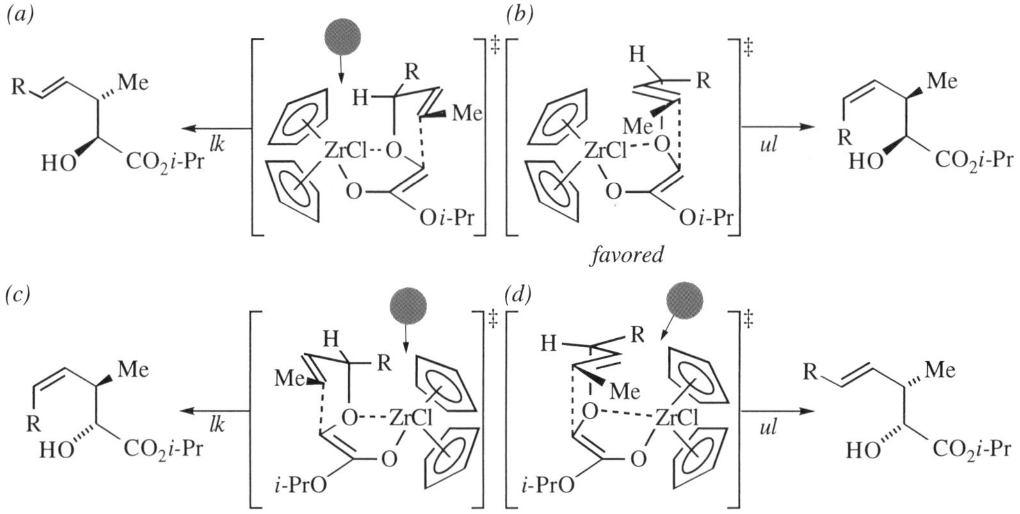

Scheme 6.14. Possible transition structures for the [2,3]-Wittig rearrangement of the *R*-allylic ester enolates shown in Scheme 6.13. For amide enolates, see Scheme 6.22.

Using a similar protocol, Marshall showed that the propargyloxy esters shown in Scheme 6.15a undergo [2,3]-Wittig rearrangements with 100% chirality transfer [81,82]. Marshall also showed that the corresponding lithium carboxylate dianions rearrange with 100% chirality transfer, and with excellent diastereoselectivity; often the yields are higher, as shown by the examples in Scheme 6.15b [81-83]. Similar to the rationale for the selective rearrangement of the α-allyloxy ester enolates in Scheme 6.13, the rationale for the asymmetric induction in the present case has the propargylic substituent on the convex face of the transition state assembly (*cf.* Scheme 6.14b).

Following the lead provided by Nakai, who showed that racemic allyloxy esters can be rearranged using trimethylsilyl triflate [84], Marshall examined similar conditions for the rearrangement of nonracemic propargyloxy esters, and reported the results tabulated in Scheme 6.16 [82]. These two reactants are identical to the two reported in Scheme 6.15a that were rearranged under strongly basic conditions. In the silyl triflate mediated rearrangement, the yields are much higher, although the selectivity is somewhat lower. Additionally, the relative configuration of the allene and the C-2 stereocenter are different. Nevertheless, the chirality transfer is

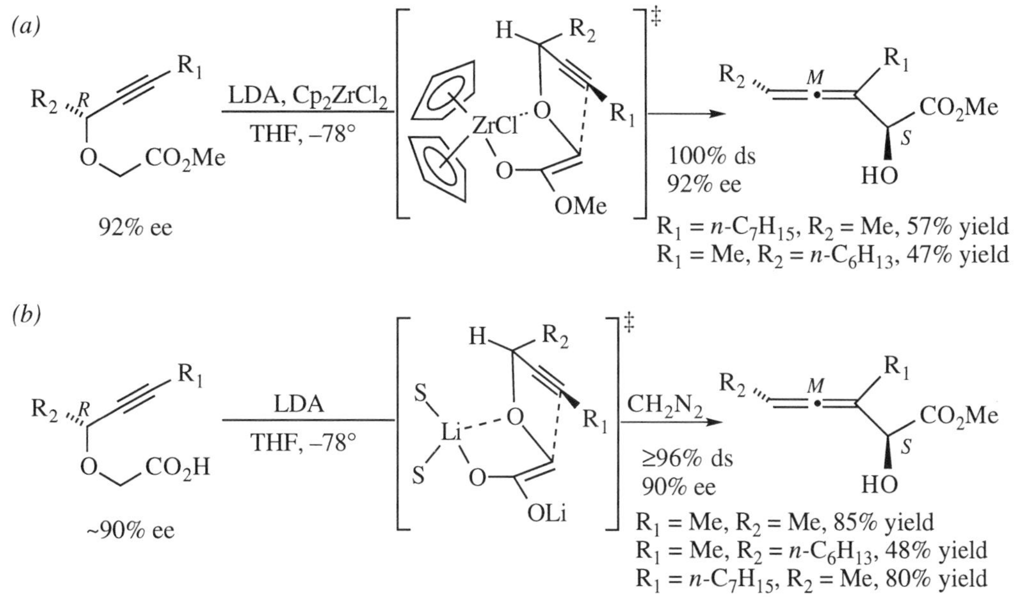

(a)

LDA, Cp$_2$ZrCl$_2$
THF, −78°

92% ee

100% ds
92% ee

S
HO

R$_1$ = n-C$_7$H$_{15}$, R$_2$ = Me, 57% yield
R$_1$ = Me, R$_2$ = n-C$_6$H$_{13}$, 47% yield

(b)

LDA
THF, −78°

CH$_2$N$_2$

≥96% ds
90% ee

S
HO

R$_1$ = Me, R$_2$ = Me, 85% yield
R$_1$ = Me, R$_2$ = n-C$_6$H$_{13}$, 48% yield
R$_1$ = n-C$_7$H$_{15}$, R$_2$ = Me, 80% yield

~90% ee

Scheme 6.15. [2,3]-Wittig rearrangements of chiral propargyloxy acetates: *(a)* Zirconium ester enolates [81,82]. *(b)* Lithium endiolates, S = solvent [81-83].

100% (*i.e.*, both the major and minor isomer have the same enantiomeric purity as the starting propargyl ether). An advantage of this procedure over the base-mediated protocol is that terminal alkynes (R$_1$ = H) survive the silicon-mediated process [82]. Nakai suggested an 'oxygen ylide' as the intermediate in these silicon-mediated [2,3]-rearrangements, with the silicon and the enolate moieties trans to each other in the 5-membered ring transition structure [84], as shown for the propargyl ether in Scheme 6.16 [82].

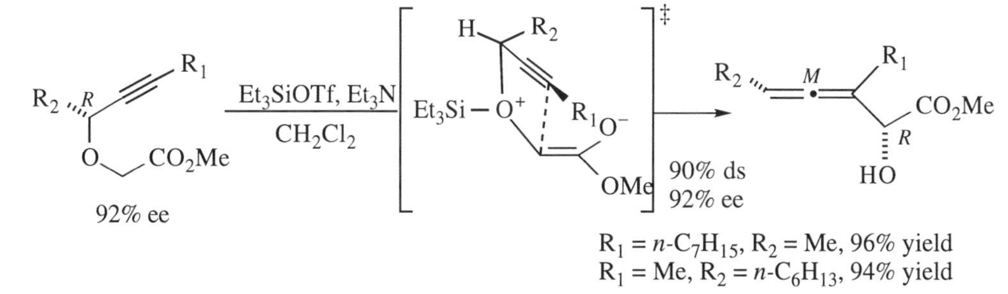

Et$_3$SiOTf, Et$_3$N
CH$_2$Cl$_2$

Et$_3$Si—O$^+$

90% ds
92% ee

R
HO

92% ee

R$_1$ = n-C$_7$H$_{15}$, R$_2$ = Me, 96% yield
R$_1$ = Me, R$_2$ = n-C$_6$H$_{13}$, 94% yield

Scheme 6.16. Silicon-mediated [2,3]-Wittig rearrangement of chiral propargyloxy acetates [82]. The minor diastereomer is the C-2 *S* hydroxyl.

Chirality transfer in the rearrangement of allyloxymethyl stannanes is complete, even in cases where the rearrangement itself is not selective for one product, as shown by the examples in Scheme 6.17 [85]. Recall from Scheme 6.7b and c that in the Still-Wittig rearrangement, one product double bond configuration is formed selectively only when the educt has the Z configuration. This is due to severe A^{1,3} strain in one of the two transition structures (*e.g.*, between the isopropyl and the methyl in Scheme 6.17a). In 1985, Midland reported that rearrangement of the Z-

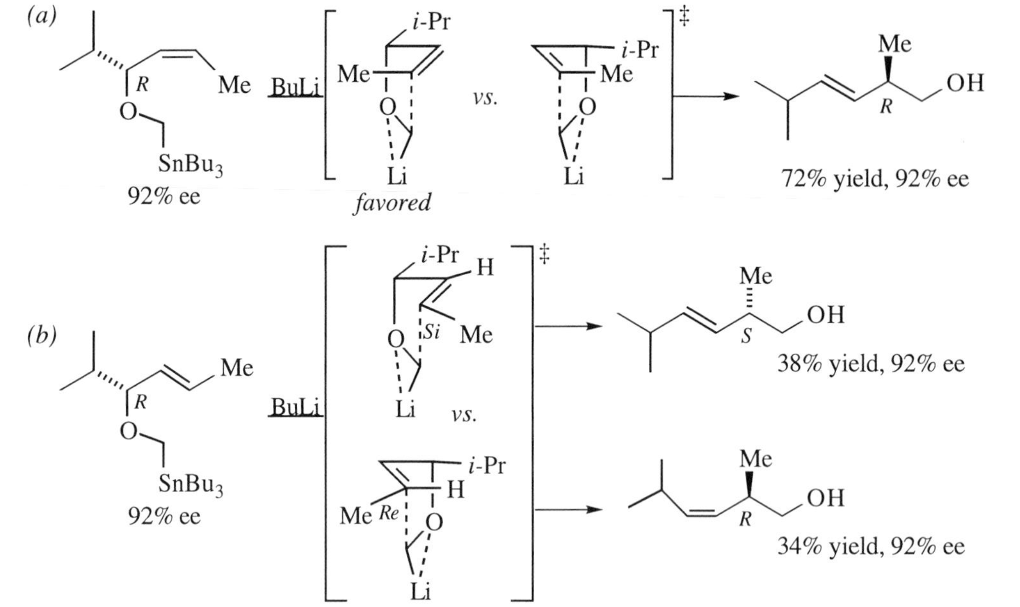

Scheme 6.17. [2,3]-Still-Wittig rearrangements of allyl ethers [85].

olefin illustrated in Scheme 6.17a is 100% selective for the *E*-double bond geometry, and that the enantiomeric purity of the product matches the enantiomeric purity of the starting material. As expected (*cf.* Scheme 6.7c), the isomeric *E*-educt affords a 53:47 mixture of *E* and *Z* products, as shown in Scheme 6.17b. However, chirality transfer for the formation of each of these products is 100%, even though the absolute configurations of the newly created stereocenter in the two products are opposite! This result may be explained by examining the two transition structures illustrated. The conformation that presents the *Si* face of the olefin to the metalated carbon (Scheme 6.17b, top) is destabilized by A^{1,2} strain (between the isopropyl and the neighboring vinyl proton) while the conformer that presents the *Re* face is destabilized by A^{1,3} strain (between the isopropyl and the other vinylic hydrogen). These two effects are approximately equal in this relatively 'loose' transition structure (*cf.* Figure 6.4a and b), so the product ratio is nearly equal.

The rearrangement of propargyloxy stannanes is highly selective, as shown by Marshall in 1989 [83]. The two examples illustrated in Scheme 6.18 show 100% chirality transfer. In this case, there is no conformational ambiguity, since neither of the carbons involved in bond formation are heterotopic.

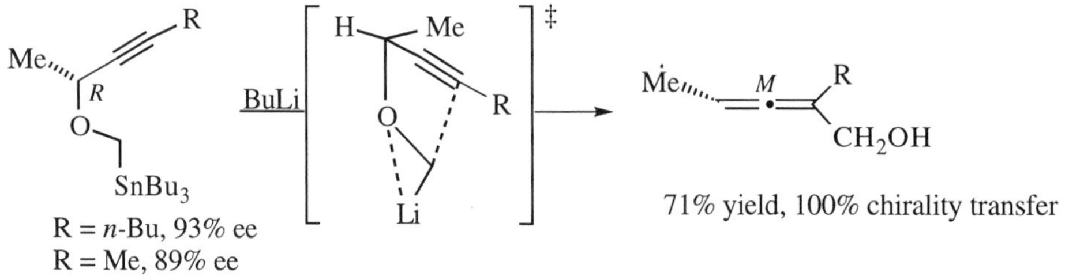

Scheme 6.18. [2,3]-Still-Wittig rearrangements of propargyl ethers [83].

Chirality transfer is also quantitative when the metalated carbon is stereogenic, as shown by the examples in Scheme 6.19 [58]. When R is hydrogen, the two faces of the terminal allylic carbon are homotopic and it does not matter which of the illustrated transition structures is involved. The only important point is that the metal-bearing carbon undergoes inversion of configuration (see also Scheme 6.6b). When R is methyl, the metal-bearing carbon still undergoes inversion, but the configuration at the second stereocenter is determined by consideration of the two illustrated transition structures. Here, the *ul* topicity is favored (the reaction is 99% diastereoselective for the anti relative configuration) because of the preference for the ethyl group to occupy the convex face of the transition structure (see Table 6.1, entry 9).

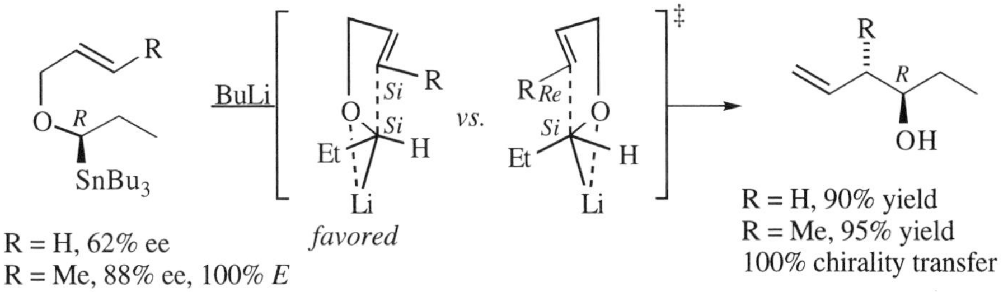

Scheme 6.19. [2,3]-Still-Wittig rearrangements of allyl ethers having stereogenic metalated carbons [58].

6.1.2.3 Chiral auxiliaries and chiral bases

The examples of [2,3]-Wittig rearrangements of allyloxy enolates listed in Table 6.2 show good to excellent simple diastereoselectivity. Chiral auxiliaries, in the form of esters of chiral alcohols and amides of C_2-symmetric chiral amines have been evaluated in these rearrangements. For example, Nakai showed that the lithium enolates of 8-phenylmenthol esters afford good simple diastereoselectivity with good asymmetric induction as well (Scheme 6.20, [86]. As before, the rationale invokes an α-alkoxyenolate that chelates the lithium metal. The inset of Scheme 6.20 illustrates the most stable conformation of the chelated enolate, and shows the

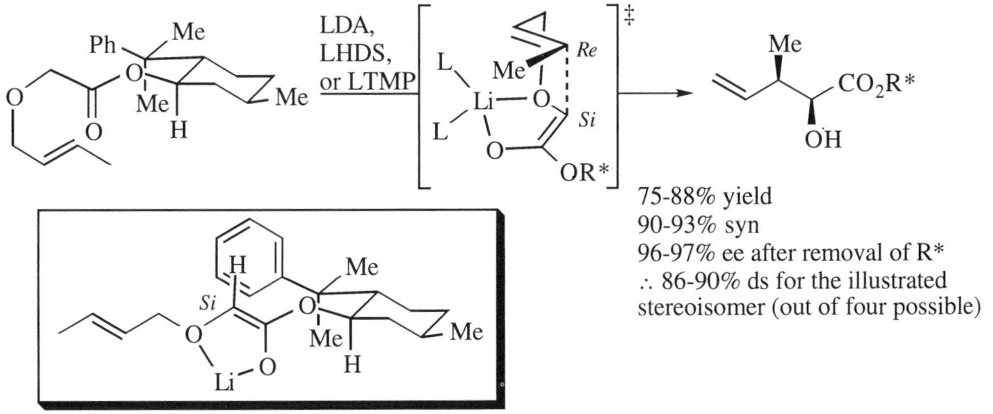

75-88% yield
90-93% syn
96-97% ee after removal of R*
∴ 86-90% ds for the illustrated stereoisomer (out of four possible)

Scheme 6.20. 8-Phenylmenthol as a chiral auxiliary in the [2,3]-Wittig rearrangement [86]. *Inset:* Rationale for the *Si*-face selectivity of the enolate.

rationale for preferential attack on the *Si* face of the enolate. The preferred topicity of an enolate is often *ul* (*cf.* Figure 5.4d, Figure 6.5c, Scheme 6.9, Scheme 6.13), which produces the syn rearrangement product, as shown in the illustrated transition structure. There is a slight dependence of the selectivity on the specific lithium amide base used, so it is likely that the amine (conjugate acid of the base) is still associated with the lithium enolate (*cf.* Section 3.1.1).

Other examples that underscore the close association of the amine with the lithium ion are examples of interligand asymmetric induction,[12] reported by Marshall and illustrated in Scheme 6.21. In Scheme 6.21a [70], Overberger's base is used to doubly deprotonate a propargyloxy acetic acid; presumably, the enolate is chelated by the α oxygen, as shown in the illustrated transition structure. Higher enantioselectivity is achieved with the 13-membered propargyl ether shown in Scheme 6.21b [87,88]. This example exhibits the highest degree of asymmetric induction for [2,3]-Wittig rearrangements using the Overberger base. Even other cyclic ethers afford only low selectivity, such as the example shown in Scheme 6.21c [89]. Nevertheless, the principle of interligand asymmetric induction is established by these examples; it then remains to improve on the observed selectivities. A rationale to explain the absolute configuration of the latter two examples may involve an enantioselective deprotonation or a mixed aggregate.

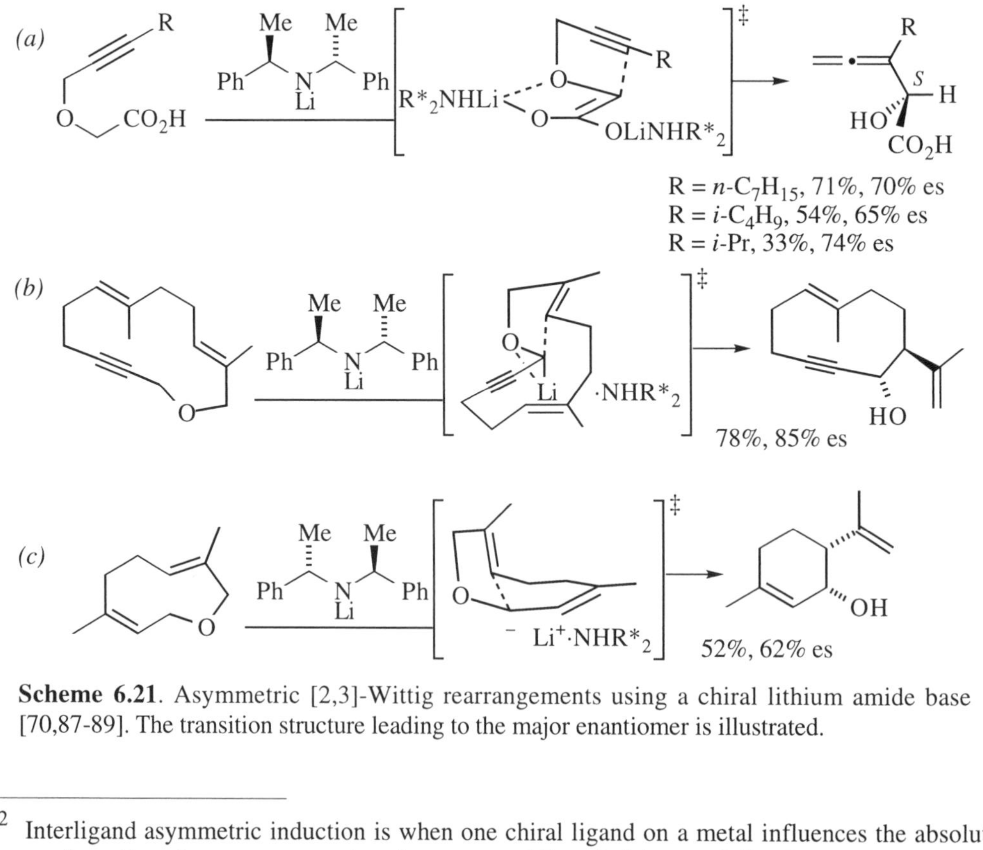

$R = n\text{-}C_7H_{15}$, 71%, 70% es
$R = i\text{-}C_4H_9$, 54%, 65% es
$R = i\text{-}Pr$, 33%, 74% es

78%, 85% es

52%, 62% es

Scheme 6.21. Asymmetric [2,3]-Wittig rearrangements using a chiral lithium amide base [70,87-89]. The transition structure leading to the major enantiomer is illustrated.

---

[12] Interligand asymmetric induction is when one chiral ligand on a metal influences the absolute configuration of a new stereogenic unit on a second ligand of the metal (Section 1.3).

As indicated by Entry 5 in Table 6.2, the lithium enolates of pyrrolidine amides show excellent simple diastereoselectivity, and rearrange in excellent yields [69]. These amides also show a slight dependence of selectivity on the structure of the amide base used [69]. Monosubstituted pyrrolidine amides were poor auxiliaries for this reaction (≤76% ds) [69], but C2-symmetric pyrrolidines are highly selective, as shown in Scheme 6.22 [90]. The *Si* facial selectivity of the lithium enolate and the illustrated zirconium enolate were comparable, but only the zirconium enolate also showed a high preference for the *ul* topicity illustrated. The two views of the transition structure rationalize both the topicity and the absolute configuration of the product. The enolate *Si* face is favored because the closer of the two pyrrolidine stereocenters blocks the *Re* face. The *ul* topicity is favored because when the enolate moiety is on the concave face of the cyclopentane envelope, a severe interaction between a pseudoaxial hydrogen and a cyclopentadiene is avoided (*cf.* Scheme 6.14 a for another illustration).

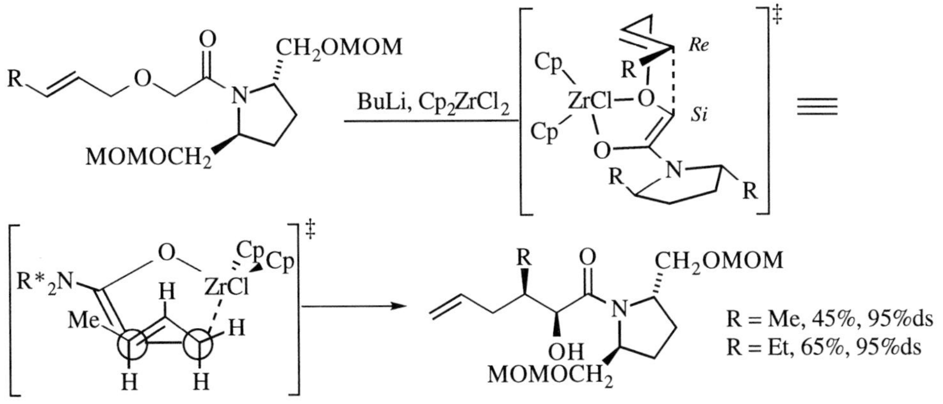

Scheme 6.22. [2,3]-Wittig rearrangement of amide zirconium enolates using Katsuki's pyrrolidine auxiliary [90].

Any reaction that forms a bond between two prochiral atoms in a stereoselective manner is a valuable synthetic method. Some of the natural products that have been made in nonracemic form using the [2,3]-Wittig rearrangement as the key step are illustrated in Figure 6.7. The stereocenters formed in the Wittig rearrangement are indicated (∗).

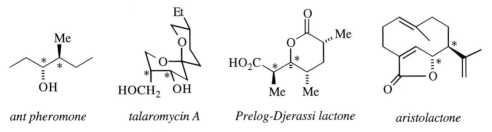

ant pheromone   *talaromycin A*   *Prelog-Djerassi lactone*   *aristolactone*

Figure 6.7. Natural products using the [2,3]-Wittig rearrangement as the key step: *(a)* ant pheromone [58]; *(b)* talaromycin A (J768); *(c)* Prelog-Djerassi lactone (J771); *(d)* aristolactone [87,88].

6.2 Cycloadditions

Cycloaddition reactions have considerable value in organic synthesis for a number of reasons, not the least of which are that two bonds are formed in one operation and that the reactions often exhibit high stereoselectivities. Even if this huge field were limited only to examples that fall into the category of asymmetric synthesis, it would take several volumes to completely do it justice. In this section, only selected [2+1]- and [4+2]-cycloadditions (and equivalent transformations) are covered, and the discussion is not limited to concerted processes.

6.2.1 [2+1]-Cyclopropanations and related processes[13]

Although the addition of carbene to a double bond to make a cyclopropane is well known, it is not particularly useful synthetically because of the tendency for extensive side reactions and lack of selectivity for thermally or photochemically generated carbenes. Similar processes involving carbenoids (species that are not free carbenes) are much more useful from the preparative standpoint [91,92]. For example, metal catalyzed decomposition of diazoalkanes usually results in addition to double bonds without the interference of side reactions such as C–H insertions. Consider the possible retrosynthetic approaches to a 1,2-disubstituted cyclopropane shown in Figure 6.8. Disconnection *a* entails the addition of a methylene across a double bond, a conversion that is often stereospecific (*e.g.,* the Simmons-Smith reaction [93]). Disconnections *b* and *c* are more problematic, since the issue of cis/trans product isomers (simple diastereoselection) arises.

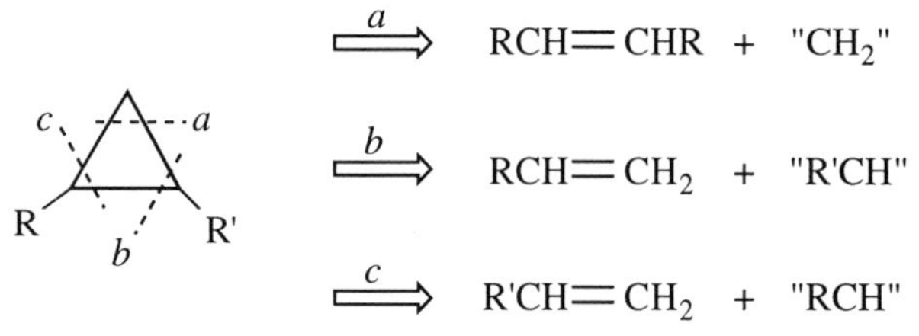

$$\xrightarrow{\ a\ } \quad RCH = CHR \ + \ "CH_2"$$

$$\xrightarrow{\ b\ } \quad RCH = CH_2 \ + \ "R'CH"$$

$$\xrightarrow{\ c\ } \quad R'CH = CH_2 \ + \ "RCH"$$

Figure 6.8. Retrosynthesis of 1,2-disubstituted cyclopropanes.

Two strategies have been taken to apply cyclopropanations to asymmetric synthesis: auxiliary based methods whereby a covalently attached adjuvant renders either the olefin or the cyclopropanating reagent chiral, and processes that utilize a chiral ligand on a metal catalyst. Scheme 6.23 illustrates these approaches as applied to the more complex case of disconnections *b* and *c* of Figure 6.8. Scheme 6.23a and b show chiral auxiliaries (R*) in the olefin and carbenoid moieties, respectively, while Scheme 6.23c shows a chiral ligand on the metal. Since the transition states of both processes still involve the metal, asymmetric syntheses using these reactions may be said to occur by intraligand or interligand asymmetric induction. Still another approach to asymmetric cyclopropanations involves reaction

---

[13] Not covered in this section are cyclopropanations that involve initial 1,3-dipolar cycloadditions of diazoalkanes to give pyrazolines, followed ring contraction and nitrogen extrusion.

sequences, such as a tandem 1,4 addition–intramolecular alkylation, that do not involve carbenes but which accomplish a similar transformation (also by intraligand asymmetric induction, Scheme 6.23d). Double asymmetric induction may be achieved by 'crossing' two methods, for example by using a chiral catalyst to promote reaction with a carbenoid and olefin that are also chiral. As will be seen, double asymmetric induction is often used in cyclopropanations of carbenes as a means of enhancing selectivity.

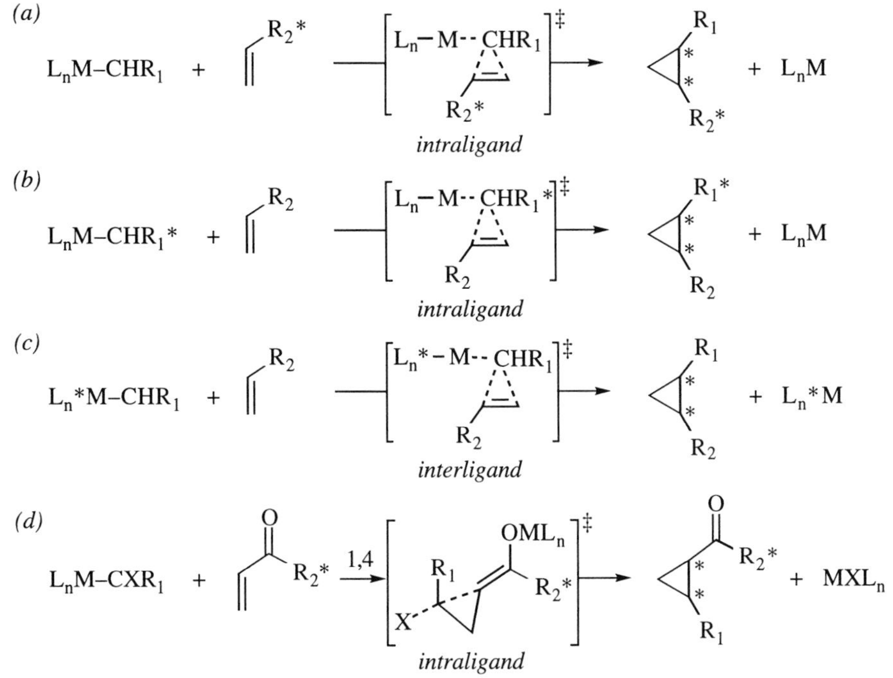

(a)

$$L_nM-CHR_1 \; + \; \overset{R_2*}{\diagup\diagdown} \quad \longrightarrow \quad \left[\begin{array}{c} L_n-M\cdots CHR_1 \\ \diagup\diagdown \\ R_2* \end{array} \right]^{\ddagger} \quad \longrightarrow \quad \overset{R_1}{\underset{R_2*}{\triangleleft^*_*}} \; + \; L_nM$$

intraligand

(b)

$$L_nM-CHR_1* \; + \; \overset{R_2}{\diagup\diagdown} \quad \longrightarrow \quad \left[\begin{array}{c} L_n-M\cdots CHR_1* \\ \diagup\diagdown \\ R_2 \end{array} \right]^{\ddagger} \quad \longrightarrow \quad \overset{R_1*}{\underset{R_2}{\triangleleft^*_*}} \; + \; L_nM$$

intraligand

(c)

$$L_n*M-CHR_1 \; + \; \overset{R_2}{\diagup\diagdown} \quad \longrightarrow \quad \left[\begin{array}{c} L_n*-M\cdots CHR_1 \\ \diagup\diagdown \\ R_2 \end{array} \right]^{\ddagger} \quad \longrightarrow \quad \overset{R_1}{\underset{R_2}{\triangleleft^*_*}} \; + \; L_n*M$$

interligand

(d)

$$L_nM-CXR_1 \; + \; \overset{O}{\underset{R_2*}{\diagdown\diagup}} \overset{1,4}{\longrightarrow} \left[\begin{array}{c} OML_n \\ \diagup\diagdown \\ R_1 \quad R_2* \\ X \end{array} \right]^{\ddagger} \quad \longrightarrow \quad \overset{O}{\underset{R_1}{\diagdown\triangleleft^*_* R_2*}} \; + \; MXL_n$$

intraligand

Scheme 6.23. General strategies for asymmetric induction in cyclopropanations.

The issue of simple stereoselectivity in cyclopropanations of the types shown in Figure 6.8, disconnections *b* and *c*, is not a trivial one, and relatively few additions of ketocarbenoids (by far the most common type of carbenoid studied) show high selectivity. The difficulty can be seen by inspection of the transition states of Scheme 6.24. The transition state leading to the trans isomer (*lk* topicity) is usually favored, but unless the COZ group is quite large, the trans-selectivities are not great. Recently, for example, Doyle showed that if Z = OEt (*i.e.*, ethyl diazoacetate), the $Rh_2(OAc)_4$ catalyzed cyclopropanation of alkenes having R = *n*-alkyl, Ph, and *i*-Pr is only about 60 - 70% trans-selective. With R = *tert*-butyl, the selectivity is 81%. If the olefin is in a ring, the selectivity is not much better [94]. If hindered esters (Z = OCMe*i*-Pr$_2$) or amides (Z = N*i*-Pr$_2$) are used, the trans-selectivity for the $Rh_2(OAc)_4$ catalyzed cyclopropanation of styrene can be improved to 71% and 98%, respectively [95]. BHT esters (Z = O-2,6-*t*-Bu-4-Me-C_6H_2) also give good trans-selectivity (71-97%) with a variety of alkenes [96]. With $Rh_2(NHCOMe)_4$ as catalyst, these selectivities can be increased further due to the decreased reactivity of the rhodium carbenoid, which results in a more selective reaction [95,96].

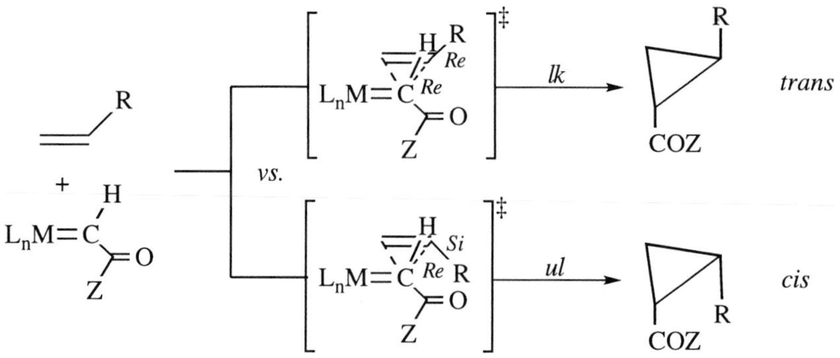

Scheme 6.24. Transition states and relative topicities for cycloaddition of ketocarbenoids and monosubstituted alkenes.

H. Davies has found that vinyl carbenoids tend to show high selectivities in Rh$_2$(OAc)$_4$ catalyzed cycloadditions, as shown by the examples in Scheme 6.25 [97]. It is also important to note that the stereoselectivity of the cyclopropanations shown in Schemes 6.24 and 6.25 are not due only to steric effects. For example, changing R$_1$ in Scheme 6.25 from *n*-butyl to *tert*-butyl lowers the selectivity from 85% to 78% (R$_2$ = CO$_2$Et), while changing R$_2$ from phenyl to CO$_2$Et (R$_1$ = Ph), lowers the selectivity from >95% to 89% [97]. Presumably there is a contribution to the relative stabilities of the transition states by both electronic and steric effects, but they have not been quantified.

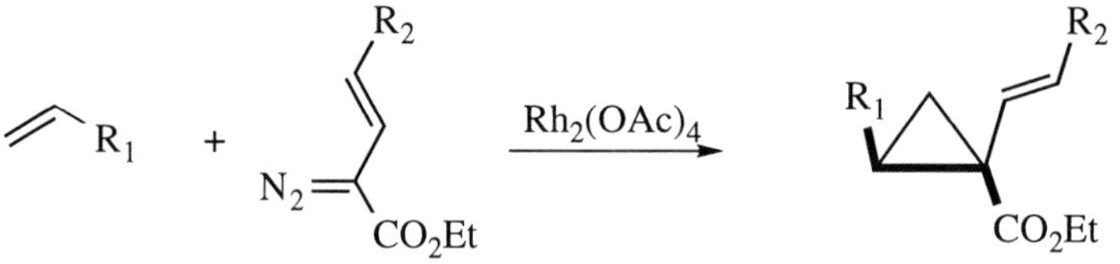

R$_1$ = Ph, 1°, 2°, 3° alkyl, AcO, EtOCH$_2$
R$_2$ = CO$_2$Et, Ph, CH=CHPh

78 - >95% ds
trans

Scheme 6.25. Diastereoselective cyclopropanations of vinyl carbenoids [97]. For disubstituted carbenes, cis/trans nomenclature is used to describe relative configuration, referring to R$_1$ relative to the carbonyl moiety, as shown in bold.

The following discussion is organized along the lines of the examples in Scheme 6.23. First, auxiliary-based methods are discussed, followed by methods using chiral catalysts, including examples of double asymmetric induction employing chiral catalysts on chiral substrates and substrates having chiral auxiliaries attached, and finally stepwise cyclopropanation sequences. Within each section, the addition of "CH$_2$" is covered first (*i.e.,* disconnection *a* in Figure 6.8), followed by examples of the addition of "RCH" (*i.e.,* disconnections *b* and *c* of Figure 6.8).

6.2.1.1 Chiral auxiliaries for carbenoid cyclopropanations

Cyclopropanations of functionalized alkenes using the Simmons-Smith reaction [93], or a similar cyclopropanation, have been developed by modifying carbonyl and hydroxyl groups with chiral auxiliaries. A single example was reported by Carrié in 1982 (Scheme 6.26a, [98]), whereby the oxazolidine derived from con-

densation of (–)-ephedrine and cinnamaldehyde was cyclopropanated with diazo-
methane using palladium acetate as catalyst. The yield was quantitative and the
selectivity was ≥95%, but no further examples were provided. More systematic
studies were undertaken by the groups of Yamamoto [99,100] and Mash [101-104].
Both of these groups used C_2-symmetric acetals as auxiliaries, as shown in Scheme
6.26b-c. Yamamoto studied the tartrate-derived acetals shown in Scheme 6.26b
while Mash examined a series of related acetals, including the two shown in Scheme
6.26c. Both groups showed that the acetal could be hydrolyzed in the normal
manner to the corresponding carbonyl compound, but Yamamoto also showed that
the acetal could be cleaved to the carboxylic acid using ozone.

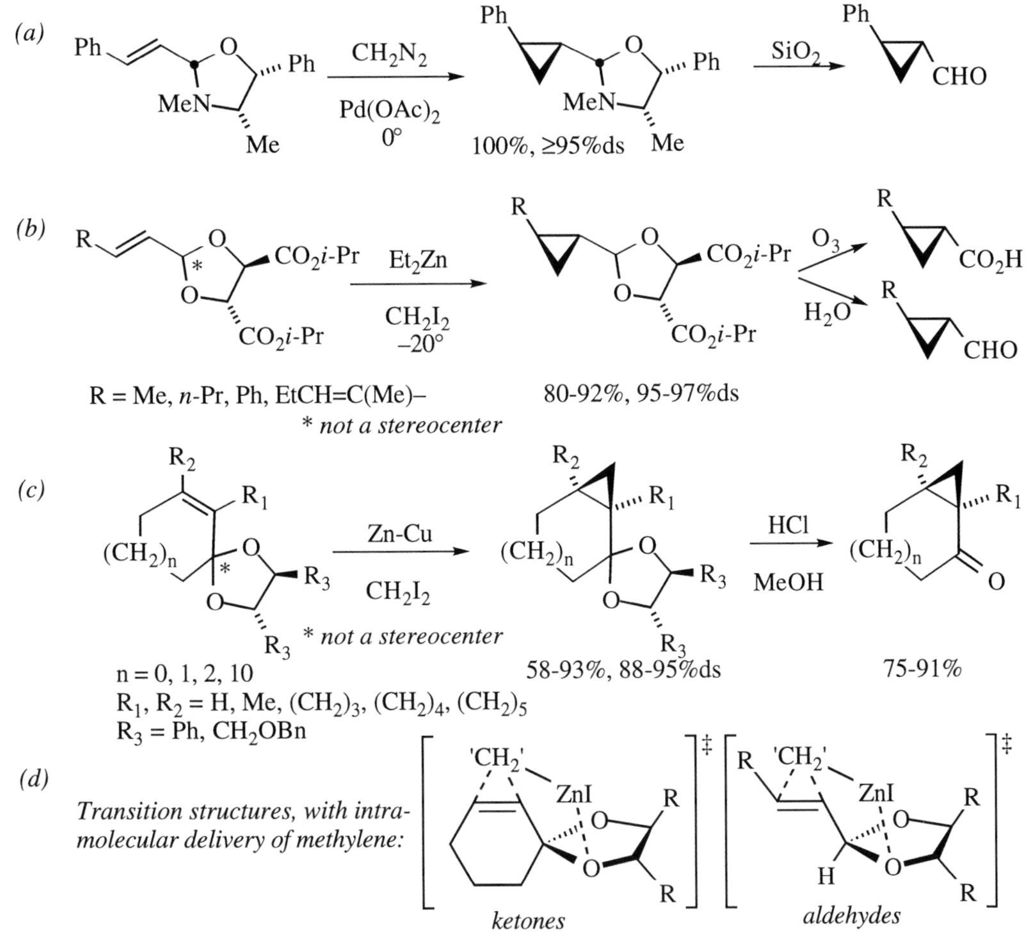

Scheme 6.26. Auxiliary-based asymmetric cyclopropanations (addition of "CH_2") of α,β-
unsaturated aldehydes and ketones. *(a)* [98]; *(b)* [99,100]; *(c)* [101-104]; *(d)* Proposed transition
structures [104]. Only one zinc and the transfer methylene are shown; other atoms associated with the
Simmons-Smith reagent are deleted for clarity.

Note that in both the aldehyde and ketone acetals, the acetal carbon is not
stereogenic, due to the C_2 symmetry of the starting diol. For the ketone acetals,
there is no conformational ambiguity, and the mechanistic rationale shown in
Scheme 6.26d was proposed to account for the selectivity of the reaction [104].

Thus, coordination of the zinc to one of the diastereotopic oxygens and oriented anti to the adjacent dioxolane substituent places the 'transfer methylene' on the face of the olefin toward the viewer, consistent with the observed absolute configuration. Note that coordination to the other oxygen and orienting anti to the adjacent substituent would place the 'transfer methylene' distal to the double bond. A similar explanation can be offered to rationalize the results of the aldehyde acetal additions, assuming that the olefin adopts the indicated conformation in the transition state.[14]

S. Davies has used an iron complex as an auxiliary for the asymmetric cyclopropanation of α,β-unsaturated carbonyls [105]. The iron acyl is most stable in the *s*-cis conformation, as illustrated in Scheme 6.27, in order to avoid severe interactions between the iron ligands and R. Coordination of the Simmons-Smith reagent to the carbonyl oxygen, anti to the iron, forces the alkene moiety out of conjugation and approximately orthogonal to the carbonyl. Because of the bulky triphenyl phosphine in the rear, this rotation can only be towards the front. Transfer of the methylene via the illustrated transition state accounts for the observed diastereoselectivity. Oxidation with bromine removes the iron acyl and derivatization with α-methyl-benzyl amine allowed evaluation of the stereoselectivity.

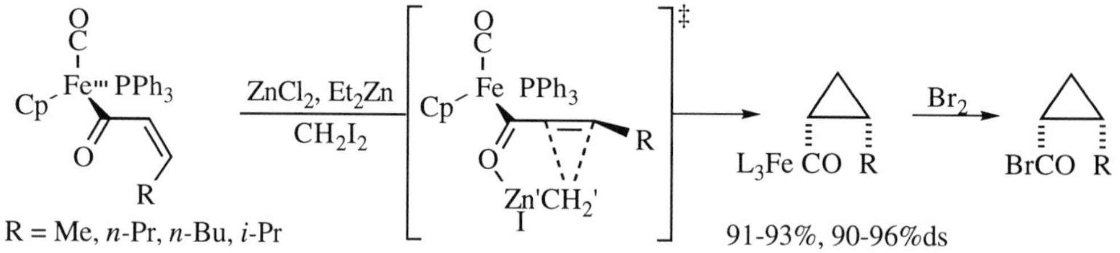

Scheme 6.27. S. Davies's asymmetric cyclopropanation of Z-iron acyls [105].

Charette has shown that allylic alcohols can be cyclopropanated by attaching a chiral auxiliary in the form of a glucose derivative [106] or *trans*-1,2-cyclohexane diol [107], as shown in Scheme 6.28. The yields are outstanding, as are the diastereoselectivities. The topicity can be rationalized by chelation of one of the zinc atoms of the Simmons-Smith reagent by the hydroxyl and the ether oxygen and intramolecular delivery of the methylene to the olefin in the conformation shown. Note however, that the conditions that are optimum for the glucose auxiliary afford very low selectivity in the cyclohexane diol system [107], which may mean that the mechanism is not so simple. Two procedures allow (destructive) removal of the auxiliary from the cyclopropane methanol. In one, the free hydroxyl of the glucose is triflated, the ring fragmented, and the resultant acylium ion hydrolyzed [106,108]. In another, the hydroxyl is converted to an iodide; halogen-lithium exchange then effects elimination of the alkoxide [107]. To get the opposite absolute configuration at the cyclopropane, a derivative of L-rhamnose may be used in place of the D-glucose [106], or the enantiomeric cyclohexane diol can be used.

---

[14] Although this explanation is self-consistent with that of the ketone acetals, a related 6-membered C_2-symmetric aldehyde acetal affords cyclopropanation products with the opposite topicity sense [100]. Also, the structure of the Simmons-Smith reagent is unknown, and aggregates may be involved. Thus, this explanation must still be regarded as tentative.

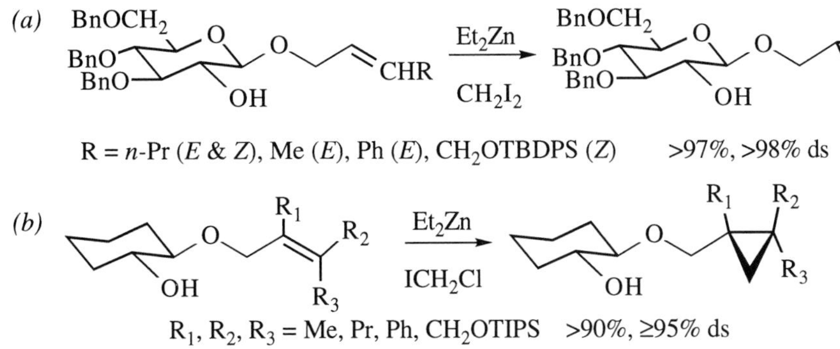

(a)

R = n-Pr (E & Z), Me (E), Ph (E), CH$_2$OTBDPS (Z)   >97%, >98% ds

(b)

R$_1$, R$_2$, R$_3$ = Me, Pr, Ph, CH$_2$OTIPS   >90%, ≥95% ds

Scheme 6.28. Asymmetric cyclopropanation of allylic alcohols: *(a)* Using a glucose-derived auxiliary [106]; *(b)* A cyclohexane diol auxiliary [107].

A process for the asymmetric cyclopropanation of the enol ethers of cyclic and acyclic ketones has been developed by Tai [109-111]. In this process, a C$_2$-symmetric acetal is isomerized to a hydroxy enol ether which serves as substrate for the Simmons-Smith cyclopropanation, as shown in Scheme 6.29. The stereo-selectivity is nearly perfect, but a mechanistic hypothesis has not been proposed. The auxiliary may be removed either by hydrolysis, to give the methyl ketone, or by oxidation of the alcohol and β-elimination [111].

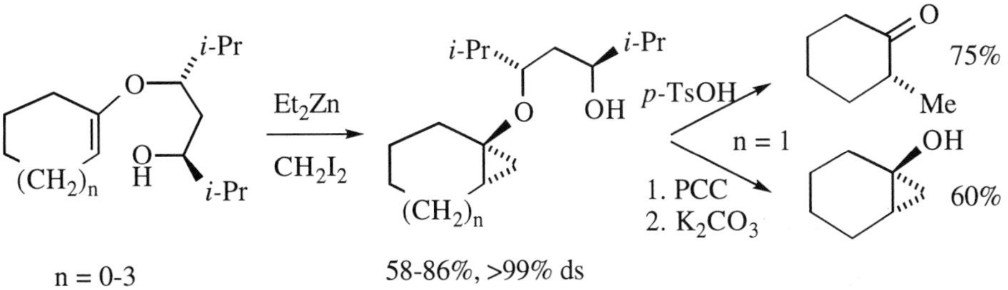

n = 0-3   58-86%, >99% ds

Scheme 6.29. Asymmetric cyclopropanation of ketone enol ethers [109-111].

Cyclopropanation reactions involving diazoalkanes and catalyzed by transition metals involve metal carbenes as intermediates. Scheme 6.30 illustrates the proposed catalytic cycle for such processes [112]. The catalyst, L$_n$M, is coordinatively unsaturated and therefore electrophilic. Loss of nitrogen from the zwitterion at the top affords the metal carbene shown at the right. Two canonical forms for the metal carbenoid are shown. For rhodium carbenes, it is thought that they tend to resemble metal stabilized carbocations, with a low barrier to rotation [112,113]. For control of the absolute configuration at the carbenoid carbon in the cyclopropanation, an auxiliary (usually the alcohol of the ester) must somehow shield one face of the trigonal carbenoid carbon in order to influence the absolute configuration at that center. Also, recall (Scheme 6.24) that the simple diastereoselectivity (relative configuration) in these processes is not high unless very bulky esters are used.

In light of the above analysis, it is perhaps not surprising that asymmetric cyclopropanations of styrene using bornyl, menthyl, and 2-phenylcyclohexyl esters of diazoacetic acid afforded both poor cis/trans selectivity and low enantioselectivity with cuprous chloride [114] or rhodium acetate [115] catalysts. On the other hand,

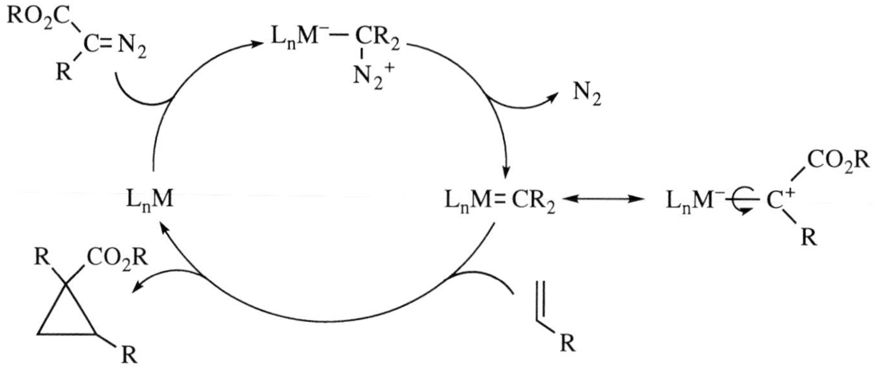

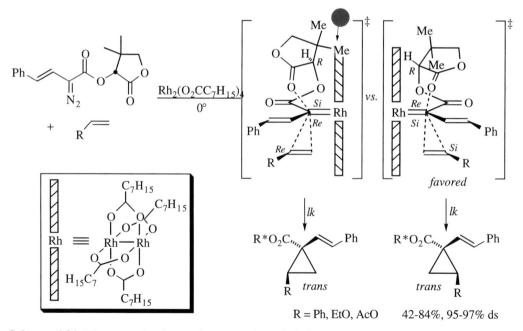

Scheme 6.30. Catalytic cycle for the transition metal-catalyzed cyclopropanation of olefins by diazoalkanes (after [112] and [113]).

vinyl carbenoids (Scheme 6.25) show good simple diastereoselection [97], and H. Davies has shown that pantolactone is an excellent chiral auxiliary, as shown in Scheme 6.31 [116-118]. The mechanistic hypothesis involves intramolecular interaction of the pantolactone carbonyl with the electrophilic carbenoid carbon, which shields the *Re* face of the carbene. Note that the conformer in which the carbene's *Si* face is shielded suffers severe steric interactions between the catalyst 'wall' and the pantolactone moiety. Approach of the alkene toward the *Si* face of the carbene, coupled with diastereoselectivity favoring *lk* relative topicity, affords a mixture containing only the two trans diastereomers. The examples in Scheme 6.25

Scheme 6.31. Diastereoselective cyclopropanation of olefins with vinyl carbenes [116]. Note that only two of the four possible stereoisomers were found in the product mixture. The trans nomenclature refers to the relative configuration of R and CO_2R^*, consistent with that of Scheme 6.24.

showed a lower cis/trans selectivity. In the examples shown in Scheme 6.31, however, only the two trans diastereomers are found. Thus, a weakness of the transition state models shown in Scheme 6.31 is that, although the absolute configuration is rationalized, it is not obvious why the cis/trans selectivity (*lk* topicity) should be 100%. This underscores the statement in the previous section which noted the presence of unquantified electronic effects contributing to the stereoselectivity of the rhodium catalyzed cyclopropanation using vinyl carbenes.

The cis relationship between the vinyl group and the R group of the olefin raises an interesting possibility: if the R group is also a vinyl substituent, the product of the cyclopropanation is a *cis*-divinylcyclopropane, precursor to a Cope rearrangement [119]. Although the Cope rearrangement destroys the stereocenters created in the cyclopropanation, it creates others, as shown by the examples in Scheme 6.32.

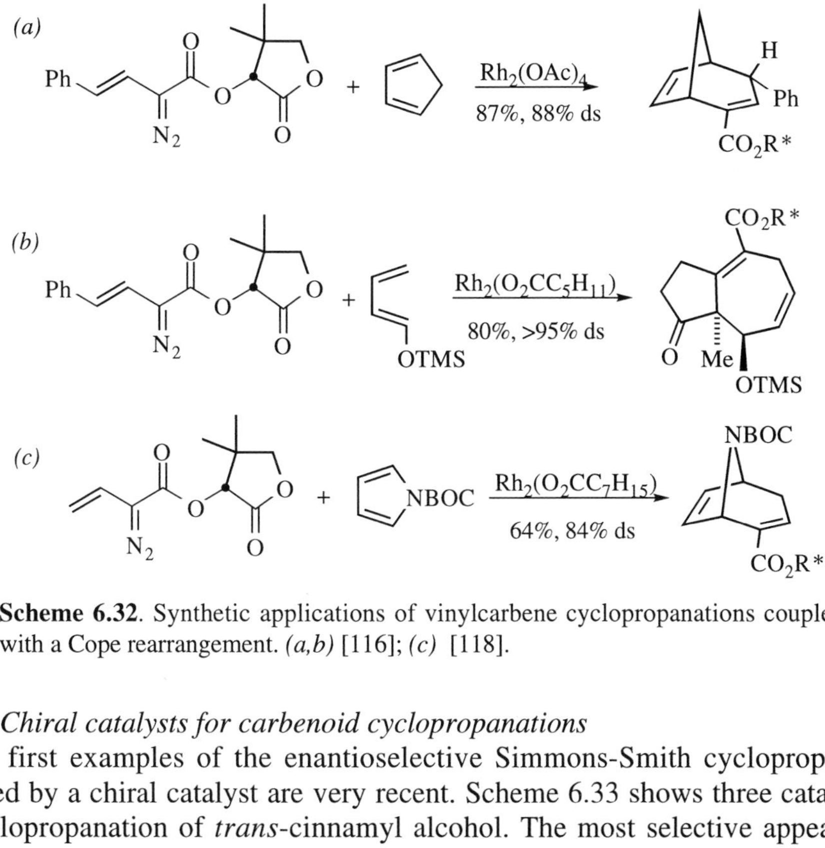

Scheme 6.32. Synthetic applications of vinylcarbene cyclopropanations coupled with a Cope rearrangement. *(a,b)* [116]; *(c)* [118].

6.2.1.2 Chiral catalysts for carbenoid cyclopropanations

The first examples of the enantioselective Simmons-Smith cyclopropanations mediated by a chiral catalyst are very recent. Scheme 6.33 shows three catalysts for the cyclopropanation of *trans*-cinnamyl alcohol. The most selective appears to be Charette's dioxaborolane (Scheme 6.33c, [120-122], which also affords the highest yield of product, although this procedure is only suitable for small scale.[15] With other olefins, such as cis and trans disubstituted alkenes and β,β-trisubstituted alkenes, the yields are nearly as good and the enantioselectivities are 96-97%. An important finding in this study [120] was that, in addition to the Lewis acid (boron) that binds the alcohol, a second atom to chelate the zinc is also necessary. In the

---

[15] Charette has noted an explosion hazard on scale-up of the original procedure [121], and has published an alternative procedure [122].

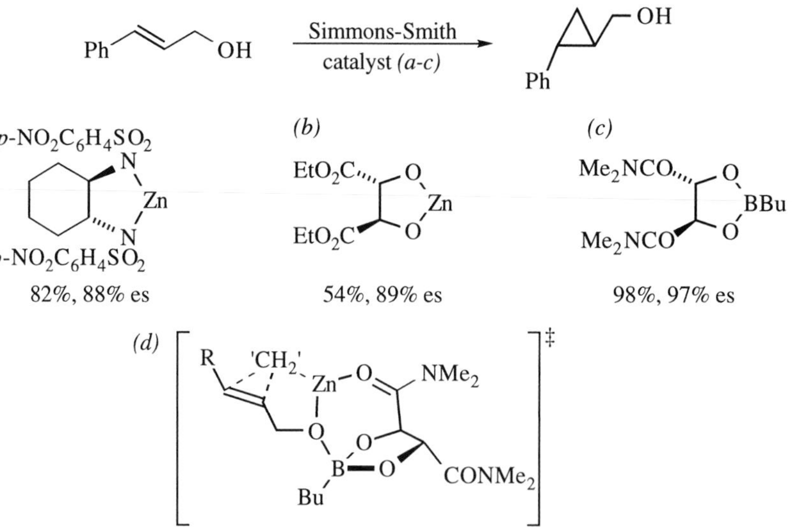

Scheme 6.33. Asymmetric catalysts for the Simmons-Smith cyclopropanation of *trans*-cinnamyl alcohol: *(a)* [123]. *(b)* [124]. *(c)* [120,122]. *(d)* Transition state model for catalyst *c* [120]. Only one zinc and the transfer methylene are shown; other atoms associated with the Simmons-Smith reagent are deleted for clarity.

Charette catalyst, this atom is the amide carbonyl oxygen (Scheme 6.33d). Evidence for this feature is that when the amide substituents are replaced by phenyl groups, the cyclopropane product is racemic.

In 1966, Nozaki, et al., reported the first example of an asymmetric cyclopropanation using a chiral copper (II) catalyst [125]. Although the enantioselectivities were low (<10% ee), the contribution is important because it was the first example of an asymmetric synthesis using a chiral, homogeneous transition metal catalyst. Subsequently, Aratani optimized the ligand design and reported a number of asymmetric cyclopropanations, as shown in Scheme 6.34 [126-128]. For symmetrical *trans*-olefins, relative configuration is not an issue, and better selectivity is achieved with *l*-menthyl (from (–) menthol) diazoacetate than with the ethyl ester (double asymmetric induction, [127]). Cyclopropanation of isobutene is used on a factory scale for the commercial manufacture of the drug cilistatin (Scheme 6.34b) [128]. With monosubstituted olefins, relative as well as absolute configuration are an issue, but trans is favored, and double asymmetric induction again increases the stereoselectivity (Scheme 6.34c, [127]). Trisubstituted, unconjugated alkenes favor the cis relative configuration, as shown by the example in Scheme 6.34d, used in the synthesis of the cis isomer of the insecticide permethric acid [127]. Dienes, on the other hand, favor the *trans*-isomer, as shown by the synthesis of chrysanthemic acid shown in Scheme 6.34e [126,128].

The mechanism that has been proposed to explain the relative and absolute configurations of these examples is illustrated in Scheme 6.35 [128]. The catalyst, shown on the left of the scheme, is coordinatively unsaturated. Reaction with the diazoalkane affords the copper carbene shown at the top. The olefin approaches from the less hindered back side (note that the absolute configuration of the carbene carbon is set at this point), such that the indicated carbon (*, which is the one most able to stabilize a cationic charge) is oriented toward the carbene carbon. This is consistent with the metal atom acting as a Lewis acid. A metallacyclobutane is thought to be a discrete intermediate (bottom), and as it is formed, the hydroxyl is released from the copper. Steric repulsion by the large aryl substitutents of the chiral ligand tends to force R_1 downward, cis to the ester function. Similarly, steric repulsion tends to favor R_2 in a position trans to the ester. Collapse of the metallacyclobutane releases the cyclopropane and regenerates the catalyst.

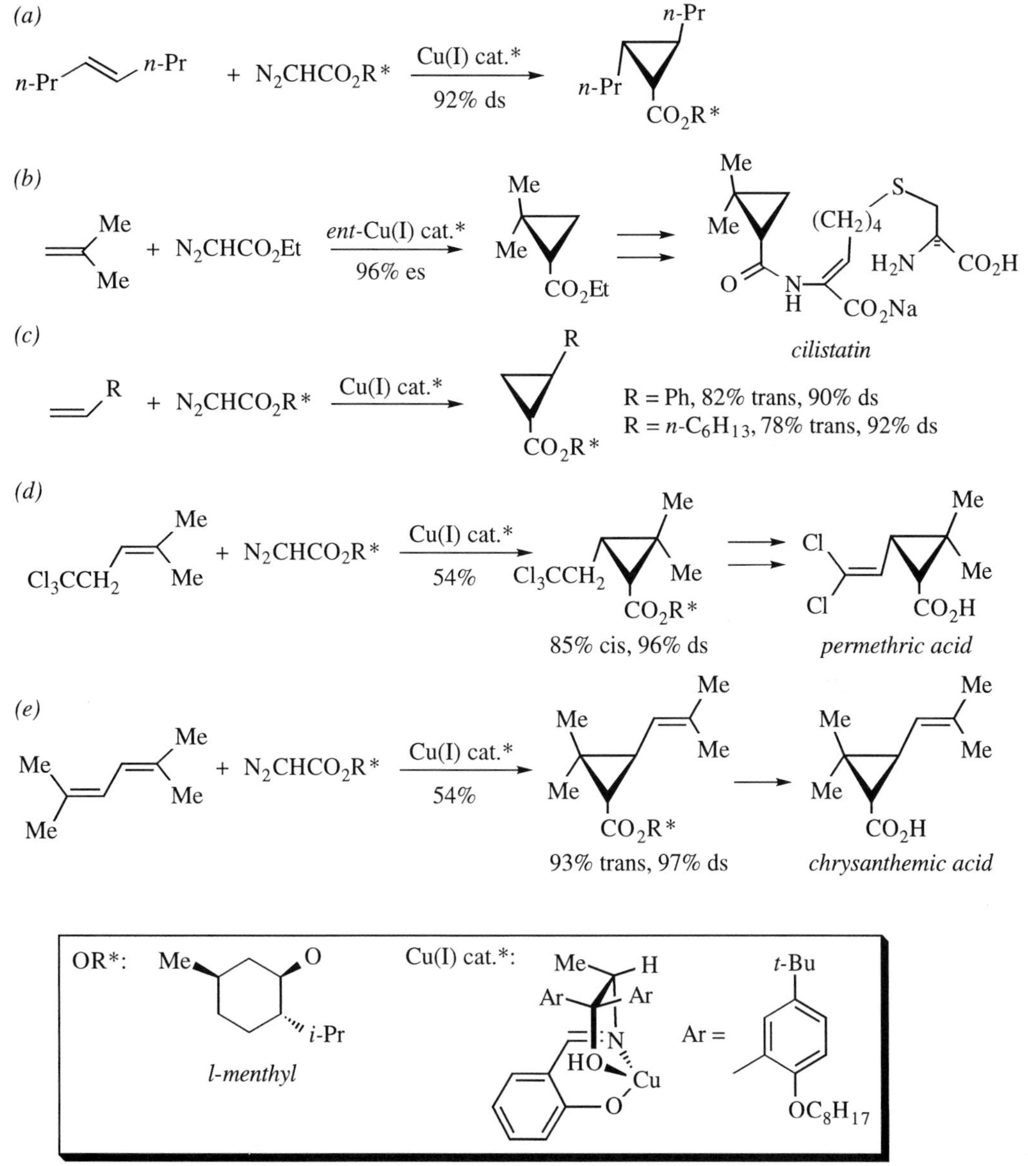

(a)
n-Pr
n-Pr
+ N2CHCO2R*
Cu(I) cat.*
92% ds
n-Pr
n-Pr
CO2R*
(b)
Me
Me
+ N2CHCO2Et
ent-Cu(I) cat.*
96% es
Me
Me
CO2Et
Me
Me
S
(CH2)4
O
N
H
H2N
CO2H
CO2Na
cilistatin
(c)
R
+ N2CHCO2R*
Cu(I) cat.*
R
CO2R*
R = Ph, 82% trans, 90% ds
R = n-C6H13, 78% trans, 92% ds
(d)
Cl3CCH2
Me
Me
+ N2CHCO2R*
Cu(I) cat.*
54%
Cl3CCH2
Me
Me
CO2R*
85% cis, 96% ds
Cl
Cl
Me
Me
CO2H
permethric acid
(e)
Me
Me
Me
Me
+ N2CHCO2R*
Cu(I) cat.*
54%
Me
Me
Me
Me
CO2R*
93% trans, 97% ds
Me
Me
Me
Me
CO2H
chrysanthemic acid
OR*:
Me
O
i-Pr
l-menthyl
Cu(I) cat.*:
Me
H
Ar
Ar
N
HO
Cu
O
Ar =
t-Bu
OC8H17

Scheme 6.34. Aratani's copper-catalyzed asymmetric cyclopropanation of olefins. *(a) trans*-1,2-disubstituted [127]. *(b)* 1,1,-disubstituted [128]. *(c)* monosubstituted, trans favored [127]. *(d)* trisubstituted, cis favored [127]. *(e)* dienes, trans favored [126,128]. *Inset:* chiral auxiliary and coordinatively unsaturated chiral catalyst.

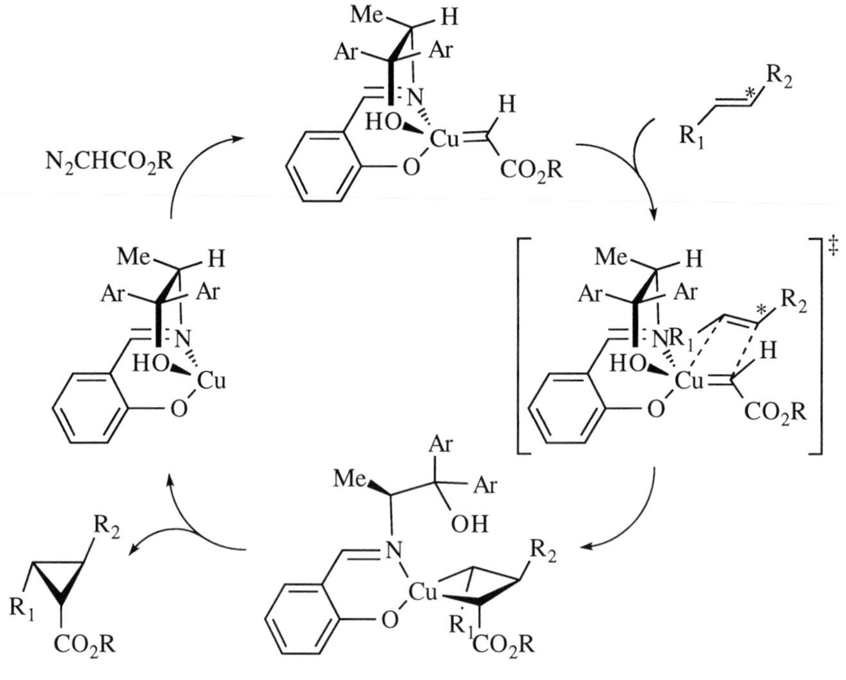

Scheme 6.35. Proposed catalytic cycle for asymmetric cyclopropanation using Aratani's copper catalyst [128].

This speculative rationale may be used to explain the apparent reversal of both relative and absolute configuration preference exhibited by the examples in Scheme 6.34d and e. In Scheme 6.34d, R_1 is Cl_3CCH_2-; attack of the copper occurs at the secondary carbon and the carbene carbon attaches to the tertiary site (∗), as shown in Scheme 6.36a. The controlling elements are the tertiary carbon of the olefin attaching to the carbene carbon, while the bulky Cl_3CCH_2- is oriented away from the nitrogen ligand. In the example in Scheme 6.34e, the more stable carbocation is

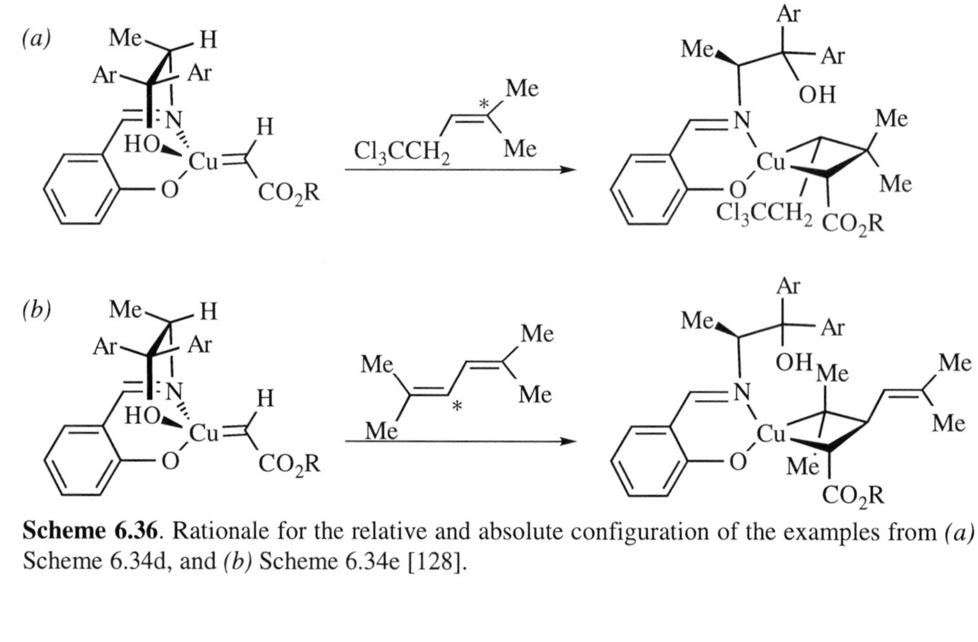

Scheme 6.36. Rationale for the relative and absolute configuration of the examples from (a) Scheme 6.34d, and (b) Scheme 6.34e [128].

allylic, so the trisubstituted olefin 'turns around' (Scheme 6.36b). Here, the controlling element is the trans orientation of the ester with respect to the isobutenyl group [128].

Several other groups have used C_2-symmetric ligands with copper and ruthenium as cyclopropanation catalysts. These ligands, shown in Figure 6.9, are generally more selective than the Aratani ligands. The first to be introduced was the semicorrin of Pfaltz (Figure 6.9a), and most of the others bear a close structural resemblance in that they all have pyrroline, oxazoline or bipyridine ligands chelating the metal. Copper(I) is the oxidation state of the active catalyst for all complexes containing copper, and the mechanism of the cyclopropanation using these catalysts is probably similar to that illustrated above (Schemes 6.35 and 6.36): electrophilic attack by copper, metallacyclobutane formation, etc. Table 6.3 lists selected examples for each ligand. It was generally found that bulky esters (*e.g.,* *tert*-butyl, BHT, menthyl) are more selective than less bulky ethyl esters (not listed). Entries 2 and 3 illustrate the effects of double asymmetric induction using the two enantiomers of menthol. Ligands c and f were also tested with both enantiomers of menthol, but there were no differences in selectivity. These examples show very high selectivity for *trans*-cyclopropanes; only one is cis-selective, but not by much (entry 18), which is >99% enantioselective for the cis product but only 62% enantioselective for the trans.

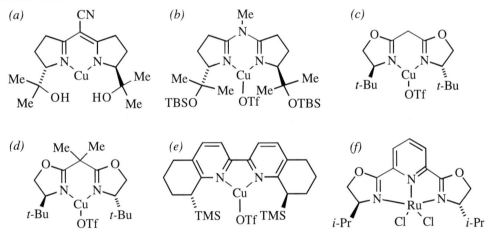

Figure 6.9. C_2-symmetric catalysts for cyclopropanation: *(a)* Pfaltz, 1988 [129]; *(b)* Pfaltz, 1992 [130]; *(c)* Masamune, 1990 [131] (see also [132]); *(d)* Evans, 1991 [133]; *(e)* Katsuki, 1993 [134]; *(f)* Nishiyama, 1994 [135].

Because of fluctuations in atom priority using the CIP sequencing rules (*i.e.,* in spite of their obvious differences, the CIP descriptor for the stereocenters in all ligands except e is *S*), we define the chirality sense of these ligands using the *P/M* nomenclature [136], applied to the R–C–N–M bond (see the inset in Figure 6.10). Thus, the ligands in Figure 6.9a and b have the *MM* configuration, while those in Figure 6.9c, d, and f have the *PP* configuration. Ligand 6.9e has an extra carbon and is not strictly definable by this system, but its symmetry features are similar to ligands 6.9a and b, so it is considered along with them.

Table 6.3. Asymmetric cyclopropanations. The "cat." column refers to the catalysts in Figure 6.9. For the structure of *l*-menthyl, see Scheme 6.34.

Entry	cat.	N_2CO_2R	alkene	% Yield	% trans	% es	Ref.
1	a	*t*-Bu	$PhCH=CH_2$	60	81	96	[129]
2	a	*l*-menth	$PhCH=CH_2$	65-75	85	95	[129]
3	a	*d*-menth	$PhCH=CH_2$	60-70	82	98	[129]
4	a	*d*-menth	$CH_2=CHCH=CH_2$	60	63	98	[129]
5	a	*d*-menth	$Me_2C=CHCH=CH_2$	77	63	98	[129]
6	a	*d*-menth	n-$C_5H_{11}CH=CH_2$	30	89	96	[129]
7	b	*t*-Bu	$PhCH=CH_2$	75	81	97	[130]
8	b	*d*-menth	$PhCH=CH_2$	75	84	99	[130]
9	c	*t*-Bu	$PhCH=CH_2$	73	80	97	[131][a]
10	c	*l*-menth	$PhCH=CH_2$	72	86	99	[131][a]
11	d	BHT	$PhCH=CH_2$	85	94	>99	[131][a]
12	d	BHT	$PhCH_2CH=CH_2$	-	-	93	[133]
13	d	BHT	$Ph_2C=CH_2$	70	-	>99	[133]
14	d	BHT	$Me_2C=CH_2$	91	-	>99	[133]
15	e	*t*-Bu	$PhCH=CH_2$	75	86	96	[134]
16	e	*t*-Bu	n-$C_6H_{13}CH=CH_2$	65	85	95	[134]
17	e	*t*-Bu	$PhCH=CHCH=CH_2$	90	70	91	[134]
18	e	*t*-Bu	E-$PhCH=CHMe$	54	40	62	[134]
					60(cis)	>99	
19	e	*t*-Bu	Z-$PhCH=CHMe$	94	>99	86	[134]
20	f	*t*-Bu	$PhCH=CH_2$	81	97	97	[135]
21	f	*l*-menth	$PhCH=CH_2$	87	95	97	[135]
22	f	*l*-menth	n-$C_5H_{11}CH=CH_2$	40	94	>99	[135]
23	f	*l*-menth	$Ph_2C=CH_2$	55	-	82	[135]
24	f	*l*-menth	$Me_2C=CHCH=CH_2$	86	79	99	[135]

[a]   The absolute configuration reported in this paper is correct (1*R*, 2*R*), but it is drawn incorrectly.

In all cases, the *MM* ligand affords the 1*S*,2*S*-trans product and the *PP* ligand affords the 1*R*,2*R* product. The sense of diastereoselectivity and enantioselectivity can be explained using the cartoons in Figure 6.10 (this scheme is a model, not a mechanism). Because of the C_2-symmetry of the ligands, the configuration of the carbene is the same whether the ester moiety is drawn up or down. Note that the vertical orientation of the carbene and the horizontal orientation of the ligand divide the reagent into four quadrants. Only in the *S,S*-trans product (from the *MM* complex) are steric interactions between the olefinic substituent and both the carbene ester *and* the ligand substituent avoided (*i.e.*, the olefin substituent is in the lower right quadrant). All other orientations produce repulsive interactions between the olefin and either the ester moiety or the ligand substituent. For ligands having the *PP* configuration, the preferred product is the *R,R-trans*-cyclopropane.

Weaknesses of the model in Figure 6.10 include the fact that there may be other ligands on the metal that are not taken into consideration here, and that it assumes a similar geometry of the carbene relative to the chelating ligand for all the com-

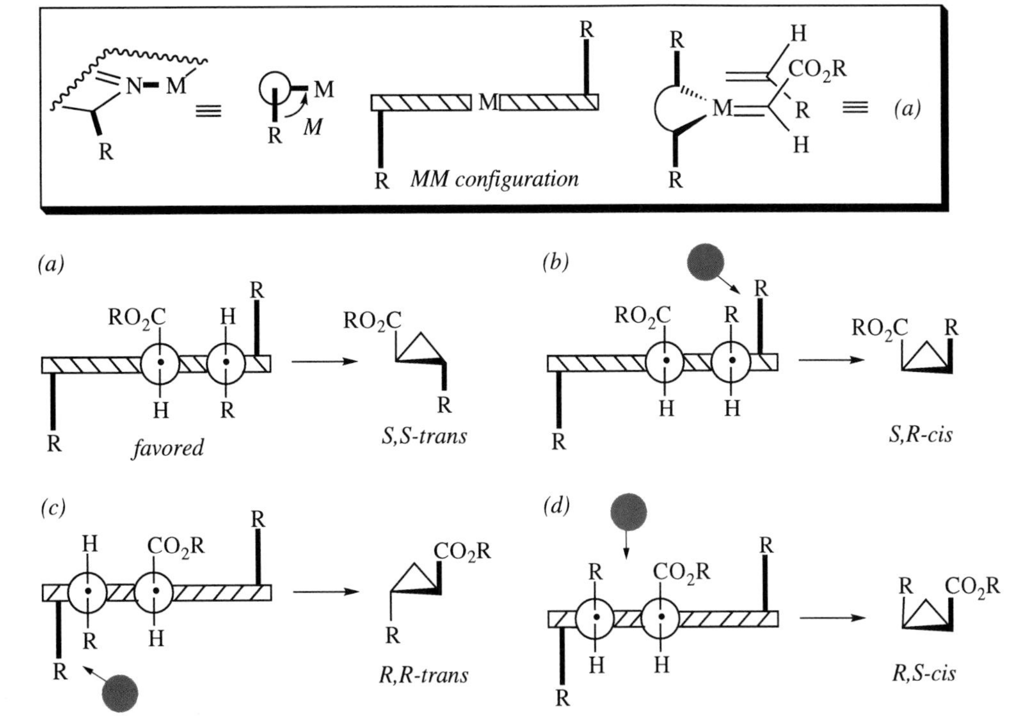

Figure 6.10. *Inset:* Definition of *M* configuration of metal complexes, and generalized side view of an *MM*-metal carbene complex with the olefin approaching from the rear (equivalent to the Newman projection shown in *a*). *(a)* Favored approach, leading to the *S,S*-trans product. *(b)* Disfavored approach, leading to *S,R*-cis product. *(c)* Disfavored approach leading to the *R,R* trans product. *(d)* Disfavored approach leading to the *R,S*-cis product. After ref. [112].

plexes. On the other hand, the formation of metallacyclobutanes in copper-catalyzed cyclopropanations appears to be an accepted hypothesis [112,133], and the consistency of these representations with an accumulating body of fact make them useful predictive models, and a good starting point for developing more detailed mechanistic hypotheses.

It was noted in the previous section that rhodium acetate catalyzed cyclopropanations of chiral diazo acetates afforded poor diastereoselectivity. Using achiral diazo acetates and methyl 2-oxopyrrolidinone-carboxylate (MEPY) as a chiral ligand on rhodium, reasonable trans selectivity and moderate enantioselectivity can be achieved, as shown by the example in Scheme 6.37a [137]. More recently, the groups of Doyle, H. Davies, and Whitesell have examined chiral esters with the Rh$_2$[MEPY]$_4$ catalyst in the hopes of improving selectivity through double asymmetric induction, but the results still leave room for improvement [115]. Intermolecular cyclopropanation of alkynes produces only two stereoisomeric products, and Doyle and Müller have found that double asymmetric induction pushes the selectivities over 90% (although the absolute configuration was not determined), as shown in Scheme 6.37b [138]. Although these menthyl esters afford higher selectivities, they offer lower yields than ethyl diazoacetates (70-85% yields) due to competitive C–H insertion reactions. H. Davies has reported that the rhodium prolinate-catalyzed addition of vinyl carbenes to alkenes is 100% selective for the

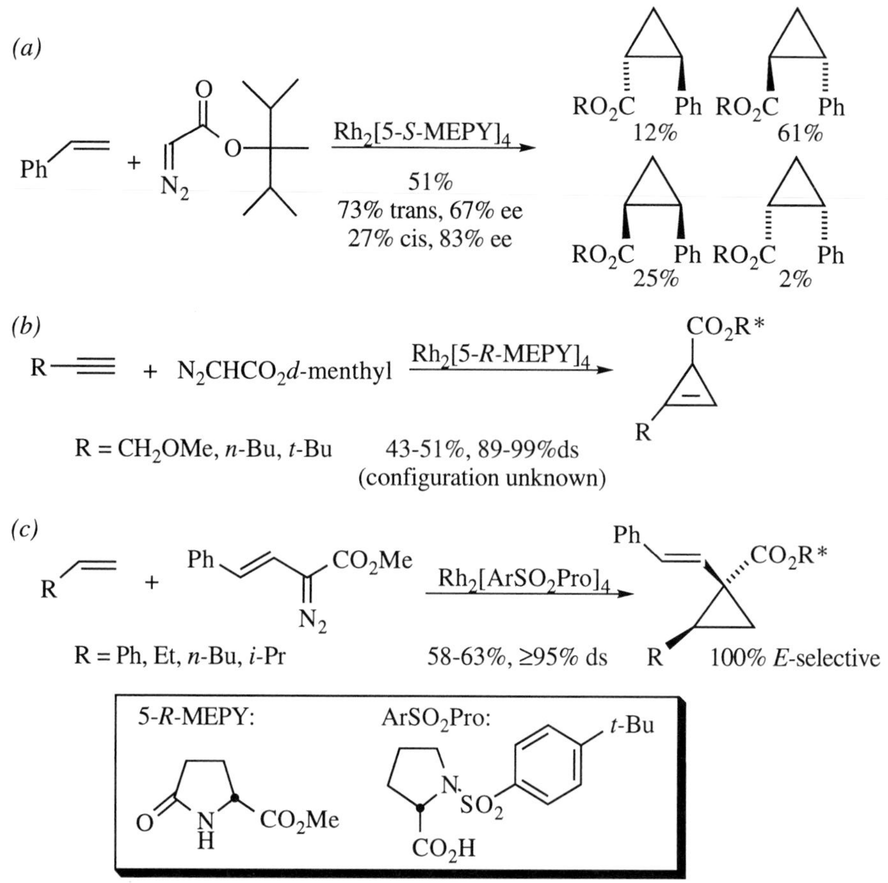

(a)

Ph⎯⎯= + N₂CHCO₂d-menthyl Rh₂[5-S-MEPY]₄

51%
73% trans, 67% ee
27% cis, 83% ee

RO₂C Ph RO₂C Ph
   12%   61%

RO₂C Ph RO₂C Ph
   25%   2%

(b)

R⎯≡ + N₂CHCO₂d-menthyl Rh₂[5-R-MEPY]₄

R = CH₂OMe, n-Bu, t-Bu   43-51%, 89-99%ds
(configuration unknown)

CO₂R*

R

(c)

R⎯= + Ph⎯⎯CO₂Me Rh₂[ArSO₂Pro]₄
          |
         N₂

R = Ph, Et, n-Bu, i-Pr   58-63%, ≥95% ds   R   100% E-selective

Ph
  \\=   ,,,CO₂R*

5-R-MEPY: ArSO₂Pro: t-Bu

O N CO₂Me   N.
   H   SO₂
                      CO₂H

Scheme 6.37. (a) Asymmetric cyclopropanation of styrene [137]. (b) Cyclopropanation of alkynes [137]. For menthyl structure, see Scheme 6.34. (c) Asymmetric cyclopropanation of alkenes with vinyl carbenes [139]. *Inset:* Ligand structures.

E-diastereomers, which are formed in an ≥95:5 ratio for several alkenes, as shown in Scheme 6.37c [139]. Surprisingly, Davies also noted that the stereoselectivity *decreased* when esters of larger alcohols were used.

Conformational considerations restrict the number of possible transition state geometries in intramolecular cyclopropanations, which are quite selective, as shown by the examples from Doyle, Martin, and Müller illustrated in Scheme 6.38a [140,141]. Intramolecular cyclopropanation of diazo esters of chiral allylic alcohols are subject to double asymmetric induction, as shown by the series of examples in Scheme 6.38b. For all of these substrates, the exo product is slightly preferred when cyclopropanation is mediated by an achiral catalyst [142], but this selectivity is reversed dramatically when the *S* ester is allowed to react with the 5-*S*-MEPY catalyst. This pronounced endo selectivity persists for both the *E* and the *Z*-alkenes, although it is higher for the *Z* alkenes. Note also that when the chirality sense of the substrate and the catalyst are mismatched (*S* substrate and *R* catalyst), the endo selectivities are low, unless R₁/R₂ are trimethylsilyl. For the matched case of double asymmetric induction, the same features that cause the endo selectivity can be used

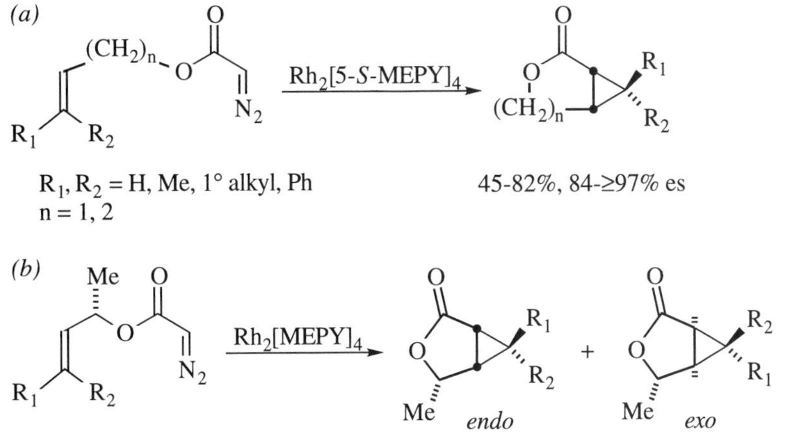

(a)

R_1, R_2 = H, Me, 1° alkyl, Ph
n = 1, 2

45-82%, 84-≥97% es

(b)

R$_1$	R$_2$	MEPY	endo:exo	% Yield
H	n-Bu	S	>95:5	80
H	n-Bu	R	40:60	39
n-Bu	H	S	86:14	77
n-Bu	H	R	50:50	42
H	TMS	S	>95:5	74
H	TMS	R	37:63	31
TMS	H	S	95:5	76
TMS	H	R	92:8	47

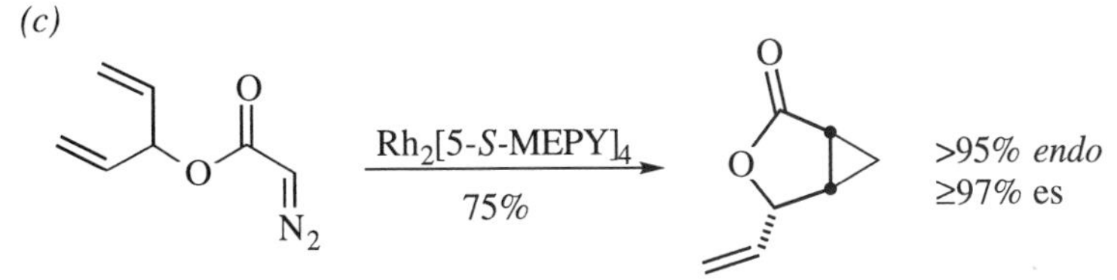

(c)

Rh$_2$[5-S-MEPY]$_4$

75%

>95% endo
≥97% es

Scheme 6.38. (a) Enantioselectivity in intramolecular cyclopropanations [140,141]. (b) Double asymmetric induction in intramolecular cyclopropanations [142]. (c) Group-selective asymmetric cyclopropanation [142].

to effect the enantioselective (group selective) cyclopropanation of the divinyl alcohol illustrated in Scheme 6.38c. The group selectivity is significantly diminished, however, if there are substituents on the double bonds [142].

The mechanism by which selectivity is induced in rhodium mediated asymmetric cyclopropanations is not clear. What is known is that the pyrrolidinone of the MEPY catalyst is bonded to the rhodiums through the carboxamide, with the nitrogens cis to each other, as shown in Figure 6.11 [113]. This arrangement places the two carbomethoxy groups cis to each other on both sides of the catalyst. With

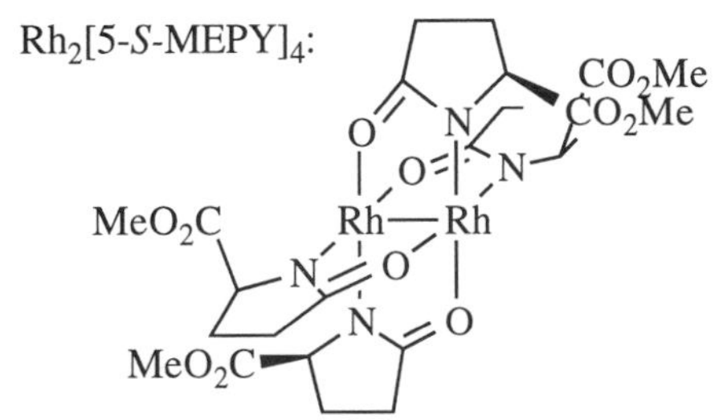

Figure 6.11. Depiction of the structure of Doyle's Rh2[5-S-MEPY]4 catalyst, showing the cis arrangement of the nitrogens [113].

the carbene bound to the rhodium on the 'right side' for example, the two carbomethoxy groups will hinder approach of the olefin from the upper right quadrant and selectivity is determined by the effects of the carbomethoxy groups on the stabilities of the various possible conformations of the transition state. These effects are quite complex and have not been fully quantified, although efforts have been made [113].

6.2.1.3 Stepwise cyclopropanations

Chiral malonate esters have been used successfully in asymmetric cyclopropanations, as shown by the example in Scheme 6.39, part of a total synthesis of steroids such as estrone [143,144]. The key step in this sequence is an intramolecular S_N2' alkylation of the monosubstituted malonate. The rationale for the diastereoselectivity is shown in the illustrated transition structure. Note that the enolate has C_2 symmetry, so it doesn't matter which face of the enolate is considered. The illustrated conformation has the ester residues syn to the enolate oxygens to relieve $A^{1,3}$ strain, with the enolate oxygens and the carbinol methines eclipsed. The allyl halide moiety is oriented away from the dimethylphenyl substituent, exposing the alkene *Re* face to the enolate. The crude selectivity is about 90% as determined by conversion to the dimethyl ester and comparison of optical rotations [143], but a single diastereomer may be isolated in 67% yield by preparative HPLC [144]. This reaction deserves special note because it was conducted on a reasonably large scale: 67.5 grams of diester (127 mmol) [144].

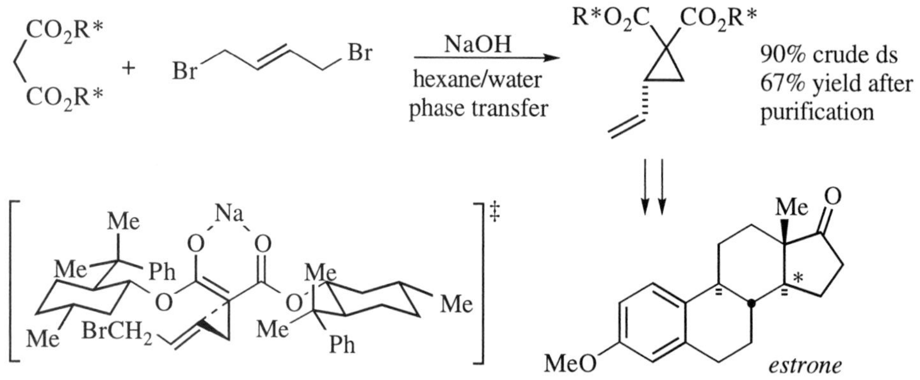

Scheme **6.39**. Asymmetric cyclopropanation of malonate enolates for steroid synthesis [143,144].

A more common strategy for stepwise asymmetric cyclopropanation is the use of chiral electrophiles. Meyers has used bicyclic lactams (*cf.* Scheme 3.19, 3.20) [145,146] as electrophilic auxiliaries in sulfur ylide cyclopropanations [147]. These auxiliaries, for reasons that are not entirely clear, are preferentially attacked from the α-face. After separation of the diastereomers, the amino alcohol auxiliary may be removed by refluxing in acidic methanol or reductively [145]. This methodology has been used in asymmetric syntheses of *cis*-deltamithrinic acid and dictyopterene C, illustrated in the inset of Scheme 6.40 [145].

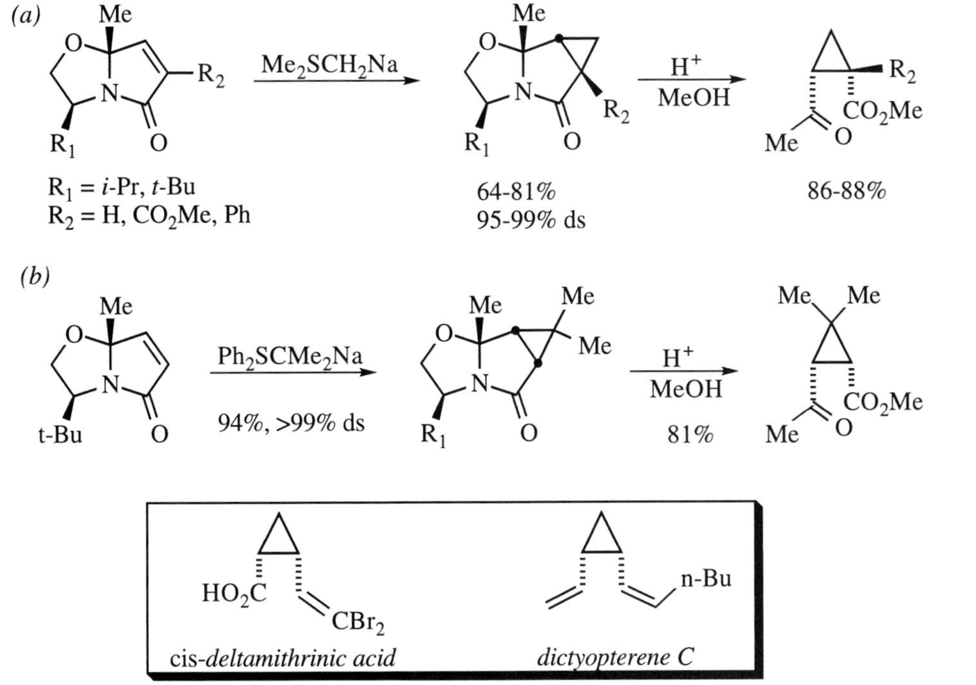
(a)
Me
O
N
R2
R1
O
Me2SCH2Na
H+
MeOH
R2
CO2Me
Me
R1 = i-Pr, t-Bu
R2 = H, CO2Me, Ph
64-81%
95-99% ds
86-88%
(b)
Ph2SCMe2Na
94%, >99% ds
t-Bu
81%
HO2C
CBr2
n-Bu
cis-deltamithrinic acid
dictyopterene C

Scheme 6.40. Meyers's asymmetric cyclopropanations using the bicyclic lactam auxiliary. *(a)* Methylene transfer. *(b)* Isopropylidine transfer. *Inset:* Synthesis targets [145].

For the synthesis of cyclopropyl amino acids, Williams has used an oxazinone auxiliary (*cf.* Scheme 3.12) as an electrophilic component in a sulfur ylide cyclopropanation using Johnson's sulfoximines, as illustrated in Scheme 6.41 [148]. Surprisingly, the sulfur ylide approaches from the β face; the authors speculate that there may be some sort of π-stacking between the phenyls on the oxazinone ring and the phenyl in the sulfoximine to account for this [149]. With Corey's [147] dimethylsulfonium methylide, the diastereoselectivity was only about 75%, but with Johnson's sulfoximines (used in racemic form), only one diastereomer could be detected for most substrates studied (with the exception of R = H, [149]). Dissolving metal reduction afforded moderate yields of the cyclopropyl amino acids.

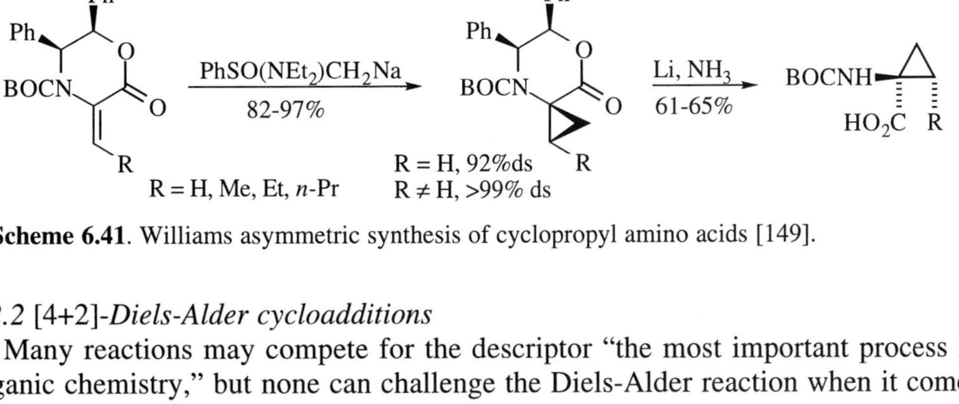
Ph
Ph
O
BOCN
O
R
PhSO(NEt2)CH2Na
82-97%
Li, NH3
61-65%
BOCNH
HO2C R
R = H, Me, Et, n-Pr
R = H, 92%ds
R ≠ H, >99% ds

Scheme 6.41. Williams asymmetric synthesis of cyclopropyl amino acids [149].

6.2.2 [4+2]-*Diels-Alder cycloadditions*

Many reactions may compete for the descriptor "the most important process in organic chemistry," but none can challenge the Diels-Alder reaction when it comes to synthetic utility in the formation of six-membered rings.[16] The enormous body of work that includes synthetic applications and mechanistic investigations of this venerable reaction cannot be adequately summarized in anything less than a monograph. Even the literature limited to the asymmetric Diels-Alder reaction is formidable,[17] and the following review is therefore selective. The discussion is limited to examples that serve to illustrate some of the methods that have been developed for the synthesis of enantiopure cyclohexenes,[18] and for which transition state models have been proposed. It is hoped that this sampling will afford the reader a taste for the breadth of the process, as well as a basic knowledge of the types of transition state assemblies that favor stereoselective cycloadditions. The historical development of the asymmetric Diels-Alder reaction begins with auxiliary-based methods for (covalently) modifying the cycloaddition reactants, and has now progressed through chiral (stoichiometric) catalysts, to true catalysts [162] that are efficient in both enantioselectivity and turnover. Thus, the development of the Diels-Alder reaction is a microcosm of the field of asymmetric synthesis itself. The following discussion is organized according to the strategy employed: auxiliaries for dienophile modification, diene auxiliaries, and chiral catalysts.

6.2.2.1 Dienophile auxiliaries

In general, cycloadditions catalyzed by Lewis acids proceed at significantly lower temperatures and with higher selectivities than their uncatalyzed counterparts. Factors that contribute to the increased selectivity of the catalyzed reactions include lower temperatures and more organized transition states. For enthalpy-controlled reactions, lowering temperatures increases selectivity (recall Section 1.4, equation 1.5). Coordination of a Lewis acid to the enone carbonyl not only activates the enone by electron withdrawal, it also restricts conformational motion and thereby reduces the number of competing transition states. Figure 6.12 illustrates several chiral auxiliaries for dienophile modification that have been used in the Diels-Alder reaction.

Principles of conformational analysis may be invoked to rationalize the face-selectivity of these compounds. Note, however, that there are two broad types of auxiliary: those that contain a second carbonyl and those that do not. The former may function by chelating the metal of the Lewis acid catalyst, while the latter can only act as monodentate ligands to the metal of the Lewis acid. Figure 6.13a illustrates the probable transition state conformations of ester dienophiles when bound as monodentate ligands to the Lewis acid catalyst, M (auxiliaries 6.12a-f). The C(=O)–O bond prefers the Z (or cis) conformation for a variety of reasons [163], but the preference is large: probably >4 kcal/mole. Because of this constraint, the C–O bond may be considered to be similar to a double bond (hence the E/Z or cis/trans designation). A subtle consequence of this constraint is the effect it has on

---

[16] Monographs reviewing the Diels-Alder reaction: [150,151]. For recent reviews with extensive references to other reviews and pertinent literature, see refs. [152-154].
[17] Reviews of the asymmetric Diels-Alder reaction: [155-157].
[18] For a monograph covering [4+2] cycloadditions that form heterocycles, see ref. [158]. For recent reviews, see ref. [159-161].

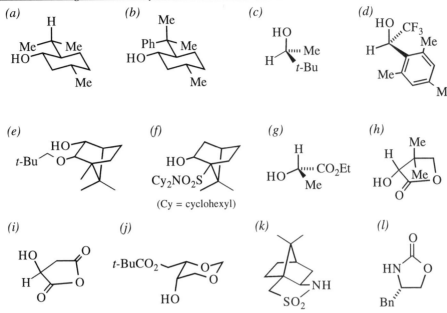

Figure 6.12. Dienophile chiral auxiliaries for the asymmetric Diels-Alder reaction. *(a)* [164]. *(b)* [165-167]. *(c)* [164,168]. *(d)* [169]. *(e)* [170], see also refs. [171,172]. *(f)* [173]. *(g)* [6]. *(h)* [174,175]. *(i)* [175]. *(j)* [176]. *(k)* [177]. *(l)* [178].

the conformation of the O–C bond (leading to the stereocenter of the chiral auxiliary). Because of $A^{1,3}$ strain, the C–H bond of the carbinol carbon eclipses the carbonyl in the lowest energy conformation, which places the other two substituents (L̲arge and S̲mall in Figure 6.13) above and below the plane of the enone. When bound to a Lewis acid, the most stable conformation about the C_1–C_2 bond of the

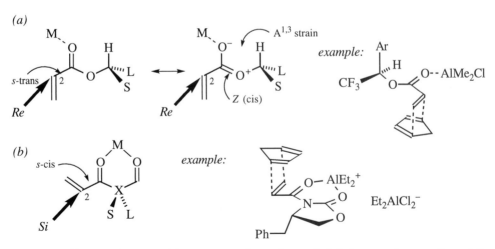

Figure 6.13. Probable transition state conformations of: *(a)* A monodentate dienophile complex such as Corey's mesityl trifluoroethanol auxiliary (Figure 6.12d, [169]). *(b)* A bidentate dienophile such as Evans's oxazolidinone (Figure 6.12l, [178]). S and L refer to the small and large substituents of the auxiliary.

acrylate is *s*-trans, as shown [179].[19] Approach of the diene from the direction of the C_2 *Re* face is favored since this is the face having the least steric interactions with the auxiliary (S *vs.* L). Note also that for an endo transition state, the diene should be oriented *toward* the ester auxiliary. The specific example shown is Corey's mesityl trifluoroethanol auxiliary (Figure 6.12d, [169]).

Figure 6.12g-l illustrates auxiliaries that may chelate the metal of the Lewis acid catalyst. In these cases, the metal is coordinated anti to the olefin and the preferred conformation of the C_1–C_2 bond is *s*-cis, as shown in Figure 6.13b. Again, the preferred approach of the diene is from the direction of the viewer, but because of the different conformation of the enone, it is now the C_2 *Si* face. The example is Evans's oxazolidinone [178]. In this example, the Lewis acid is Et$_2$AlCl, but more than one molar equivalent is required for optimum results [178]. Castellino has shown by NMR that Et$_2$AlCl initially binds in a monodentate fashion, but excess acid creates a bidentate dione·AlEt$_2^+$ complex having a Cl$_2$AlEt$_2^-$ gegenion [180].

Acrylates. Cyclopentadiene is often used to evaluate selectivity in asymmetric Diels-Alder reactions. Table 6.4 lists the selectivities found for acrylate cyclo-additions using the auxiliaries shown in Figure 6.13 under conditions that are opti-mized for each auxiliary. Note that there are four possible norbornene stereo-isomers, two endo and two exo. In accord with Alder's endo rule, the endo is heavily favored in all these examples. Although several authors report selectivities in these reactions in terms of selectivity for one endo adduct over the other, the selectivities indicated in the table reflect the *total* diastereoselectivity of the major adduct over the other three, if this information could be deduced from the information provided in the paper.

Because of the different conditions (Lewis acid, temperature, solvent) used for each of these auxiliaries, it is difficult to determine the "most selective" auxiliary. Indeed, considerations such as ease of separability and reaction scale are important factors in selecting an auxiliary for any given application. Our concern is the factors governing selectivity. An analysis of auxiliary 6.12e and two close relatives illustrate how structural changes can affect the selectivity of the cycloaddition and how conformational principles can explain the effects. Scheme 6.42 shows three acrylate/cyclopentadiene cycloadditions with three very similar auxiliaries, run with the same catalyst at similar temperatures, but which exhibit markedly different stereoselectivities. All of these auxiliaries were designed to place the acrylate and a shielding neopentyl group on a rigid scaffolding (camphor skeleton) such that the enone and a *tert*-butyl group lie (more or less) parallel, and they are thought to react via a nonchelated conformation analogous to Figure 6.13a. Scheme 6.42a duplicates the data listed in Table 6.4, entry 5 [170]. This auxiliary, developed by Oppolzer, shows outstanding selectivity at –20°, but its close counterpart, shown in Scheme 6.42b, exhibits significantly lower (although still useful) selectivity [170]. The only difference is the relationship of the bridgehead methyl to the neopentyl. In the absence of the bridgehead methyl, the *tert*-butyl can rotate away from the acrylate, leaving the *Si* face more accessible. The auxiliary in Scheme 6.42c was

---

[19] In the absence of a Lewis acid catalyst, both *s*-cis and *s*-trans conformers are present.

Table 6.4. Asymmetric Diels-Alder reactions of cyclopentadiene and acrylates. The X_c column refers to the auxiliaries in Figure 6.12; the probable transition state (TS) conformations of the dienophile are illustrated in Figure 6.13; the % ds refers to the formation of one of the four possible products (two endo and two exo isomers).

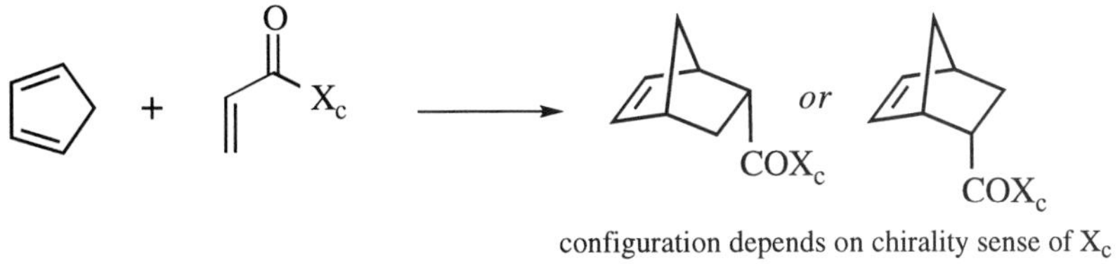

configuration depends on chirality sense of X_c

Entry	X_c	Lewis Acid	Probable TS	Temp.	% yield	% ds	Ref.
1	a	BF$_3$·OEt$_2$	non-chelated	−70°	80	91	[164]
2	b	SnCl$_4$	non-chelated	0°	95	75	[165,166]
3	c	BF$_3$·OEt$_2$	non-chelated	−70°	75	87	[164,168]
4	d	Me$_2$AlCl	non-chelated	−78°	96	97	[169]
5	e	TiCl$_2$(Oi-Pr)$_2$	non-chelated	−20°	96	96	[170]
6	f	TiCl$_2$(Oi-Pr)$_2$	non-chelated	−20°	97	89	[173]
7	g	TiCl$_4$	chelated[a]	−63°	88	91	[6]
8	h	TiCl$_4$	chelated[a]	−64°	81	95	[174,175]
9	i	TiCl$_4$	chelated[a]	−78°	86	97	[175]
10	j	Et$_2$AlCl	chelated[b]	−70°	88	99	[176]
11	k	Et$_2$AlCl	chelated	−130°	96	95	[177]
12	l	Et$_2$AlCl	chelated	−100°	94	78	[178]

a In this case, the TiCl$_4$ is thought to shield one face of the enone [181].
b Chelation is postulated to occur at a ring oxygen.

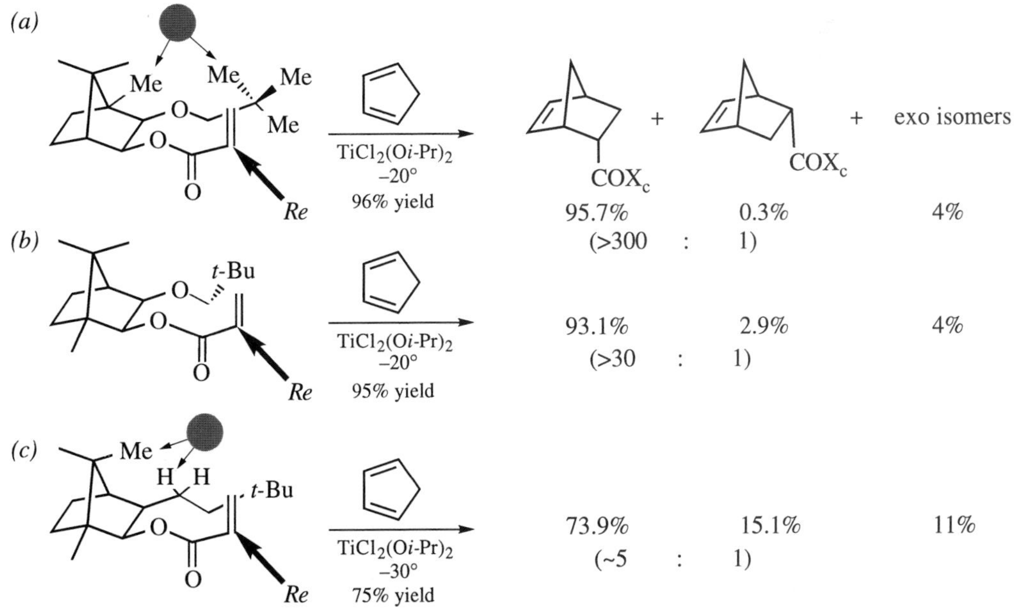

Scheme 6.42. Camphor-derived auxiliaries for asymmetric Diels-Alder cycloadditions. (*a,b*) [170]. (*c*) [171]. The auxiliary illustrated in (*a*) is the enantiomer of that reported in ref. [170].

prepared to further probe the effects of conformation on selectivity [171]. In this case, an oxygen has been replaced by a methylene. The most likely rationale for the further lowering of selectivity (compare Scheme 6.42b and c) is that the protons of the methylene experience unfavorable van der Waals repulsion with the indicated methyl in the conformation which most shields the acrylate *Si* face. Population of other (unspecified) conformations results in both lower endo selectivity, and lower *Re* facial selectivity within the endo manifold. It is interesting to recall the discussion in Chapter 1 on selectivity (*cf.* Figure 1.3 and the accompanying discussion) which emphasized the small energetic differences that can result in large effects on selectivity. In Scheme 6.42, the selectivities for the three examples correspond to differences in energies of activation ($\Delta\Delta G^{\ddagger}$) of 2.9, 1.7, and 0.8 kcal/mole for examples 6.42a-c, respectively, for the two endo isomers. In each case, an increment of approximately 1 kcal/mole (about the same as the energy difference between gauche and anti butane) has a profound effect on the observed selectivity.

The presence of a potential chelating functional group in an auxiliary does not necessarily mean that chelation occurs. For example, in his optimization studies of *S*-ethyl lactate as a chiral auxiliary (Figure 6.12g), Helmchen noted a marked dependence on the identity and amount of the Lewis acid added [6]. For example, excess $TiCl_4$ or $SnCl_4$ induced cyclopentadiene addition to the C_2 *Si* face of the acrylate, whereas excess $BF_3 \cdot OEt_2$ or $AlCl_3$ catalyzed addition to the C_2 *Re* face [6]. Moreover, a dependence of selectivity on the stoichiometry (acrylate/acid) was also noted [174]. Figure 6.14 shows three (among many) conformations that could be important in the transition state. Figure 6.14a illustrates a chelating conformation that was found in the X-ray crystal structure of a $TiCl_4$ complex (Figure 6.14d, [181]), while Figure 6.14b and c show possible monodentate conformers: Figure 6.14b shows monodentate coordination to one Lewis acid, while Figure 6.14c shows monodentate coordination to two Lewis acids. The differing face-selectivity of the Lewis acids mentioned above was attributed to a chelated transition structure in the case of $TiCl_4$ and $SnCl_4$ (*i.e.,* Figure 6.14a), and reactive intermediates such as shown in Figure 6.14b and c were postulated for $BF_3 \cdot OEt_2$ and $AlCl_3$ catalysts [174].

Without any further information, it is not obvious what facial selectivity might be expected from any of these conformations. However, the crystal structure (Figure 6.14d) of the $TiCl_4$ complex of *O*-acryloyl ethyl lactate reveals the probable origin of the observed stereoselectivity [181]. Interestingly, the coordination of the two carbonyl oxygens to titanium shows appreciable π-character, which produces a geometry in which a chlorine on the titanium shields the C_2 *Re* face of the acrylate [181]. Additional notable features of the crystal structure include a small (~40°) H–C–O–C(=O) torsion angle (*cf.* Figure 6.13a) and an even smaller O–C(=O)–C–Me angle (20°). Comparing the conformations of Figure 6.14a-c suggests that an entropic price must be paid in order to populate conformation 6.14a. But the small torsion angle observed between the ethoxy and the methyl suggest that this price might be avoided if these functional groups were constrained in a ring. Scheme 6.43 shows a comparison of the data of Table 6.4, entries 7-9. Substitution of pantolactone (Figure 6.12h) for ethyl lactate as the auxiliary under

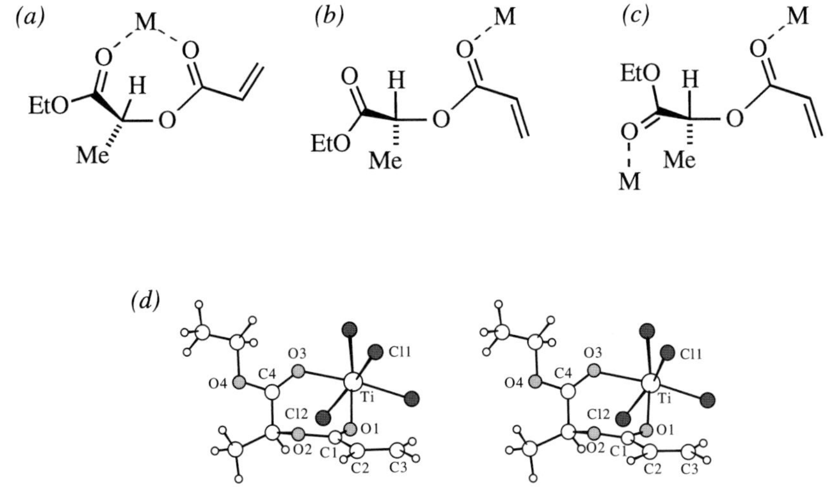

Figure 6.14. (a-c) Possible conformations of O-acryloyl lactates coordinated to one or more Lewis acids (after ref. [174]). (d) Stereoview of the crystal structure of O-acryloyl ethyl lactate · TiCl$_4$ complex (reprinted with permission from ref. [181]).

otherwise identical conditions (Scheme 6.43b) yields an increase in selectivity from 91% to 95%, corresponding to an increase in relative rates from 13:1 to 49:1 for formation of the two endo isomers. This corresponds to free energies of activation ($\Delta\Delta G^{\ddagger}$) of 1.1 and 1.6 kcal/mole, respectively. Note also that the *gem* dimethyls of the pantolactone are not important contributors to the selectivity, as shown by

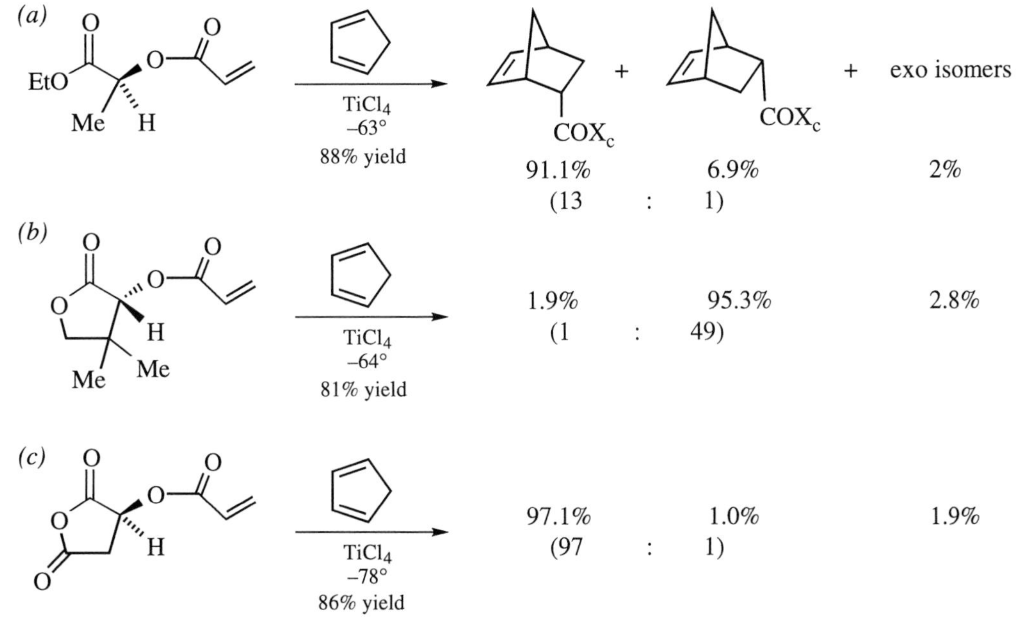

Scheme 6.43. Acrylate ester cycloadditions using chelating auxiliaries: (a) [6]; (b) [174,175]; (c) [175].

comparison of the selectivities in Scheme 6.43b and c, respectively. The latter auxiliary (Figure 6.12i, Table 6.4, entry 9) exhibits still higher selectivity, but at lower temperature. The relative rate corresponds to $\Delta\Delta G^{\ddagger} = 1.8$ kcal/mole, which would give a relative rate of about 70:1 at $-63°$, not significantly different from auxiliary 6.12h at that temperature (Scheme 6.43b and Table 6.4, entry 8). The steric shielding for all three auxiliaries is thought to be a chlorine on titanium.

Several of these auxiliaries were also tested with acyclic dienes. Table 6.5 lists the stereoselectivities found. Here again, Lewis acid catalysis was found to be advantageous in each case. The cycloadditions in entries 1 and 2 are thought to proceed by monodentate coordination to the catalyst (Figure 6.13a), while entries 3-6 proceed through a chelated intermediate. For the auxiliaries in Figure 6.12h, k, and l, high selectivities are also observed with E-crotonates and E-2-bromoacrylates, as would be expected by examination of the position of an E-β-substituent in the chelated transition structures of Figure 6.13b.

Table 6.5. Examples of asymmetric cycloadditions of acrylates with acyclic dienes. The X_c column refers to Figure 6.12.

Entry	X_c	Lewis Acid	Diene	% yield	% ds	Ref.
1	b	TiCl$_4$	butadiene	70	93	[167]
2	e	TiCl$_4$	butadiene	98	98	[172]
3	h	TiCl$_4$	butadiene	73[a]	93	[174]
4	k	EtAlCl$_2$	butadiene	93	97	[177]
5	h	TiCl$_4$	2-methylbutadiene	76[a]	97	[174]
6	l	Et$_2$AlCl	2-methylbutadiene	85	95	[178]

a Yield of a single diastereomer after 3 recrystallizations.

Intramolecular Cycloadditions. Diels-Alder reactions[20] having diene and dienophile connected by three or four atom carbon chains are selective (for trans-fused bicyclic adducts) only when the dienophile is trans and when Lewis acid catalysis is employed. The competing transition states are illustrated in Scheme 6.44a [182]. The auxiliaries illustrated in Figure 6.12a, k, and l have been used to modify the dienophile fragment for asymmetric intramolecular Diels-Alder reactions for trienes having these attributes. The examples shown in Scheme 6.44b-d reveal that the facial selectivity dictated by chelating and non-chelating auxiliaries as rationalized in Figure 6.13 determine the chirality sense of the trans-fused product.[21] Thus, the absolute configuration of the product obtained using the menthyl auxiliary (Scheme 6.44b) is consistent with an s-trans C_1–C_2 conformation (*cf.* Figure 6.13a) and an anti transition state. The camphor sultam (Scheme 6.44c)

---

[20] For reviews of the intramolecular Diels-Alder reaction, see ref. [151,153,182,183].

[21] For ease of comparison, the chirality sense of the camphor sultam is inverted from that reported in the literature [184] so that the favored approach at C_2 is toward the *Si* face for all three examples.

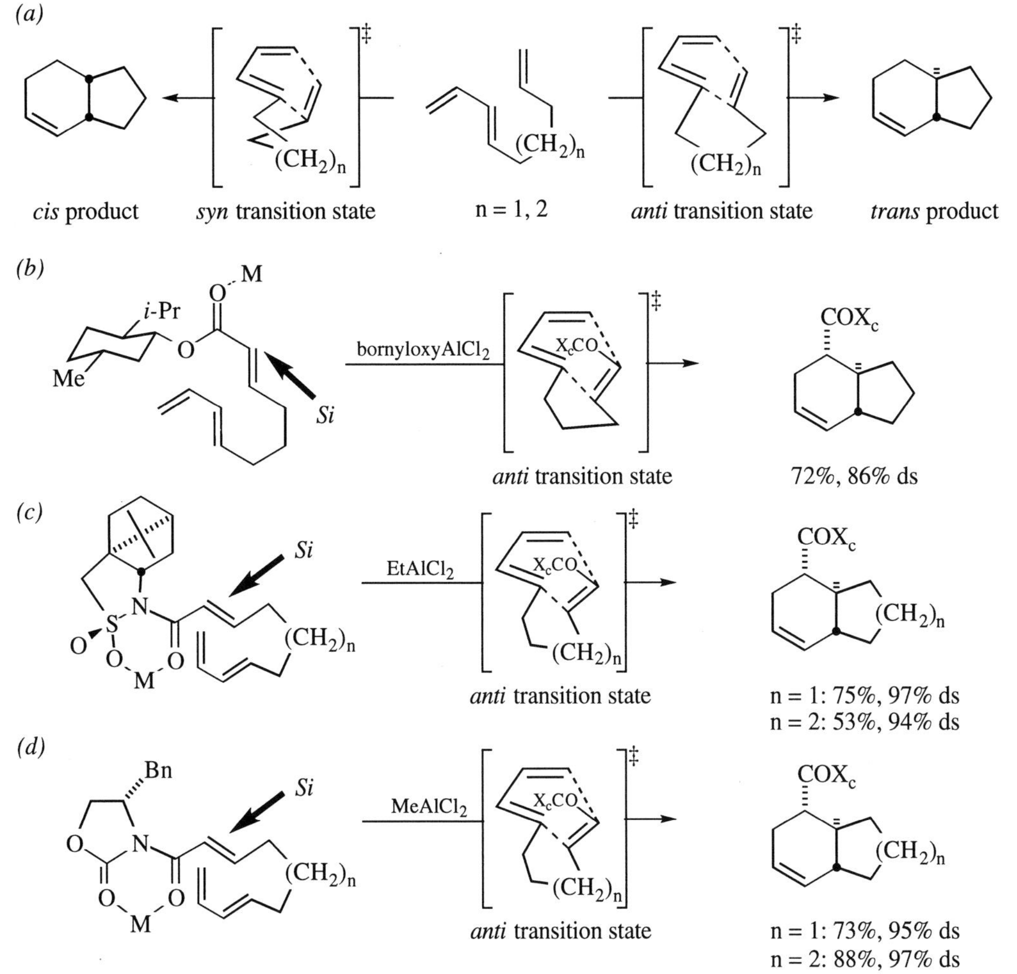

Scheme 6.44. *(a)* Syn and anti transition states for the intramolecular Diels-Alder reaction [182]. *(b)* The contribution of the chiral Lewis acid to the stereoselectivity was neglible [185]. *(c)* [184]; the illustrated examples are enantiomeric to those reported. *(d)* [178].

and oxazolidinone imide (Scheme 6.44d) appear to react through an *s*-cis C_1–C_2 conformation (*cf.* Figure 6.13b) in the anti transition state.

Fumarates. Asymmetric cycloaddition to fumarates has been accomplished by modification of either one or both ends of the diacid. In fact, addition of butadiene to dimenthyl fumarate, reported by Walborsky in 1961, was the first highly selective (89% ds) asymmetric Diels-Alder reaction ever recorded [186,187]. Scheme 6.45 shows examples of cycloadditions of several dienes to dimenthyl fumarate [186-189]. Scheme 6.45a illustrates the presumed reactive conformation of dimenthyl fumarate This conformation features (*cf.* Figure 6.13a) an *s*trans conformation at C_1–C_2, cis orientation of the ester ligand relative to the carbonyl oxygen, and orientation of the menthyl moiety to relieve $A^{1,3}$ strain. In this conformation, preferred approach of the diene is from the (rear) C_2 *Si* face. In addition to menthol (Figure 6.12a) 1-mesityl trifluoroethanol (Figure 6.12d) has been used as a *bis*-auxiliary [169].

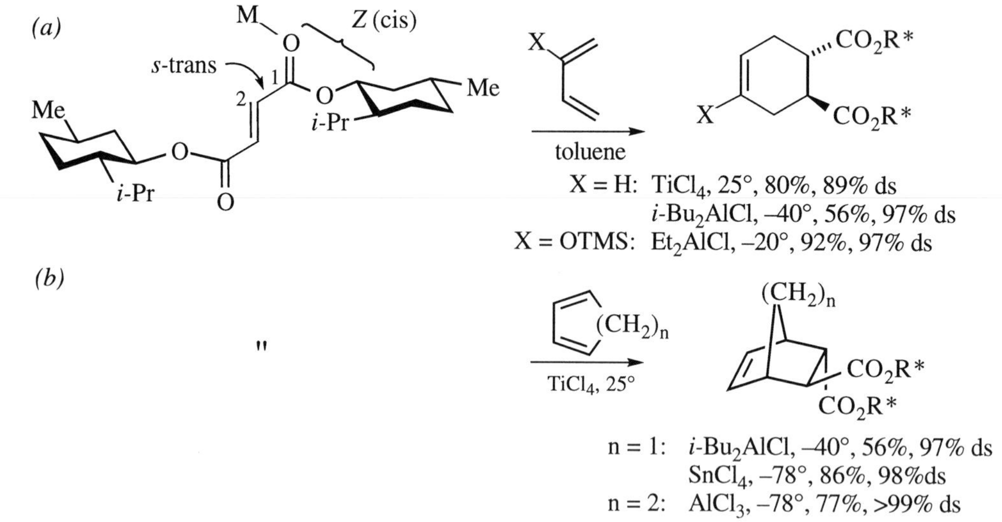

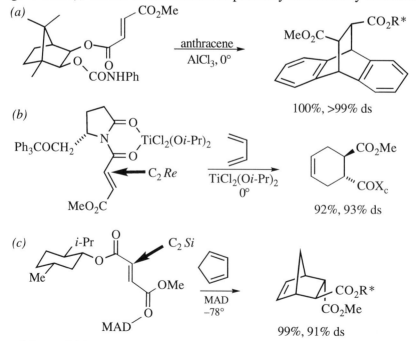

(a) s-trans

M. O
Z (cis)
Me

Me
i-Pr
O
i-Pr

X

toluene

X = H: TiCl$_4$, 25°, 80%, 89% ds
            i-Bu$_2$AlCl, –40°, 56%, 97% ds
X = OTMS: Et$_2$AlCl, –20°, 92%, 97% ds

,,,CO$_2$R*
X
CO$_2$R*

(b)

"

(CH$_2$)$_n$

TiCl$_4$, 25°

(CH$_2$)$_n$

CO$_2$R*
CO$_2$R*

n = 1: i-Bu$_2$AlCl, –40°, 56%, 97% ds
         SnCl$_4$, –78°, 86%, 98%ds
n = 2: AlCl$_3$, –78°, 77%, >99% ds

Scheme 6.45. Asymmetric cycloadditions to doubly modified fumarates. *(a)* X = H: ref. [186-188]. X = OTMS: ref. [188]. *(b)* n = 1: ref. [188,189]. n = 2: ref. [189].

Another approach to fumarate modification is to attach the chiral auxiliary to only one of the two carboxylate groups. One auxiliary for this purpose was introduced by Helmchen, as illustrated in Scheme 6.46a [190]. The only diene reported for this auxiliary was anthracene, probably because unsymmetrical dienes would introduce additional stereoisomers into the product mixture. In this case, the conformation of the ester is similar to that presented in previous examples (*cf.,* Scheme 6.45a, Figure 6.13a), but the conformation is probably additionally constrained by

(a)

CO$_2$Me

O O
CONHPh

anthracene
AlCl$_3$, 0°

CO$_2$R*
MeO$_2$C

100%, >99% ds

(b)

Ph$_3$COCH$_2$
N
O
O
TiCl$_2$(O*i*-Pr)$_2$

C$_2$ *Re*

MeO$_2$C

TiCl$_2$(O*i*-Pr)$_2$
0°

CO$_2$Me
,,,COX$_c$

92%, 93% ds

(c)

i-Pr
O
Me
O
OMe
O
MAD

C$_2$ *Si*

MAD
–78°

CO$_2$R*
CO$_2$Me

99%, 91% ds

Scheme 6.46. Asymmetric Diels-Alder reactions to fumarates having only one auxiliary. *(a)* ref. [190]. *(b)* ref. [191]. *(c)* ref. [192].

coordination of the metal to the phenylurethane carbonyl. Again, the favored approach is from the C_2 (or C_3) *Si* (rear) face. Later, Koga studied the pyrrolidinone auxiliary shown in Scheme 6.46b [191] and Yamamoto examined methyl menthyl fumarate in Scheme 6.46c [192]. Koga's auxiliary showed excellent selectivity with a titanium catalyst in cycloadditions with butadiene. Yamamoto found that methylaluminum bis(2,6-di-*tert*butyl-4-methylphenoxide (MAD) is 91% selective for the illustrated isomer [192]. The face-selectivity of the Koga auxiliary can be rationalized by titanium chelation of the two carbonyls as shown in Figure 6.13b. For the Yamamoto auxiliary, it is thought that the MAD binds to the methyl ester in favor of the menthyl ester, but that the face-selectivity is determined by the menthyl auxiliary (*cf.* Figure 6.13a). In addition to the face-selectivity, the reaction is also selective for the isomer having the menthyl in the exo position, due to the diene orienting away from the MAD and menthyl moieties, and toward the methoxy, as illustrated.

Maleates. Cycloaddition of a symmetrical diene such as butadiene or cyclopentadiene to maleic acid or a symmetrical derivative affords achiral (meso) adducts (Scheme 6.46a). To break the symmetry, either the diene or the dienophile must be unsymmetrical. For example, cycloaddition of an unsymmetric diene would give a chiral adduct, and Scheme 6.47b shows one such approach. Maleimide having an α-methylbenzyl auxiliary on nitrogen is highly selective when there is a large substituent at the diene 2-position [193]. A second tactic is the same as the fumarate approach in Scheme 6.46: attach an auxiliary to only one carboxyl group. After considerable experimentation, Yamamoto showed that 2-phenylcyclohexanol is an excellent auxiliary for *tert*-butyl maleate, as shown in Scheme 6.47c [194]. In this case, the catalyst is thought to chelate the two carbonyls with the phenyl group interacting with the double bond in a π-stacking arrangement.

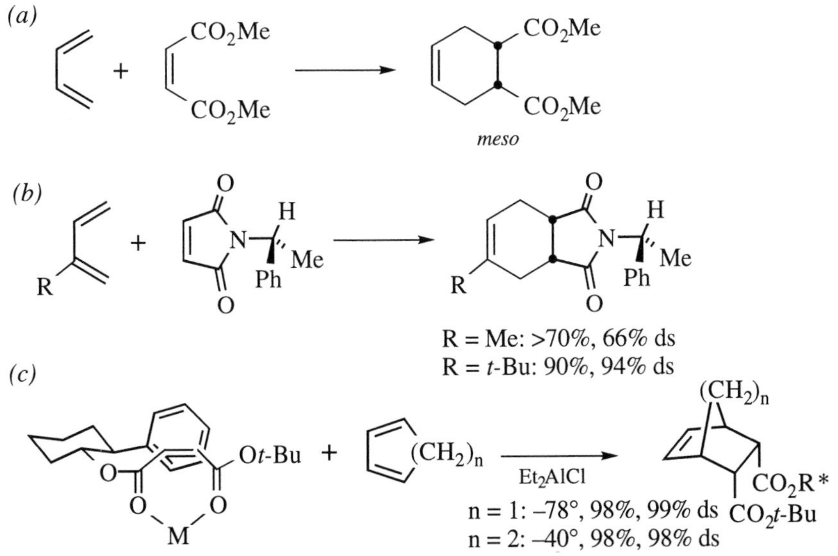

Scheme 6.47. (*a*) Cycloadditions of symmetrical dienes to maleates gives achiral products. (*b*) Asymmetric cycloadditions to chiral maleimide [193]. (*c*) Asymmetric cycloadditions to chiral maleic ester [194].

Figure 6.15 shows some natural products synthesized using the asymmetric Diels-Alder reaction. It is interesting to note that none of these compounds are cyclohexenes, even though that is the structural unit formed in the key step! In fact, only in yohimbine is the 6-membered ring formed by the Diels-Alder reaction preserved.

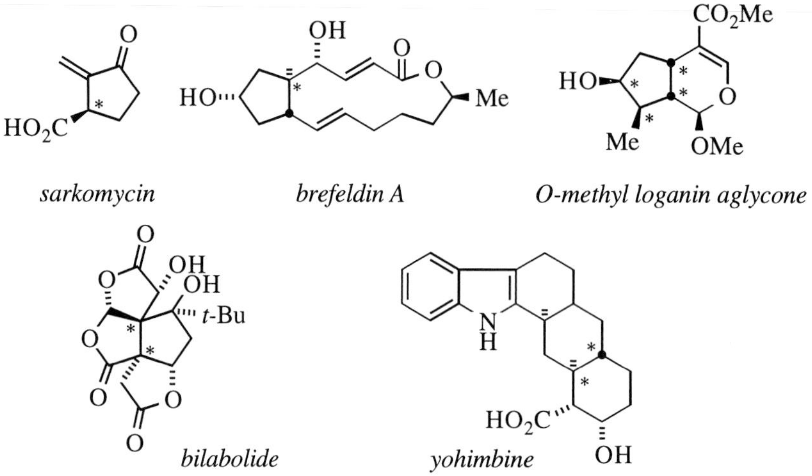

sarkomycin   *brefeldin A*   *O-methyl loganin aglycone*

bilabolide   *yohimbine*

Figure 6.15. Natural products synthesized using asymmetric cycloadditions to chiral dienophiles as the key stereodifferentiating step: sarkomycin [167,195]; brefeldin A [168] (of the 3 stereocenters formed in the asymmetric acrylate/cyclopentadiene cycloaddition, the indicated stereocenter is the only one retained); *O*-methyl loganin aglycone [196]; bilabolide [197]; yohimbine [198].

6.2.2.2 Diene auxiliaries

In comparison to the large amount of work on chiral dienophiles for the asymmetric Diels-Alder reaction, there have been very few reports of chiral auxiliaries for the diene component. This may be due, in part, to the lack of convenient methods for the synthesis of modified dienes, but it may also be due to the inherent complexity of the problem. A general analysis of the magnitude of the challenge is shown in Scheme 6.48 for the "simple" case of a monosubstituted diene and a monosubstituted dienophile. If the diene and dienophile are not C_2 symmetric, two constitutional isomers may be produced as cycloadducts. If a substituent is present at the 2-position of the diene, only one stereocenter is formed in the cycloaddition (Scheme 6.48a), so that two stereoisomers are possible for each constitutional isomer (referred to as *meta* and *para* for simplicity). On the other hand, if the substituent is at C_1 of the diene, two stereocenters are formed for each of the two regioisomeric products, for a total of 8 possible products from a single pair of reactants (Scheme 6.48b: 4 from each of the regioisomers, again labeled as *ortho* and *meta* for convenience). Of course, a great deal is known about the regioselectivity of the Diels-Alder reaction of unsymmetric dienes and dienophiles [152], and good regioselectivity is often possible. Nevertheless the primary and secondary molecular orbital considerations that govern regioselectivity constitute a limiting factor in auxiliary design. Regiochemical issues such as these undoubtedly

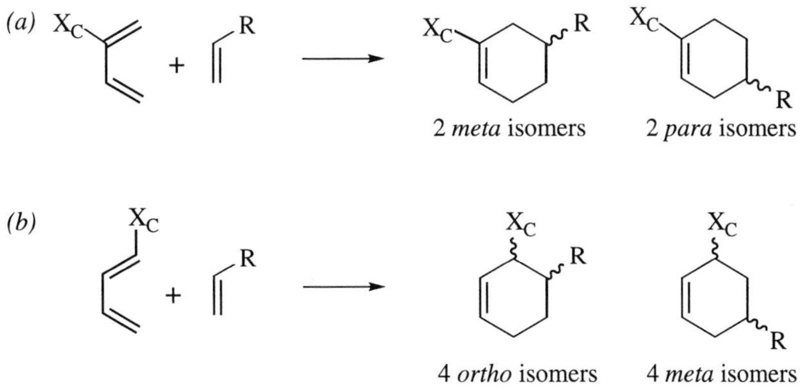

(a)

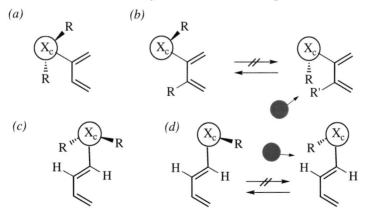

2 *meta* isomers   2 *para* isomers

(b)

4 *ortho* isomers   4 *meta* isomers

Scheme 6.48. Constitutional isomers (regioisomers) and stereoisomers possible from the Diels-Alder cycloaddition of a monosubstituted diene with a monosubstituted dienophile.

contribute to the paucity of examples of unsymmetrical dienes reported in the previous section, but they are more important here because unsymmetrical dienes are unavoidable if the diene is to be modified with a chiral auxiliary.

For a diene with the auxiliary at C_2, C_2–X_c bond rotation will populate two rotational isomers unless the auxiliary is C_2-symmetric, in which case the two rotamers are identical (Figure 6.16a) or unless there is a substituent either at C_1, cis to the auxiliary, or at C_3, in which case one of the conformers may be destabilized by repulsive van der Waals interactions (Figure 6.16b, R' ≠ H). When the auxiliary is at C_1, a similar situation exists: a C_2-symmetric auxiliary has only one conformational isomer possible (Figure 6.16c), but an otherwise unsubstituted diene will have two rotational isomers of unequal energy (Figure 6.16d). Thus, for a diene with the auxiliary at C_2, a C_2-symmetric auxiliary would seem to have an advantage [199]. When the auxiliary is at C_1, the two conformers are unequally populated as long as there is no other substituent at C_1. A similar analysis could be made for other substitution cases, but this analysis covers the examples which follow.

(a)   (b)

(c)   (d)

Figure 6.16. Generalized conformational considerations for chiral auxiliaries attached to butadiene: *(a)* C_2-symmetric auxiliary at position 2; *(b)* C_s-symmetric auxiliary at position 2; *(c)* C_2-symmetric auxiliary at position 1; *(d)* C_s-symmetric auxiliary at position 1.

Scheme 6.49 illustrates an asymmetric cycloaddition of an enamino diene developed in the Enders laboratory [200]. In this case the auxiliary, 2-methoxy-methyl pyrrolidine, has C_S symmetry, and excellent selectivity is achieved with β-nitrostyrenes as dienophiles, although the yields are modest. The diastereoselectivity in the cycloaddition is ≥98% in each case, however hydrolysis of the enamine on workup affords a mixture of 2-methyl diastereomers with 75-95% ds. The proposed transition state for the cycloaddition is shown in the inset, although an alternative two step mechanism (Michael addition followed by aldol cyclization) has not been ruled out [200].

A more generally useful chiral auxiliary was introduced by Trost in 1980 [201]. This auxiliary, derived from mandelic acid, is available as either enantiomer. The original diastereoselectivity reported for addition to acrolein was 82% at –20° (Scheme 6.50a), but Thornton later reported 94% ds at –78° [202]. The other examples in Scheme 6.50 illustrate similarly high selectivities and yields, although the Thornton paper does not report specific yields for each example [202]. For the *S* auxiliary, addition to the C_1 *Si* face of the diene is preferred (relative topicity *lk*). Figure 6.17 illustrates two conformational models that have been proposed to rationalize this preference. Trost suggested that π-stacking of the diene over the face of the phenyl group shields the C_1 *Re* face, as shown in Figure 6.17a [201]. A weakness of this model is that reduction of the phenyl to a cyclohexyl group afforded an auxiliary that is equally selective [201,202]. This prompted Thornton to propose that the cycloaddition took place in a diene conformation in which the bond from the α-carbon to the phenyl (or cyclohexyl) is perpendicular to the plane of the ester, as shown in Figure 6.17b [202]. Thornton asserts that the conformation in which the methoxy is nearest the carbonyl is preferred, but no explanation for this preference was offered. Nevertheless, crystal structures of three cycloadducts exhibit this conformation, which is similar to one proposed by Mosher to rationalize chemical anisotropies (*cf.,* Figure 2.4, ref. [203].

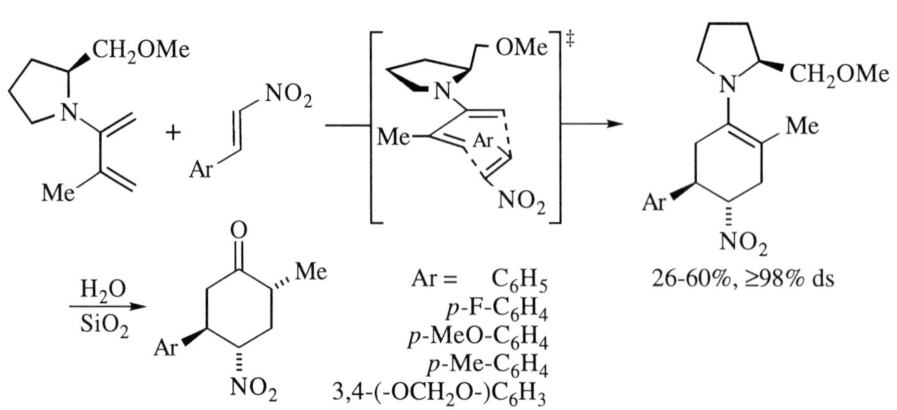

Scheme 6.49. Asymmetric Diels-Alder reaction of dieneamine [200].

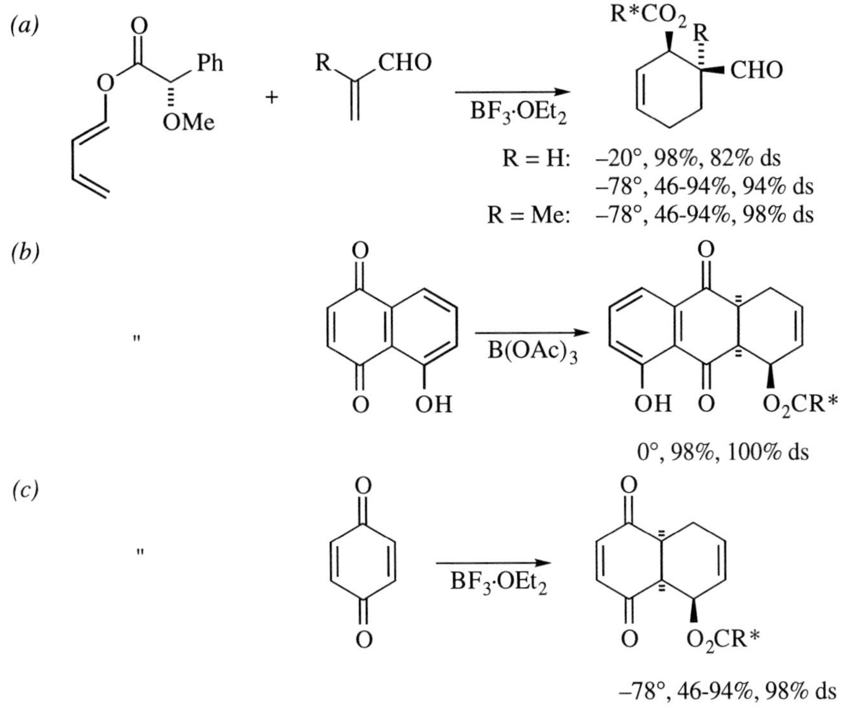

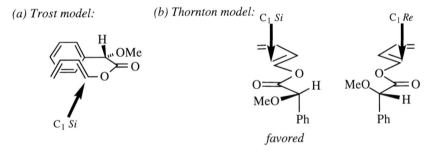

Scheme 6.50. Asymmetric Diels-Alder reactions of *O*-methylmandelate esters: *(a)* R = H [201,202]; R = Me [202]. *(b)* ref. [201]. *(c)* ref. [202].

(a) Trost model: *(b) Thornton model:*

favored

Figure 6.17. Conformational models to explain the relative topicity of the Trost auxiliary for asymmetric Diels-Alder reactions. *(a)* Trost model [201]. *(b)* Thornton model [202].

6.2.2.3 Chiral catalysts

Quite a number of ligand/metal combinations have been evaluated as chiral catalysts for the Diels-Alder reaction, with several being very successful. Much of the effort has been occupied in ligand synthesis and design, but the effort has largely been empirically driven (*i.e.*, trial and error). Figure 6.18 shows several complexes that have been tested as catalysts in the Diels-Alder reaction and which show both high diastereoselectivity and high enantioselectivity.[22] Among the metals,

---

[22] Cyclopentadiene addition to an acrylate gives two diastereomers (*endo* and *exo*), each of which has two enantiomers. In the presence of a chiral auxiliary, these four stereoisomers are all

the most commonly used are boron and titanium, but copper [204,205], magnesium [206], and lanthanides [207,208] have also found some use. In this section, detailed analysis is presented for only a few of these catalysts, chosen to illustrate current levels of understanding. Additional references and other Lewis acid catalysts can be found in recent reviews [152,157,209,210].

Monodentate dienophiles. The first chiral Lewis acid catalyst to show high selectivity (86% es in the cycloaddition of cyclopentadiene to 2-methylacrolein), is a dichloroaluminum alkoxide derived from menthol (Figure 6.18a, [211]). This catalyst, as well as several others (*e.g.,* Figure 6.18b-d) have C_s symmetry, but most of the catalysts shown in Figure 6.18 are C_2-symmetric. This feature reduces the number of competing transition states, which is especially important when the ligand sphere is greater than 4-coordinate. Because of fewer possible coordination sites, the binding and face-selectivities of catalysts containing boron, aluminum, or other tetravalent metals are better understood than those of octahedral complexes, and these are examined first.

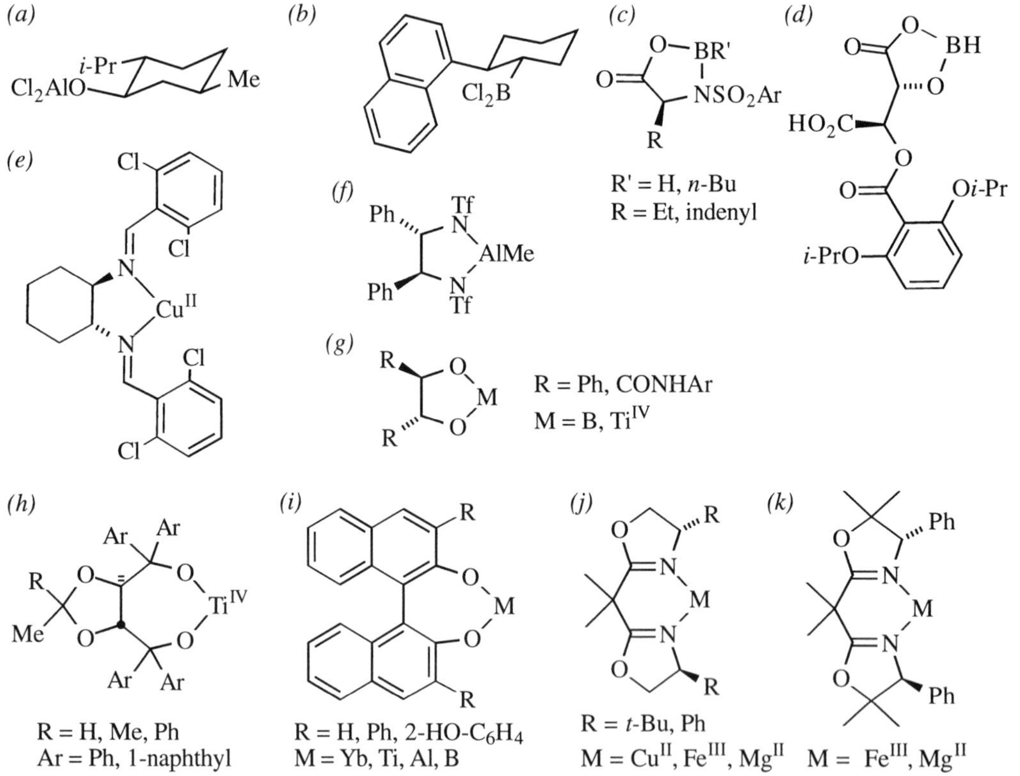

Figure 6.18. Selected catalysts for the asymmetric Diels-Alder reaction. (*a*) [211]. (*b*) [212,213]. (*c*) R = Et [214], R = CH$_2$indenyl [215-217]. (*d*) [218-220]. (*e*) [204]. (*f*) [221-223]. (*g*) R = Ph [224], R = CONHAr [225]. (*h*) [226-229]. (*i*) R = H [207,208,230], R = Ph [231,232], R = 2-HO-C$_6$H$_4$ [233]. (*j*) R = Ph [205,234], R = *t*-Bu [205]. (*k*) [206]

diastereomers, so that selectivity for one of the four can be expressed as percent diastereoselectivity, % ds. But in the present case it is necessary to express selectivity in terms of both diastereoselectivity (*endo/exo*) and enantioselectivity (% es for the major diastereomer).

Figure 6.19 shows Hawkins's (2-aryl)cyclohexylboron dichloride catalyst (Figure 6.18b [212,213]) coordinated to methyl acrylate (*cf.* Figure 6.13). Note that monodentate coordination of the ester is thought to occur trans to the alkoxy group, which forces the enone into an *s*-trans conformation, similar to that seen previously (*cf.* Figure 6.13 and Scheme 6.45). The geometry shown in Figure 6.19 has been observed in the crystal of five related catalyst·crotonate complexes [213], and NMR studies show that this conformation persists in solution [212]. Comparison of these five structures indicates that, as the polarizability of the aryl group increases, a dipole-induced dipole attraction draws the polar ester group of the boron-bound crotonate towards the arene (the five complexes are, in increasing order of polarizability: Ar = phenyl, 3,5-dimethylphenyl, 3,5-dichlorophenyl, 3,5-dibromo-phenyl, and 1-naphthyl). Since this effect correlates with enantioselectivity, Hawkins concluded that the effect is operative in the transition state [213]. In this conformation, the rear (C$_2$ *Re*) face of the dienophile is blocked by the aryl group, and approach of the diene toward the face of the crotonate that is not blocked by the aryl moiety is favored. The 2-(1-naphthyl)-cyclohexyl boron catalyst produces ≥95% enantioselectivities in the addition of cyclopentadiene to methyl acrylate, methyl crotonate, and dimethyl fumarate [212].

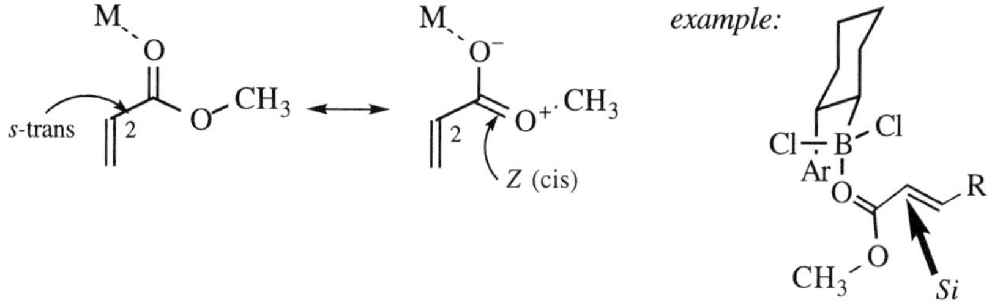

Figure 6.19. Methyl acrylate coordinated to Hawkins's 2-arylcyclohexyl boron catalyst [212,213].

Scheme 6.51 shows the reaction of 2-bromoacrolein and cyclopentadiene catalyzed by the indenyl oxazaborolidine shown in Figure 6.18c [215,216]. This reaction, which is both highly diastereoselective and enantioselective, is thought to react *via* the *s*-cis conformation shown in the inset of Scheme 6.51. This catalyst conformation is suggested by nuclear Overhauser effects in the NMR spectrum of the catalyst-dienophile complex, and by chemical shift changes upon complexation to boron trifluoride [216]. Also, a 1:1 complex of the catalyst and 2-bromoacrolein is orange-red at 210° K, a color that is attributed to charge-transfer complexation between the indene ring and the boron-bound aldehyde [216]. Similar catalysts with different substituents on the nitrogen [216] or the carbon of the oxazaborolidine [216,217] show significantly lower selectivities. For example the oxazaborolidine having a 2-naphthyl group (comparable in size, but not as good a π-donor) in place of the indene exhibits only 88% enantioselectivity. Phenyl, cyclohexyl, or isopropyl groups give only about 65% enantioselectivity with *opposite* topicity [216]. Oxazaborolidine auxiliaries having donor *atoms* in the side chain also show improved selectivities [217].

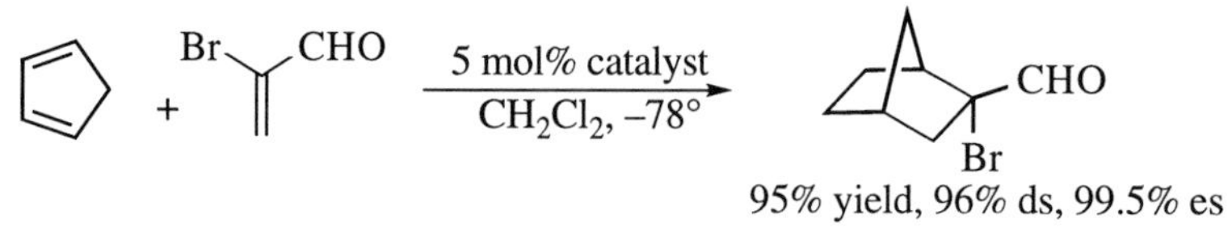

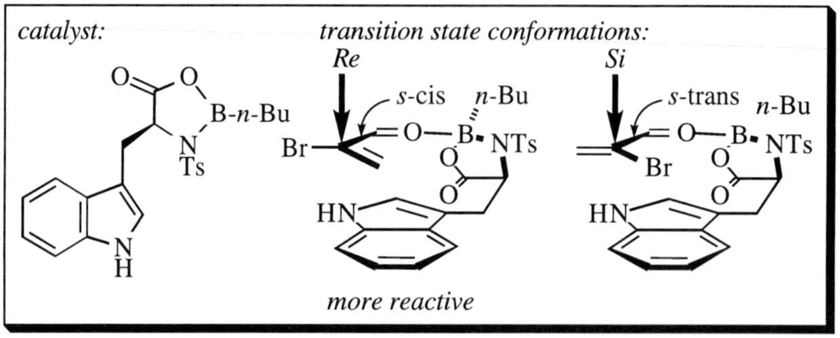

Scheme 6.51. Asymmetric cycloaddition of 2-bromoacrolein and cyclopenta-diene using Corey's indenyl oxazaborolidine catalyst [215,216]

Note that the illustrated conformation has the acrolein oriented in an *s*-cis con-formation. This is in contrast to the usual *s*-trans conformation of acroleins co-ordinated to a Lewis acid (Figure 6.13a), but it is supported by the fact that cyclo-pentadiene adds to the opposite face of acrolein itself [216]. It is likely that both *s*-cis and *s*-trans dienophile conformers are present, and that the *s*-cis conformer is more reactive. In other words, Curtin-Hammett kinetics [235] are operative. The rationale for this increased reactivity is as follows: the *s*-trans conformation of 2-bromoacrolein would place the bromine above the indene ring. Cycloaddition to the top (*Si*) face of the *s*-trans conformer would force the bromine into closer proximity to the indene as C_2 rehybridizes from sp^2 to sp^3, a situation that is avoided in cycloaddition to the top (*Re*) face of the *s*-cis conformer.

Bidentate dienophiles. When a dienophile such as *N*-acryloyloxazolidinone coordinates the metal in a bidentate fashion (*cf.* Figure 6.13b), $A^{1,3}$ strain between the enone β-carbon and the oxazolidinone C-4 methylene forces the enone into an *s*-cis conformation, as shown in Figure 6.20a. Interactions between the other ligands on the metal, the coordinated dienophile, and the approaching diene then determine the topicity of the cycloaddition. The exact nature of the interactions will depend on the coordination sphere of the metal.

For example, Figure 6.20b and c shows examples of similar C_2-symmetric ligands (Figure 6.18j,k) coordinated to metals having diffent tetravalent geometries and which result in enantiomeric cycloadducts, but with excellent selectivity in both cases. The explanation for the topicity of the two catalysts is revealed by examination of the proposed arrangements of the catalyst/dienophile complexes, as shown in Scheme 6.20d. The tetrahedral magnesium [206] complex facilitates addition to the C_2 *Si* face because the rear phenyl is blocking the *Re* face [206]. In contrast, the square-planar copper complex facilitates C_2 *Re* addition because the *Si* face is blocked by the *tert*-butyl group [205]. It should be noted, however, that in these two examples, the geometry of the coordination complex appears to be inferred (at least partly) from the topicity of the cycloaddition (note the absence of any anionic ligands in these models).

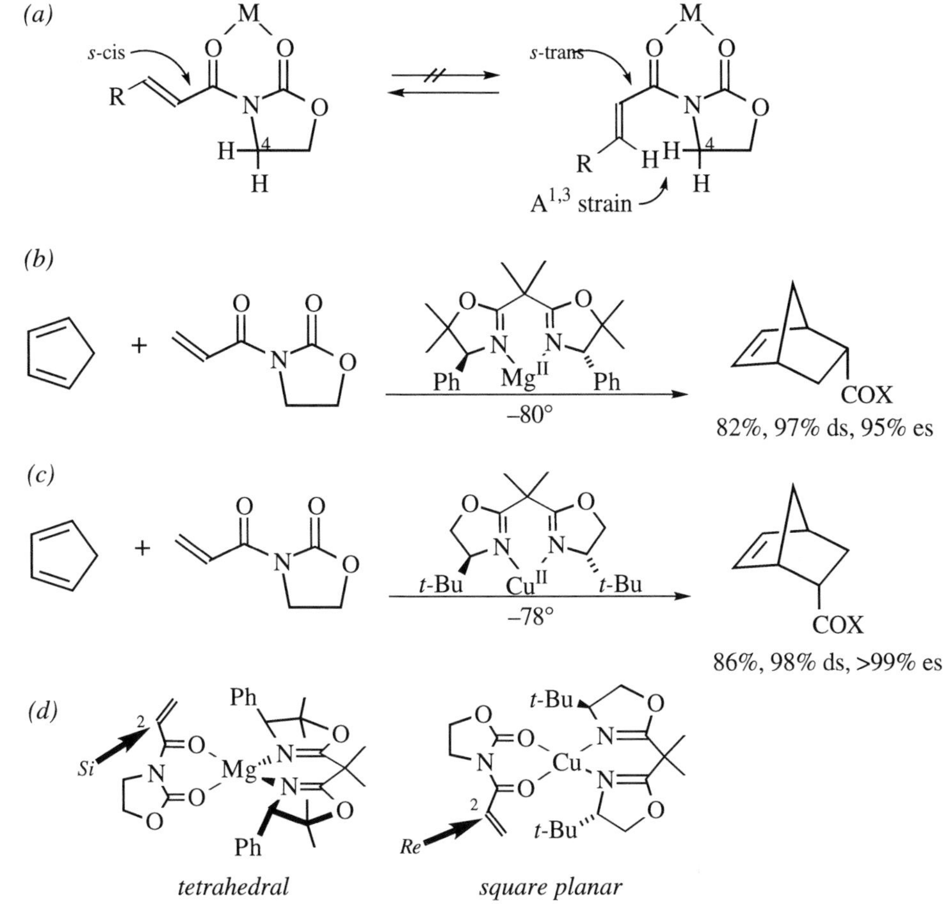

Figure 6.20. *(a)* Acryloyloxazolidinone in bidentate coordination. $A^{1,3}$ strain favors the *s*-cis conformation. *(b)* Cycloaddition of C_2-symmetric bisoxazoline-magnesium complex [206]. *(c)* Cycloaddition of C_2-symmetric bisoxazoline-copper complex [205]. *(d)* Rationale for the different topicities of the bisoxazoline complexes, even though both ligands have the same absolute configuration. The dienophile is drawn in the plane of the paper, and the favored approach is from the direction of the viewer.

Ligands having C_2-symmetry have also been used with metals that are undoubtedly octahedral, however the analysis of facial selectivity in octahedral complexes is complicated by several possible competing coordination modes of the dienophile. One class of ligand that has been well studied are the TADDOLs (TADDOL is an acronym for $\alpha,\alpha,\alpha',\alpha'$-tetraaryl-1,3-dioxolane-4,5-dimethanol). Both the Narasaka [236] and the Seebach [228] groups have evaluated a number of TADDOLs as ligands for titanium in the asymmetric Diels-Alder reaction. Table 6.6 lists selected data from two extensive reports, which illustrates not only the utility of the titanium TADDOLate complex as an asymmetric catalyst, but which also illustrates some subtle differences that are not readily explained. For example, Narasaka found that the tetraphenyl dimethyldioxolane ligand (R = R' = Me; Ar = Ph) promoted the reaction (88% yield) when used in stoichiometric quantities

Table 6.6. Asymmetric cycloadditions of crotyloxazolidinones and cyclopentadiene catalyzed by titanium TADDOLate complexes.

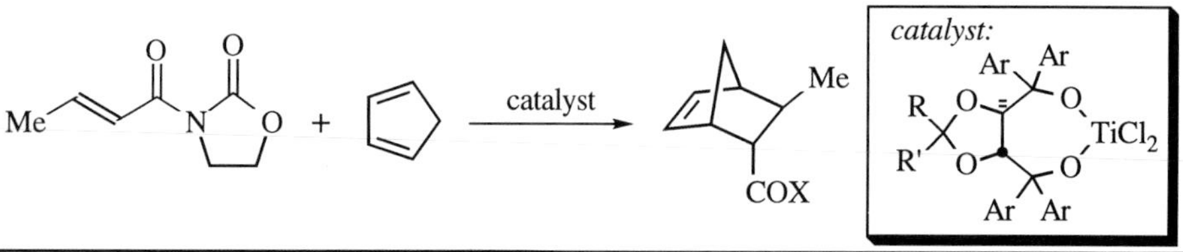

Entry	R/R'	Ar	Temp.	Eq. cat.	% yield	% ds	% es	Ref.
1	Me/Me	Ph	−15	1	88	93	77	[236]
2	Me/Me	Ph	−15	0.15	25	83	72	[228]
3	Me/Me	2-naphthyl	−15	0.15	96a	90	94	[228]
4	Me/Ph	Ph	−15	2	93	90	96	[236]
5	Me/Ph	Ph	0	0.10	87	92	95	[236]

a This experiment done on a >4g (crotyloxazolidinone) scale.

(entry 1), but Seebach found that a catalytic amount was not as effective (25% yield) under similar conditions (entry 2). Note the difference in diastereo- and enantioselectivity for these two entries, as well. In contrast, replacing the phenyl group with a 2-naphthyl group affords an outstanding catalyst (entry 3), that gives excellent yields and selectivities on a multigram scale [228]. Entries 4 and 5 illustrate the tetraphenyl methyl-phenyl dioxolane catalyst (R = Me, R' = Ph; Ar = Ph), which affords outstanding yields and selectivities in either stoichiometric or catalytic modes [236]. Comparison of entry 2 with entry 5 is particularly puzzling: replacement of one the dioxolane substituents (a position remote from the catalytic site) results in an amazing improvement in catalyst efficiency and selectivity.[23]

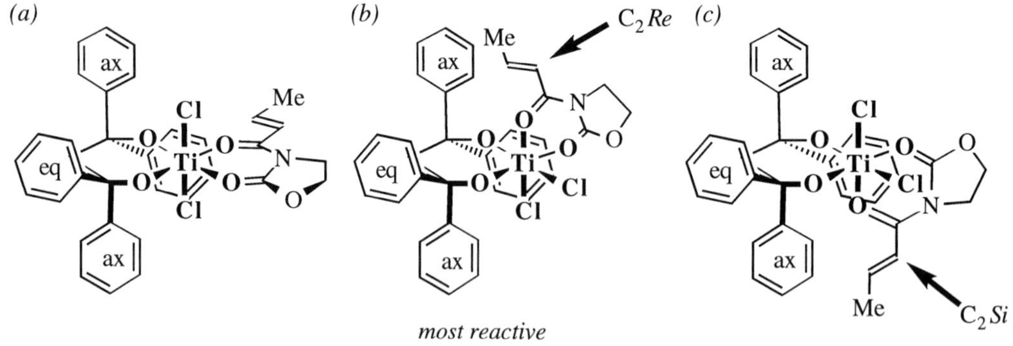

Figure 6.21. Titanium TADDOLate - crotyloxazolidinone complexes. The dioxolane ring of the chiral ligand (Figure 6.18h) is deleted for clarity, and the phenyl groups are labelled as axial (ax) or equatorial (eq). *(a)* Symmetrical complex found by NMR to be the predominant species in solution [237], and also characterized crystallographically [238]. *(b)* Complex judged to be most likely to be responsible for the asymmetic cycloaddition [228,237]. *(c)* This complex is probably less reactive, since approach of the dienophile is hindered by the axial phenyl [228].

---

[23] Although entries 2 and 5 are from different laboratories, Seebach's group has reported results similar to those of entry 5: 99% conversion, 88% ds, and 94% es using 15 mol% catalyst at −5° [228].

NMR studies have shown that at least three hexacoordinate catalyst oxazolidinone complexes exist in solution [237]. The most abundant has been assigned a structure that has the oxazolidinone oxygens trans to the TADDOL oxygens and the chlorines trans to each other, as shown in Figure 6.21a. This species has also been characterized crystallographically [238]. There are four other possible complexes, two of which are illustrated in Figure 6.21b and c.[24] It is not known whether these two complexes are the ones that are observed in the NMR [237], but these two are judged to be more reactive, since in these structures, the enone oxygen is trans to the weaker π-donor ligand (chlorine) and may therefore experience a higher degree of Lewis acid activation. NMR studies show that one of the axial phenyls undergoes restricted rotation when bidentate ligands are bound to the titanium TADDOLate [237]. When the oxazolidinone ligand is oriented as shown in Figure 6.21b, the dienophile and the axial phenyl are in close proximity and approach of the diene from the direction of the viewer (toward the C_2 *Re* face) is unhindered, and would result in cycloadduct with the observed absolute configuration [228]. The alternative geometry, shown in Scheme 6.21c, is judged to be less reactive, since the diene must approach either from the direction of the viewer (toward the C_2 *Si* face), where it may encounter the nearby axial phenyl, or from the rear, where it is blocked by the equatorial phenyl [228].[25]

This explanation is described as a "mnemonic rule" [228], which can only be taken as a first approximation of reality. The same rule can be used to rationalize the topicity of other asymmetric Diels-Alder reactions, such as those employing titanium BINOLate catalysts (Figure 6.18i, [230]), or iron bisoxazoline catalysts (Figure 6.18j,k [206,215]). Although the explanation seems reasonable, the picture is not complete, since it does not account for a number of observations, including the fact that the dioxolane substituents exert an extraordinary effect on catalyst efficiency (*cf.* Table 6.6, entries 2 and 5). Additionally, both titanium TADDOLate [228] and BINOLate [230] complexes show a nonlinear relationship between enantiomeric purity of the catalyst and that of the product, which suggests that some sort of dimerization phenomenon is involved.[26]

6.2.2.4 Prostaglandins: A case study in the synthesis of enantiopure compounds

Efficiency in the synthesis of prostanoids has been an important aspect of organic chemistry for over two decades. Because the prostaglandins are chiral, synthesis of enantiopure drugs is highly desirable for clinical applications. The following examples of prostaglandin synthesis are taken from the work of Corey, much of which is summarized in Chapter 11 of his recent book [239]. The hydroxy acid shown in Scheme 6.52a has been used as a key intermediate in a number of

---

[24] The other two have the oxazolidinone transposed such that the enone oxygen is trans to a TADDOL oxygen.

[25] Jørgensen has proposed another rationale, based on the geometry observed in the crystal structure [238].

[26] One possibility is that heterochiral dimerization of the ligand or the titanium complex produces an inactive catalyst; this tends to sequester the minor enantiomer (*cf.* Scheme 4.6). Another is that the catalyst is a dinuclear species, which is more reactive when homochiral.

prostaglandin syntheses, two of which are shown. To control the *relative* configuration of the stereocenters in the cyclopentane ring, the lactone was synthesized by a Diels-Alder strategy employing a substituted cyclopentadiene, as shown in Scheme 6.52b. Baeyer-Villiger oxidation of the key bicycloheptenone and hydrolysis afforded a hydroxy acid that was initially (in the early 1970s) separated into its enantiomers by resolution [240]. Asymmetric synthesis was then applied to the problem. For example in 1975, 8-phenylmenthol (Figure 6.12b) was used as an acrylate auxiliary to provide an endo bicycloheptene carboxylate [165] that was oxidatively cleaved to the ketone, and carried on to the hydroxy acid as before (Scheme 6.52c). Then in 1989, the first of two chiral catalysts (Figure 6.18f) was

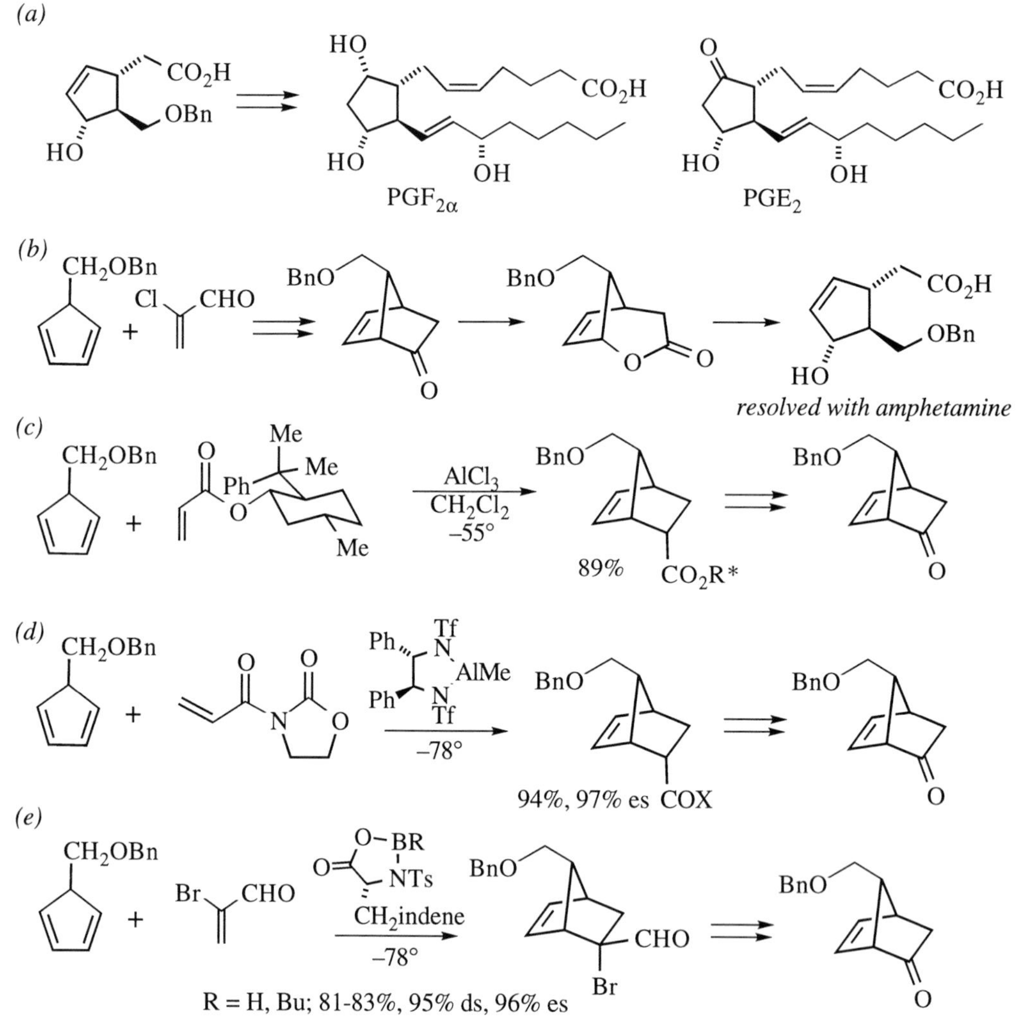

(a)

(b)

resolved with amphetamine

(c)

89% CO_2R*

(d)

94%, 97% es COX

(e)

R = H, Bu; 81-83%, 95% ds, 96% es

Scheme 6.52. Corey's synthetic approaches to prostaglandins (see also ref. [239], chapter 11): *(a)* Key hydroxy acid intermediate for the synthesis of $PGF_{2\alpha}$ and PGE_2. *(b)* Early synthesis that relied on resolution for obtaining enantiopure products [240]. *(c)* 8-Phenylmenthol as a chiral auxiliary [165]. *(d)* Acryloyl oxazolidinone as dienophile with a chiral catalyst [221,222]. *(e)* 2-Bromoacrolein as dienophile with a chiral catalyst [215].

applied to the problem: with an acryloyl oxazolidinone as the dienophile, a bicycloheptene carboximide similar to the ester obtained previously was obtained, as shown in Scheme 6.52 [221,222]. Development of the oxazaborolidine catalyst (Figure 6.18c) and 2-bromoacrolein as a dienophile provided a means for streamlining the preparation even further (Scheme 6.52e, [215]). Thus, the development of an efficient synthetic plan has been continually improved as progress in asymmetric synthesis has taken the route from classical chemical resolution, through auxiliary-based methods, to efficient chiral catalysts.

6.3 References

1. S. J. Rhoads; N. R. Raulins *Org. React.* **1975**, *22*, 1-252.
2. F. E. Ziegler *Acc. Chem. Res.* **1977**, *10*, 227-232.
3. K. Mori; H. Nomi; T. Chuman; M. Kohno; K. Kato; M. Noguchi *Tetrahedron* **1982**, *38*, 3705-3711.
4. R. E. Ireland *Aldrichimica Acta* **1988**, *21*, 59-69.
5. S. Blechert *Synthesis* **1989**, 71-82.
6. P. Wipf In *Comprehensive Organic Synthesis. Selectivity, Strategy, and Efficiency in Modern Organic Chemistry*; B. M. Trost, I. Fleming, Eds.; Pergamon: Oxford, 1991; Vol. 5, p 827-873.
7. R. K. Hill In *Comprehensive Organic Synthesis. Selectivity, Strategy, and Efficiency in Modern Organic Chemistry*; B. M. Trost, I. Fleming, Eds.; Pergamon: Oxford, 1991; Vol. 5, p 785-826.
8. S. Pereira; M. Srebnik *Aldrichimica Acta* **1993**, *26*, 17-29.
9. E. J. Corey; B. E. Roberts; B. R. Dixon *J. Am. Chem. Soc.* **1995**, *117*, 193-196.
10. S. G. Davies *Organotransition Metal Chemistry Applications to Organic Synthesis*; Pergamon: Oxford, 1982.
11. S. Otsuka; K. Tani *Synthesis* **1991**, 665-680.
12. A. J. Birch; I. D. Jenkins In *Transition Metal Organometallics in Organic Synthesis*; H. Alper, Ed.; Academic: New York, 1976; Vol. 1, p 1-82.
13. H. M. Colquhoun; J. Holton; D. J. Thompson; M. V. Twigg In *New Pathways for Organic Synthesis. Practical Applications of Transition Metals*; Plenum: New York, 1984, p 173-193.
14. H. Kumobayashi; S. Akutagawa; S. Otsuka *J. Am. Chem. Soc.* **1978**, *100*, 3949-3950.
15. K. Tani; T. Yamagata; S. Otsuka; S. Akutagawa; H. Kumobayashi; T. Taketomi; H. Takaya; A. Miyashita; R. Noyori In *Asymmetric Reactions and Processes in Organic Chemistry. ACS Symposium Series 185*; E. L. Eliel, S. Otsuka, Eds.; American Chemical Society: Washington, 1982, p 187-193.
16. S. Akutagawa; K. Tani In *Catalytic Asymmetric Synthesis*; I. Ojima, Ed.; VCH: New York, 1993, p 41-61.
17. R. A. Sheldon In *Chirotechnology. Industrial Synthesis of Optically Active Compounds*; Marcel Dekker: New York, 1993, p 304.
18. V. Prelog; G. Helmchen *Angew. Chem. Int. Ed. Engl.* **1982**, *21*, 567-583.
19. K. Tani; T. Yamagata; S. Akutagawa; H. Kumobayashi; T. Taketomi; H. Takaya; A. Miyashita; R. Noyori; S. Otsuka *J. Am. Chem. Soc.* **1984**, *106*, 5208-5217.
20. T. Ohta; H. Takaya; R. Noyori *Inorg. Chem.* **1988**, *27*, 566-569.
21. K. Mashima; K. Kusano; T. Ohta; R. Noyori; H. Takaya *J. Chem. Soc., Chem. Commun.* **1989**, 1208-1210.
22. K. Tani *Pure Appl. Chem.* **1985**, *57*, 1845-1854.
23. M. Kitamura; K. Manabe; R. Noyori; H. Takaya *Tetrahedron Lett.* **1987**, *28*, 4719-4720.

24. K. Tani; T. Yamagata; S. Otsuka; H. Kumobayashi; S. Akutagawa *Organic Syntheses* **1993**, *Coll. Vol. VIII*, 183-188.

25. H. Takaya; S. Akutagawa; R. Noyori *Organic Syntheses* **1993**, *Coll. Vol. VIII*, 57-63.

26. P. Schorigen *Chem. Ber.* **1925**, *58*, 2028-2036.

27. P. Schorigin *Chem. Ber.* **1924**, *57*, 1634-1637.

28. W. Schlenk; E. Bergmann *Liebigs Ann. Chem.* **1928**, *464*, 35-42.

29. G. Wittig; L. Löhmann *Liebigs Ann. Chem.* **1942**, *550*, 260-268.

30. G. Wittig; L. Löhmann *Liebigs Ann. Chem.* **1947**, *557*, 205-220.

31. C. R. Hauser; S. W. Kantor *J. Am. Chem. Soc.* **1951**, *73*, 1437-1441.

32. W. C. Still; A. Mitra *J. Am. Chem. Soc.* **1978**, *100*, 1927-1928.

33. P. T. Lansbury; V. A. Pattison; J. D. Sidler; J. B. Bieber *J. Am. Chem. Soc.* **1966**, *88*, 78-84.

34. K. Tomooka; T. Igrashi; T. Nakai *Tetrahedron Lett.* **1993**, *34*, 8139-8142.

35. R. B. Woodward; R. Hoffmann *The Conservation of Orbital Symmetry*; Academic: New York, 1970.

36. G. Wittig; H. Döser; I. Lorenz *Liebigs Ann. Chem.* **1949**, *562*, 192-205.

37. U. Schöllkopf; K. Feldenberger *Liebigs Ann. Chem.* **1966**, *698*, 80-85.

38. Y. Makisumi; S. Notzumoto *Tetrahedron Lett.* **1966**, 6393-6397.

39. J. E. Baldwin; J. DeBernardis; J. E. Patrick *Tetrahedron Lett.* **1970**, 353-356.

40. V. Rautenstrauch *J. Chem. Soc., Chem. Commun.* **1970**, 4-6.

41. J. E. Baldwin; J. E. Patrick *J. Am. Chem. Soc.* **1971**, *93*, 3556-3558.

42. R. Hoffmann *Angew. Chem. Int. Ed. Engl.* **1979**, *18*, 563-640.

43. R. K. Hill In *Asymmetric Synthesis*; J. D. Morrison, Ed.; Academic: Orlando, 1984; Vol. 3, p 503-572.

44. T. Nakai; K. Mikami *Chem. Rev.* **1986**, *86*, 885-902.

45. K. Mikami; T. Nakai *Synthesis* **1991**, 594-604.

46. J. A. Marshall In *Comprehensive Organic Synthesis. Selectivity, Strategy, and Efficiency in Modern Organic Chemistry*; B. M. Trost, I. Fleming, Eds.; Pergamon: Oxford, 1991; Vol. 3, p 975-1014.

47. R. Brückner In *Comprehensive Organic Synthesis. Selectivity, Strategy, and Efficiency in Modern Organic Chemistry*; B. M. Trost, I. Fleming, Eds.; Pergamon: Oxford, 1991; Vol. 6, p 873-908.

48. K. Mikami; K. Azuma; T. Nakai *Tetrahedron* **1984**, *40*, 2303-2308.

49. K. Mikami; T. Maeda; T. Nakai *Tetrahedron Lett.* **1986**, *27*, 4189-4190.

50. M. Koreeda; J. I. Luengo *J. Am. Chem. Soc.* **1985**, *107*, 5572-5573.

51. J. I. Luengo; M. Koreeda *J. Org. Chem.* **1989**, *54*, 5415-5417.

52. L. Castedo; J. R. Granja; A. Mouriño *Tetrahedron Lett.* **1985**, *26*, 4959-4960.

53. M. Koreeda; D. J. Ricca *J. Org. Chem.* **1986**, *51*, 4090-4092.

54. W. C. Still; C. Sreekumar *J. Am. Chem. Soc.* **1980**, *102*, 1201-1202.

55. Y.-D. Wu; K. N. Houk; J. A. Marshall *J. Org. Chem.* **1990**, *55*, 1421-1423.

56. E. J. Verner; T. Cohen *J. Am. Chem. Soc.* **1992**, *114*, 375-377.

57. R. Hoffmann; R. Brückner *Angew. Chem. Int. Ed. Engl.* **1992**, *31*, 647-649.

58. K. Tomooka; T. Igarashi; M. Watanabe; T. Nakai *Tetrahedron Lett.* **1992**, *33*, 5795--5798.

59. H. Eichenauer; E. Friedrich; W. Lutz; D. Enders *Angew. Chem. Int. Ed. Engl.* **1978**, *17*, 206-208.

60. D. Seebach; J. Golinski *Helv. Chim. Acta* **1981**, *64*, 1413-1423.

61. K. Mikami; Y. Kimura; N. Kishi; T. Nakai *J. Org. Chem.* **1983**, *48*, 279-281.

62. E. Nakai; T. Nakai *Tetrahedron Lett.* **1988**, *29*, 5409-5412.

63. M. Uchikawa; T. Katsuki; M. Yamaguchi *Tetrahedron Lett.* **1986**, *27*, 4581-4582.

64. H. B. Mekelburger; C. S. Wilcox In *Comprehensive Organic Synthesis. Selectivity, Strategy, and Efficiency in Modern Organic Chemistry*; B. M. Trost, I. Fleming, Eds.; Pergamon: Oxford, 1991; Vol. 2, p 99-131.

65. S. D. Burke; W. F. Fobare; G. J. Pacofsky *J. Org. Chem.* **1983**, *48*, 5221-5228.

66. J. Kallmerten; T. J. Gould *Tetrahedron Lett.* **1983**, *24*, 5177-5180.

67. J. K. Whitesell; A. M. Helbing *J. Org. Chem.* **1980**, *45*, 4135-4139.

68. R. E. Ireland; S. Thaisrivongs; N. Vanier; C. S. Wilcox *J. Org. Chem.* **1980**, *45*, 48-61.

69. K. Mikami; O. Takahashi; T. Kasuga; T. Nakai *Chem. Lett.* **1985**, 1729-1732.

70. J. A. Marshall; X. Wang *J. Org. Chem.* **1992**, *57*, 2747-2750.

71. T. Nakai; K. Mikami; S. Taya; Y. Kimura; T. Mimura *Tetrahedron Lett.* **1981**, *22*, 69-72.

72. J. A. Marshall; W. Y. Gung *Tetrahedron Lett.* **1988**, *29*, 1657-1660.

73. P. C.-M. Chan; J. M. Chong *J. Org. Chem.* **1988**, *53*, 5584-5586.

74. J. A. Marshall; G. S. Welmaker; B. W. Gung *J. Am. Chem. Soc.* **1991**, *113*, 647-656.

75. J. M. Chong; E. K. Mar *Tetrahedron Lett.* **1991**, *32*, 5683-5686.

76. D. S. Matteson; P. B. Tripathy; A. Sarkur; K. N. Sadhu *J. Am. Chem. Soc.* **1989**, *111*, 4399-4402.

77. A. F. Thomas; R. Dubini *Helv. Chim. Acta* **1974**, *57*, 2084-2087.

78. D. J.-S. Tsai; M. M. Midland *J. Org. Chem.* **1984**, *49*, 1842-1843.

79. N. Sayo; K. Azuma; K. Mikami; T. Nakai *Tetrahedron Lett.* **1984**, *25*, 565-568.

80. J. A. Marshall; T. M. Jenson *J. Org. Chem.* **1984**, *49*, 1707-1712.

81. J. A. Marshall; X. Wang *J. Org. Chem.* **1990**, *55*, 2995-2996.

82. J. A. Marshall; X. Wang *J. Org. Chem.* **1991**, *56*, 4913-4918.

83. J. A. Marshall; E. D. Robinson; A. Zapata *J. Org. Chem.* **1989**, *54*, 5854-5855.

84. K. Mikami; O. Takahashi; T. Tabei; T. Nakai *Tetrahedron Lett.* **1986**, *27*, 4511-4514.

85. M. M. Midland; Y. C. Kwon *Tetrahedron Lett.* **1985**, *26*, 5013-5016.

86. O. Takahashi; K. Mikami; T. Nakai *Chem. Lett.* **1987**, 69-72.

87. J. A. Marshall; J. Lebreton *Tetrahedron Lett.* **1987**, *28*, 3323-3326.

88. J. A. Marshall; J. Lebreton *J. Am. Chem. Soc.* **1988**, *110*, 2925-2931.

89. J. A. Marshall; J. Lebreton *J. Org. Chem.* **1988**, *53*, 4108-4112.

90. M. Uchikawa; T. Hanemoto; T. Katsuki; M. Yamaguchi *Tetrahedron Lett.* **1986**, *27*, 4577-4580.

91. P. Helquist In *Comprehensive Organic Synthesis. Selectivity, Strategy, and Efficiency in Modern Organic Chemistry*; B. M. Trost, I. Fleming, Eds.; Pergamon: Oxford, 1991; Vol. 4, p 951-997.

92. H. M. L. Davies In *Comprehensive Organic Synthesis. Selectivity, Strategy, and Efficiency in Modern Organic Chemistry*; B. M. Trost, I. Fleming, Eds.; Pergamon: Oxford, 1991; Vol. 4, p 1031-1067.

93. H. E. Simmons, Jr.; T. L. Cairns; S. A. Vladuchick; C. M. Hoiness *Org. React.* **1973**, *20*, 1-131.

94. M. P. Doyle; R. L. Dorow; W. E. Buhro; J. H. Griffin; W. H. Tamblyn; M. L. Trudell *Organometallics* **1984**, *3*, 44-52.

95. M. P. Doyle; R. L. Dorow; W. H. Tamblyn; W. E. Buhro *Tetrahedron Lett.* **1982**, *23*, 2261-2264.

96. M. P. Doyle; V. Bagheri; T. J. Wandless; N. K. Harn; D. A. Brinker; C. T. Eagle; K. L. Loh *J. Am. Chem. Soc.* **1990**, *112*, 1906-1912.

97. H. M. L. Davies; T. J. Clark; L. A. Church *Tetrahedron Lett.* **1989**, *30*, 5057-5060.

98. H. Abdallah; R. Grée; R. Carrié *Tetrahedron Lett.* **1982**, *23*, 503-506.

99. I. Arai; A. Mori; H. Yamamoto *J. Am. Chem. Soc.* **1985**, *107*, 8254-8256.

100. A. Mori; I. Arai; H. Yamamoto; H. Nakai; Y. Arai *Tetrahedron* **1986**, *42*, 6447-6458.

101. E. A. Mash; K. A. Nelson *J. Am. Chem. Soc.* **1985**, *107*, 8256-8258.

102. E. A. Mash; K. A. Nelson *Tetrahedron* **1987**, *43*, 679-692.

103. E. A. Mash; D. S. Torok *J. Org. Chem.* **1989**, *54*, 250-253.

104. E. A. Mash; S. B. Hemperly; K. A. Nelson; P. C. Heidt; S. V. Deusen *J. Org. Chem.* **1990**, *55*, 2045-2055.

105. P. W. Ambler; S. G. Davies *Tetrahedron Lett.* **1988**, *29*, 6979-6982.

106. A. B. Charette; B. Côté; J.-F. Marcoux *J. Am. Chem. Soc.* **1991**, *113*, 8166-8167.

107. A. B. Charette; J.-F. Marcoux *Tetrahedron Lett.* **1993**, *34*, 7157-7160.

108. A. B. Charette; B. Côté *J. Org. Chem.* **1993**, *58*, 933-936.

109. T. Sugimura; T. Futugawa; A. Tai *Tetrahedron Lett.* **1988**, *29*, 5775-5778.

110. T. Sugimura; T. Futugawa; M. Yoshikawa; A. Tai *Tetrahedron Lett.* **1989**, *30*, 3807-3810.

111. T. Sugimura; M. Yoshikawa; T. Futugawa; A. Tai *Tetrahedron* **1990**, *46*, 5955-5966.

112. M. P. Doyle *Rec. Trav. Chim. Pays-Bas* **1991**, *110*, 305-316.

113. M. P. Doyle; W. R. Winchester; J. A. A. Hoorn; V. Lynch; S. H. Simonsen; R. Ghosh *J. Am. Chem. Soc.* **1993**, *115*, 9968-9978.

114. P. A. Krieger; J. A. Landgrebe *J. Org. Chem.* **1978**, *43*, 4447-4452.

115. M. P. Doyle; M. N. Protopopova; B. D. Brandes; H. M. L. Davies; N. J. S. Hruby; J. K. Whitesell *Synlett* **1993**, 151-153.

116. H. M. L. Davies; N. J. S. Huby; W. R. Cantrell, Jr.; J. L. Olive *J. Am. Chem. Soc.* **1993**, *115*, 9468-9479.

117. H. M. L. Davies *Tetrahedron Lett.* **1991**, *32*, 6509-6512.

118. H. M. L. Davies; N. J. S. Huby *Tetrahedron Lett.* **1992**, *33*, 6935-6938.

119. H. M. L. Davies *Tetrahedron* **1993**, *49*, 5203-5223.

120. A. B. Charette; H. Juteau *J. Am. Chem. Soc.* **1994**, *116*, 2651-2652.

121. A. B. Charette *Chem. Eng. News* **1995**, *Feb 6*, p. 2.

122. A. B. Charette; S. Prescott; C. Brochu *J. Org. Chem.* **1995**, *60*, 1081-1083.

123. H. Takayashi; M. Yoshioka; M. Ohno; S. Kobayashi *Tetrahedron Lett.* **1992**, *33*, 2575-2578.

124. Y. Ukaji; M. Nishimura; T. Fujisawa *Chem. Lett.* **1992**, 61-64.

125. H. Nozaki; S. Moriuti; H. Takaya; R. Noyori *Tetrahedron Lett.* **1966**, 5239-5242.

126. T. Aratani; Y. Yoneyoshi; T. Nagase *Tetrahedron Lett.* **1977**, 2599-2602.

127. T. Aratani; Y. Yoneyoshi; T. Nagase *Tetrahedron Lett.* **1982**, *23*, 685-688.

128. T. Aratani *Pure Appl. Chem.* **1985**, *57*, 1839-1844.

129. H. Fritschi; U. Leutenegger; A. Pfaltz *Helv. Chim. Acta* **1988**, *71*, 1553-1565.

130. U. Leutenegger; G. Umbricht; C. Fahrni; P. von Matt; A. Pfaltz *Tetrahedron* **1992**, *48*, 2143-2156.

131. R. E. Lowenthal; A. Abiko; S. Masamune *Tetrahedron Lett.* **1990**, *31*, 6005-6008.

132. R. E. Lowenthal; S. Masamune *Tetrahedron Lett.* **1991**, *32*, 7373-7376.

133. D. A. Evans; K. A. Woerpel; M. M. Hinman; M. M. Faul *J. Am. Chem. Soc.* **1991**, *113*, 726-728.

134. K. Ito; T. Katsuki *Tetrahedron Lett.* **1993**, *34*, 2662-2664.

135. H. Nishiyama; Y. Itoh; H. Matsumoto; S.-B. Park; K. Itoh *J. Am. Chem. Soc.* **1994**, *116*, 2223-2224.

136. W. Klyne; V. Prelog *Experientia* **1960**, *16*, 521-523.

137. M. P. Doyle; B. D. Brandes; A. P. Kazala; R. J. Pieters; M. B. Jartsfer; L. M. Watkins; C. T. Eagle *Tetrahedron Lett.* **1990**, *31*, 6613-6616.

138. M. N. Protopopova; M. P. Doyle; P. Müller; D. Ene *J. Am. Chem. Soc.* **1992**, *114*, 2755-2757.

139. H. M. L. Davies; D. K. Hutcheson *Tetrahedron Lett.* **1993**, *34*, 7243-7246.

140. M. P. Doyle; R. J. Pieters; S. F. Martin; R. E. Austin; C. J. Oalmann; P. Müller *J. Am. Chem. Soc.* **1991**, *113*, 1423-1424.

141. S. F. Martin; C. J. Oalmann; S. Liras *Tetrahedron Lett.* **1992**, *33*, 6727-6730.

142. S. F. Martin; M. R. Spaller; S. Liras; B. Hartmann *J. Am. Chem. Soc.* **1994**, *116*, 4493-4494.

143. G. Quinkert; U. Schwartz; H. Stark; W.-D. Weber; H. Baier; F. Adam; G. Dürner *Angew. Chem. Int. Ed. Engl.* **1980**, *19*, 1029-1030.

144. G. Quinkert; U. Schwartz; H. Stark; W.-D. Weber; F. Adam; A. Baier; G. Frank; G. Dürner *Liebigs Ann. Chem.* **1982**, 1999-2040.

145. D. Romo; J. L. Romine; W. Midura; A. I. Meyers *Tetrahedron* **1990**, *46*, 4951-4994.

146. D. Romo; A. I. Meyers *J. Org. Chem.* **1992**, *57*, 6265-6270.

147. E. J. Corey; M. J. Chaykovsky *J. Am. Chem. Soc.* **1965**, *87*, 1353-1364.

148. C. R. Johnson *Aldrichimica Acta* **1985**, *18*, 3-11.

149. R. M. Williams; G. J. Fegley *J. Am. Chem. Soc.* **1991**, *113*, 8796-8806.

150. Wasserman *Diels-Alder Reactions*; Elsevier: New York, 1965.

151. D. F. Taber *Intramolecular Diels-Alder and Alder Ene Reactions*; Springer: Berlin, 1984.

152. W. Oppolzer In *Comprehensive Organic Synthesis. Selectivity, Strategy, and Efficiency in Modern Organic Chemistry*; B. M. Trost, I. Fleming, Eds.; Pergamon: Oxford, 1991; Vol. 5, p 315-399.

153. W. R. Roush In *Comprehensive Organic Synthesis. Selectivity, Strategy, and Efficiency in Modern Organic Chemistry*; B. M. Trost, I. Fleming, Eds.; Pergamon: Oxford, 1991; Vol. 5, p 513-550.

154. J. March In *Advanced Organic Chemistry, 4th ed.*; Wiley-Interscience: New York, 1992, p 839-852.

155. W. Oppolzer *Angew. Chem. Int. Ed. Engl.* **1984**, *23*, 876-889.

156. G. Helmchen; R. Karge; J. Weetman In *Modern Synthetic Methods*; R. Scheffold, Ed.; Springer: Berlin, 1986; Vol. 4, p 262-306.

157. M. Taschner In *Organic Synthesis. Theory and Applications*; T. Hudlicky, Ed.; JAI: Greenwich, CT, 1989; Vol. 1, p 1-101.

158. D. L. Boger; S. M. Weinreb *Hetero Diels-Alder Methodology in Organic Synthesis*; Academic: New York, 1987.

159. S. M. Weinreb In *Comprehensive Organic Synthesis. Selectivity, Strategy, and Efficiency in Modern Organic Chemistry*; B. M. Trost, I. Fleming, Eds.; Pergamon: Oxford, 1991; Vol. 5, p 401-449.

160. D. L. Boger In *Comprehensive Organic Synthesis. Selectivity, Strategy, and Efficiency in Modern Organic Chemistry*; B. M. Trost, I. Fleming, Eds.; Pergamon: Oxford, 1991; Vol. 5, p 451-512.

161. H. Waldmann *Synthesis* **1994**, 535-551.

162. H. B. Kagan; O. Riant *Chem. Rev.* **1992**, *92*, 1007-1019.

163. K. B. Wiberg; K. E. Laidig *J. Am. Chem. Soc.* **1987**, *109*, 5935-5943.

164. J. Sauer; J. Kredel *Tetrahedron Lett.* **1966**, 6359-6364.

165. E. J. Corey; H. E. Ensley *J. Am. Chem. Soc.* **1975**, *97*, 6908-6909.

166. W. Oppolzer; M. Kurth; D. Reichlin; F. Moffatt *Tetrahedron Lett.* **1981**, *22*, 2545-2548.

167. R. K. Boeckman, Jr.; P. C. Naegley; S. D. Arthur *J. Org. Chem.* **1980**, *45*, 752-754.

168. C. LeDrain; A. E. Greene *J. Am. Chem. Soc.* **1982**, *104*, 5473-5483.

169. E. J. Corey; X.-M. Cheng; K. A. Crimprich *Tetrahedron Lett.* **1991**, *32*, 6839-6842.

170. W. Oppolzer; C. Chapuis; G. M. Dao; D. Reichlin; T. Godel *Tetrahedron Lett.* **1982**, *23*, 4781-4784.

171. W. Oppolzer; C. Chapuis *Tetrahedron Lett.* **1984**, *25*, 5383-5386.

172. W. Oppolzer In *Selectivity, a Goal for Synthetic Efficiency: Proceedings of the 14th Workshop Conference Hoechst*; W. Bartmann, B. M. Trost, Eds.; Verlag Chemie: Weinheim, 1984; Vol. 14, p 137-167.

173. W. Oppolzer; C. Chapuis; G. Bernardinelli *Tetrahedron Lett.* **1984**, *25*, 5885-5888.

174. T. Poll; A. Sobczak; H. Hartmann; G. Helmchen *Tetrahedron Lett.* **1985**, *26*, 3095-3098.

175. T. Poll; A. F. A. Hady; R. Karge; G. Linz; J. Weetman; G. Helmchen *Tetrahedron Lett.* **1989**, *30*, 5595-5598.

176. J.-L. Gras; H. Pellissier *Tetrahedron Lett.* **1991**, *32*, 7043-7046.

177. W. Oppolzer; C. Chapuis; G. Bernardinelli *Helv. Chim. Acta* **1984**, *67*, 1397-1401.

178. D. A. Evans; K. T. Chapman; J. Bisaha *J. Am. Chem. Soc.* **1988**, *110*, 1238-1256.

179. R. J. Loncharich; T. R. Schwarz; K. N. Houk *J. Am. Chem. Soc.* **1987**, *109*, 14-23.

180. S. Castellino; W. J. Dwight *J. Am. Chem. Soc.* **1993**, *115*, 2986-2987.

181. T. Poll; J. O. Metter; G. Helmchen *Angew. Chem. Int. Ed. Engl.* **1985**, *24*, 112-114.

182. E. Ciganek *Org. React.* **1984**, *32*, 1-374.
183. P. Deslongchamps *Pure Appl. Chem.* **1992**, *64*, 31-47.
184. W. Oppolzer; D. Dupuis *Tetrahedron Lett.* **1985**, *26*, 5437-5440.
185. W. R. Roush; H. R. Gillis; A. I. Ko *J. Am. Chem. Soc.* **1982**, *104*, 2269-2283.
186. H. M. Walborsky; L. Barash; T. C. Davis *J. Org. Chem.* **1961**, *26*, 4778-4779.
187. H. M. Walborsky; L. Barash; T. C. Davis *Tetrahedron* **1963**, *19*, 2333-2351.
188. K. Furuta; K. Iwanaga; H. Yamamoto *Tetrahedron Lett.* **1986**, *27*, 4507-4510.
189. Y. N. Ito; A. K. Beck; A. Boháč; C. Ganter; R. E. Gawley; F. N. M. Kühnle; J. A. Piquer; J. Tuleja; Y. M. Wang; D. Seebach *Helv. Chim. Acta* **1994**, *77*, 2071-2110.
190. G. Helmchen; R. Schmierer *Angew. Chem. Int. Ed. Engl.* **1981**, *20*, 205-207.
191. K. Tomioka; N. Hamada; T. Suenaga; K. Koga *J. Chem. Soc., Perkin Trans. 1* **1990**, 426-428.
192. K. Maruoka; S. Saito; H. Yamamoto *J. Am. Chem. Soc.* **1992**, *114*, 1089-1090.
193. S. W. Baldwin; P. Greenspan; C. Alaimo; A. T. McPhail *Tetrahedron Lett.* **1991**, *32*, 5877-5880.
194. K. Maruoka; M. Akakura; S. Saito; T. Ooi; H. Yamamoto *J. Am. Chem. Soc.* **1994**, *116*, 6153-6158.
195. G. Linz; J. Weetman; A. F. Abdel-Hady; G. Helmchen *Tetrahedron Lett.* **1989**, *30*, 5599-5602.
196. M. Vandewalle; J. V. d. Eyken; W. Oppolzer; C. Vullioud *Tetrahedron* **1986**, *42*, 4035-4043.
197. E. J. Corey; W. Su *Tetrahedron Lett.* **1988**, *29*, 3423-3426.
198. J. Aubé; S. Ghosh; M. Tanol *J. Am. Chem. Soc.* **1994**, *116*, 9009-9018.
199. D. Enders; O. Meyer; G. Raabe; J. Runsink *Synthesis* **1994**, 66-72.
200. D. Enders; O. Meyer; G. Raabe *Synthesis* **1992**, 1242-1244.
201. B. M. Trost; D. O'Krongly; J. L. Belletire *J. Am. Chem. Soc.* **1980**, *102*, 7595-7596.
202. C. Siegel; E. R. Thornton *Tetrahedron Lett.* **1988**, *29*, 5225-5228.
203. J. A. Dale; H. S. Mosher *J. Am. Chem. Soc.* **1973**, *95*, 512-519.
204. D. A. Evans; T. Lectka; S. J. Miller *Tetrahedron Lett.* **1993**, *34*, 7027-7030.
205. D. A. Evans; S. J. Miller; T. Lectka *J. Am. Chem. Soc.* **1993**, *115*, 6460-6461.
206. E. J. Corey; K. Ishihara *Tetrahedron Lett.* **1992**, *33*, 6807-6810.
207. S. Kobayashi; I. Hachiya; H. Ishitani; M. Araki *Tetrahedron Lett.* **1993**, *34*, 4535-4538.
208. S. Kobayashi; H. Ishitani *J. Am. Chem. Soc.* **1994**, *116*, 4083-4084.
209. K. Tomioka *Synthesis* **1990**, 541-549.
210. K. Narasaka *Synthesis* **1991**, 1-11.
211. S. Hashimoto; N. Komeshima; K. Koga *J. Chem. Soc., Chem. Commun.* **1979**, 437-438.
212. J. M. Hawkins; S. Loren *J. Am. Chem. Soc.* **1991**, *113*, 7794-7795.
213. J. M. Hawkins; S. Loren; M. Nambu *J. Am. Chem. Soc.* **1994**, *116*, 1657-1660.
214. M. Takasu; H. Yamamoto *Synlett* **1990**, 194-196.
215. E. J. Corey; T.-P. Loh *J. Am. Chem. Soc.* **1991**, *113*, 8966-8967.
216. E. J. Corey; T.-P. Loh; T. D. Roper; M. D. Azimioara; M. C. Noe *J. Am. Chem. Soc.* **1992**, *114*, 8290-8292.
217. J.-P. G. Seerden; H. W. Scheeren *Tetrahedron Lett.* **1993**, *34*, 2669-2672.
218. K. Furuta; S. Shimizu; Y. Miwa; H. Yamamoto *J. Org. Chem.* **1989**, *54*, 1481-1483.
219. K. Furuta; A. Kanamatsu; H. Yamamoto; S. Takaoka *Tetrahedron Lett.* **1989**, *30*, 7231-7232.
220. K. Ishihara; Q. Gao; H. Yamamoto *J. Am. Chem. Soc.* **1993**, *115*, 10412-10413.
221. E. J. Corey; R. Imwinkelried; S. Pikul; Y. B. Xiang *J. Am. Chem. Soc.* **1989**, *111*, 5493-5495.
222. E. J. Corey; N. Imai; S. Pikul *Tetrahedron Lett.* **1991**, *32*, 7517-7520.
223. E. J. Corey; S. Sarshar; J. Bordner *J. Am. Chem. Soc.* **1992**, *114*, 7938-7939.
224. P. N. Devine; T. Oh *J. Org. Chem.* **1992**, *57*, 396-399.
225. K. Maruoka; M. Sakurai; J. Fujiwara; H. Yamamoto *Tetrahedron Lett.* **1986**, *27*, 4895-4898.
226. K. Narasaka; M. Inoue; N. Okada *Chem. Lett.* **1986**, 1109-1112.

227. D. Seebach; A. K. Beck; R. Imwinkelried; S. Roggo; A. Wonnacott *Helv. Chim. Acta* **1987**, *70*, 954-974.

228. D. Seebach; R. Dahinden; R. E. Marti; A. K. Beck; D. A. Plattner; F. N. M. Kühnle *J. Org. Chem.* **1995**, *60*, 1788-1799.

229. T. A. Engler; M. A. Letavic; K. O. Lynch, Jr.; F. Takusagawa *J. Org. Chem.* **1994**, *59*, 1179-1183.

230. K. Mikami; Y. Motoyama; M. Terada *J. Am. Chem. Soc.* **1994**, *116*, 2812-2820.

231. T. R. Kelly; A. Whiting; N. S. Chandrakumar *J. Am. Chem. Soc.* **1986**, *108*, 3510-3512.

232. C. Chapuis; J. Jurczak *Helv. Chim. Acta* **1987**, *70*, 436-440.

233. K. Ishihara; H. Yamamoto *J. Am. Chem. Soc.* **1994**, *116*, 1561-1562.

234. E. J. Corey; N. Imai; H.-Y. Zhang *J. Am. Chem. Soc.* **1991**, *113*, 728-729.

235. J. I. Seeman *Chem. Rev.* **1983**, *83*, 83-134.

236. K. Narasaka; N. Iwasawa; M. Inoue; T. Yamada; M. Nakashima; J. Sugimori *J. Am. Chem. Soc.* **1989**, *111*, 5340-5345.

237. C. Haase; C. R. Sarko; M. DiMare *J. Org. Chem.* **1995**, *60*, 1777-1787.

238. K. V. Gothelf; R. G. Hazell; K. A. Jørgensen *J. Am. Chem. Soc.* **1995**, *117*, 4435-4436.

239. E. J. Corey; X.-M. Cheng *The Logic of Chemical Synthesis*; Wiley: New York, 1989.

240. E. J. Corey; S. M. Albonico; U. Koelliker; T. K. Schaaf; R. K. Varma *J. Am. Chem. Soc.* **1971**, *93*, 1491-1493.

Chapter 7

Reductions and Hydroborations

Addition of a hydrogen atom to a trigonal (sp^2) carbon atom is the theme of this chapter. Within this scope are additions of dihydrogen, hydrides, and hydroborations. For the latter, the product boranes may be converted to a number of useful functional groups, but this chemistry is not covered here (reviews: [1,2]). The chapter is divided into three parts: reduction of carbon-heteroatom double bonds, reduction of carbon-carbon double bonds, and hydroborations. Several books have been written on these topics, so the present coverage is necessarily selective. As in previous chapters, the coverage is intended to highlight particularly important and selective reagents, with an emphasis on understanding the factors that influence stereoselectivity.

7.1 Reduction of carbon-heteroatom double bonds

Larock's *Comprehensive Organic Transformations* lists over fifty reagents in the section "Asymmetric Reduction of Aldehydes and Ketones" [3]. The nonenzymatic entries can be divided into several categories based on reagent type and/or mechanism: lithium aluminum hydrides modified with chiral ligands, borohydrides modified (sometimes catalytically) with chiral ligands, chiral boranes that reduce carbonyls in a self-immolative chirality transfer process, and chiral transition metal complexes that catalyze hydrogenation or hydrosilylation. Each of these involves interligand asymmetric induction (Section 1.3). Only selected examples from each category will be presented in detail; the objective being to analyze the factors that determine enantioselectivity for each reaction. A judgement of which reducing agent is most selective and/or convenient depends on the substrate, but an attempt at a comprehensive evaluation of 10 ketone classes with available reducing agents was made a few years ago [4]. Highly selective reductions of the carbon-nitrogen bond have been achieved only recently. Examples of azomethine reduction are included in the following sections as appropriate.

7.1.1 Modified lithium aluminum hydride

The first efforts to modify lithium aluminum hydride (LAH) with a chiral ligand were by Bothner-by in 1951 [5]. Although the result was later challenged, the seed was planted and many attempts have been made to produce an efficient chiral reducing agent using this strategy (reviews: [6-9]). Of these, we will examine the binaphthol-LAH-ROH reagent (BINAL-H) introduced by Noyori in 1979 [10-13]. Binaphthol is a popular ligand (like its cousin BINAP) for asymmetric synthesis because it has a pleasing C_2 symmetry which, when bound in a bidentate fashion to a metal, often affords excellent differentiation between heterotopic faces of a bound ligand.

Noyori's reagent is prepared by addition of binaphthol to a <u>solution</u> of LAH in THF, then adding another equivalent of an alcohol such as ethanol or methanol to form the reagent (Scheme 7.1).[1] The ethanol or methanol is a pragmatic necessity, as the reagent having two (presumed) active hydrides shows poor enantioselectivity in asymmetric reductions [12]. The exact nature of the reagent is not known, since aluminum hydrides may disproportionate and/or aggregate, processes that may continue as the product of the reduction (an alkoxide) accumulates during the reaction. Perhaps because of such processes, optimal selectivity is achieved with a 3-fold excess of the hydride reagent. Under these conditions, the reagent is highly enantioselective in reductions of certain classes of ketones. Some examples are listed in Table 7.1.

Entries 1 and 2 show the reagent's ability to reduce deuterated aldehydes to afford primary alcohols that are chiral by virtue of isotopic substitution. Note that the rest of the examples showing high selectivity (entry 13 being the exception) have one ketone substituent that is unsaturated and one that is not. Note also that in the saturated substituent, branching at the α-position lowers enantioselectivity significantly (compare entries 4/5 and 7/8). The fact that 3-octyn-2-one (entry 9) is reduced with 92% enantioselectivity (84% ee) whereas 2-octanone (entry 13) is reduced with only 62% enantioselectivity (24% ee) is curious. The authors submit that this comparison (among others) suggests that the facial discrimination involves more than just steric effects.

The rationale offered by the Noyori group to explain the chirality sense of the observed products is predicated on the 6-membered ring transition structures shown in Figure 7.1a and b. These structures differ only in the orientation of the two ketone substituents. Another pair, in which the 6-membered ring is flipped, is destabilized by a steric repulsion between the alkoxy methyl (or ethyl) and the C-3 position of the binaphthol. Figure 7.1c shows this interaction, which is (note the bold lines) a "gauche pentane-like" conformation (*cf.* Figure 5.5 and accompanying

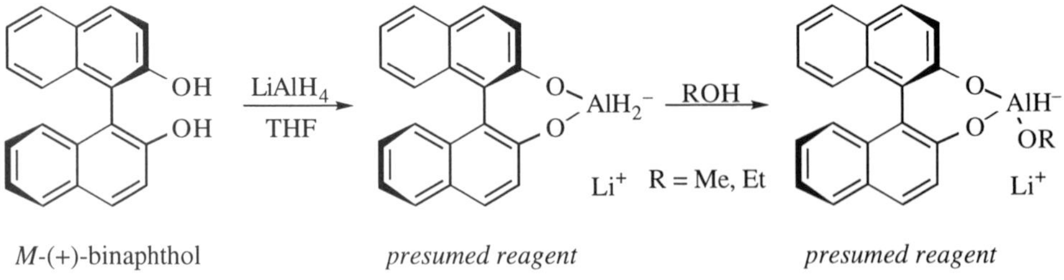

M-(+)-binaphthol   *presumed reagent*   *presumed reagent*

Scheme 7.1. Preparation of Noyori's BINAL-H reagents [12]. The aluminum complexes shown are postulated structures that may represent "time averages" of several equilibrating species.

---

1   For those wishing to use this reagent, care should be taken to follow the Noyori experimental procedure exactly. Precipitous drops in enantioselectivity result from very minor changes in protocol. Note that a "milk-white" or "cloudy" reagent solution is OK; but when there is "extensive precipitation", the reagent should not be expected to perform as advertised [12] (see also ref. [14,15]).

Table 7.1. Asymmetric reductions using BINAL-H. The reactions were conducted by initial reaction at $-100°$ for 3 hours, followed by several hours at $-78°$ C. All examples favor *ul* relative topicity (see Figure 7.1a). Thus, the *M* reagent adds to the *Si* face to give the *R* product, and vice versa for the *P* reagent.

Entry	R_1	R_2	RO	% Yield	% es	Ref.
1	Ph	D	EtO	59	93	[13]
2	*(structure)*	CDO	EtO	91	92	[13]
3	Ph	Me	EtO	61	97	[12]
4	Ph	*n*-Pr	EtO	92	100	[12]
5	Ph	*i*-Pr	EtO	68	85	[12]
6	α-tetralone		EtO	91	87	[12]
7	HC≡C	n-C_5H_{11}	MeO	87	92	[13]
8	HC≡C	*i*-Pr	MeO	84	79	[13]
9	n-C_4H_9C≡C	Me	MeO	79	92	[13]
10	n-C_4H_9C≡C	n-C_5H_{11}	MeO	85	95	[13]
11	E-n-C_4H_9CH=CH	Me	EtO	47	89	[13]
12	E-n-C_4H_9CH=CH	n-C_5H_{11}	EtO	91	95	[13]
13	n-C_6H_{13}	Me	EtO	67	62	[12]
14	*(structure)*		EtO	87	100	[13]

discussion).[2] With respect to the 6-membered ring in Figures 7.1a and b, note that one of the ketone substituents is equatorial and one is axial. The interaction of the latter with the axial naphthyloxy ligand is postulated to account for the enantio-selectivity. This interaction is suggested to be one of two types: steric interactions, which are repulsive, and electronic, which may (in principle) be either repulsive or attractive, but which are repulsive for all the examples in Table 7.1 (other substrates are suggested to have dominant *attractive* electronic interactions [12]). For the examples in Table 7.1, it is observed that the *P* BINAL-H reagent selectively adds hydride to the *Re* face (*ul* relative topicity - see Glossary, section 1.6), as shown in the transition structure of Figure 7.1a. In this structure, the saturated ligand (R_{sat}) bears a 1,3-diaxial relationship to the naphthloxy ligand on

---

2   Note that in Figure 7.1a-c, the alkoxy "R group" is always axial. The authors point out that structures in which the R group occupies an equatorial position would be further destabilized by repulsive interactions between R and the BINOL moieties [12]. It may be useful to note that the configuration of the alkoxy oxygen in the favored chairs (Figure 7.1a,b) having the *P* BINOL ligand is *R*. The configuration of the oxygen in the disfavored chair (Figure 7.1c, *P* BINOL ligand) is *S*.

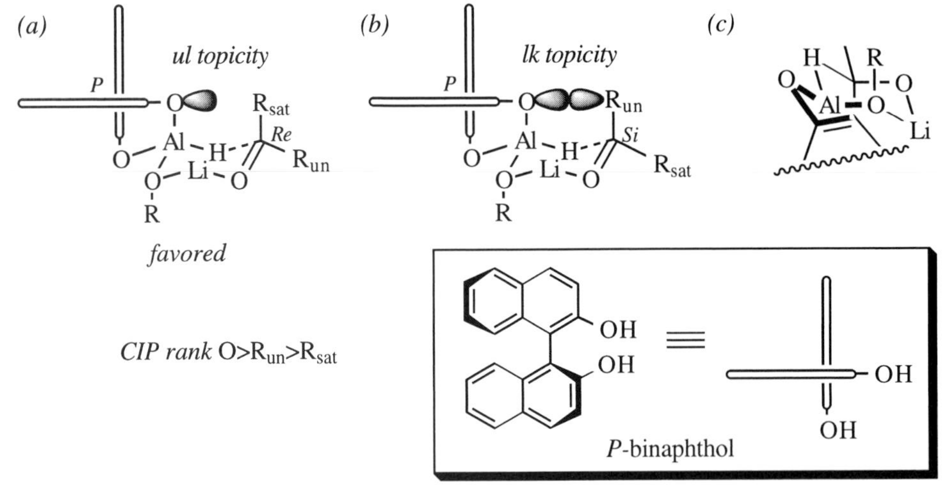

Figure 7.1. Postulated transition structures for the asymmetric reduction of unsaturated ketones by BINAL-H [12]. Structures *(a)* and *(b)* differ in the orientation of R_{sat} and R_{un}, the saturated and unsaturated ketone ligands, respectively. *(a) Ul* topicity: *P* reagent attacking *Re* face of ketone. *(b) Lk* topicity: *P* reagent attacking *Si* face of ketone. *(c)* Alternate chair that is destabilized by the "gauche pentane" conformation accented by the bold lines (*cf.* Figure 5.5). Transition structures containing this conformation were considered by Noyori to be unimportant [12].

aluminum. Since an alkene or alkyne ligand is generally considered to be "smaller" than an *n*-alkyl ligand,[3] this situation is somewhat counterintuitive. Noyori suggests that the reason for this topicity has to do with an unfavorable repulsive electronic interaction between the unpaired electrons on the axial naphthyl oxygen and the π orbital of the unsaturated ligand (R_{un}) in the transition structure having *lk* topicity, shown in Figure 7.1b, and that this interaction causes greater repulsion than that of an axial saturated ligand.

These reductions distinguish the enantiotopic *faces* of aldehydes and ketones. An interesting extension of the use of this reagent is the enantioselective reduction of *meso* anhydrides [17]. In this application, the reagent distinguishes enantiotopic *ligands,* not faces. A generic example of the process, along with yields and enantioselectivities of several substrates, is shown in Scheme 7.2.

7.1.2 Modified borane

The first attempt to use a chiral ligand to modify borane was Kagan's attempt at enantioselective reduction of acetophenone using amphetamine-borane and desoxy-ephedrine-borane in 1969 [18]. However, both reagents afforded 1-phenyl ethanol in <5% ee. The most successful borane-derived reagents are oxazaborolidines, introduced by Hirao in 1981, developed by Itsuno, and further developed by Corey several years later (reviews: [19,20]). Figure 7.2 illustrates several of the Hirao-Itsuno and Corey oxazaborolidines that have been evaluated to date. All of these examples are derived from amino acids by reduction or Grignard addition. Hirao

---

3   The *A* values of $-CH_2CH_3$, $-CH=CH_2$, and $-C\equiv CH$ are ~1.75, 1.7, and 0.41 kcal/mole, respectively [16].

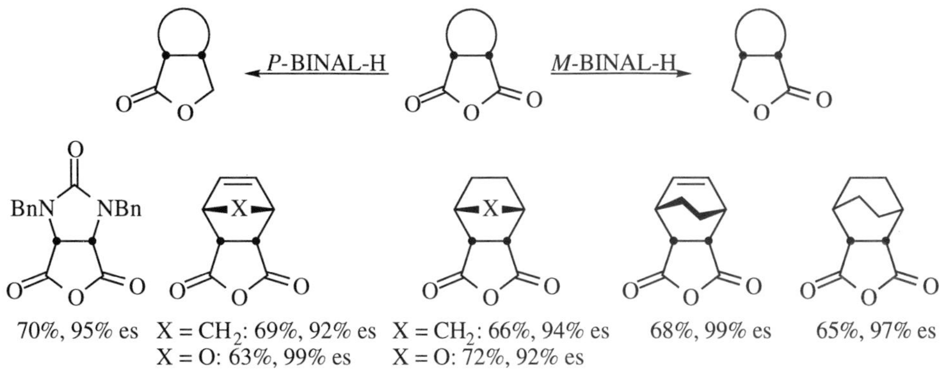

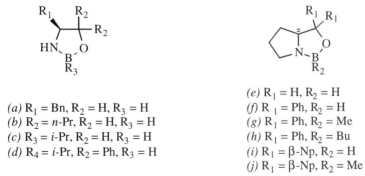

70%, 95% es   X = CH$_2$: 69%, 92% es   X = CH$_2$: 66%, 94% es   68%, 99% es   65%, 97% es
              X = O: 63%, 99% es   X = O: 72%, 92% es

Scheme 7.2. Yields and enantioselectivities of reduction of *meso* anhydrides using BINAL-H [17].

originally investigated the reagent derived from condensation of amino alcohols such as valinol and prolinol with borane (Figure 7.2*a-c, e*), and found enantio-selectivities in the neighborhood of 70-80% es [21]. Optimization studies revealed that enantioselectivities of ~85% es (for the reduction of acetophenone) could be obtained in THF solvent at 30° C, using amino alcohol:borane ratios of 1:2 [22]. In 1983, Itsuno found that the reagent was much more selective (96-100% es with acetophenone) if tertiary alcohols derived from addition of phenyl magnesium bromide to valine (Figure 7.2*d*) were used [23,24]. Additionally, Itsuno found that a polymer-bound amino alcohol could be used for the process with equal facility [25]. Reduction of aliphatic ketones was not quite as selective, affording reduction products in 77-87% es [24,26]. Itsuno [24] and Corey [27] demonstrated the synthesis of oxiranes by asymmetric reduction of α-halo ketones followed by cyclization. In 1985, Itsuno showed that oxime ethers (but not oximes) could be enantioselectively reduced to primary amines (84-99% es) using the valinol-derived reagent (Figure 7.2*d*, [24]), and in 1987 showed that this process could be catalytic in oxazaborolidine: acetophenone *O*-methyloxime was reduced to α-methylbenzyl amine in 90% yield and 100%es [28]. In 1987, Corey characterized the Itsuno reagent (Figure 7.2*d*) and showed that the diphenyl derivative (Figure 7.2*f*) of the Hirao reagent (Figure 7.2*e*) afforded excellent enantioselectivities (≥95%) when

(*a*) R$_1$ = Bn, R$_2$ = H, R$_3$ = H
(*b*) R$_2$ = *n*-Pr, R$_2$ = H, R$_3$ = H
(*c*) R$_3$ = *i*-Pr, R$_2$ = H, R$_3$ = H
(*d*) R$_4$ = *i*-Pr, R$_2$ = Ph, R$_3$ = H

(*e*) R$_1$ = H, R$_2$ = H
(*f*) R$_1$ = Ph, R$_2$ = H
(*g*) R$_1$ = Ph, R$_2$ = Me
(*h*) R$_1$ = Ph, R$_2$ = Bu
(*i*) R$_1$ = β-Np, R$_2$ = H
(*j*) R$_1$ = β-Np, R$_2$ = Me

Figure 7.2. Oxazaborolidines for the asymmetric reduction of ketones: (*a-c*) [21,22]. (*d*) [23-26,28]. (*e*) [21]. (*f*) [29]. (*g*) [30]. (*h*) [27]. (*i-j*) [31].

used in catalytic amounts [29]. In the same year, the Corey group reported that *B*-methyl oxazaborolidines (Figure 7.2*g,h*) were easier to prepare, could be stored at room temperature, could be weighed and transferred in air, and afforded enantioselectivities comparable to the *B*-H reagents [27,30]. In 1989, Corey found that the β-naphthyl derivative of prolinol afforded a reagent with still higher enantioselectivities than either the *B*-H (Figure 7.2*i*) or *B*-Me (Figure 7.2*j*) derivative (*e.g.*, 99% es with acetophenone [31]).

X-ray crystal structures of the oxazaborolidine reagent [32] and a derivative [33] have been published, and a mechanistic hypothesis has been formulated [29]. Heterocycles such as the boranes shown in Figure 7.2*a-f,i* do not, by themselves, reduce carbonyls; but in the presence of excess borane, they catalyze the reduction by the mechanism shown in Scheme 7.3 for the *B*-methyl catalyst of Figure 7.2*g*. In the first step, borane coordinates to the nitrogen of the oxazaborolidine on the less hindered convex face of the fused bicyclic system; the ketone then coordinates to the convex face. From the perspective of the ketone, the Lewis acid (boron atom) is trans to the larger ketone substituent [34]. Hydride transfer occurs *via* a 6-membered chair transition structure [35,36] having *lk* relative topicity (the *R* enantiomer of the catalyst favoring the *Re* face of the carbonyl carbon). Elimination of the alkoxy borane completes the catalytic cycle [37]. Table 7.2 lists representative examples of oxazaborolidine reductions. Entry 4 is one example (among several) of the asymmetric reduction [38] of trichloromethyl ketones [39]. Corey's group has shown that the resulting carbinols are versatile intermediates for the preparation of α-amino acids [38], α-hydroxy and α-aryloxy acids [40], and terminal epoxides [41].

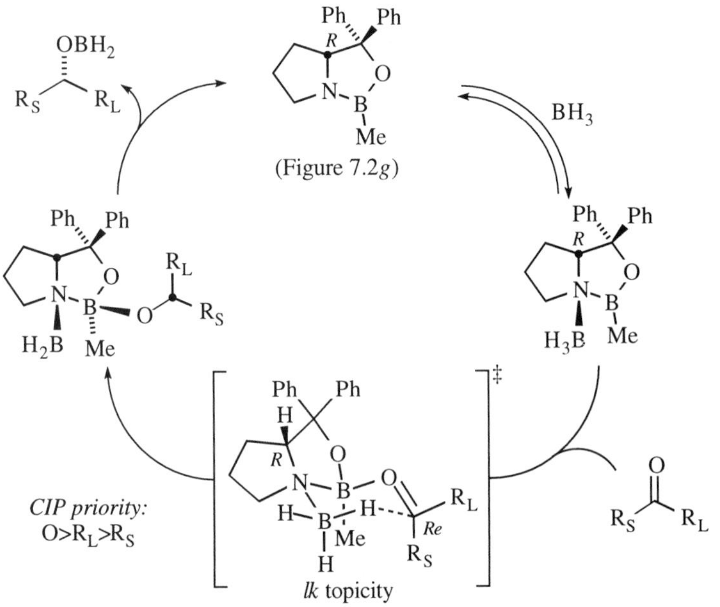

Scheme 7.3. Catalytic cycle for the asymmetric reduction of a ketone with an oxazaborolidine catalyst [29,35,36].

Table 7.2. Examples of ketone reductions mediated by oxazaborolidines. The "Cat." column refers to the catalysts in Figure 7.2. The reductant is borane, unless otherwise noted. For entries 3 and 9, the product may spontaneously cyclize. The products of entries 16 and 17 are primary amines.

Entry	Ketone	Cat.	T, °C	% Yield	% es	Ref.
1	PhCOMe	*c, f-j*	2-30	95-100	≥97	[23,24,29-31,42]
2	PhCOEt	*c, f, g, j*	−10-30	100	≥94	[23,24,29-31]
3	PhCOCH$_2$Cl	*c, f, g*	25-32	100	98	[24,27,29,30]
4	*t*-BuCOCCl$_3$	*h*	−20	96	99	[38]
5	α-tetralone	*f, g, i, j*	−10-31	100	≥93	[29-31]
6	*t*-BuCOMe	*f, g, j*	−10-25	100	≥96	[29-31]
7	*cyclo*-C$_6$H$_{11}$COMe	*g, j*	−10-0	100	91-92	[30,31]
8	*i*-PrCOMe	*c*	30	100	80	[24]
9	*n*-C$_6$H$_{13}$COMe	*c*	30	100	79	[24,26]
10	PhCO(CH$_2$)$_n$CO$_2$Me n = 2, 3	*g, j*	0	100	97-98	[30,31]
11	![ketone with Br]	*g, i*	23-36	100	95	[30,31]
12	![cyclohexyl methyl ketone]	*h*	−78[a]	>95	90	[43]
13	*E*-PhCH=CHCOMe	*h*	−78[a]	>95	96	[43]
15	![cyclohexenyl methyl ketone]	*h*	−78[a]	>95	96	[43]
16	![NOMe acetophenone oxime ether]	*c*	30	90-100	99-100	[24,28]
17	![NOMe tetralone oxime ether]	*c*	30	100	84	[24]

[a] Catechol borane as reductant

Operationally, these reagents are effective at or near room temperature, which may be of significant benefit to large-scale employment. The preparation of the *S*-diphenylprolinol ligand (*cf.* Figure 7.2*f-h*) is most easily accomplished by addition of a phenyl Grignard reagent to *L*-proline *N*-carboxyanhydride (73% yield, 99%

ee, [33]). The *R* enantiomer of the amino alcohol may be made by a similar addition to *D*-proline, but may also be made by enantioselective lithiation of BOC-pyrrolidine and addition to benzophenone (70% yield, 99% ee, as illustrated in Scheme 3.33 [44]).[4] The catalysts may be made by condensation of the amino alcohol with methyl boronic acid [30,31,33] or trimethylboroxine [33] with simultaneous water removal. *B*-Methyl or *B*-butyl catalysts can be made by condensation of the amino alcohol with bis(trifluoroethyl) alkylboronate and removal of trifluoroethanol *in vacuo* [42].

The catalysts may be used in 5-10 mol% concentrations, with either borane or catechol borane [43] as the stoichiometric reductant. Use of the more reactive catechol borane allows one to conduct the reduction at lower temperature, a feature that may be advantageous in cases where selectivity at room temperature is not high enough. The reductions are sensitive to moisture: Jones, et al. [45] found that the presence of 1 mg of water per gram of ketone lowered the enantioselectivity from 97% to 75% es.

7.1.3 Chiral organoboranes[5]

The reaction of a chiral alkene with borane in the proper stoichiometry may afford alkyl boranes R^*BH_2 or dialkyl boranes R^*_2BH, where R^* is a chiral ligand. Attempts to achieve highly selective reductions of ketones using such reagents have met with little success, however.[6] Trialkyl boranes R_3B were first reported to reduce aldehydes and ketones (under forcing conditions) in 1966 by Mikhailov [50]. Mechanistic studies (summarized in ref. [46]) showed that there are two limiting mechanisms for the reduction of a carbonyl compound by a trialkylborane, as shown in Scheme 7.4: a pericyclic process reminiscent of the Meerwein-Pondorf-Verley reaction (Scheme 7.4a), and a two step process that involves dehydro-

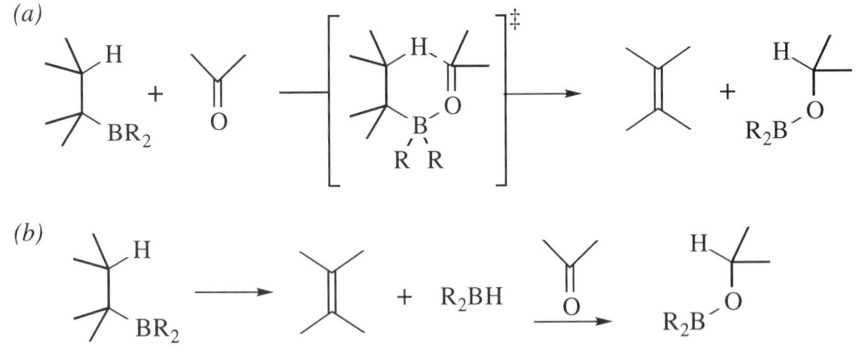

Scheme 7.4. Limiting mechanisms for carbonyl reduction of carbonyls by a trialkylborane: *(a)* pericyclic mechanism. *(b)* Two step mechanism involving dehydroboration of a trialkylborane followed by carbonyl reduction by the resultant dialkylborane.

---

[4] This procedure will be published in *Organic Syntheses*, probably in volume 74, 1996 (P. Beak, personal communication).

[5] Reviews: ref. [46-48].

[6] For a notable exception, see ref. [49].

boration to a dialkylborane plus olefin, followed by carbonyl reduction by the dialkylborane (Scheme 7.4b). With unhindered carbonyl compounds such as aldehydes, the reaction is bimolecular and appears to proceed by the pericyclic pathway [51]. With ketones, the rate is independent of ketone concentration, indicating a switch to the dehydroboration-reduction pathway.

In 1979-80, Midland showed that the trialkyl borane formed by hydroboration of α-pinene by 9-borabicyclononane (9-BBN), known as *B*-isopinocampheyl-9-borabicyclo[3.3.1]nonane or Alpine-borane™, efficiently reduces aldehydes [52,53] and propargyl ketones [54,55] with a high degree of enantioselectivity, as shown in Scheme 7.5. The mechanism was shown to be a self-immolative chirality transfer process (Scheme 7.4a), proceeding through the 6-membered ring boat transition structure shown in Scheme 7.5b and c [46]. This reduction is probably the method of choice for the production of enantiomerically enriched primary alcohols that are chiral by virtue of isotopic substitution, provided enantiopure α-pinene is used [56]. Most ketones other than propargyl ketones are not readily reduced by trialkylboranes, making this process highly chemoselective for aldehydes and propargyl ketones in the presence of other ketones, esters, acid chlorides, alkyl halides, alkenes and alkynes. Under forcing conditions, Alpine-borane dehydroborates (the reverse of Scheme 7.5a) with a half-life 500 min in refluxing THF [46], and non-selective reduction by 9-BBN becomes competitive (*cf.* Scheme 7.4b).

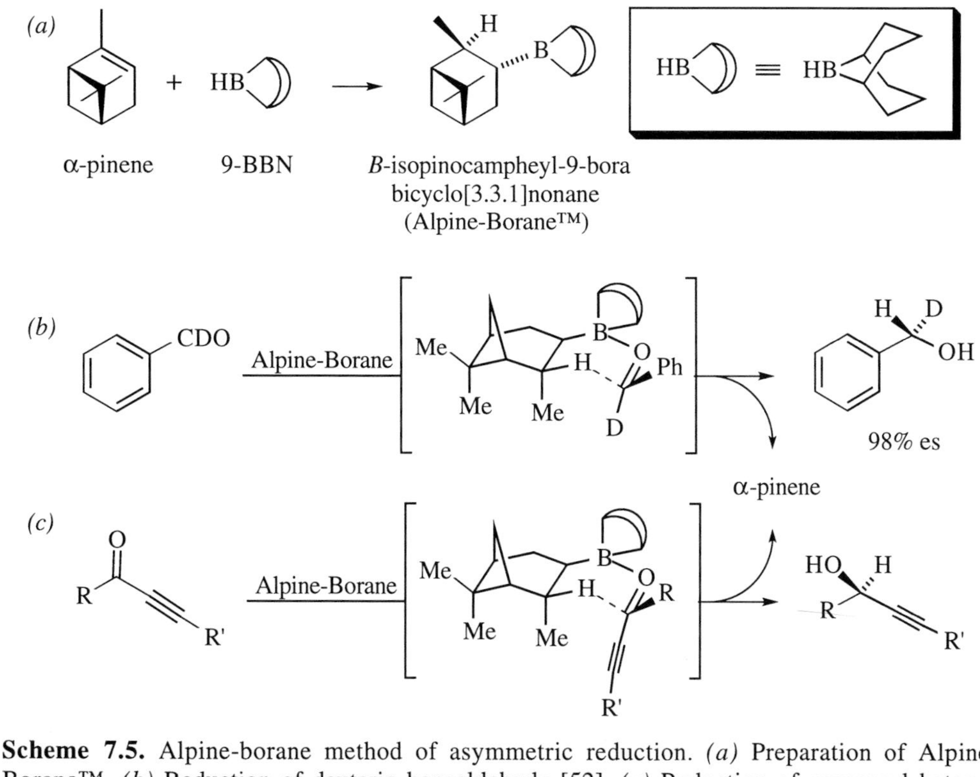

Scheme 7.5. Alpine-borane method of asymmetric reduction. *(a)* Preparation of Alpine-Borane™. *(b)* Reduction of deuterio benzaldehyde [52]. *(c)* Reduction of propargyl ketones [54,55].

To circumvent the problem of competitive dehydroboration with ketones, the Alpine-borane reductions can be conducted in neat (excess) reagent [57] or at high pressure (6000 atm, [58]). Experiments done in neat reagent take several days to go to completion, and afford enantioselectivities of 70-98% [57]. At pressures of 6000 atmospheres, the reactions are faster and dehydroboration is completely suppressed. Ketones are reduced with slightly higher enantioselectivities (75-100% es) under these conditions [58].

A better solution to asymmetric ketone reduction is to make a more reactive borane. Brown showed that hindered dialkylchloroboranes (R_2BCl) are less prone to dehydroboration than hindered trialkylboranes (R_3B) such as Alpine-borane and are excellent reagents for the reduction of aldehydes and ketones. Inductive electron withdrawal by the chlorine also increases the Lewis acidity of the boron. *B*-Chloro-diisopinocampheylborane (Ipc_2Cl, DIP-chloride™) is such a reagent, and is an excellent reagent for the asymmetric reduction of aryl-alkyl ketones [59,60]. Scheme 7.6 shows the preparation of Ipc_2Cl and the postulated transition structure to rationalize the chirality sense of the products. Table 7.3 lists several examples. Note that dialkyl ketones and alkynyl-alkyl ketones are reduced with low selectivity unless one of the substituents is tertiary. For a summary of other pinene-based self-immolative reducing agents, see Brown's reviews [47,48].

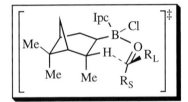

Scheme 7.6. Preparation of Ipc_2Cl. *Inset:* Proposed transition structure for asymmetric reductions using Ipc_2Cl [59].

Table 7.3. Asymmetric reduction of ketones, $R_1C(=O)R_2$, with Ipc_2Cl.

Entry	R_1	R_2	% yield	% es	Ref.
1	Me	Et	–	52	[59]
2	Me	*i*-Pr	–	66	[59]
3	Me	*t*-Bu	50	93	[59]
4	2,2-dimethylcyclopentanone		71	98	[59]
5	2,2-dimethylcyclohexanone		60	91	[59]
6	1-indanone		62	97	[59]
7	α-tetralone		70	86	[59]
8	HC≡C	Me	83	58	[60]
9	PhC≡C	Me	92	60	[60]
10	PhC≡C	*i*-Pr	85	92	[60]
11	PhC≡C	*t*-Bu	80	>99	[60]
12	*cyclo*-$C_5H_{11}C≡C$	*i*-Pr	81	69	[60]
13	*cyclo*-$C_5H_{11}C≡C$	*t*-Bu	76	98	[60]
14	*n*-$C_8H_{17}C≡C$	*i*-Pr	86	63	[60]
15	*n*-$C_8H_{17}C≡C$	*t*-Bu	77	99	[60]

7.1.4 Chiral transition metal catalysts

Enantioselective reduction of simple ketone carbonyls is possible, but catalysts that deliver consistently high selectivities in such reactions have been elusive [61-64]. More success has been recorded in the asymmetric reduction of functionalized ketones and imines (reviews: [65,66]). Two types of stoichiometric reductants are used: dihydrogen and dihydrosilanes (reviews: ref. [67,68]), but as the mechanism of hydrosilylation is "highly controversial" [68], we will discuss only the former.

Ketone reductions. For the asymmetric hydrogenation of functionalized ketones, a team led by Noyori in Nagoya and Akutagawa in Tokyo introduced ruthenium(II) BINAP catalysts that produce excellent enantioselectivities for a number of functionalized ketones [69-75] (review: [76]; for a recent reference to a more reactive catalyst see ref. [77]). The topicity of the reduction is illustrated in Scheme 7.7, and is suggestive of a mechanism in which the heteroatom X and the carbonyl oxygen chelate the metal (*vide infra*). The catalyst is thought to be a monomeric BINAP ruthenium(II) dichloride, which was originally prepared by a tedious process using Schlenk techniques [69]; however, improved procedures have since been developed [71-73].

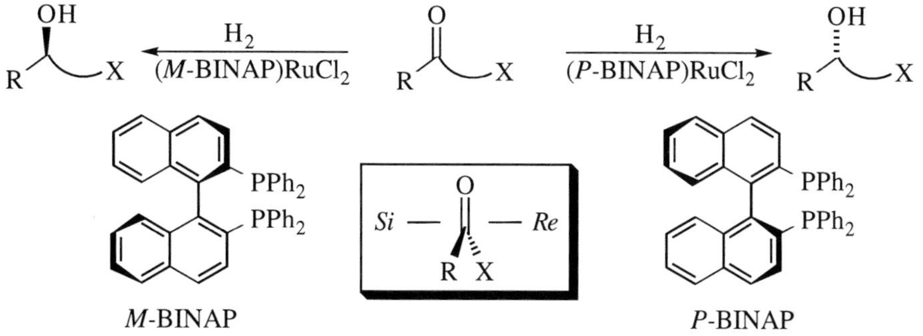

Scheme 7.7. *Ul* relative topicity (*e.g.,* *P*-BINAP/*Re* face) is uniformly observed for ruthenium BINAP catalyzed asymmetric reduction of functionalized ketones [70].

Selected examples that afford high selectivity are listed in Table 7.4. Several β-keto esters are reduced with excellent enantioselectivity (entries 1, 3-6); however, α-keto esters are reduced with somewhat diminished enantioselectivity [70]. β-Keto amides and thioesters (entry 2) are good substrates, as are α- and β-hydroxy ketones (entry 7) and α-amino ketones (entries 7 and 8). Particularly striking is the chemoselectivity observed when the reductions are conducted at low pressures: isolated double bonds are left intact (entry 6). Bifunctional ketones may be problematic, since chelation might occur by more than one functional group. For example, a ketone such as $HOCH_2COCH_2CO_2Et$ could chelate *via* either the hydroxyl or the ester oxygen, and this competition would lower the enantioselectivity. However, protection of hydroxyl as its triisopropylsilyl (TIPS) ether prevents chelation by the hydroxyl oxygen and excellent enantioselectivity results (entry 5). Competition is less of a worry if chelation forms a 6-membered ring, and protection as a benzyl ether suffices (entry 5).[7]

---

[7] For an example of the effect of chelation on regioselectivity, see Scheme 4.3 and the accompanying discussion.

Table 7.4. Selected examples of asymmetric ketone reductions using $Ru^{II}Cl_2$·BINAP. Reactions were run at room temperature and 50-100 atm unless otherwise noted. Yields were determined spectroscopically unless noted.

Entry	Ketone	% Yield	% es	Ref.
1	Me—C(=O)—CH2—CO_2R R = Me, Et, *i*-Pr, *t*-Bu	≥97	≥99	[69-71]
2	Me—C(=O)—CH2—C(=O)—X			
	X = NMe_2	100	98	[70]
	X = SEt	42^a	96	
3	R—C(=O)—CH2—CO_2Me R = Me, *n*-Bu, *i*-Pr	99	≥99	[69]
4	Ph—C(=O)—CH2—CO_2Et	99	92	[69]
5	RO—$(CH_2)_n$—C(=O)—CH2—CO_2Et			
	R = TIPS, n = 1	100	97	[70]
	R = BnO, n = 2	94	99	
6	(structure)—CO_2Me^b			
	R = H	73	99	[72]
	R = Me	96	99	
7	Me—C(=O)—CH2—X			
	X = NMe_2	72	98	[70]
	X = OH	100	96	
	X = CH_2OH	100	99	
8	R—C(=O)—CH2—NMe_2			
	R = *i*-Pr	83	97	[70]
	R = Ph	85	97	

^{a} Isolated yield.
^{b} 50 psi, 80°

These catalytic reductions are relatively slow, requiring high pressures or high temperatures, and chiral β-ketoesters racemize faster than they can be reduced. As it happens, reduction of one enantiomer is considerably faster than reduction of the other. This is a case of double asymmetric induction (see Section 1.5) applied to a

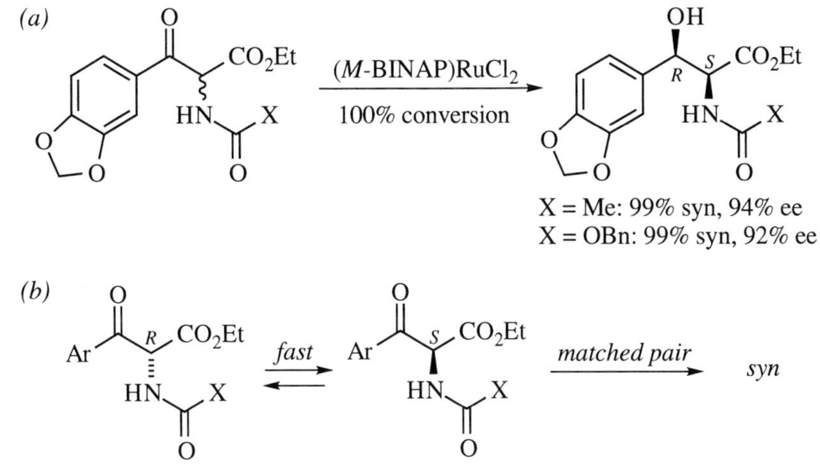

(a)

$$\xrightarrow[\text{100\% conversion}]{(M\text{-BINAP})RuCl_2}$$

X = Me: 99% syn, 94% ee
X = OBn: 99% syn, 92% ee

(b)

fast

matched pair

syn

Scheme 7.8. Asymmetric reduction of chiral β-keto esters may be used in an asymmetric transformation of the first kind (dynamic kinetic resolution) [78].

kinetic resolution. Since the enantiomers racemize rapidly, the ruthenium BINAP catalyst can be used to effect an asymmetric transformation of the first kind (see Glossary, section 1.6), as shown in Scheme 7.8a [78]. In this example, the *racemic* β-keto ester is completely converted to the syn amino alcohol with a diastereoselectivity (syn:anti) of 99:1. The syn product is obtained in 94% ee, indicating that of the four possible stereoisomeric products (syn and anti enantiomers), the major product is 96% of the mixture. The simple explanation for this beautiful result is shown in Scheme 7.8b: racemization under the reaction conditions is fast compared to reduction of either enantiomer, but reduction of the *S*-enantiomer by the *M*-BINAP catalyst (matched pair, addition to the ketone *Si* face) is itself faster than reduction of the *R*-enantiomer (mismatched pair, not shown), so the net result is a draining of the fast racemization equilibrium (Curtin-Hammett principle [79,80]).

The proposed catalytic cycle for these reductions is shown in Scheme 7.9 [76]. In this scheme, it is assumed that the polymeric catalyst precursor [(BINAP)RuCl2]n is dissociated to monomer by the methanolic solvent. Reduction and loss of hydrogen afford the putative catalyst, (BINAP)RuHCl(MeOH)2. Displacement of the two methanols by the bidentate substrate then sets the stage for the hydrogen transfer step (*vide infra*). Exchange of the alkoxide product with the methanolic solvent and reaction with hydrogen to regenerate the catalyst completes the catalytic cycle. Deuterium labeling experiments showed that the mechanism involves C=O reduction, and not the alternative C=C reduction of an enol tautomer [78].

The X-ray crystal structures of two ruthenium BINAP complexes have been determined [74,81]. Figure 7.3a illustrates the structural features that are thought to influence stereoselectivity (see also Figure 6.3 and the accompanying discussion). In both crystal structures, the 7-membered chelate ring formed by the *P*-enantiomer of the BINAP ligand and the metal adopt similar conformations and have the pseudoequatorial phenyl groups occupying the lower left and upper right quadrants, as viewed from the P–Ru–P plane with the BINAP to the rear. The pseudoaxial

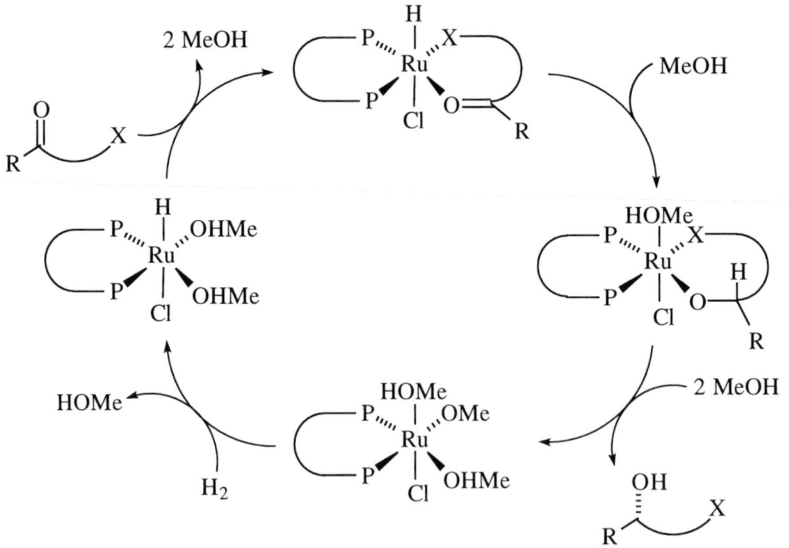

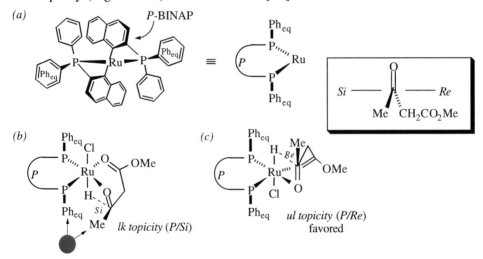

Scheme 7.9. Catalytic cycle proposed for the asymmetric reduction of functionalized ketones by ruthenium BINAP catalyst (after ref. [76]).

phenyls are slanted to the rear and would not significantly interact with a ligand bound trans to either phosphorous. For reduction of methyl acetoacetate, *ul* relative topicity is observed (*P*-BINAP catalyst preferentially attacking the *Re* face of the ketone). Assuming that the catalyst is a mononuclear monohydride complex having the hydrogen and chlorine trans, with the substrate chelated to the ruthenium (each carbonyl oxygen being trans to a phosphorous), the chirality sense may be rationalized by the two transition structures illustrated in Figure 7.3b and c. A four-membered transition structure having *lk* topicity (Figure 7.3b) would force the C-4 methyl into the crowded lower left quadrant, while the transition structure with *ul* topicity (Figure 7.3c) is less hindered [76].

Figure 7.3. (*a*) Conformation of *P*-BINAP in two crystal structures [74,81]. (*b*) *lk* Topicity transition structure for asymmetric reduction of methyl acetoacetate. (*c*) *ul* Topicity transition structure. (After ref. [76]). *Inset:* definition of *Re* and *Si* faces of ketone.

Figure 7.4 illustrates three natural products that have been synthesized using ruthenium(II)·BINAP-mediated ketone reduction as the key step. For pyrenophorin [82] and gloesporone [83], the secondary carbinol is retained, but for indolizidine 223AB, the Mitsunobu reaction is employed to convert the alcohol to an amine [73].

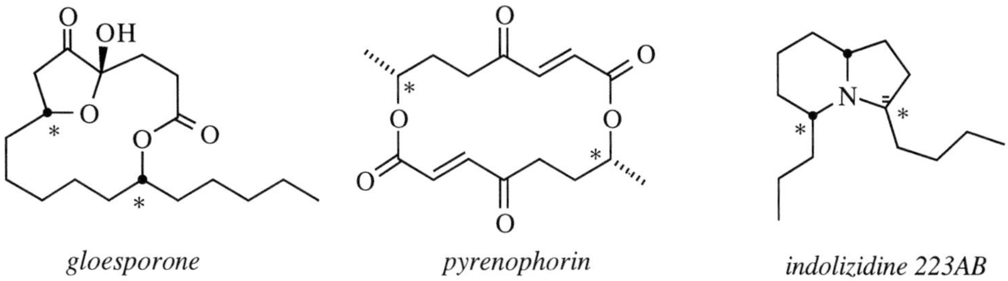

gloesporone   *pyrenophorin*   *indolizidine 223AB*

Figure 7.4. Ruthenium(II)·BINAP catalysts have been used as a key step in the asymmetric synthesis of gloesporone [83], pyrenophorin [82], and indolizidine 223AB [73]. Stereocenters formed by asymmetric reduction are indicated (*).

Imine reductions. The asymmetric reduction of carbon–nitrogen double bonds is not possible using ruthenium(II) catalysts, but Buchwald has recently shown that a titanocene catalyst (Scheme 7.10) exhibits good to excellent enantioselectivity in the reduction of imines [84-86] (review: ref. [87]). The reaction can be highly stereoselective for both acyclic and cyclic imines, but since acyclic imines are usually a mixture of *E* and *Z* isomers, and since the imine isomerization is catalyzed by the titanocene, the reaction is not always preparatively useful for acyclic substrates. Examples are listed in Table 7.5. For the cyclic imines (entries 1-8), the enantioselectivities indicated were obtained under hydrogen pressure of 80 psi, at temperatures of 45-65°; higher pressures (500-2000 psi) gave slightly higher enantioselectivities, although reduction of side-chain double bonds occurs.

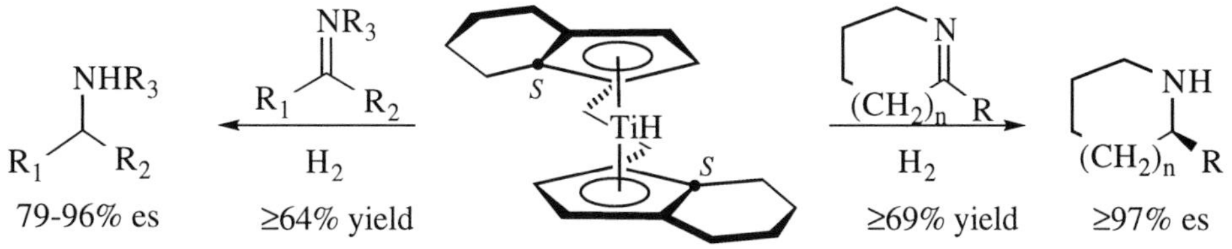

Scheme 7.10. Titanocene catalyzed asymmetric reduction of imines [85]. In the accompanying discussion, the catalyst shown is designated the *S,S* enantiomer, in accord with the *CIP* rules for describing metal arenes [88]. This is a different designation than that used by Buchwald, however.[8]

---

[8] In the original paper describing the preparation of the titanocene catalyst precursor [89], Brintzinger specified the chirality sense of the ansa metallocene by referring to the absolute configuration at C-1 of the indene (the carbon bearing the ethylene bridge), and Buchwald has adopted this usage. However, the *CIP* system states that the chirality sense of the <u>complex</u> should be assigned with reference to <u>the arene ring atom</u> (or in general, the π-complexed atom of any ligand) <u>having the highest *CIP* rank</u> [88]. In this case, the highest-ranking atom is the C7a (indicated by •), which has the opposite *CIP* designation of C-1. For rules on assigning a *CIP* descriptor to π-complexes, see ref. [90-93]. For another method (Ω+/Ω–), see ref. [94].

Table 7.5. Examples of asymmetric imine reduction using Buchwald's chiral titanocene catalyst. Reactions were run at 45° and 80 psi, with 5 mol% *S,S* catalyst, unless noted otherwise.

Entry	Imine	Amine	% Yield	% es	Ref.
1	Ph–N=(CH$_2$)$_n$ n = 1-3	Ph–NH–(CH$_2$)$_n$	71-83	≥99	[84,85, 95]
2	MeO, MeO isoquinoline, Me	MeO, MeO, NH, Me	79	97	[84,85, 95]
3	Bn pyrrole dihydropyrrole	Bn pyrrole pyrrolidine	72	>99	[85,95]
4	prenyl dihydropyrrole	prenyl pyrrolidine	79	>99	[85,95]
5	R–(CH$_2$)$_4$–N R = H R = TMS	R–(CH$_2$)$_4$–NH	72 73^a	>99 >99	[85,95]
6	(CH$_2$)$_5$–N	(CH$_2$)$_5$–NH	69^b	>99	[85]
7	R–(CH$_2$)–N R = TBSOCH$_2$ R = ethylenedioxy-CH	R–NH	82 82	>99 >99	[85,95]
8	HO–(CH$_2$)$_7$–N	HO–(CH$_2$)$_7$–NH	84	>99	[85,95]
9	cyclohexyl–C(=NR)Me^c R = Me (92% E) R = Bn (92% E)	cyclohexyl–CH(NHR)Me	85 85	96 71	[84,85]
10	NBn^d (75% E)	NHBn	64	81	[85]
11	RC(=NBn)Me^d R = i-Pr (93% E) R = Ph (94% E) R = 2-naphthyl (98% E)	R–CH(NHBn)Me	66 81 82	88 88 85	[84,85]

^{a} Yield includes 5-8% of product having a saturated side chain.
^{b} Yield includes 13-18% of product having a saturated side chain and 14-18% E-olefin.
^{c} 500 psi H$_2$.
^{d} 2000 psi H$_2$.

Examination of the enantioselectivities in Table 7.5 indicates a striking difference in selectivity achieved in the reduction of cyclic (entries 1-8) *vs.* acyclic imines (entries 9-11). The former is very nearly 100% stereoselective. The simple reason for this is that the acyclic imines are mixtures of *E* and *Z* stereoisomers, which reduce to enantiomeric amines (*vide infra*). The mechanism proposed for this reduction is shown in Scheme 7.11 [86]. The putative titanium(III) hydride catalyst is formed in situ by sequential treatment of the titanocene BINOL complex with butyllithium and phenylsilane. The latter reagent serves to stabilize the catalyst. Kinetic studies show that the reduction of cyclic imines is first order in hydrogen and first order in titanium but zero order in imine. This (and other evidence) is consistent with a fast 1,2-insertion followed by a slow hydrogenolysis (σ-bond metathesis), as indicated [86]. Although β-hydride elimination of the titanium amide intermediate is possible, it appears to be slow relative to the hydrogenolysis.

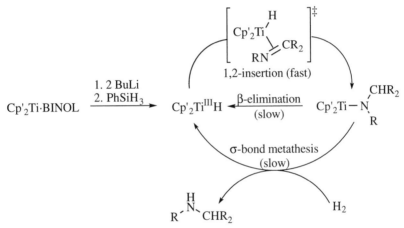

Scheme 7.11. Proposed catalytic cycle for the titanocene catalyzed reduction of imines [86].

Note the η^2 bonding of the imine to the titanium at the transition state for insertion. The geometry of this complex is critical to the stereoselectivity of the reaction, since it is in this step that the stereocenter in the product is created. A dichlorotitanocene is tetrahedral around titanium, as indicated by the X-ray crystal structure shown in Figure 7.5 [89]. Note the C_2 symmetry of the complex, the orientation of the two cyclohexane moieties in the upper left and lower right quadrants (Figure 7.5b), and the placement of the two chlorines with respect to the cyclohexanes, especially as viewed from the "top" (Figure 7.5d). Based on valence orbital calculations of olefin complexes that are isolobal to the titanium-imine transition structure shown in Scheme 7.11, Buchwald has proposed that the configuration of the titanium in the transition state is similar to that of the dichlorotitanium complex, with one chloride being replaced by a hydride, the other by the η^2 imine ligand, as shown in Figure 7.6 [86]. In a tetrahedral geometry, the imine can only coordinate to the titanium as shown, with the *N*-methylene oriented to the lower left quadrant of the drawing in Figure 7.6a. That this can be the only possible orientation is shown clearly by the top view in Figure 7.6b. This view also

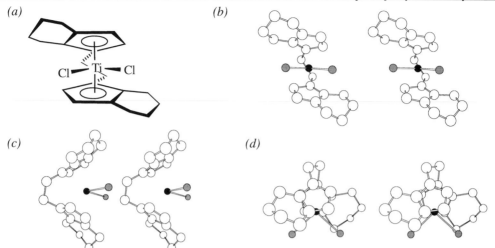

Figure 7.5. Crystal structure of *S,S* ethylene-bis(tetrahydroindenyl)titanium chloride [89]: *(a)* Perspective drawing of complex. *(b)* Front stereoview. *(c)* Side stereoview. *(d)* Top stereoview.

illustrates positioning of the phenyl in the vacant upper right quadrant, with a minor interaction taking place between C-3 of the heterocycle and the cyclohexyl in the lower right quadrant. This aspect of binding in the transition structure is important in the analysis of the reduction of acyclic imines, as shown in Figure 7.6c and d.

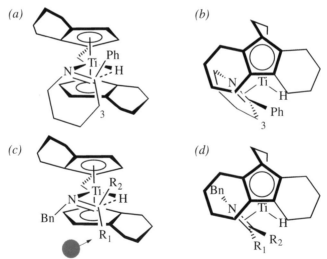

Figure 7.6. Transition structures for titanocene hydride imine reduction [86]: *(a)* Front view of heterocycle reduction. *(b)* Top view of heterocycle reduction. *(c)* Front view of acylic imine reduction. *(d)* Top view of imine reduction.

For acyclic imines, note that interchange of R_1 and R_2 in the transition structure is equivalent to an *E/Z* isomerization of the educt. Reduction of cyclohexyl methyl *N*-benzyl imine, using a stoichiometric amount of catalyst affords a 92:8 *R/S* enantiomer ratio that is identical to the 92:8 *E/Z* ratio of the educt (*i.e.*, the reaction is stereospecific). This is interpreted as follows: the major imine isomer is *E* (R_2 = cyclohexyl, R_1 = methyl). Addition to the *Si* face gives the *R* enantiomer of the amine. With the *Z* imine, R_2 is methyl and R_1 is cyclohexyl. Addition to the *Re* face gives the *S* amine. Entry 9 of Table 7.5 (R = Bn) is the same reaction, but using only 5 mol% of catalyst. Under catalytic conditions, the reaction is no longer stereospecific for two reasons: first, the *E* and *Z* isomers interconvert slowly under the reaction conditions (probably catalyzed by the titanium), and second, the *Z* isomer is reduced faster than the *E* isomer. If the hydrogen pressure is reduced from 2000 psi to 500 psi, the enantioselectivity drops to 71%, consistent with a slower rate of reduction relative to *E/Z* isomerization [86].[9]

The titanocene catalyzed asymmetric imine reduction may be used in kinetic resolutions of racemic pyrrolines [96]. The most efficient kinetic resolution was observed for 5-substituted pyrrolines, and the mechanistic postulate outlined above readily accomodates the experimental results, as shown by the matched pair transition structure in Scheme 7.12 [96].[10] Pyrrolines substituted at the 3- and 4-positions were reduced with excellent enantioselectivity, but kinetic resolution of the starting material was only modest [96].

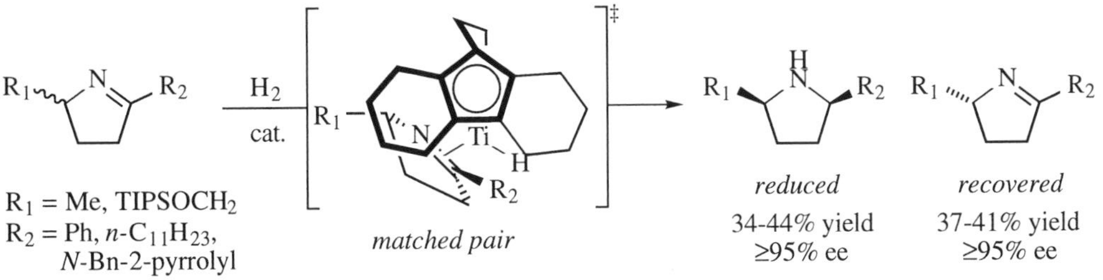

R_1 = Me, TIPSOCH$_2$
R_2 = Ph, *n*-C$_{11}$H$_{23}$,
  N-Bn-2-pyrrolyl

matched pair

reduced
34-44% yield
≥95% ee

recovered
37-41% yield
≥95% ee

Scheme 7.12. Kinetic resolution of 5-substituted 1-pyrrolines by asymmetric reduction using the *S,S* chiral titanocene catalyst [96].

7.2 Reduction of carbon-carbon bonds

Reduction of a carbon-carbon double bond will produce a chiral product if the olefin is (unsymmetrically) geminally disubstituted. Although hundreds of catalysts having chiral ligands have been synthesized and screened with a number of alkene structural types (reviews: ref. [65,97-107]), the present discussion will focus on only one: the reduction of acetamido cinnamates using soluble rhodium catalysts (reviews: ref. [97,100,108-110]). The development of chiral bisphosphine ligands and the herculean effort that led to the elucidation of the mechanism of this reaction make it an important example for study, since we now know that the major enantiomer of the product arises from a minor (often invisible) component of a pre-equilibrium [109,111]. This aspect of chemical reactivity is an important lesson whose importance cannot be overemphasized: when we strive to understand the

---

[9] In contrast, the enantioselectivity of cyclic imine reduction is independent of hydrogen pressure.
[10] In reference [96] the *R,R* enantiomer of the catalyst (*cf* Scheme 7.10) was employed. To maintain consistency with Scheme 7.10 and Figure 7.6, we illustrate the *S,S* catalyst.

forces that govern reactivities and selectivities, we must never overlook the fact that an observable intermediate in a chemical process may not be the one responsible for the observed products.

Following Wilkinson's detailed studies of tris-triphenylphosphine rhodium chloride as a soluble catalyst for hydrogenations [112], it did not take long for chemists to realize that chiral phosphines could be substituted for triphenylphosphine so as to effect an asymmetric reduction [113]. Following Mislow's development of a synthesis of chiral phosphine oxides [114], the groups of Knowles [115] and Horner [113] tested methyl phenyl *n*-propyl phosphine in the Wilkinson catalyst system, but found only low selectivities in the reduction of substrates such as α-phenylacrylic acid. These efforts were predicated on the (very reasonable) assumption that the chiral rhodium complex should contain chirality centers at phosphorous (since they are close to the metal). However, this assumption was proven wrong in 1971 when Morrison [116] and Kagan [117] independently showed that ligands such as R*–PPh$_2$ (Morrison) and Ph$_2$P–R*–PPh$_2$ (Kagan) (where R* contains a chirality center) were capable of reducing substituted cinnamic acids with enantioselectivities in the 80% (es) range, as shown by the "record-setting" examples in Scheme 7.13. The Kagan ligand, derived from tartaric acid, later became known as "DIOP", and served as the prototype for many more chiral, chelating diphosphine ligands. The Kagan example also demonstrates the utility of an asymmetric reduction protocol for the synthesis of α-amino acids. A similar reaction is now used industrially for the enantioselective production of dihydroxyphenylalanine (DOPA, a drug for treating Parkinson's disease) and aspartame (an artificial sweetener). It can be fairly stated that these spectacular early successes served to heighten optimism for the prospects of asymmetric synthesis in general, and asymmetric catalysis in particular, hopes that have been well rewarded in the interim.

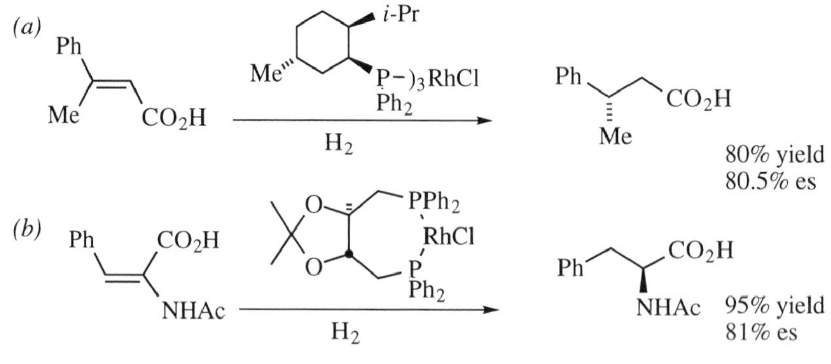

Scheme 7.13. *(a)* Morrison's asymmetric reduction of β-methyl cinnamic acid [116]. *(b)* Kagan's asymmetric reduction of *N*-acetyl dehydrophenylalanine and the debut of the DIOP ligand [117].

These examples were followed with a continuous stream of ligands (that continues to this day: *cf.* ref. [66,105,107,108,118-120]) that were tested with rhodium and other metals in asymmetric reductions and other reactions catalyzed by transition metals [102-104,121]. Simultaneously, studies of the mechanism of the asymmetric hydrogenation were pursued, most agressively in the labs of Halpern and Brown. The currently accepted mechanism is shown in Scheme 7.14 [111]. The substrate (methyl *Z*-acetamidocinnamate, middle left) displaces two solvent molecules from the cationic rhodium catalyst (center) in an equilibrium that favors the diastereomer in which the rhodium is bound to the *Si* face (at C-2) of the alkene [122]. This equilibrium defines the "major" and "minor" manifolds of the reaction. The sequence of oxidative addition of dihydrogen, migratory insertion, and reductive elimination, completes the cycle in both manifolds. With some bisphosphines, both of the initially formed diastereomeric complexes are visible by NMR; with others, signals from the minor diastereomer are lost in the noise. Each subsequent reaction is irreversible, but at low temperature the rhodium alkyl hydride product of migratory insertion can be intercepted and characterized spectroscopically [123,124]. Surprisingly, the intercepted complex (leading to the *S* product) has the metal on the *Re* face of C-2! *Thus, the major product of the reduction is produced by oxidative addition of dihydrogen to the minor diastereomer of the catalyst-substrate complex* [111,123].

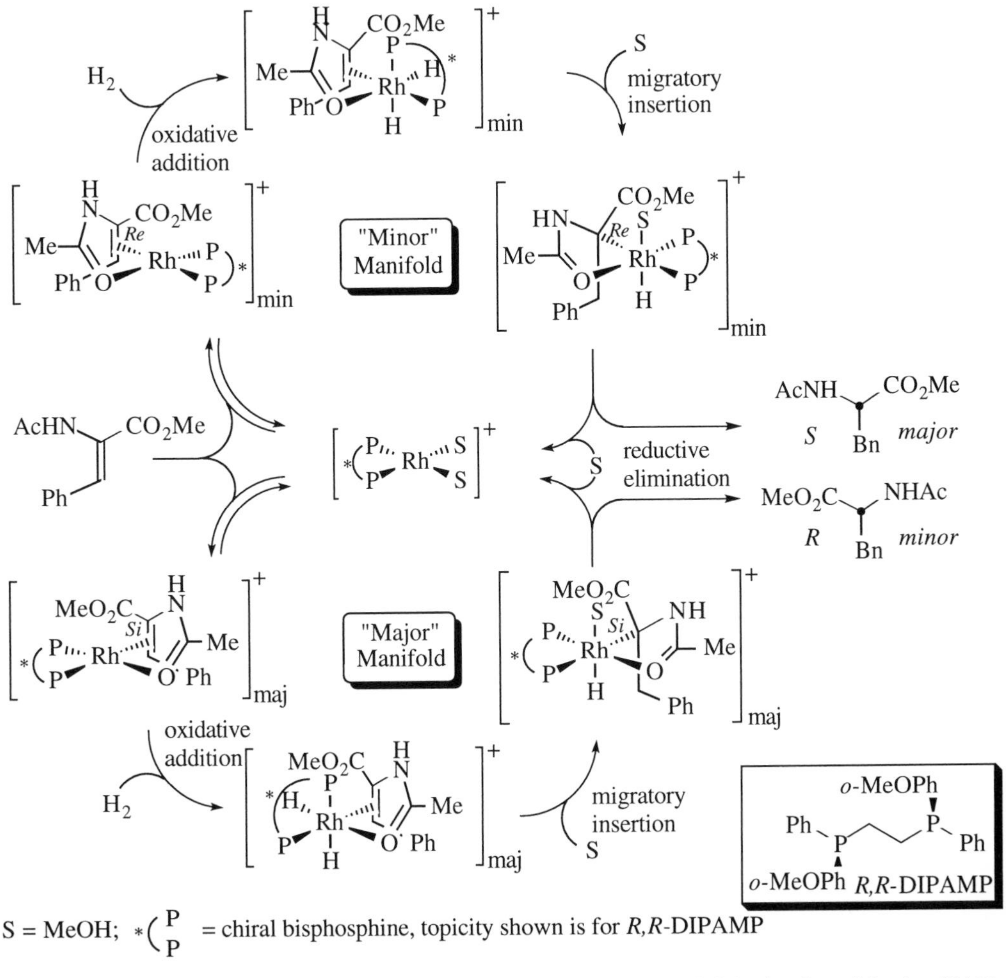

S = MeOH; $*\binom{P}{P}$ = chiral bisphosphine, topicity shown is for *R,R*-DIPAMP

Scheme 7.14. Mechanism of asymmetric hydrogenation of *N*-acetyl dehydrophenylalanine [111].

At temperatures above $-40°$, the rate-determining step of the reaction is the oxidative addition of hydrogen to the catalyst-substrate complex. Because the interconversion of the two diastereomeric catalyst-substrate complexes is fast relative to the rate of oxidative addition [111], and because the migratory insertion and reductive elimination steps are kinetically invisible, the complex equilibria in Scheme 7.14 reduce to a classic case of Curtin-Hammett kinetics [80], whereby the relative rate of the formation of the two enantiomers is determined solely by the relative energy of the two transition states, as illustrated in Figure 7.7. This energy difference is 2.3 kcal/mole, corresponding to a relative rate of about 50:1 for formation of the S product over the R (from R,R-DIPAMP catalyst at 25°), or 98% es (*cf.* Figure 1.3, p. 9).

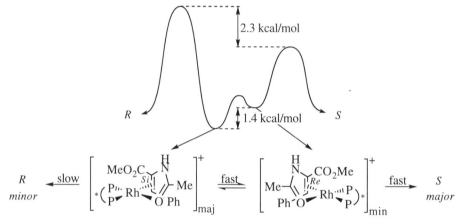

Figure 7.7. The asymmetric hydrogenation of N-acetyl dehydrophenylalanine ester as an example of Curtin-Hammett kinetics. Energy values taken from ref. [111].

Still, the question remains: why is the one diastereomer of the catalyst-substrate complex so much more reactive than the other? In 1977, Knowles suggested, based on examination of the crystal structures of several metal complexes having chiral bisphosphine ligands, that the orientation of the P-phenyl groups could be the source of the enantioselectivity of these reactions [125]. The common feature Knowles observed (Figure 7.8) is that the two equatorial phenyls are oriented such that – with the bisphosphine in the horizontal plane and the metal in front – their

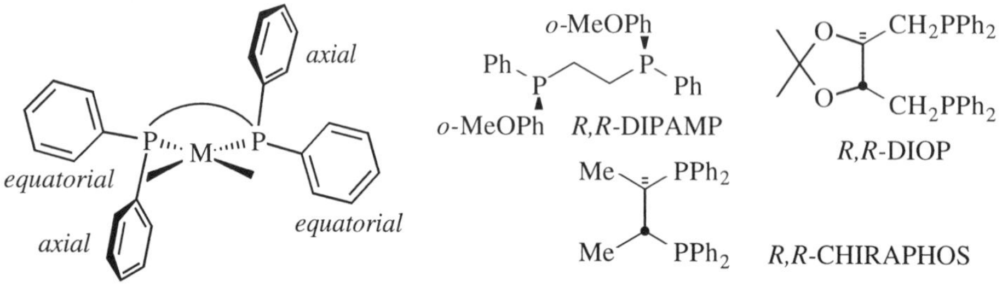

Figure 7.8. Bisphosphine ligands and the common structural feature that affects steric crowding of ligands bound trans to phosphorus [97,125]. Note vacant upper left and lower right quadrants in the generalized structure on the left.

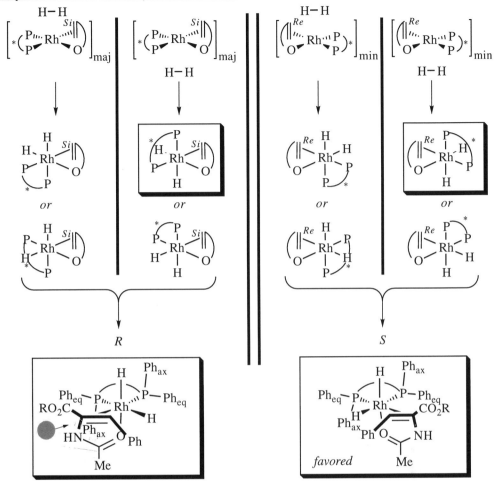

Scheme 7.15. Possible orientations for oxidative addition of dihydrogen to the major (left) and minor (right) diastereomers of the catalyst-substrate complex (for simplicity, the linkages connecting the atoms bonded to the metal are indicated with a curved line). The boxed structures are the only octahedral structures that are not encumbered by severe non-bonded interactions; they are redrawn at the bottom with the bisphosphine to the rear and in the horizontal plane. The topicity illustrated is for ligands having the structure of Figure 7.8, such as *R,R* DIPAMP, *R,R*-DIOP, or *R,R* CHIRAPHOS (see Figure 7.8).

'faces' are exposed to a ligand coordinated at a site towards the viewer, while the two axial phenyls expose 'edges'. This conformation produces two crowded quadrants (those having the axial phenyls) and two vacant quadrants, as shown in Figure 7.8. Structural features similar to this have turned up in the interim in the structures of numerous other ligands, and can be conveniently used to explain the stereoselectivity of a number of metal-catalyzed reactions (*cf.* Figures 4.18, 6.3, 6.10, 6.20, 6.21, 7.3, 7.6).[11]

For the asymmetric hydrogenation, both substrate-catalyst complexes are square-planar, and hydrogen could, in principle, add from either the top or the bottom of

---

[11] For a recent leading reference to these structural features, see ref. [126,127]

either complex, as illustrated in Scheme 7.15. From a molecular orbital standpoint, there is no reason to expect any one of these four possibilities to dominate. Indeed, it is likely that the oxidative addition of dihydrogen occurs stepwise, and that the hydrogen binds "edgewise" initially (making a square pyramid complex), followed by H–H cleavage to form the octahedral dihydride product. Further, it is likely that the dihydrogen associates with the metal many times before H–H bond cleavage occurs.[12] Lacking an electronic rationale, the only other possible explanation is that the movement of the ligands (as the oxidative addition proceeds) is determinant.

Scheme 7.15 illustrates the eight possible octahedral complexes that could arise by addition of dihydrogen to either the top or the bottom of the two square complexes. Each is drawn so that the orientation of the substrate remains unchanged, and one of the phosphines is moved trans to the incoming hydrogen. A molecular mechanics investigation [129] indicates that only two of the eight structures are viable (boxed, redrawn at the bottom of Scheme 7.15); all the others suffer severe nonbonded interactions between ligands. Note the similarity of the two boxed structures: in both, a hydride is trans to the chelating oxygen and cis to both phosphorous atoms. The double bond and the second hydride are then meridonal with respect to the two phosphorous atoms. The reason for the difference in stability can be seen by examining the orientation of the substrate relative to the axial phenyls: in the favored configuration, the substrate occupies the less crowded "lower right" quadrant. It is implicit in this analysis that the energetic consequences of these various structural features must be felt in the competing transition states for oxidative addition.

Two factors contribute to the success of this reaction: the outstanding enantio-selectivity achieved, and efficiency of the catalyst (*i.e,* high turnover). The above analysis emphasizes only the former, but the latter also varies with the nature of the chiral bisphosphine ligand and the structure of the substrate. The structural features of the substrate and the catalyst are mutually optimal in the example cited above. Perturbation of any of these features usually lowers either the enantioselectivity or the turnover rate. The range of substrates that are amenable to asymmetric hydro-genation with this catalyst system is, therefore, limited. Figure 7.9 illustrates the classes of substrate that can be accomodated by cationic rhodium bisphosphine catalysts [104]. For a more extensive summary, see ref. [110].

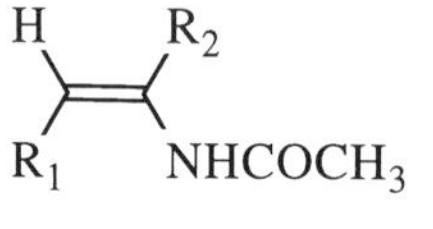

$$R_1 = H, \text{ alkyl, aryl}$$
$$R_2 = CO_2R, Ph$$

Figure 7.9. Substrate tolerance in the asymmetric hydrogenation (after ref. [104]).

---

[12] Hoff has shown that the rate of dissociation of a tungsten-dihydrogen complex is at least one order of magnitude faster than the rate of oxidative addition (H–H bond cleavage) to a dihydride complex [128].

7.3 Hydroborations

The first asymmetric synthesis to achieve >90% optical yield was Brown's hydroboration of cis alkenes with diisopinocampheylborane (Ipc$_2$BH, Figure 7.10) in 1961 [130,131]. The reagent was prepared by hydroboration of α-pinene of ~90% ee; 2-butanol obtained from hydroboration/oxidation of *cis*-2-butene had an optical purity of 87%, indicating an optical yield of 90%. *cis*-3-Hexene was hydroborated in ~100% optical yield. Since then, simple methods for the enantiomer enrichment of Ipc$_2$BH (and IpcBH$_2$) have been developed [132-134], and enantioselectivities have been evaluated more carefully with the purified material. For example, Ipc$_2$BH of 99% ee[13] affords 2-butanol (from *cis*-2-butene) in 98% ee and 3-hexanol (from *cis*-3-hexene) in 93% ee, both determined by rotation (see Table 7.6, entries 1 and 5) [132].[14]

)$_2$BH   BH$_2$ BH

diisopinocampheyl borane
Ipc$_2$BH

monoisopinocampheyl borane
IpcBH$_2$

trans 2,5-dimethylborolane
DMB

Figure 7.10. Chiral hydroborating reagents: IpcBH$_2$ [130-134], IpcBH$_2$ [133,135]; DMB [136].

Today, Ipc$_2$BH is still as good a reagent as any for achieving enantioselective hydroboration of cis alkenes (Table 7.6, entries 1, 2, 6, 7, 10, 12-17; reviews: [2,137-139]). However, it does not afford good enantioselectivities with *trans*-disubstituted or trisubstituted alkenes. For these classes of compounds, monoisopino-campheylborane, IpcBH$_2$ (Figure 7.10), gives good selectivities, as indicated by the examples in Table 7.6, entries 4, 8, and 11 [135]; recrystallization of the intermediate borane may be used to purify the major borane diastereomer in some cases (Table 7.6, entries 2, 18, 20, 22, 24, and 26), and serves to improve the overall enantioselectivity of the process [133].

In 1985 [136], Masamune introduced *trans*-2,5-dimethylborolane (Figure 7.10) as a chiral hydroborating agent that works well for cis and trans disubstituted and trisubstituted alkenes (Table 7.6, entries 3, 5, 7, 9, 19, 21, and 23). Although this reagent is the most versatile yet invented, its preparation is sufficiently cumbersome that its synthetic utility is not great. On the other hand, the conformational rigidity of this reagent allows us to postulate a reasonable transition structure to account for the topicity of the hydroboration (Scheme 7.16). Specifically, when R$_1$ ≠ H, and either R$_2$ or R$_3$ = H, the boron of the *R,R*-borolane adds preferentially to the *Si* face of the alkene carbon. Good stereoselectivity will result when either R$_2$ or R$_3$ (or both) are ≠ H, since the carbon which is attacked by boron determines the stereoselectivity. Conversely, if R$_1$ = H, there is little difference in energy between

---

[13] This is the enantiomeric purity of isopinocampheol produced by oxidation of the purified Ipc$_2$BH, measured by rotation [132].

[14] The discrepancy between the optical yields using enantiopure α-pinene and that of ~90% ee is probably due experimental error in the measurement of rotations.

Table 7.6. Examples of enantioselective hydroborations. The "Reagent" column refers to the structures in Figure 7.10. The "% es" column reflects the overall enantioselectivity of the process, including any diastereomeric enrichment, and is corrected for the enantiomeric purity of the borane.

Entry	Alkene	Reagent	% Yield	% es	Ref.
1	*cis*-2-butene	Ipc₂BH	74	99	[130-132]
2	"	IpcBH₂	78	99	[133]
3	"	DMB	75	97	[136]
4	*trans*-2-butene	IpcBH₂	73	86	[135]
5	"	DMB	71	100	[136]
6	*cis*-3-hexene	Ipc₂BH	68-81	95-96	[130-132]
7	"	DMB	83	100	[136]
8	*trans*-3-hexene	IpcBH₂	83	87	[135]
9	"	DMB	83	>99	[136]
10	norbornene	Ipc₂BH	62	91	[130-132]
11	*trans*-stilbene	IpcBH₂	69	82	[135]
12	(ring, X)	Ipc₂BH X = O	74	>99	[140]
13		X = NCO₂Bn	85	>99	[140]
14	(ring, X)	Ipc₂BH X = O	87	>99	[140]
15		X = S	80	>99	[140]
16	(ring, X)	Ipc₂BH X = O	81-85	91-93	[140]
17		X = S	68	83	[140]
18	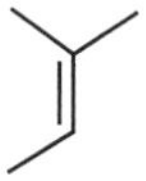 (alkene)	IpcBH₂	78	99	[133]
19		DMB	90	94	[136]
20	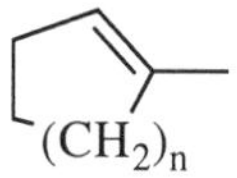 (CH₂)ₙ	n = 1: IpcBH₂	65	100	[133]
21		n = 1: DMB	79	98	[136]
22		n = 2: IpcBH₂	75	99	[133]
23		n = 2: DMB	60	96	[136]
24	(CH₂)ₙ—Ph	IpcBH₂ n = 1	72-92	100	[133,135]
25		n = 2	79	94	[135]
26	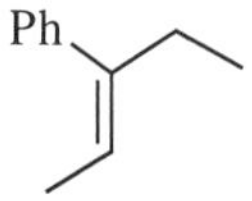 Ph (alkene)	IpcBH₂	77	100	[133]

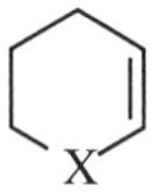

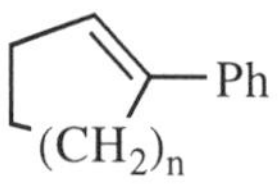

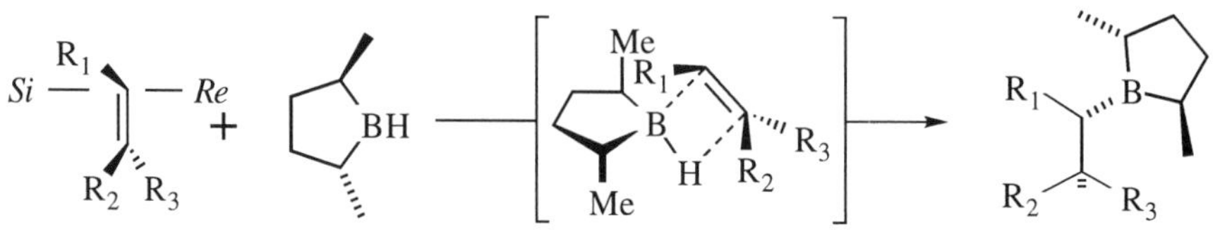

Scheme 7.16. Favored transition structure for asymmetric hydroboration by Masamune's borolane [136].

the illustrated transition structure and an alternative one in which R_2 and R_3 are interchanged. In fact, for 1,1-disubstituted alkenes, none of the reagents of Figure 7.10 affords products in greater than 10% ee (≤55% es).

The conformational mobility around the B–C bond(s) in $IpcBH_2$ and Ipc_2BH complicate the analysis for these terpene-derived boranes, but Figure 7.11 gives a simplified picture. Using ab initio techniques, Houk and coworkers [141] located the transition structures for the hydroboration of simple alkenes, and found that the most consistent feature of the most stable transition structures has the auxiliary (R*) and the substituent on carbon (R) anti to each other, as shown in Figure 7.11a. Analysis of the conformational motion of the B–R* bond revealed that the substituents on boron prefer to be staggered with respect to the forming C–B bond. Furthermore, the most stabile position is anti to this bond. The so-called *"outside"* position is less encumbered sterically than the *"inside"* position, and the difference in energy between these two is affected by whether the alkene is cis or trans (Figure 7.11b). In Figure 7.11c the pinene substituent is reduced to a shorthand notation of **S**mall (H), **M**edium, and **L**arge substituents on the carbon bearing the boron.

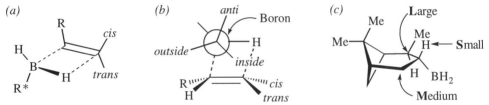

Figure 7.11. Terminology definitions for hydroboration transition structures [141]: *(a)* The auxiliary may be either syn or anti to the alkene substituents, but anti to the substituent (R) on the nearest carbon. *(b)* A stereocenter attached to boron, in a staggered conformation with respect to the forming C–B bond, has substituents in anti, *inside,* and *outside* positions. *(c)* Definition of the Large, Medium, and Small substituents of $IpcBH_2$.

With these generalizations in mind, it is possible to qualitatively[15] rationalize the results with $IpcBH_2$ and Ipc_2BH. The more easily understood example, of course, is $IpcBH_2$, since there is only one pinene moiety involved. This reagent is most selective with trans alkenes, so this olefin-type is illustrated first. The lowest energy (molecular mechanics) transition structure for addition of boron to the *Si* face of the alkene (Figure 7.12a) has the pinene anti to R (methyl), and has the small, medium, and large ligands in the most stable positions relative to the newly forming C–B bond: **L**-*anti*, **M**-*outside,* and **S**mall(H)-*inside*. In contrast, the transition structure for addition to the *Re* face (Figure 7.12b) has **L** in the less favorable *outside* position [141]. Note that in both of these structures the *inside* position is in close proximity to the second methyl (R) group, which increases the destabilization of any conformer in which either **M** or **L** occupy the *inside* position. $IpcBH_2$ is also fairly selective with trisubstituted alkenes (Table 7.6), and the transition structures of Figure 7.12 show why this should be the case: an additional substituent in the cis

---

[15] Houk, et al, note that the magnitude of the experimentally observed selectivities do not correspond to the energy differences their molecular mechanics calculations indicate, so this analysis and the calculated transition structures may only be taken as a first approximation [141].

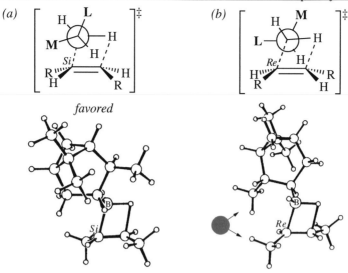

Figure 7.12. Transition structures for the asymmetric hydroboration of *trans*-2-butene with IpcBH$_2$. Reprinted with permission from ref. [141], copyright 1984 Elsevier Science, Ltd.

position (*cf.* Figure 7.11a,b) imposes no additional crowding on the transition structure. On the other hand, IpcBH$_2$ is much less selective with cis alkenes. Here, the position of the alkyl group is moved away from its close proximity to the *inside* position, and a number of other transition structures become feasible [141].

For hydroborations with Ipc$_2$BH, there are two pinene moieties to consider. Ipc$_2$BH is only selective for cis alkenes, and the alkene substituents (R in Figure 7.13a) must be near one of them (R* in Figure 7.13a). Houk, et al., find that there is only one conformation that the two pinenes may adopt relative to each other, and that is shown schematically in Figure 7.13b [141]. In the conformation shown, the olefin can align itself with the B–H bond and have the two R groups oriented either toward the proximal or distal pinene. Note that the proximal pinene has the **S**mall hydrogen in the *inside* position, whereas the **L**arge substituent of the distal pinene is

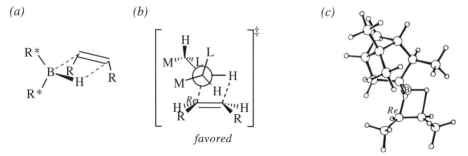

Figure 7.13. Transition structure for hydroboration of a cis alkene with Ipc$_2$BH. (*a*) The alkene substituents <u>must</u> be syn to one of the pinenes (R*). (*b*) Schematic representation of the lowest energy conformation. (*c*) Molecular mechanics - derived structure, with the rear (distal) pinene deleted for clarity. Reprinted with permission from ref. [141], copyright 1984, Elsevier Science, Ltd.

in the crowded *inside* position. The alkene is least hindered in the orientation shown in Figure 7.13b. Figure 7.13c shows the transition structure with the distal pinene deleted for clarity [141]. Note that in this structure, the alkene substituent is oriented toward the proximal pinene (with respect to the 4-membered ring), and it is therefore clear why Ipc$_2$BH preferentially attacks the *Re* face (Figure 7.13). This is in contrast to IpcBH$_2$, which prefers the *Si* face (Figure 7.12), because the alkene substituent is anti to the pinene.

7.4 References

1. H. C. Brown; P. K. Jadhav; A. K. Mandal *Tetrahedron* **1981**, *37*, 3574-3587.
2. H. C. Brown; B. Singaram *Pure Appl. Chem.* **1987**, *59*, 879-894.
3. R. C. Larock *Comprehensive Organic Transformations. A Guide to Functional Group Preparations*; VCH: New York, 1989.
4. H. C. Brown; W. S. Park; B. T. Cho; P. V. Ramachandran *J. Org. Chem.* **1987**, *52*, 5406-5412.
5. A. A. Bothner-by *J. Am. Chem. Soc.* **1951**, *73*, 846.
6. J. D. Morrison; H. S. Mosher In *Asymmetric Organic Reactions*; Prentice-Hall: Englewood Cliffs, NJ, 1971, p 202-215.
7. E. R. Grandbois; S. I. Howard; J. D. Morrison In *Asymmetric Synthesis*; J. D. Morrison, Ed.; Academic: Orlando, 1983; Vol. 2, p 71-79.
8. M. Nishizawa; R. Noyori In *Comprehensive Organic Synthesis. Selectivity, Strategy, and Efficiency in Modern Organic Chemistry*; B. M. Trost, I. Fleming, Eds.; Pergamon: Oxford, 1991; Vol. 8, p 159-182.
9. J. Seyden-Penne *Reductions by the Alumino- and Borohydrides in Organic Synthesis*; VCH: New York, 1991.
10. R. Noyori; R. Tomino; I. Tomino; Y. Tanimoto *J. Am. Chem. Soc.* **1979**, *101*, 3129-3131.
11. M. Nishiwaza; M. Yamada; R. Noyori *Tetrahedron Lett.* **1981**, *22*, 247-250.
12. R. Noyori; I. Tomino; Y. Tanimoto; M. Nishizawa *J. Am. Chem. Soc.* **1984**, *106*, 6709-6716.
13. R. Noyori; I. Tomino; M. Yamada; M. Nishizawa *J. Am. Chem. Soc.* **1984**, *106*, 6717-6725.
14. D. Seebach; A. K. Beck; R. Dahinden; M. Hoffmann; F. N. M. Kühnle *Croatia Chem. Acta* **1996**, in press.
15. A. K. Beck; R. Dahinden; F. N. M. Kühnle In *ACS Symposium Series. Reduction in Organic Chemistry*; A. F. Abdel-Magid, Ed.; American Chemical Society: Washington, D. C., 1996.
16. J. March In *Advanced Organic Chemistry, 4th ed.*; Wiley-Interscience: New York, 1992, p 145.
17. K. Matsuki; H. Inoue; M. Takeda *Tetrahedron Lett.* **1993**, *34*, 1167-1170.
18. J. C. Fiaud; H. B. Kagan *Bull. Soc. Chim. Fr.* **1969**, 2742-2743.
19. S. Wallbaum; J. Martens *Tetrahedron Asymmetry* **1992**, *3*, 1475-1504.
20. L. Deloux; M. Srebnik *Chem. Rev.* **1993**, *93*, 763-784.
21. A. Hirao; S. Itsuno; S. Nakahama; N. Yamazaki *J. Chem. Soc., Chem. Commun.* **1981**, 315-317.
22. S. Itsuno; A. Hirao; S. Nakahama; N. Yamazaki *J. Chem. Soc., Perkin Trans. 1* **1983**, 1673-1676.
23. S. Itsuno; K. Ito; A. Hirao; S. Nakahama *J. Chem. Soc., Chem. Commun.* **1983**, 469-470.
24. S. Itsuno; M. Nakano; K. Miyazaki; H. Masuda; K. Ito *J. Chem. Soc., Perkin Trans. 1* **1985**, 2039-2044.
25. S. Itsuno; K. Ito; A. Hirao; S. Nakahama *J. Chem. Soc., Perkin Trans. 1* **1984**, 2887-2893.
26. S. Itsuno; K. Ito; A. Hirao; S. Nakahama *J. Org. Chem.* **1984**, *49*, 555-557.
27. E. J. Corey; R. K. Bakshi; S. Shibata *J. Org. Chem.* **1988**, *53*, 2861-2863.

28. S. Itsuno; Y. Sakurai; K. Ito; A. Hirao; S. Nakahama *Bull. Chem. Soc. Jpn.* **1987**, *60*, 395-396.

29. E. J. Corey; R. K. Bakshi; S. Shibata *J. Am. Chem. Soc.* **1987**, *109*, 5551-5553.

30. E. J. Corey; R. K. Bakshi; S. Shibata; C.-P. Chen; V. K. Singh *J. Am. Chem. Soc.* **1987**, *109*, 7925-7926.

31. E. J. Corey; J. O. Link *Tetrahedron Lett.* **1989**, *30*, 6275-6278.

32. E. J. Corey; M. Azomioara; S. Sarshar *Tetrahedron Lett.* **1992**, *33*, 3429-3430.

33. D. J. Mathre; T. K. Jones; L. C. Xavier; T. J. Blacklock; R. A. Reamer; J. J. Mohan; E. T. T. Jones; K. Hoogsteen; M. W. Baum; E. J. J. Grabowski *J. Org. Chem.* **1991**, *56*, 751-762.

34. V. Nevalainen *Tetrahedron Asymmetry* **1991**, *2*, 429-435.

35. D. K. Jones; D. C. Liotta; I. Shinkai; D. J. Mathre *J. Org. Chem.* **1993**, *58*, 799-801.

36. G. J. Quallich; J. F. Blake; T. M. Woodall *J. Am. Chem. Soc.* **1994**, *116*, 8516-8525.

37. V. Nevalainen *Tetrahedron Asymmetry* **1992**, *3*, 921-932.

38. E. J. Corey; J. O. Link *J. Am. Chem. Soc.* **1992**, *114*, 1906-1908.

39. E. J. Corey; J. O. Link; Y. Shao *Tetrahedron Lett.* **1992**, *33*, 3435-3438.

40. E. J. Corey; J. O. Link *Tetrahedron Lett.* **1992**, *33*, 3431-3434.

41. E. J. Corey; C. J. Helal *Tetrahedron Lett.* **1993**, *34*, 5227-5230.

42. E. J. Corey; J. O. Link *Tetrahedron Lett.* **1992**, *33*, 4141-4144.

43. E. J. Corey; R. K. Bakshi *Tetrahedron Lett.* **1990**, *31*, 611-614.

44. P. Beak; S. T. Kerrick; S. Wu; J. Chu *J. Am. Chem. Soc.* **1994**, *116*, 3231-3239.

45. T. K. Jones; J. J. Mohan; L. C. Xavier; T. J. Blacklock; D. J. Mathre; P. Sohar; E. T. T. Jones; R. A. Reamer; F. E. Roberts; E. J. J. Grabowski *J. Org. Chem.* **1991**, *56*, 763-769.

46. M. M. Midland *Chem. Rev.* **1989**, *89*, 1553-1561.

47. H. C. Brown; P. V. Ramachandran *Pure Appl. Chem.* **1991**, *63*, 307-316.

48. H. C. Brown; P. V. Ramachandran *Acc. Chem. Res.* **1992**, *25*, 16-24.

49. T. Imai; T. Tamura; A. Yamamuro; T. Sato; T. A. Wollmann; R. M. Kennedy; S. Masamune *J. Am. Chem. Soc.* **1986**, *108*, 7402-7404.

50. B. M. Mikhailov; Y. N. Bubnov; V. G. Kiselev *J. Gen. Chem. USSR (Engl. Transl.)* **1966**, *36*, 65-69.

51. M. M. Midland; S. A. Zderic *J. Am. Chem. Soc.* **1982**, *104*, 525-528.

52. M. M. Midland; S. Greer; A. Tramontano; S. A. Zderic *J. Am. Chem. Soc.* **1979**, *101*, 2352-2355.

53. M. M. Midland; S. Greer *Synthesis* **1978**, 845-846.

54. M. M. Midland; D. C. McDowell; R. L. Hatch; A. Tramontano *J. Am. Chem. Soc.* **1980**, *102*, 867-869.

55. M. M. Midland; R. S. Graham *Organic Syntheses* **1990**, *Coll. Vol. VII*, 402-406.

56. M. M. Midland; A. Tramontano; A. Kazubski; R. Graham; D. J.-S. Tsai; D. B. Cardin *Tetrahedron* **1984**, *40*, 1371-1380.

57. H. C. Brown; G. G. Pai *J. Org. Chem.* **1985**, *50*, 1384-1394.

58. M. M. Midland; J. I. McLoughlin; J. Gabriel *J. Org. Chem.* **1989**, *54*, 159-165.

59. H. C. Brown; J. Chandrasekharan; P. V. Ramachandran *J. Am. Chem. Soc.* **1988**, *110*, 1539-1546.

60. P. V. Ramachandran; A. V. Teodorovic; M. V. Rangaishenvi; H. C. Brown *J. Org. Chem.* **1992**, *57*, 2379-2386.

61. G. Zassinovich; R. Betella; G. Mestroni; N. Bresciani-Pahor; S. Geremia; L. Randaccio *J. Organomet. Chem.* **1989**, *370*, 187-202.

62. S. Gladiali; L. Pinna; G. Deloga; S. D. Martin; G. Zassinovich; G. Mestroni *Tetrahedron Asymmetry* **1990**, *1*, 635-648.

63. C. Bolm *Angew. Chem. Int. Ed. Engl.* **1991**, *30*, 542-543.

64. X. Zhang; T. Taketomi; T. Yoshizumi; H. Kumobayashi; S. Akutagawa; K. Mashima; H. Takaya *J. Am. Chem. Soc.* **1993**, *115*, 3318-3319.

65. H. Takaya; T. Ohta; R. Noyori In *Catalytic Asymmetric Synthesis*; I. Ojima, Ed.; VCH: New

York, 1993, p 1-39.

66. R. Noyori *Asymmetric Catalysis in Organic Synthesis*; Wiley-Interscience: New York, 1994.

67. H. Brunner; H. Nishiyama; K. Itoh In *Catalytic Asymmetric Synthesis*; I. Ojima, Ed.; VCH: New York, 1993, p 303-322.

68. R. Noyori In *Asymmetric Catalysis in Organic Synthesis*; Wiley-Interscience: New York, 1994, p 124-131.

69. R. Noyori; T. Ohkuma; M. Kitamura; H. Takaya; N. Sayo; H. Kumobayashi; S. Akutagawa *J. Am. Chem. Soc.* **1987**, *109*, 5856-5858.

70. M. Kitamura; T. Ohkuma; S. Inoue; N. Sayo; H. Kumobayashi; S. Akutagawa; T. Ohta; H. Takaya; R. Noyori *J. Am. Chem. Soc.* **1988**, *110*, 629-631.

71. M. Kitamura; M. Tokunaga; T. Ohkuma; R. Noyori *Tetrahedron Lett.* **1991**, *32*, 4163-4166.

72. D. F. Taber; L. J. Silverberg *Tetrahedron Lett.* **1991**, *32*, 4227-4230.

73. D. F. Taber; P. B. Deker; L. J. Silverberg *J. Org. Chem.* **1992**, *57*, 5990-5994.

74. K. Mashima; K. Kusano; T. Ohta; R. Noyori; H. Takaya *J. Chem. Soc., Chem. Commun.* **1989**, 1208-1210.

75. K. Mashima; K. Kusano; N. Sato; Y. Matsumura; K. Nozaki; H. Kumobayashi; N. Sayo; Y. Hori; T. Ishikazi; S. Akutagawa; H. Takaya *J. Org. Chem.* **1994**, *59*, 3064-3076.

76. R. Noyori In *Asymmetric Catalysis in Organic Synthesis*; Wiley-Interscience: New York, 1994, p 63-66.

77. M. J. Burk; T. G. P. Harper; C. S. Kalberg *J. Am. Chem. Soc.* **1995**, *117*, 4423-4424.

78. R. Noyori; T. Ikeda; T. Ohkuma; M. Widhalm; M. Kitamura; H. Takaya; S. Akutagawa; N. Sayo; T. Saito; T. Taketomi; H. Kumobayashi *J. Am. Chem. Soc.* **1989**, *111*, 9134-9135.

79. D. Y. Curtin *Rec. Chem. Progr.* **1954**, *15*, 111-128.

80. J. I. Seeman *Chem. Rev.* **1983**, *83*, 83-134.

81. T. Ohta; H. Takaya; R. Noyori *Inorg. Chem.* **1988**, *27*, 566-569.

82. J. E. Baldwin; R. M. Adlington; S. H. Ramcharitar *Synlett* **1992**, 875-877.

83. S. L. Schreiber; S. E. Kelly; J. A. Porco, Jr.; T. Sammakia; E. M. Suh *J. Am. Chem. Soc.* **1988**, *110*, 6210-6218.

84. C. A. Willoughby; S. L. Buchwald *J. Am. Chem. Soc.* **1992**, *114*, 7562-7564.

85. C. A. Willoughby; S. L. Buchwald *J. Am. Chem. Soc.* **1994**, *116*, 8952-8965.

86. C. A. Willoughby; S. L. Buchwald *J. Am. Chem. Soc.* **1994**, *116*, 11703-11714.

87. C. Bolm *Angew. Chem. Int. Ed. Engl.* **1993**, *32*, 232-233.

88. R. S. Cahn; C. K. Ingold; V. Prelog *Angew. Chem. Int. Ed. Engl.* **1966**, *5*, 385-415, 511.

89. F. R. W. P. Wild; L. Zsolnai; G. Huttner; H. H. Brintzinger *J. Organomet. Chem.* **1982**, *232*, 233-247.

90. K. Schlögl In *Topics in Stereochemistry*; E. L. Eliel, N. L. Allinger, Eds.; Wiley-Interscience: New York, 1967; Vol. 1, p 39-91.

91. T. E. Sloan In *Topics in Stereochemistry*; G. L. Geoffroy, Ed.; Wiley-Interscience: New York, 1981; Vol. 12, p 1-36.

92. E. L. Eliel; S. H. Wilen; L. N. Mander In *Stereochemistry of Organic Compounds*; Wiley-Interscience: New York, 1994; Vol. Ch. 14, p 1121-2.

93. G. Helmchen In *Stereoselective Synthesis*; G. Helmchen, R. W. Hoffmann, J. Mulzer, E. Schaumann, Eds.; Georg Thieme: Stuttgart, 1995; Vol. E21a, p 1-74.

94. K. Hortmann; H.-H. Brintzinger *New J. Chem.* **1992**, *16*, 51-55.

95. C. A. Willoughby; S. L. Buchwald *J. Org. Chem.* **1993**, *58*, 7627-7629.

96. A. Viso; N. E. Lee; S. L. Buchwald *J. Am. Chem. Soc.* **1994**, *116*, 9373-9374.

97. H. B. Kagan *Pure Appl. Chem.* **1975**, *43*, 401-421.

98. J. D. Morrison; W. F. Masler; M. K. Neuberg *Adv. in Catal.* **1976**, *25*, 81-124.

99. V. Caplar; G. Comisso; V. Sunjic *Synthesis* **1981**, 85-116.

100. W. S. Knowles *Acc. Chem. Res.* **1983**, *16*, 106-112.

101. H. Takaya; R. Noyori In *Comprehensive Organic Synthesis. Selectivity, Strategy, and Efficiency in Modern Organic Chemistry*; B. M. Trost, I. Fleming, Eds.; Pergamon: Oxford,

1991; Vol. 8, p 443-469.

102. H. Brunner *Synthesis* **1988**, 645-654.

103. S. L. Blystone *Chem. Rev.* **1989**, *89*, 1663-1679.

104. H. B. Kagan *Bull. Soc. Chim. Fr.* **1988**, 846-853.

105. R. Noyori; H. Takaya *Acc. Chem. Res.* **1990**, *23*, 345-350.

106. R. Noyori In *Asymmetric Catalysis in Organic Synthesis*; Wiley-Interscience: New York, 1994, p 16-94.

107. K. Inoguchi; S. Sakuraba; K. Achiwa *Synlett* **1992**, 169-178.

108. H. B. Kagan In *Asymmetric Synthesis*; J. D. Morrison, Ed.; Academic: Orlando, 1985; Vol. 5, p 1-39.

109. J. Halpern In *Asymmetric Synthesis*; J. D. Morrison, Ed.; Academic: Orlando, 1985; Vol. 5, p 41-69.

110. K. E. Koenig In *Asymmetric Synthesis*; J. D. Morrison, Ed.; Academic: Orlando, 1985; Vol. 5, p 71-101.

111. C. R. Landis; J. Halpern *J. Am. Chem. Soc.* **1987**, *109*, 1746-1754.

112. J. A. Osborn; F. H. Jardine; J. F. Young; G. Wilkinson *J. Chem. Soc. (A)* **1966**, 1711-1732.

113. L. Horner; H. Büthe; H. Siegel *Tetrahedron Lett.* **1968**, 4023-4026.

114. O. Korpium; R. A. Lewis; J. Chickos; K. Mislow *J. Am. Chem. Soc.* **1968**, *90*, 4842-4846.

115. W. S. Knowles; M. J. Sabacky *J. Chem. Soc., Chem. Commun.* **1968**, 1445-1446.

116. J. D. Morrison; R. E. Burnett; A. M. Aguiar; C. J. Morrow; C. Phillips *J. Am. Chem. Soc.* **1971**, *93*, 1301-1303.

117. T. P. Dang; H. B. Kagan *Chem. Commun.* **1971**, 481.

118. M. J. Burk; J. E. Feaster; R. L. Harlow *Tetrahedron Asymmetry* **1991**, *2*, 569-592.

119. *Catalytic Asymmetric Synthesis*; I. Ojima, Ed.; VCH: New York, 1993.

120. M. J. Burk; M. F. Gross; J. P. Martinez *J. Am. Chem. Soc.* **1995**, *117*, 9375-9376.

121. R. Noyori *Science* **1990**, *248*, 1194-1199.

122. A. S. C. Chan; J. J. Pluth; J. Halpern *J. Am. Chem. Soc.* **1980**, *102*, 5952-5954.

123. J. M. Brown; P. A. Chaloner *J. Chem. Soc., Chem. Commun.* **1980**, 344-346.

124. A. S. C. Chan; J. Halpern *J. Am. Chem. Soc.* **1980**, *102*, 838-840.

125. B. D. Vineyard; W. S. Knowles; M. J. Sabacky; G. L. Bachman; D. J. Weinkauff *J. Am. Chem. Soc.* **1977**, *99*, 5946-5952.

126. P. Barbaro; P. S. Pregosin; R. Salzmann; A. Albinati; R. W. Kunz *Organometallics* **1995**, *14*, 5160-5170.

127. D. Seebach; E. Devaquet; A. Ernst; M. Hayakawa; F. N. M. Kühnle; W. B. Schweizer; B. Weber *Helv. Chim. Acta* **1995**, *78*, 1636-1650.

128. K. Zhang; A. A. Gonzalez; C. D. Hoff *J. Am. Chem. Soc.* **1989**, *111*, 3627-3632.

129. J. M. Brown; P. L. Evans *Tetrahedron* **1988**, *44*, 4905-4916.

130. H. C. Brown; G. Zweifel *J. Am. Chem. Soc.* **1961**, *83*, 486-487.

131. H. C. Brown; N. R. Ayyangar; G. Zweifel *J. Am. Chem. Soc.* **1964**, *86*, 397-403.

132. H. C. Brown; M. C. Desai; P. K. Jadhav *J. Org. Chem.* **1982**, *47*, 5065-5069.

133. H. C. Brown; B. Singaram *J. Am. Chem. Soc.* **1984**, *106*, 1797-1800.

134. H. C. Brown; U. P. Dhotke *J. Org. Chem.* **1994**, *59*, 2365-2369.

135. H. C. Brown; P. K. Jadhav; A. K. Mandal *J. Org. Chem.* **1982**, *47*, 5074-5083.

136. S. Masamune; B. M. Kim; J. S. Peterson; T. Sato; S. J. Veenstra; T. Imai *J. Am. Chem. Soc.* **1985**, *107*, 4549-4551; 5832.

137. H. C. Brown; P. K. Jadhav In *Asymmetric Synthesis*; J. D. Morrison, Ed.; Academic: Orlando, 1983; Vol. 5, p 1-43.

138. H. C. Brown; P. K. Jadhav; M. C. Desai *Tetrahedron* **1984**, *40*, 1325-1332.

139. H. C. Brown; B. Singaram *Acc. Chem. Res.* **1988**, *21*, 287-293.

140. H. C. Brown; J. V. N. V. Prasad *J. Am. Chem. Soc.* **1986**, *108*, 2049-2054.

141. K. N. Houk; N. G. Rondan; Y.-D. Wu; J. T. Metz; M. N. Paddon-Row *Tetrahedron* **1984**, *40*, 2257-2274.

Chapter 8

Oxidations

8.1 Introduction and scope

Some of the most effective and commonly used techniques in asymmetric synthesis utilize oxidation reactions, especially epoxidation and (increasingly) dihydroxylation reactions. The reasons for this begin with the general utility of the products in organic synthesis. Because of ring strain, epoxides are excellent partners for substitution reactions by a very wide variety of nucleophiles. Epoxides can also be readily converted to allylic alcohols by elimination or ketones by rearrangement. Although less important historically, the chemistry of 1,2-diols (as obtained by hydration of epoxides or directly by dihydroxylation) has received more and more attention, largely driven by the increasing availability of simple enantioselective methods for their synthesis.

Some of the most pertinent virtues of asymmetric epoxidations and dihydroxylations were already present in their classical versions. Both reactions are highly chemo-selective and can be carried out in the presence of many other functional groups. More important with respect to stereochemistry, each reaction is stereospecific in that the product faithfully reflects the E or Z configuration of the starting olefin (the nucleophilic epoxidation of α,β-unsaturated carbonyl compounds is an important exception). And one should not underestimate the importance of experimental simplicity: in most cases, one can carry out these reactions by simply adding the often commercially available reagents to a substrate in solvent, without extravagant precautions to avoid moisture or air.

This chapter summarizes several types of asymmetric oxidation reactions. Since most of the these reactions have been thoroughly reviewed, coverage is selective. Once again, the emphasis is on utility and rationales of stereoselectivity.

8.2 Epoxidations and related reactions

8.2.1 Early approaches and relevant issues of diastereoselectivity

Most early approaches to the incorporation of enantioselectivity into oxidation chemistry utilized straightforward chiral variants of the peracids so popular in standard epoxidation reactions; the essential aspects of this work have been summarized [1]. The main difficulties arose from the nature of the transition state in peracid-mediated epoxidations, as illustrated for a simple *trans* alkene (Scheme 8.1). Regardless of the size differential of the ligands in a chiral peracid R*-CO₃H, the stereogenic center(s) on R* are too far away from the developing stereogenic centers in the epoxide to exert much influence between the two possible transition structures shown in Scheme 8.1. This is true whether the transition structure has the peracid functional group and the developing epoxide in a plane (the butterfly arrangement, shown) or within planes perpendicular to each other (the spiro arrangement). Clearly,

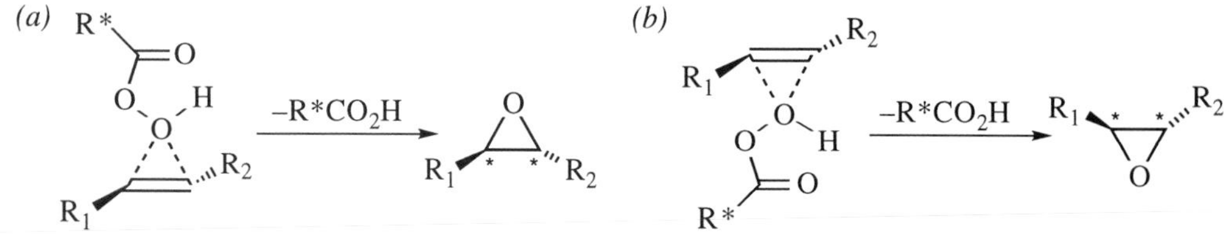

Scheme 8.1. Generalized illustration of epoxidation of a *trans*-alkene using a chiral peracid; R* = a generic chiral substituent (in early work, monoperoxycamphoric acid was often used).

a transition state in which the chirality in the reagent is closer to the reacting olefin is required.

An important clue as to how this could be done came from work done by Henbest and coworkers [2]. This group compared the diastereoselectivity of peracid oxidation reactions of 3-hydroxy and 3-acyloxycyclohex-2-enes (Scheme 8.2). When the alcohol was capped by an acetate group, the *trans* addition product predominated. Better selectivity was later obtained by placing a larger trimethylsilyl group on the allylic alcohol [3]. In both cases, the source of the selectivity could be ascribed to the approach of the reagent from the least hindered side of the molecule (*anti* to OR); put another way, the approach from one face was slowed relative to the other (Scheme 8.2a).

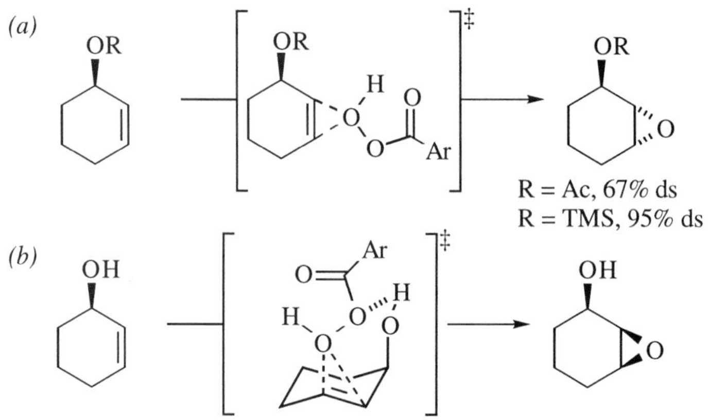

Scheme 8.2. *(a)* Addition of *m*-CPBA from the face opposite to the allylic acyloxy or trimethylsilyloxy ligand. *(b)* Proposed delivery of peracid to the β-face of the substrate mediated by the allylic alcohol group. Other modes of hydrogen bonding have been proposed for this type of reagent delivery [4,5].

In contrast, attack was found to occur *syn* to an allylic hydroxy group; obviously, simple steric effects do not account for this result. Instead, it appears that the alcohol is hydrogen bonded to the peracid in the transition state. One possible transition structure for this is shown in Scheme 8.2b; note that the allylic alcohol must occupy a pseudoaxial position to "deliver" the reagent to the olefin. In addition to this stereochemical feature, such an intrasupramolecular delivery of reagent might be expected to lower the activation barrier of the reaction due to favorable entropic considerations. Thus, rather than achieving selectivity by blocking an unfavorable path relative to an achiral model system, one might effect facial selectivity by enhancing the rate of attack from one face relative to the other. Similar directing effects have been observed in a wide variety of oxidations [6] and other reactions [7].

This idea was later extended by Sharpless and his group to include epoxidation reactions mediated by transition metals, notably those based on vanadium [5]. More than any other system, these diastereoselective epoxidation reactions laid the groundwork for the development of the first truly catalytic asymmetric epoxidation reactions. Thus, soluble metal complexes such as VO(acac)$_2$ react with simple organic peroxides, such as *tert*-butylhydroperoxide, to form a potent oxidizing system in situ. However, an allylic alcohol is *essential* for the oxidation reaction to proceed: isolated alkenes do not react under similar conditions. Accordingly, a mechanism involving intimate contact between all three components of the reaction around the transition metal was proposed. The various components of the oxidizing system seemed to be close to the reacting olefin in the transition state, as reflected in higher diastereoselectivities relative to peracid oxidations. Some outstanding results were obtained; several chemo- and stereoselective examples are depicted in Scheme 8.3 [5].

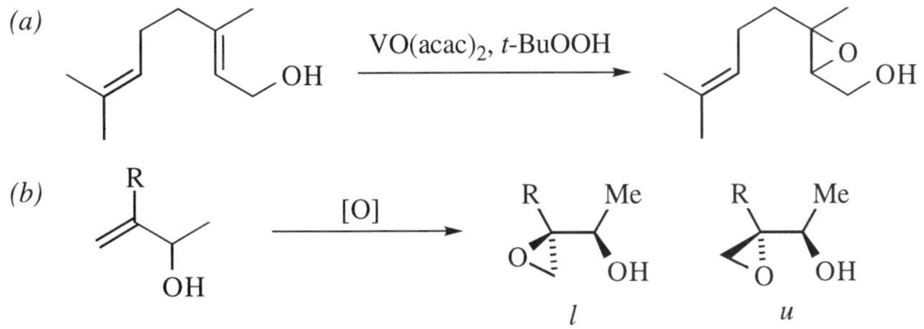

R	[O]	*l:u* ratio
H	*m*-CPBA	60 : 40
H	VO(acac)$_2$, *t*-BuOOH	20 : 80
CH$_3$	*m*-CPBA	45 : 55
CH$_3$	VO(acac)$_2$, *t*-BuOOH	5 : 95

Scheme 8.3. Some examples of V^{+5}-mediated reactions of allylic alcohols with *t*-BuOOH. *(a)* A chemoselective reaction [8]. *(b)* Stereoselective reactions of acyclic allylic alcohols, compared to results obtained using *m*-CPBA [9]. Note that better selectivity is usually obtained using the metal-based oxidation system, but not always with the same relative topicity as observed using a peracid.

The requirement for coordination of an allylic alcohol to the metal and the lack of epoxidation by *t*-BuOOH in the absence of metal guaranteed a potent rate acceleration for suitable substrates. In addition, this phenomenon allowed the very useful chemoselective differentiation between allylic alcohols and unsubstituted olefins. These experiments set the stage for the development of an efficient asymmetric epoxidation reaction.

8.2.2 Epoxidations

Katsuki–Sharpless asymmetric epoxidation. Since its introduction in 1980 [10], the Katsuki–Sharpless asymmetric epoxidation (AE) reaction of allylic alcohols has been one of the most popular methods in asymmetric synthesis ([11-14]). In this work, the metal-catalyzed epoxidation of allylic alcohols described in the previous section was rendered asymmetric by switching from vanadium catalysts to titanium ones and by the addition of various tartrate esters as chiral ligands. Although subject to some technical improvements (most notably the addition of molecular sieves, which allowed the use of catalytic amounts of the titanium–tartrate complex), this recipe has persisted to this writing.

In general, the reaction accomplishes the efficient asymmetric synthesis of hydroxymethyl epoxides from allylic alcohols (Scheme 8.4). Operationally, the catalyst is prepared by dissolving titanium isopropoxide, diethyl or diisopropyl tartrate (DET or DIPT, respectively), and molecular sieves in CH_2Cl_2 at -20 °C, followed by addition of the allylic alcohol or *t*-BuOOH. After a brief waiting period (presumably to allow the ligand equilibration to occur on titanium), the final component of the reaction is added.

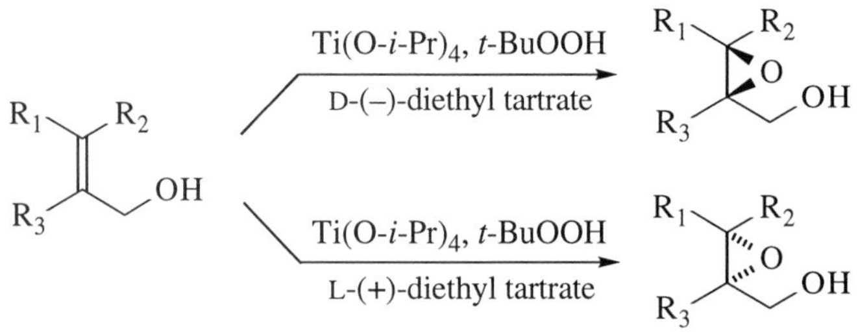

Scheme 8.4. The asymmetric epoxidation reaction of allylic alcohols. As usually carried out in CH_2Cl_2 at -20 °C, the reaction generally affords the product epoxides in excellent yields (>70%) and enantioselectivities (>95%). In addition, the reaction is predictable with respect to the predominant enantiomer obtained according to the above scheme.

The virtues of the AE are obvious. In each case, the components are commercially available at reasonable cost. The availability of tartrate esters in both enantiomeric forms is especially fortunate, allowing the synthesis of either enantiomer of a desired product. A key feature in this regard is the predictability of the process; no exceptions to the trend shown in Scheme 8.4 have been noted in reactions using achiral substrates. And the simplicity of standard epoxidation reactions has been effectively retained, especially considering that the chiral catalyst system is prepared in situ.

A simplified version of the mechanism proposed by Sharpless is given in Scheme 8.5. Early work on the mechanism of this useful and important reaction has been reviewed [11], and references to more recent mechanistic studies have been collected [13]. To date, evidence in support of this mechanism has included extensive kinetic studies, spectroscopy, and molecular weight determinations [15,16].[1]

---

[1] An alternative mechanism involving a monomeric complex has also appeared [17].

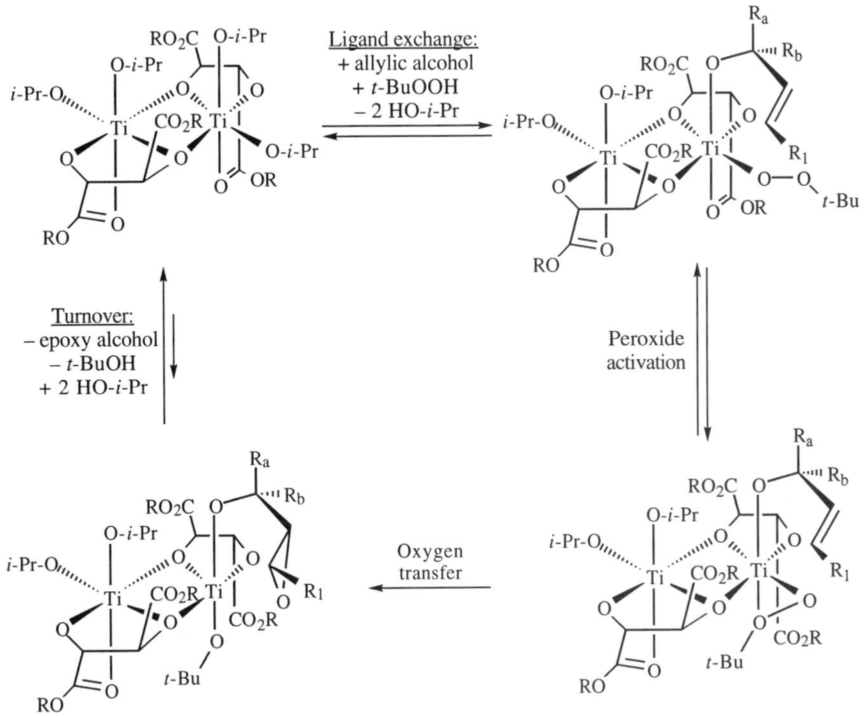

Scheme 8.5. Proposed mechanism for the Sharpless asymmetric epoxidation reaction of allylic alcohols, shown here for a simple *trans*-allylic alcohol. For the AE reaction, $R_a = R_b = H$. When one (or occasionally both) of these substituents are alkyl groups, the Scheme pertains to the kinetic resolution sequence described in the next section.

A very important aspect of this mechanism is not shown in the scheme. This is the formation of the titanium–tartrate species from its commercially available precursors, Ti(O-*i*-Pr)$_4$ and the dialkyl tartrate. The equilibrium in this step lies far toward the formation of the chiral complex formed; this is critical because the enantioselectivity of the process depends on the absence of any active *achiral* catalyst. Note that the complex as drawn (in the upper left of Scheme 8.5) is dimeric and has a C_2 axis of symmetry. This structure has not been isolated in the solid state, but is based in part on an X-ray structure of a related tartramide complex [18]. The situation is undoubtedly complicated by dynamic equilibria between this form and other species in solution.

Without specifying the order of events, two isopropoxide ligands must be replaced by one molecule of peroxide and one molecule of allylic alcohol to give the species shown in the upper right of Scheme 8.5 (recall that, in reality, the peroxide and allylic alcohol are added at different times). The ease of such ligand exchange reactions in these titanium complexes largely accounts for their utility here. The other function of the titanium is to activate the distal oxygen of the peroxide for transfer.

At this point (lower right of Scheme 8.5), the complex is fully loaded and ready for oxygen transfer to the alkene. In this mechanism, the allylic alcohol occupies a position *cis* to the reactive peroxide oxygen. In the AE reaction ($R_a = R_b = H$), the diastereofacial selectivity of the olefin in the complex results from the avoidance of the allylic carbon and a carboxylic ester (Figure 8.1b). After oxygen transfer, the final step is the exchange of the reaction products, epoxy alcohol and *t*-BuOH, with other ligands to give either the starting complex or some other species on the way to the loaded catalyst. The importance of turnover must not be underappreciated, for without it one may have a reagent but never a catalyst.

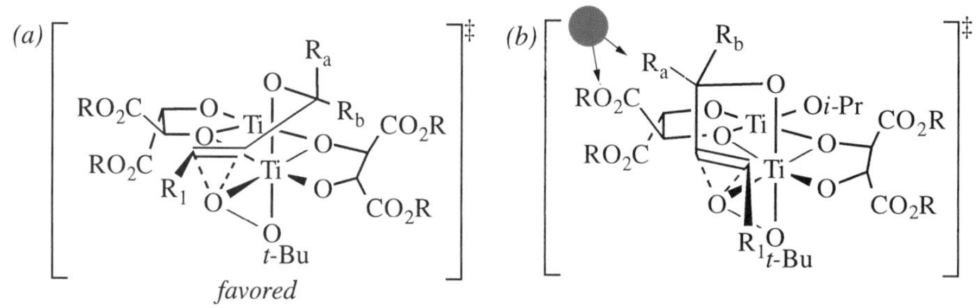

Figure 8.1. Proposed steric interactions leadering to enantioselectivity in the Sharpless AE reaction.

This model is consistent with much that is known about the scope of the Sharpless AE. The most common and best-behaved substrates are simple *trans*-allylic alcohols; their reactions are generally fast and reliably give products with very good enantioselectivity (>95% es). Inspection of the loaded complex in Figure 8.1 might suggest that substrates with an alkyl group *cis* to the hydroxymethyl substituent (*i.e.*, where $R_2 \neq H$ in Scheme 8.4) may be less stable due to steric interactions with the main portion of the catalyst. Indeed, such compounds are the slowest-reacting and subject to the most variation in enantioselectivity. However, there are examples of excellent results using alkenes of every conceivable type, although some work may need to be invested in optimizing reaction conditions (Table 8.1).

AE reactions of simple olefins. The Sharpless AE reaction has been supplemented by other approaches to asymmetric epoxide synthesis; the most evident goal being to obviate the need for an allylic alcohol. Attempts to carry out asymmetric epoxidation reactions on simple olefins have utilized transition-metal-containing catalysts such as porphyrins as well as stoichiometric chiral reagents (peroxides, dioxiranes, and oxaziridines). These approaches have been summarized [19].

The most promising procedure so far was introduced by Jacobsen and coworkers in 1990 [20] and has been reviewed [19]. The method uses chiral, C_2-symmetric (salen)Mn complexes, such as shown in Scheme 8.6. Such materials are very easily prepared by the condensation of a chiral diamine with a substituted salicylaldehyde, followed by coordination of the metal. The ready availability of both components and the swift synthesis of the target complexes permits easy access to a great many catalyst variations, which facilitates reaction optimization. The starting Mn(III) complex is subjected to *in situ* oxidation with the stoichiometric oxidant, usually NaOCl (bleach!). The use of this inexpensive and relatively safe oxidant is another virtue of this system.

Although some outstanding results have been obtained, there are some limitations to the scope of this process (see examples in Table 8.2). The reaction works best with *cis*-olefins, in contrast to the situation with the Sharpless reaction. This can be accommodated with a side-on approach of reagent to the catalyst system, as depicted

Table 8.1. Examples of Sharpless AE Reactions. These reactions were carried out under catalytic conditions (<10 mol % of Ti(OR)4 and tartrate), except for entry 8 (done using stoichiometric catalyst).

Entry	Product	Tartrate	% Yield	% es	Ref.
1		(–)-DIPT	50-60	94-96	[21]
2		(+)-DET	85	97	[21]
3		(+)-DET	54	83	[22]
4		(+)-DIPT	63	>90	[21]
5		(+)-DET	88	97	[21]
6		(–)-DIPT	87	95	[23]
7		(+)-DET	77	96	[21]
8		(+)-DET	80	94	[10]
9		(+)-DET	95	95	[21]
10		(+)-DET	*not reported*	>95	[24]

in Scheme 8.6b. However, differences in catalyst structure can lead to reversal of the sense of selectivity. This observation has been attributed to attack from different sides of the complex [25].

The reaction affords the highest selectivities with conjugated, preferentially cyclic olefins. Acyclic *cis* olefins are subject to various amounts of isomerization, one observation that led to the radical mechanism proposed (Scheme 8.6c). This isomerization can be facilitated by the addition of chiral quaternary ammonium salts, leading to synthetically useful (>10:1 *trans:cis*, >80% ee) conversions of *cis* olefins to *trans*-dialkyl epoxides (*cf.* entries 2 and 3 in Table 8.2) [26]. Further improvements have resulted in a substantial broadening of this profile, obtaining some good-to-excellent selectivities from styrene [27] and tri- and tetra-substituted olefins that are not subject to isomerization (either due to symmetry or by constraining the double bond in a ring) [28,29].

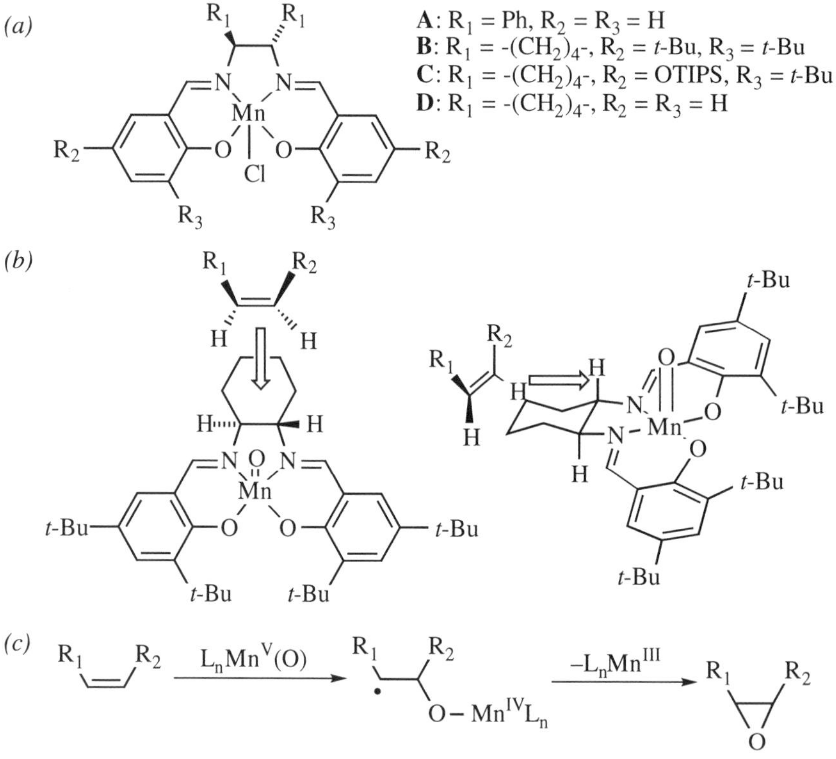

(a)

A: R_1 = Ph, R_2 = R_3 = H
B: R_1 = -(CH$_2$)$_4$-, R_2 = *t*-Bu, R_3 = *t*-Bu
C: R_1 = -(CH$_2$)$_4$-, R_2 = OTIPS, R_3 = *t*-Bu
D: R_1 = -(CH$_2$)$_4$-, R_2 = R_3 = H

(b)

(c)

Scheme 8.6. Jacobsen's approach to epoxidation of simple olefins. *(a)* A few examples of (salen)Mn(III) epoxidation catalysts prior to reaction with NaOCl. *(b)* Two views of the proposed side-on approach of a generic *cis* olefin to the loaded catalyst. Different approach vectors have been proposed depending on the catalyst structure [25]. *(c)* Proposed stepwise mechanism of the reaction [26].

In sharp contrast to the oxidation reactions of electron-rich olefins just described, attempts to carry out nucleophilic epoxidation reactions of α,β-unsaturated carbonyl compounds have enjoyed only limited success (Scheme 8.7) [19]. The most successful attempts have been with chalcones, using standard basic peroxidation conditions with additives such as a quinine-derived phase-transfer catalyst first

Table 8.2. Examples of Jacobsen AE Reactions. See Scheme 8.6(a) for catalyst structures.

Entry	Olefin	Catalyst	% Yield (*cis*/*trans*)	% es	Ref.
1	Ph / Me	A	71 (*trans* only)	10	[20]
2	Ph / Me	B	84 (92:8)	96 (*cis*) 91.5 (*trans*)	[19,25]
3	Ph / Me	C	not reported (5:95)	90.5 (*trans*)	[26]
	Used a quinine-derived additive				
4	NC ... O	B	96	98.5	[19,25]
5	Me₃Si	B	65 (16:84)	82 (*cis*) 99 (*trans*)	[30]
6		B	73	82	[31]
7	Me, Ph, Ph	B	87	94	[28]
8	Br ... O, Me, Me, Me, Ph	D	72	90.5	[29]
9	Ph, Ph, Ph, Me	A	12	72	[29]

reported by Wynberg in 1976 [32] or poly-L-leucine [33]. Although seemingly limited to this substrate type, the products have been converted to the corresponding α,β-epoxy esters via a regioselective Baeyer-Villiger oxidation in at least one case [34]. More recently, a glimmer of success in applying organometallic catalysis to this problem has been seen in a platinum-based approach, although the products have yet to be isolated in synthetically useful yields [35].

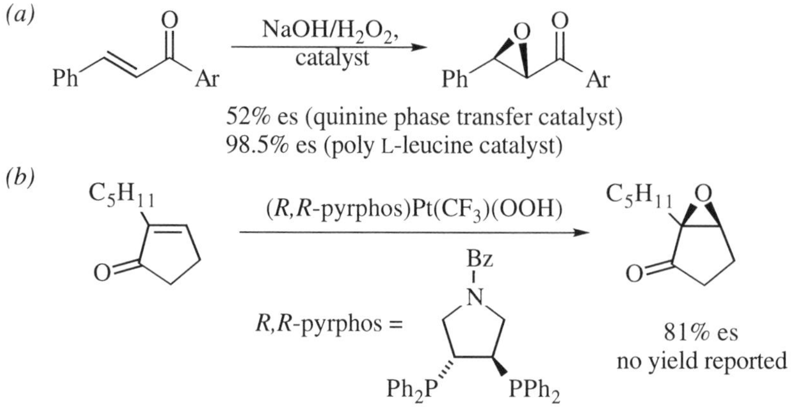

(a)

52% es (quinine phase transfer catalyst)
98.5% es (poly L-leucine catalyst)

(b)

81% es
no yield reported

Scheme 8.7. Nucleophilic epoxidation reactions of enones. *(a)* Epoxidation of chalcone using phase-transfer [32] or polymeric amino acid [33] catalysis. *(b)* Platinum-based epoxidation method [35].

8.2.3 Sharpless kinetic resolution

Inspection of the mechanism in Scheme 8.5 suggests that the Sharpless epoxidation should be relatively insensitive to configuration of any stereocenter in an alkene substituent with one very important exception: the allylic carbon bearing the alcohol. Indeed, good diastereoselectivity was often obtained in reactions of various chiral allylic alcohols with achiral epoxidizing agents (Scheme 8.3). Substitution at this particular position is important because of its proximity to the bulk of the catalyst. Thus, one might expect substitution at R_a to be well-tolerated because this group points away from the catalyst, whereas R_b should be much more sterically encumbered (Figure 8.2). Some experimental observations that address this issue and ultimately led to the application of the Katsuki–Sharpless catalyst to kinetic resolution reactions are shown in Scheme 8.8.

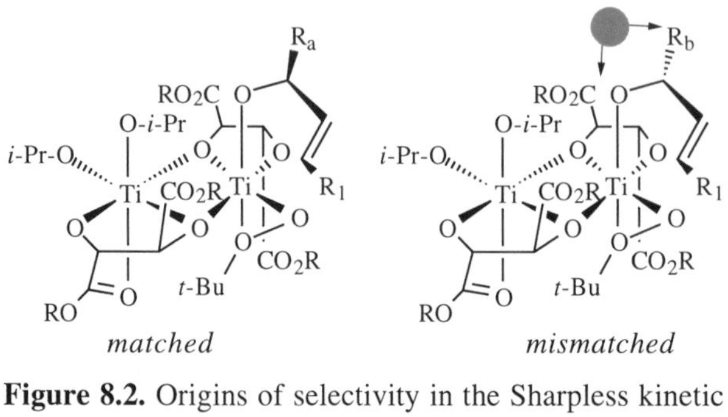

matched   *mismatched*

Figure 8.2. Origins of selectivity in the Sharpless kinetic resolution procedure.

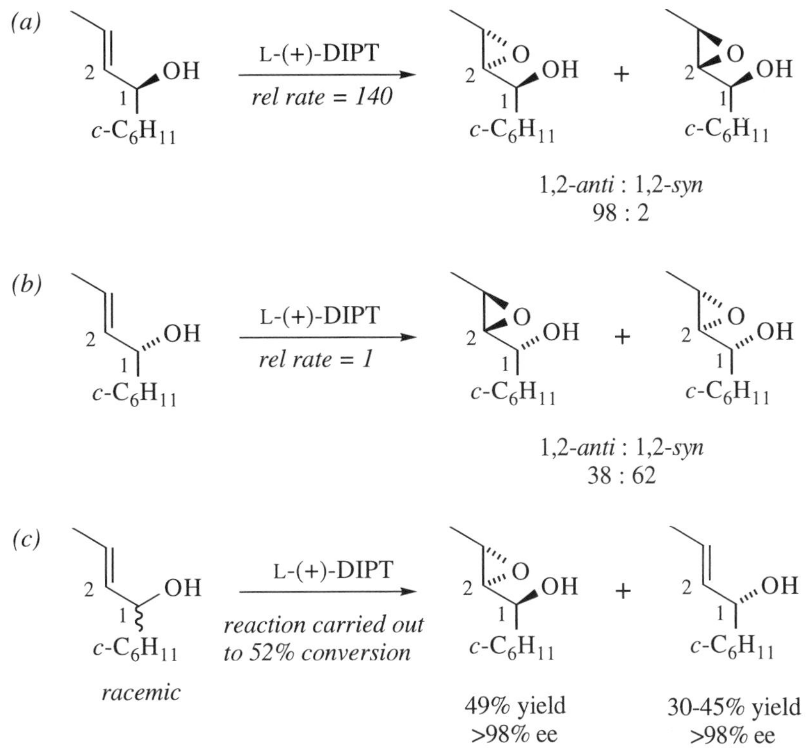

(a)

L-(+)-DIPT

rel rate = 140

1,2-*anti* : 1,2-*syn*
98 : 2

(b)

L-(+)-DIPT

rel rate = 1

1,2-*anti* : 1,2-*syn*
38 : 62

(c)

racemic

L-(+)-DIPT

*reaction carried out
to 52% conversion*

49% yield   30-45% yield
>98% ee   >98% ee

Scheme 8.8. Reactions of a chiral allylic alcohol under Sharpless epoxidation conditions (Ti(O-i-Pr)$_4$, *t*-BuOOH) using the chiral tartrates given (DIPT = diisopropyltartrate). *(a)* The "matched" case, in which the preferred approach of the asymmetric catalyst and the diastereoselectivity of the substrate are the same. *(b)* The "mismatched" case. *(c)* An example of a Sharpless kinetic resolution (KR).

Like the vanadium-based catalysts, the Sharpless AE system intrinsically favors 1,2-*anti* products; this is because the cyclohexyl group in Scheme 8.8a occupies the position denoted by group R$_a$ in Figure 8.2, away from the catalyst. In fact, this diastereoselectivity is somewhat amplified relative to achiral titanium catalysts. When the *S* allylic alcohol is used with (+)-DIPT, a matched pair results (Scheme 8.8a). The strong enantiofacial selectivity of the L-(+)-DIPT catalyst clashes with the *R* substrate's resident chirality (this is the case shown in Figure 8.2 with R$_b$ = cyclohexyl). In this mismatched pair, the preference of the chiral catalyst for α attack moderately exceeds that of the allylic alcohol for 1,2-*anti* product (Scheme 8.8b). The most important consequence is that *the latter reaction is 140 times slower than the former*.

Using a racemic allylic alcohol, one can take advantage of this rate differential to selectively epoxidize the more reactive *S* isomer in the presence of its antipode. This procedure is known as a Sharpless kinetic resolution (KR) [13,36]. The KR has very wide applicability for the preparation of both 1,2-*anti* epoxy alcohols and the unreacted allylic alcohol, often with very high enantioselectivities (note that the diastereomeric 1,2-*syn* series is not generally available by this technique). In general terms, carrying out the reaction to lower conversions will maximize the yield and enantiomeric purity of the epoxy alcohol, with longer conversions sometimes allowing very high (>99%) enantiomeric purities of allylic alcohols, albeit in reduced yields.[2] Scheme 8.8c shows an example of what is possible under optimized conditions with a favorable substrate.

The KR procedure is not limited to making simple epoxides bearing an adjacent stereogenic center. Figure 8.3 depicts several interesting classes of molecules that have been resolved using KR procedures. Although results have been spotty, alternative sites of oxidation have included attempts with alkynes, furans, and β-amino alcohols. Of particular interest to stereochemistry buffs are procedures that result in different classes of enantiomerically pure compounds, such as those with axial chirality (cycloalkylidenes or allenes) or planar chirality. And, although not as far along as the now-standard reactions utilizing allylic alcohols, some progress has been made in extending both the AE and KR procedures to homoallylic alcohols [37].

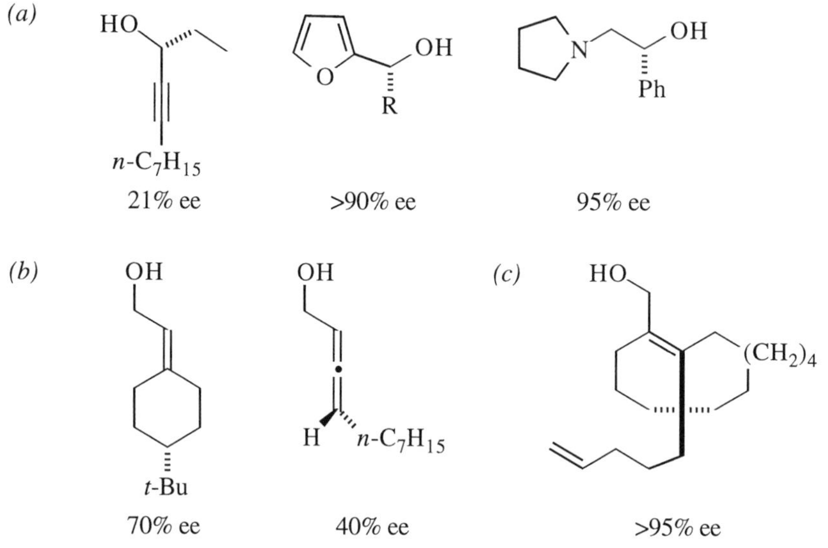

Figure 8.3. Examples of molecules prepared in enantiomerically enriched form using Sharpless KR procedure. (*a*) Compounds having alternative sites of oxidation: acetylene [38], furan [39], and amine [40]. (*b*) Compounds bearing axial chirality [38]. (*c*) An alkene with planar chirality [41].

8.2.4 Applications of asymmetric epoxidation and kinetic resolution procedures

The importance of the Sharpless AE and KR procedures is best measured by the speed with which they have become a part of the synthetic chemist's "bag of tricks". Although a measure of their utility can be gleaned from examples given in sections 8.2.2 and 8.2.3, their influence has been far too pervasive to allow even a partial representative listing here. However, a few examples where these reactions have been used to illustrate principles of more general stereochemical interest will be summarized in this section.

Carbohydrate synthesis. Save the all-important hydroxymethyl group that the titanium reagent uses as a handle, the Sharpless AE is remarkably insensitive to

---

2   The reader is directed to the original literature for a quantitative treatment [13,36].

stereogenic centers extant in the substrate. This has lead to the wide use of this system for *reagent-based stereocontrol*, wherein the chirality of a new stereocenter is determined simply by pulling the appropriate reagent off of the shelf (as opposed to *substrate control*, in which a new element of chirality is installed under the influence of those already in the reactant; see Section 1.5). This strategy was nicely illustrated by the synthesis of all eight isomeric hexoses in their unnaturally-occurring L-series, summarized for L-allose in Scheme 8.9 [42,43].

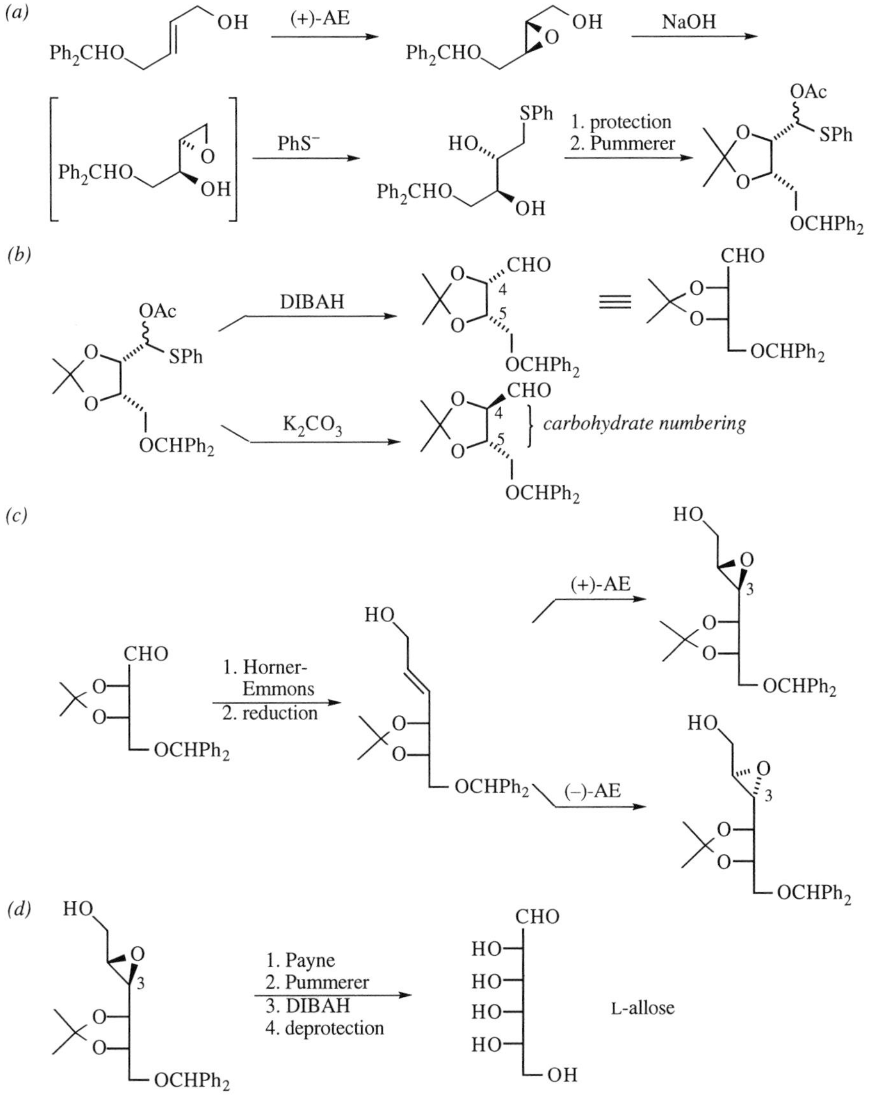

Scheme 8.9. Reagent-controlled synthesis of L-allose ((+)-AE = Sharpless AE using L-(+)-DIPT; (–)-AE = Sharpless AE using D-(–)-DIPT). *(a)* A Sharpless AE followed by Payne rearrangement and oxidation. *(b)* Stereodifferentiation of the C-4 and C-5 stereocenters. *(c)* Chain extension followed by reagent-controlled oxidation of the olefin. *(d)* Completion of the synthesis.

This is an excellent example of an iterative sequence that takes full advantage of the stereochemical versatility of the Sharpless AE reaction. The synthesis prepares the target carbohydrate in the C-6 → C-1 direction and starts with a readily prepared *trans*-allylic alcohol. The first AE directly sets the C-5 stereogenic center (carbohydrate numbering), now requiring that the epoxide be opened in a regio- and stereoselective manner and that the primary alcohol be converted to the oxidation state of an aldehyde. Both tasks were accomplished by a Payne rearrangement in base, which isomerizes the epoxy alcohol with strict inversion at the C-4 center. The new epoxide thus formed is monosubstituted and therefore suffers a kinetically favored attack by an external nucleophile, in this case the thiophenolate anion.[3,4]

Next, the researchers took advantage of the acetonide protecting group to control the relative configuration between C-4 and C-5 (the use of a protecting group for this kind of stereochemical finesse is *ancillary stereocontrol* [46]). A mild, nonbasic unraveling of the aldehyde by reduction at the acetate carbonyl group was accomplished with diisobuytyl aluminum hydride, which left the target in its initial *cis* configuration about the five-membered ring. Alternatively, basic deprotection led to epimerization to the *trans* isomer. Two interesting points:

1. Isomerization is only possible because of poor overlap between the enolate leading to epimerization and the C-5 carbon-oxygen bond (which doesn't allow for β-elimination).
2. This maneuver was used instead of making the same compound via a similar sequence using the *cis* allylic alcohol.

This is a good example of stereochemical divergence from a single precursor and obviates the necessity of preparing and working with the less reactive *cis* olefins. Overall, the conversions indicated in Schemes 8.9a and b constitute a single iteration of the synthesis.

Scheme 8.9c shows how the aldehyde could be homologated to a new allylic alcohol and how simple choice of tartrate ligand afforded the diastereomeric epoxides shown, since the AE process effectively ignores the resident stereocenter in the new substrate. This is the essence of reagent-controlled synthesis: the utilization of a tool for enantioselective elaboration to permit the selective synthesis of diastereomeric compounds. Once prepared, the utilization of the diisobutyl aluminum hydride variant of the iterative sequence followed by final deprotection steps led to the synthesis of L-allose. A useful exercise is to arbitrarily draw an isomer of allose and synthesize it using this technique (on paper, of course), or to imagine a modification that would lead to the corresponding pentoses [47].

Group selective reactions of divinyl carbinols. It is important to remember that the reagent control strategy is inapplicable to situations where the resident chirality is on the allylic position bearing the hydroxyl "handle" for the catalyst. However, the pref-

---

3   The regiochemistry of ring opening in epoxy alcohols has been more generally examined [44].

4   Although known prior to the discovery of the Sharpless AE, this use of the Payne rearrangement is a good example of how the availability of a particular functional array by asymmetric synthesis provoked a reaction's further development [45]. In this case, the product sulfide allows the chemoselective conversion of this carbon to the oxidation state of the aldehyde, in the guise of an acetoxy sulfide.

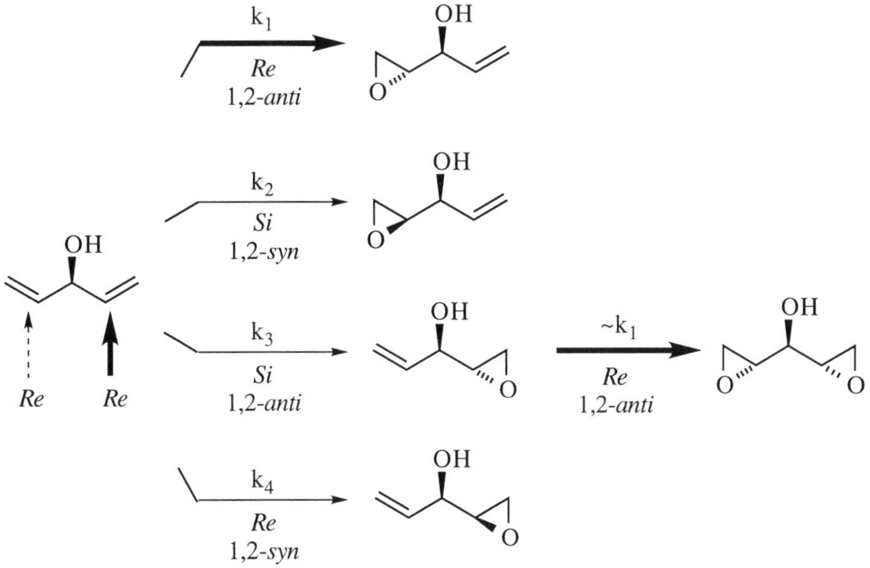

Scheme 8.10. Reaction of divinyl carbinol under (+)-AE conditions as an example of enantiotopic group selectivity in epoxidation chemistry. Matched cases of enantiofacial selectivity are shown with bold arrows. Qualitative rate differences are on the order $k_1 \gg k_2$, $k_3 \gg k_4$ (without specifying an order for k_2 *vs.* k_3 (however, *cf.* Scheme 8.8b). Note that the products arising from the pairs k_1/k_3 and k_2/k_4 are enantiomers.

erence for 1,2-*anti* product has been cleverly applied to a problem in diastereotopic group selectivity (Scheme 8.10) [48-52].

The two olefins carry a total of two enantiomeric pairs of diastereotopic faces. When a tartrate-titanium epoxidation system is allowed to react with this substrate, approach to only one of these four faces simultaneously satisfies the requirements of both the catalyst (which prefers the *Re* face) and the substrate (which prefers 1,2-*anti* addition). To the extent that the rate of epoxidation at this face exceeds that of the others (k_1 in Scheme 8.10), one product predominates. Minor diastereomers result from pathways k_2 and k_4. However, note that the pathway with a rate of k_3 (mismatched: 1,2-*anti* diastereoselectivity combined with disfavored *Si* enantiofacial attack) affords the enantiomer of the major isomer.

Schreiber has noted that this group-selective process may nonetheless be expected to provide products with very high enantioselectivity because the disfavored enantiomer resulting from pathway k_3 *still has the most favorable face available for reaction* [51]. Thus, to the extent that product from pathway k_3 accumulates, it is rapidly siphoned off at a rate comparable to k_1. Thus, the problem of enantiomer separation at the end of the reaction can largely be replaced by the problem of diastereomer and side product separation (although itself never an issue to be taken lightly; see Section 2.1).

The availability of a path for selective destruction of the unwanted enantiomer means that the desired product can be obtained with very high enantioselectivity when the reaction is pushed to higher conversions. However, this will come at the expense of overall yield because some of the desired product will also react further under such conditions [51,53]. Provided that one is able to distinguish either end of a developing chain, such reactions have promise in applications involving two-dimension chain elongation strategies [54].

Epoxide-opening reactions. The most common use of epoxides is in S_N2 ring-opening reactions leading to 1,2-difunctionalized compounds. Clearly, the availability of enantiomerically enriched epoxides, when combined with appropriate regiochemical control of their opening, has enhanced the applicability of this approach to the preparation of enantiomerically pure compounds. An alternative approach is to begin with a *meso* epoxide, and then follow this reaction with a sequence able to distinguish between the enantiotopic carbons of the epoxide (Scheme 8.11a).

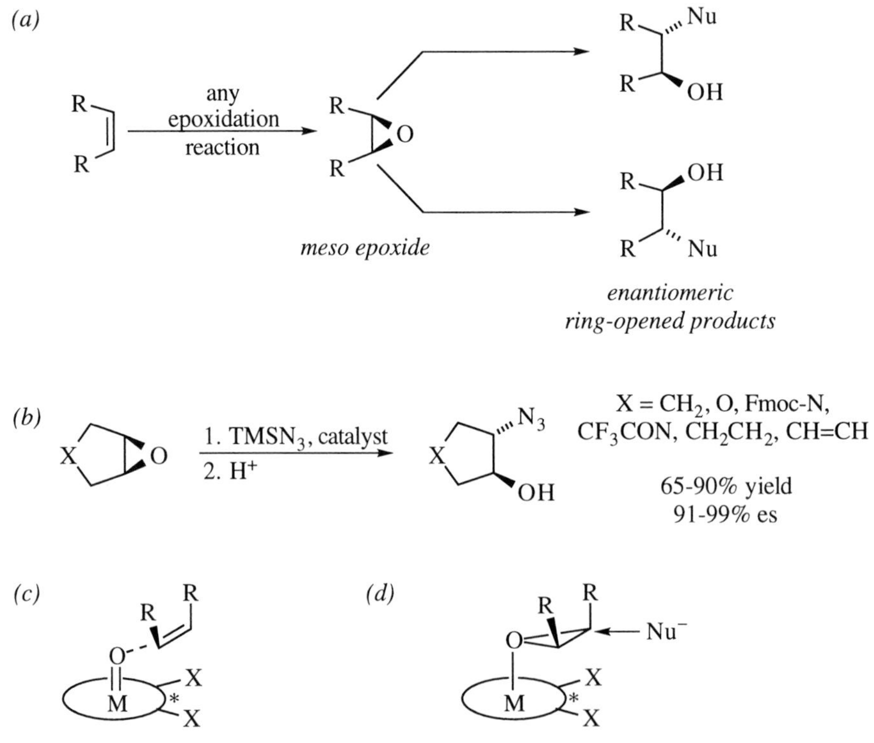

Scheme 8.11. *(a)* Group-selective ring-opening of meso epoxides by nucleophiles leads to enantioselective syntheses of 1,2-difunctionalized compounds. *(b)* Azido alcohol synthesis from epoxides and trimethylsilyl azide as catalyzed by (salen)CrCl complexes (see Scheme 8.6a for general structures of salen catalysts) [55]. Comparison of proposed ensembles for *(c)* asymmetric epoxidation and *(d)* Lewis-acid activation of epoxides for nucleophilic attack.

Several groups have addressed this problem using chiral aminating reagents [56] or, more recently, chiral catalysts [57]. Jacobsen has reported the efficient use of (salen)Cr(III) complexes for such conversions (Scheme 8.11b; *cf.* Section 8.2.2 and Scheme 8.6) [55]. These authors point out the similarities between transition structures for oxygen transfer in an epoxidation reaction (Scheme 8.11c) and epoxide activation (Scheme 8.11d), suggesting that the nonbonded interactions leading to selectivity ought to be similar in both cases.

8.2.5 Aziridinations

Although less commonly investigated, several catalytic, enantioselective aziridination reactions have also been developed. As an example, copper complexed to a chiral bisoxazoline ligand such as shown in Scheme 8.12a has been shown to catalyze the addition of *N*-(*p*-toluenesulfonylimino)phenyliodinane across a double bond [58]. Some promising results have been obtained (Scheme 8.12b), but work to fully define and optimize the range of olefins susceptible to this process is still ongoing.

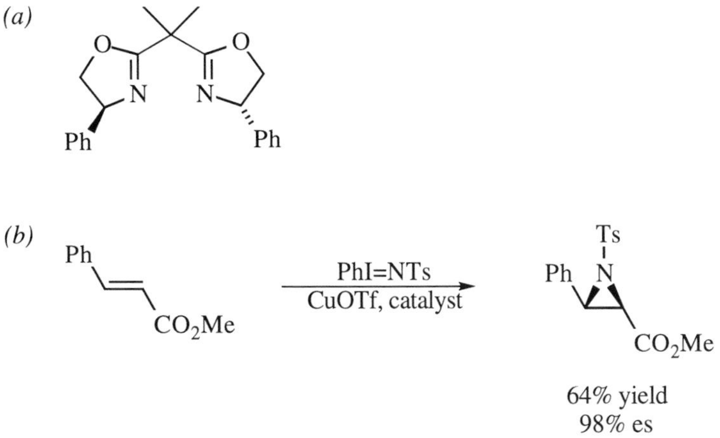

(a)

(b)

64% yield
98% es

Scheme 8.12. *(a)* An example of a bidentate catalyst for copper salts used in asymmetric aziridination reactions. *(b)* An asymmetric aziridination reaction [58].

In addition, chiral salen complexes of Mn [59] and Cu [60] have been found to catalyze similar reactions with moderate enantioselectivities. In contrast to the salen complexes used in Mn-mediated epoxidation reaction, which use tetradentate complexes, the best results were obtained with bidentate ligands. Mechanistic work has implicated a metal-bound nitrene species in the reaction [61].

8.3 Dihydroxylations[5]

The synthesis of vicinal diols from olefins using OsO_4 complements epoxidation/hydrolysis as a route to 1,2-diols (Scheme 8.13 [65]). Both reagents effect *cis* difunctionalization of an olefin, but since the epoxide-opening step involves an inversion of configuration, the two routes afford opposite diastereomers beginning with a single olefin geometry. The development of an efficient asymmetric dihydroxylation process, again pioneered by the Sharpless laboratory, has become the most general single method for the oxygenation of unactivated olefins (*i.e.*, those without an allylic alcohol). The rapid acceptance of the Sharpless asymmetric dihyroxylation (AD) reaction by the organic chemistry community is a lesson in the value of a method able to convert a simple, readily available functional group into another common moiety, particularly when the reaction can be done stereoselectively.

---

[5] Reviews: ref. [62-64].

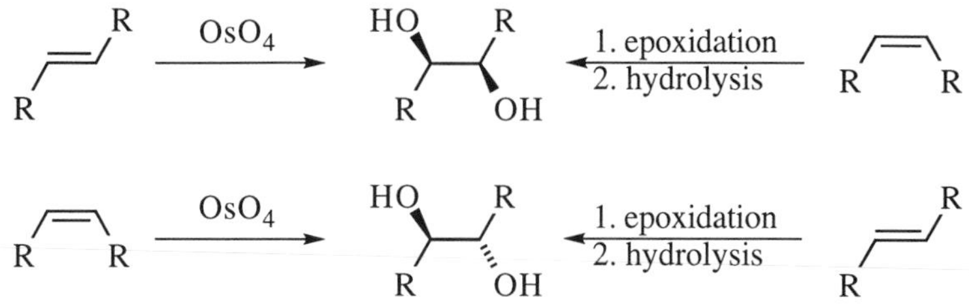

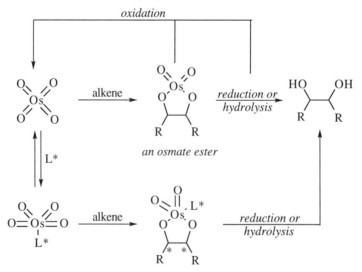

Scheme 8.13. 1,2-Diol synthesis from alkenes via direct osmylation or epoxidation followed by hydrolysis.

8.3.1 Development , scope, and mechanism

Three things were required to realize a useful catalytic, asymmetric dihydroxylation reaction:

1. An efficient osmylation reaction using achiral catalysts.
2. Appropriate chiral ligands to effect face selectivity in the addition of the osmylation reagent to an alkene.
3. Some way to enforce the participation of the chiral osmium catalyst in the reaction to the exclusion of non-stereoselective pathways.

These issues are summarized in the *extremely* simplified mechanism given in Scheme 8.14. All osmylation reactions ultimately afford the osmate ester shown, although the mechanism for this step has been controversial.[6] Note that the osmium is reduced over the course of the reaction. In efforts to minimize the amount of toxic osmium used in these reactions, catalytic methods using a variety of stoichiometric reoxidants for osmium were introduced, beginning with the introduction of KClO4 in 1917 by Hofmann [66] and including the convenient Upjohn process, which uses *N*-methylmorpholine-*N*-oxide for this purpose [67].

Scheme 8.14. Simplified mechanism for the dihydroxylation of olefins. A more complete description is available [63].

---

6   Both [3 + 2] concerted mechanisms and schemes involving the formation of a metallooxetane intermediate by [2 + 2] cycloaddition followed by rearrangement have been proposed.

An asymmetric dihydroxylation reaction requires some form of "chiral osmium". The first evidence that amine ligands could affect the chemistry of this dihydroxylation reaction was published by Criegee, who found that the reaction was accelerated by the addition of pyridine [68]. Sharpless later showed that useful enantioselectivities (up to 97% es) could be realized when chiral amines were added to OsO$_4$-mediated oxidation reactions. The ligands used by the Sharpless group were representatives of the cinchona alkaloid family, dihydroquinidine (DHQD) and dihydroquinine (DHQ) (Figures 8.4a and b). Note that although these ligands are nearly enantiomeric with respect to the quinuclidine base and aromatic side chain, their mirror symmetry is spoiled by the placement of the ethyl group on the bicyclic portion of the molecules. The existence of these alkaloids in the chiral pool would prove to be fortuitous indeed, as modifications of the alcohol group would be the key to the development of effective *catalytic* complexes. In addition, other workers have reported a variety of totally synthetic ligands able to effect highly selective stoichiometric dihydroxylation reactions; these ligands very often incorporate C_2 symmetry into their design (*e.g.*, Figures 8.4c and d).

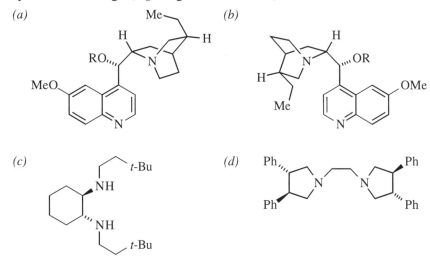

Figure 8.4. Representative ligands used in stoichiometric, asymmetric dihydroxylation reactions. *(a)* Dihydroquinidine (DHQD) and *(b)* dihydroquinine (DHQ) are used for stoichiometric osmylation reactions when R = H; effective dihydroxylation catalysts result from appropriate modifications at this position (*e.g.*, see Figure 8.5 below). Also, *(c)* [69] and *(d)* [70] are examples of C_2 symmetrical ligands used in stoichiometric reactions.

Sharpless reported the first generally useful catalytic version of the reaction in 1987 [71]. This landmark paper showed that the reaction could be rendered catalytic by combining modified cinchona ligands with the Upjohn process (Scheme 8.4). This use of *ligand-accelerated catalysis* is critical to the success of a catalytic AD reaction because of the preequilibrium present between OsO$_4$ and OsO$_4$L* in solution (Scheme 8.14). Unless the equilibrium lies so far to the latter species as to effectively lower the concentration of OsO$_4$ to zero, the nonselective reaction of OsO$_4$ with the

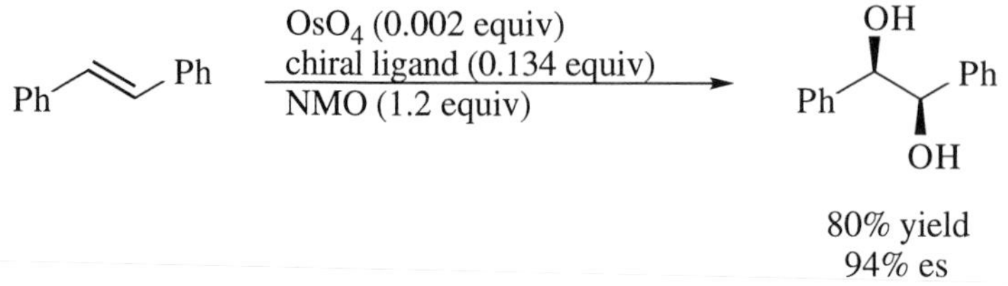

Scheme 8.15. An early example of a catalytic dihydroxylation reaction [71]. The chiral ligand used here is the *p*-chlorobenzoate ester of DHQD (Figure 8.4a).

olefin would compete with that of OsO$_4$L*, lowering the enantioselectivity of the overall process. The ligand acceleration effect provided by the chiral amine sidesteps this issue by ensuring that the OsO$_4$L* pathway is also the most kinetically competent.

An interesting contrast exists between the development of the Sharpless asymmetric epoxidation reaction and the asymmetric dihydroxylation process. In the former case, the original reagents and protocol for carrying out the reaction have basically survived in their original form. However, the AD has been subjected to a great deal of optimization since its introduction, both in terms of ligand design and modification of conditions. In particular, protocols that cut down on interference by non-selective pathways have helped raise the utility of the overall procedure to its current high level. For example, the intrusion of a second catalytic cycle was proposed to lower overall stereoselectivity of the AD (Scheme 8.16). In this second cycle, the osmate ester formed by the reaction of one olefin with the chiral Os-cinchona complex was proposed to undergo oxidation and become itself a reactive dihydroxylation reagent, albeit one that had little enantiofacial selection. This pathway could be minimized by mandating slow addition of the alkene (allowing the osmate ester time to undergo hydrolysis and reoxidation [72]), through the use of K$_3$Fe(CN)$_6$ as the reoxidant in place of NMO [73], or by increasing the rate of hydrolysis by adding MeSO$_2$NH$_2$ to the reaction mixture [74]. In particular, the use of the iron-based reoxidant remands the job of Os reoxidation to the aqueous portion

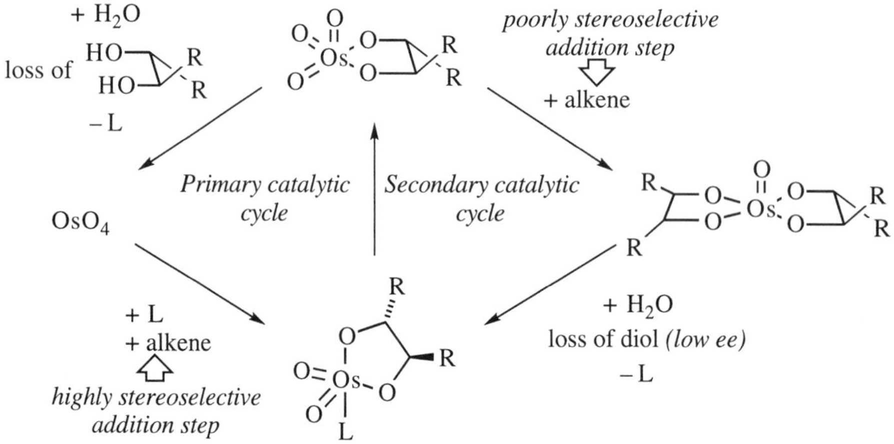

Scheme 8.16. The two catalytic cycles proposed for the Sharpless AD reaction [72].

of a biphasic reaction mixture, thus "protecting" the organic osmate ester from inopportune oxidation prior to hydrolysis. The addition of sulfonamide is doubly useful because it increases the turnover rate of the reaction and facilitates the dihydroxylation of otherwise sluggish substrates.

Finally, a mind-boggling number of analogs of the original catalysts have been prepared and tested. Although the progression through this series makes for interesting reading [62-64], only the most generally efficacious catalysts are cited in Figure 8.5. The most striking advance is the use of dimeric species featured in PHAL and PYR, the most general of the catalysts (Figures 8.5a and b). The former ligand has been formulated along with $K_2OsO_2(OH)_4$ (a non-volatile source of Os), $K_3Fe(CN)_6$, and either DHQ or DHQD, respectively; these stable, storable powders contain all of the necessary ingredients for AD reactions and are known as AD-mix-α or AD-mix-β. These mixtures are commercially available. Interestingly, although the hydroxyl substituent on the cinchona alkaloid platforms for these catalysts tolerates and benefits from a great many variations, the rest of the alkaloid has proven much less flexible [64], and this portion of the catalytic system can (conveniently) be left alone.

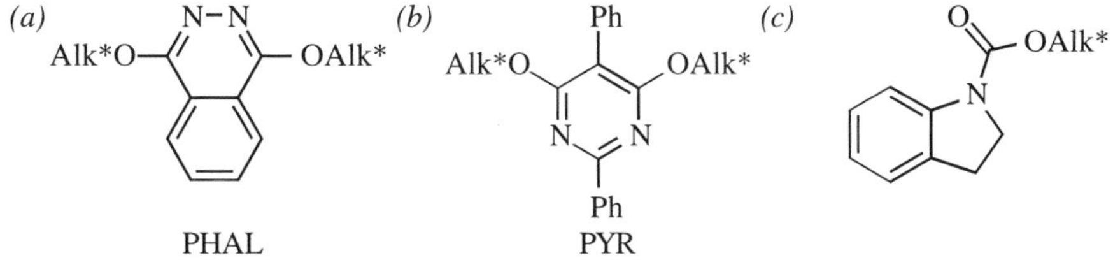

Figure 8.5. Ligands for the Sharpless AD process. *(a)* The phthalazine ligand (PHAL class), recommended for most substitution types [74,75]. *(b)* The diphenylpyrimidine ligand (PYR class), used for mono- and tetrasubstituted olefins [76,77] (along with PHAL ligands). *(c)* Indoline ligand (IND class), best for *cis*-disubstituted olefins [78]. For each, the Alk* bound to each position is DHQD or DHQ (Figure 8.4).

The results of many dihydroxylation reactions have resulted in the compilation of a mnemonic device for the prediction of the direction of attack with catalysts based on each alkaloid (Scheme 8.17). Although this model is very useful, there can be some ambiguity as to which group is the large one and which is the medium (especially with *trans*-disubstituted olefins) and electronic characteristics cannot be ignored [63]. This is a byproduct of the lack of an unambiguous group to orient the molecule (*cf.* the AE reaction, Scheme 8.1).

Some useful generalizations are that the "large" group is very often aromatic, and indeed, aromatic olefins make up some of the best substrates for this reaction. Results from modifying other sites of substitution have led to the suggestion that the loaded catalyst is very forgiving for *trans* olefins (the best substrates), but that it begins to experience some interference at the R_S position. That the binding site is even less favorable toward substituents *cis* to the R_L position (H in Scheme 8.17) is surmised by the difficulty of carrying out AD reactions with fully substituted [77] and *cis* olefins (there are only a few really good examples, and those top out at about 90% es [78]). However, very good to excellent results have been wrestled from all alkene

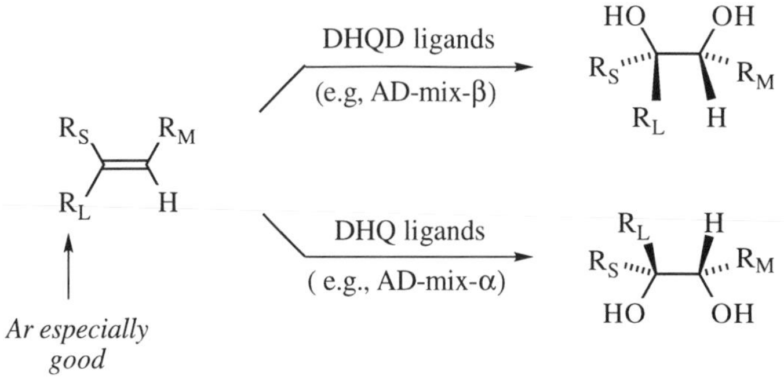

Scheme 8.17. The predicted enantioselectivity of Sharpless AD reactions using DHQD or DHQ ligands. This model is used by orienting the substrate so that the large (often aromatic), medium, and small substituents match up best with the R_L, R_M, and R_S positions. Application of this model to some alkenes will inevitably result in some compromises in placing the groups.

types and the reaction has to be considered one of the most reliable and easily performed asymmetric transformations available (Table 8.3). In many cases, the enantiomeric purity of the diol products can be increased by simple recrystallization, which increases the practicality of the method.

A preliminary picture of the mechanism has begun to emerge through examination of the reaction's reactivity profile and kinetics [79], molecular modeling [80], and structural studies of the chiral ligands [75]. One view of an intermediate is given in Figure 8.6, depicting the osmate ester of the reaction between styrene and the (DHQD)$_2$PHAL-derived complex. In the transition state leading to this product, one DHQD binds the active osmium species and the other is a bystander ligand. Overall, the binding pocket is proposed to have an "L" shape, with the floor made up by the flat PHAL heterocycle (the better to accommodate aromatic alkenes) and one wall coming from the bystander DHQD ligand. Favorable aromatic stacking interactions along with the minimization of steric interactions between the alkene group and the methoxyquinone wall result in the observed diastereoselectivity. The mechanistic picture of this reaction is still evolving, however, and alternative mechanistic proposals and transition structures have appeared [81,82].

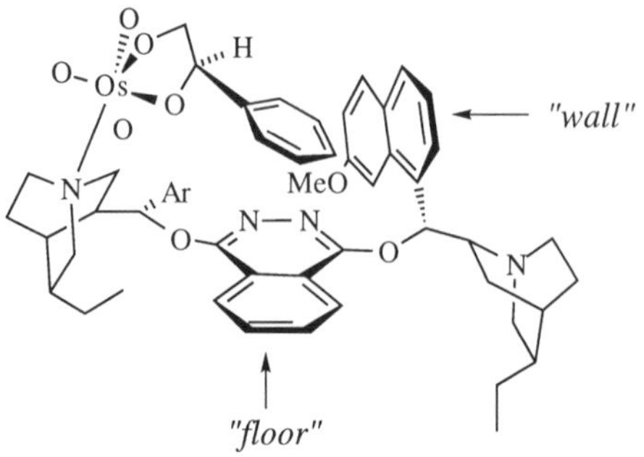

Figure 8.6. One view of a proposed intermediate in the Sharpless AD reaction of the (DHQD)$_2$PHAL-derived osmium species reacting with styrene [80].

Table 8.3. Examples of Asymmetric Dihydroxylation Reactions. See Figure 8.5 for catalyst structures.

Entry	Diol	Catalyst	% es	Ref.
1	OH, Ar-O-...-OH Ar = p-OMe-C$_6$H$_4$-	(DHQD)$_2$-PHAL	95	[83]
2	OH, Ph-CH-CH$_2$-OH	(DHQ)$_2$-PHAL	98.5	[74,76]
3	OBn, OH, Ph-...-OH	(DHQD)$_2$-PHAL	89	[84]
4	OH, C$_5$H$_{11}$-...-CO$_2$Et, OH	(DHQ)$_2$-PHAL	97.5	[74]
5	HO, R, OH (cyclohexane)	(DHQD)$_2$-PHAL	79 (R = Me) >99 (R = phenyl)	[64,74]
6	OH, R-...-OH, Me	DHQD-IND	78 (R = c-C$_6$H$_{11}$) 86 (R = Ph)	[78]
7	OH, Me, Me, C$_5$H$_{11}$-...-Me, OH	(DHQD)$_2$-PYR	61	[77]
8	OH, Me, OH, Me (tetralin)	(DHQD)$_2$-PYR	92.5	[77]

8.3.2 Applications of enantioselective dihydroxylations

As in the asymmetric epoxidation reaction, the development of such a powerful tool for the enantioselective preparation of diols spurred its application to a very wide variety of synthetic problems and the invention of new methods of manipulating the diol products [64]. A few examples confer some of the flavor of this work.

Like the AE, the AD has obvious utility to those contemplating the synthesis or use of carbohydrates as synthons. For one, glyceraldehyde and its acetonides have found very wide acceptance as chirons for asymmetric synthesis [85]. A clever utilization of the AD affords a building block that nicely complements the use of the naturally occurring material (Scheme 8.18) [86].

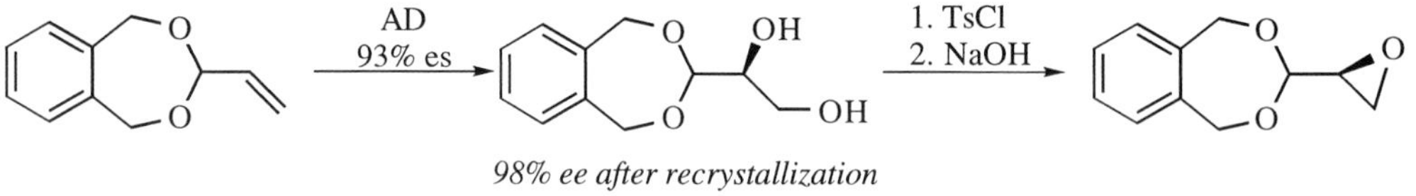

98% ee after recrystallization

Scheme 8.18. Synthesis of a glyceraldehyde equivalent and its conversion to an epoxide [86]. This AD used the 5-phenanthryl ether of DHQD as the ligand.

Note that:

1. The AD technique allows the synthesis of either enantiomer of the diol by simply switching catalyst.
2. The formation of the epoxide occurs with retention due to specific tosylation of the primary alcohol. (A variety of conditions have been developed for the conversion of 1,2-diols into epoxides [64].)
3. The use of the aromatic protecting group of the aldehyde both increasing the efficiency of the AD reaction and allows for its ready removal by hydrogenation.

Instead of activating diols by converting them to epoxides, an alternative is to activate the diols themselves to nucleophilic attack; this has been accomplished by converting them into cyclic sulfates (Scheme 8.19) [64,87]. These highly reactive species are subject to substitution by many nucleophiles, including halides, azides, reducing agents, and sulfur and carbon nucleophiles. Scheme 8.19b depicts a strategy involving irreversible epoxide formation (*cf.* the Payne rearrangement (section 8.2.4)) [88].

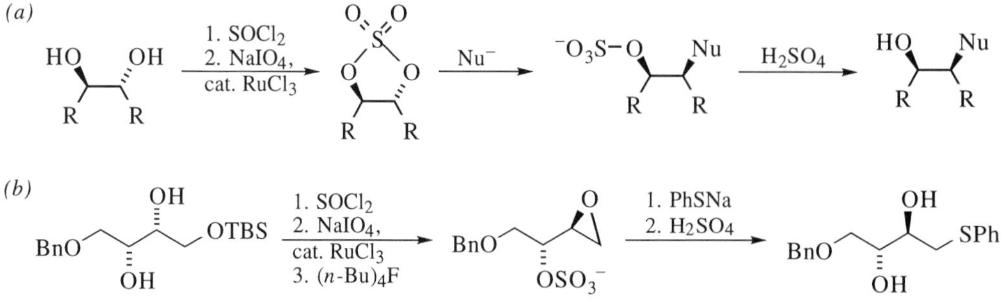

(a)

(b)

Scheme 8.19. *(a)* Formation and reactivity of cyclic sulfates. *(b)* Application of cyclic sulfates to the synthesis of erythrose [88]. Note that the epoxide formation is irreversible because the sulfate leaving group is no longer nucleophilic.

The reliably high selectivity of *trans*-olefins makes these substrates particularly amenable to synthetic schemes that depend on reagent control for the installation of new stereocenters. Unlike the AE reaction, there are no strict limitations on which diastereoisomers can be prepared. However, issues of double asymmetric induction [89] often arise because the reactions of alkenes bearing allylic electron-withdrawing groups can be highly diastereoselective. Prior to the development of this asymmetric dihydroxylation, the dependence of diastereofacial selection in alkenes bearing allylic substitution had been catalogued by Kishi (Scheme 8.20) [90,91]. When AD reactions are carried out on substrates already bearing stereogenic centers, matched *vs.* mismatched situations develop (Section 1.5), with the former affording very high selectivity. However, the ability of the AD system to induce enantiofacial selectivity is often high enough that varying levels of selectivity in either direction can be obtained. Although a few cases have been published (and summarized [64]), relatively little work has addressed the use of the AD reaction for kinetic resolution.

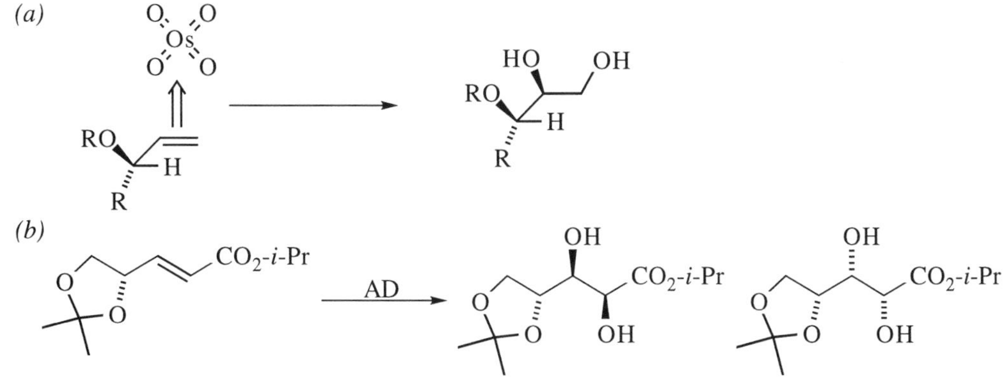

ligand	ratio
(DHQD)₂PHAL	98 : 2
(DHQ)₂PHAL	43 : 57

Scheme 8.20. *(a)* Kishi model for acyclic control in osmylation reactions [90]. *(b)* Double diastereoselectivity in an AD reaction [92].

Of the many further examples that could be provided here, one that utilizes the AD reaction to gain access to a series of molecules bearing axial chirality will be discussed [93]. In this case, the *trans* biaryl alkene shown in Scheme 8.21 was subjected to a highly efficient AD. Cyclization of the diol restricts the motion of the

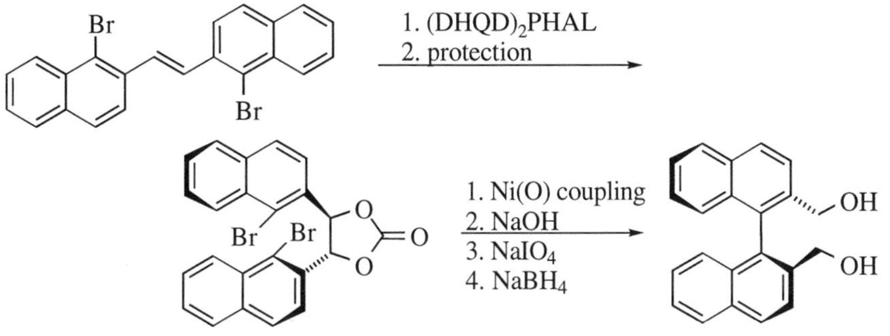

Scheme 8.21. Use of the Sharpless AD reaction in the synthesis of an enantiomerically enriched (98.5% ee) biaryl diol [93].

aryl groups so that they can only undergo intramolecular biaryl coupling to give one configuration about the newly formed single bond. Diol oxidation removed the original stereocenters installed by the AD reaction.

8.4 Oxidation of enolates and enol ethers

The asymmetric α-functionalization of carbonyl compounds with OR and NR$_2$ through their enolates has become another standard method for the synthesis of chiral 1,2-dioxygenated compounds. Most such methods utilize the rich chemistry developed for the asymmetric alkylation of carbonyls (Chapter 3), although one important class of chiral reagents has been developed for just this purpose.

8.4.1 Hydroxylations

In cases where chiral auxiliaries are used to differentiate the two faces of an enolate, oxygen-transfer agents produce α-hydroxy ketones or esters after cleavage of the auxiliary. Some examples are shown in Scheme 8.22. In each case, the chiral auxiliary had been previously developed for reaction with alkyl halide electrophiles, and is applied to oxidation chemistry in these examples. For example, the metalloenamine shown in Scheme 8.22a was originally developed by Koga [94]; its use in the illustrated oxidation was reported by Snyder [95]. Scheme 8.22b illustrates the application of Evans's imides to hydroxylations [96] (*cf.* Scheme 3.17). Enders's SAMP hydrazone enolates are also amenable to oxidation, as shown in Scheme 8.22c (*cf.* Scheme 3.21 and 3.22) [97]. All three of these reactions presumably involve mechanisms for diastereofacial discrimination similar to those involved in carbon-carbon bond-forming reactions (Chapters 3 and 5). In addition to reagents such as MoOPh or benzoyl peroxide, *N*-sulfonyl oxaziridines have become especially useful for this purpose (Scheme 8.22b and c) [98]. Note that, although this oxaziridine is chiral, its configuration is not important, and it is generally used in racemic form for this purpose.[7]

The utility of oxaziridines in asymmetric α-hydroxylation also extends to reactions with achiral enolates. This has been made possible by the discovery that certain chiral *N*-sulfonyl oxaziridines can react with enolates to afford α-hydroxy carbon compounds in excellent yield and enantioselectivity. An application of a highly selective sulfonyloxaziridine derived from camphor to the synthesis of daunomycin is shown in Scheme 8.23. Attack of the oxaziridine presumably occurs such that the enolate ester avoids nonbonded interactions with the *exo* methoxy group on the bicyclic ring system (*cf.* Schemes 8.23c and d). This is a very useful reaction of wide scope, and can be carried out on both stabilized enolates derived from keto esters (shown) and simple ketone enolates [99].

---

[7]   These reactions are, therefore, examples of double asymmetric induction whereby the selectivity of one chiral reactant is overwhelmed by the facial bias of another (*cf.* Section 1.5).

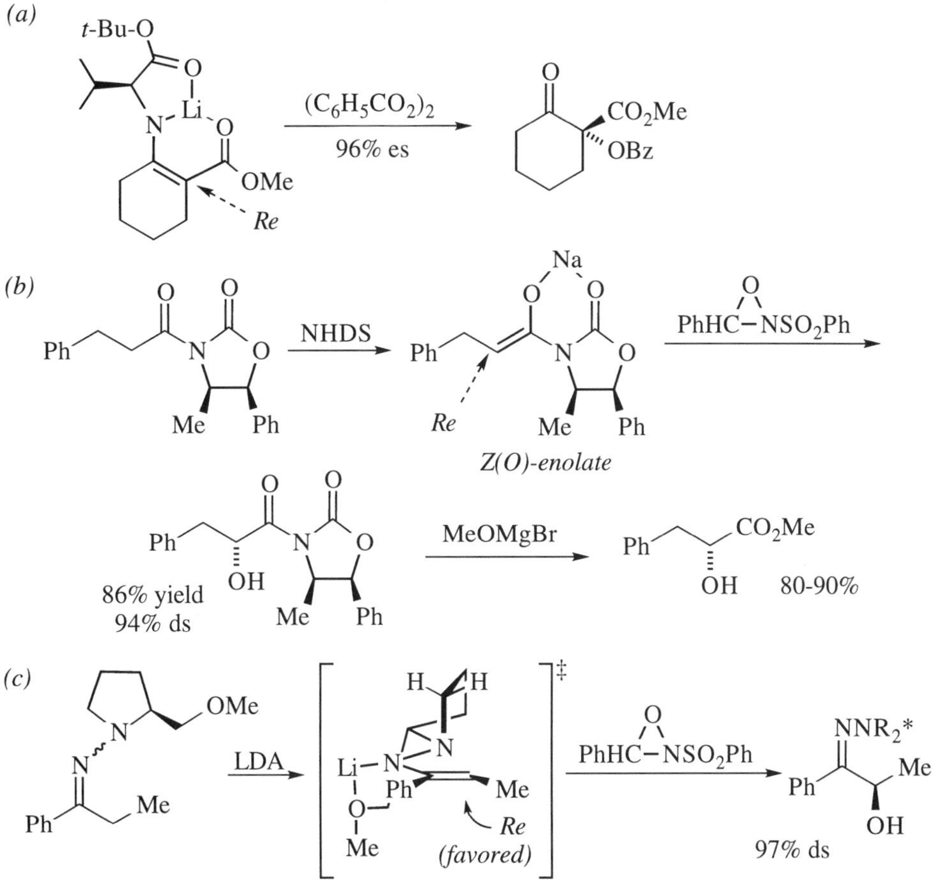

(a)

(b)

$Z(O)$-enolate

(c)

97% ds

Scheme 8.22. Representative hydroxylation reactions of chiral enolates, using *(a)* metalloenamine [95] (*cf.* Scheme 5.33), *(b)* oxazolidinone [96] (*cf.* Scheme 3.17), and *(c)* hydrazone [97] (*cf.* Scheme 3.21 and 3.22) auxiliaries.

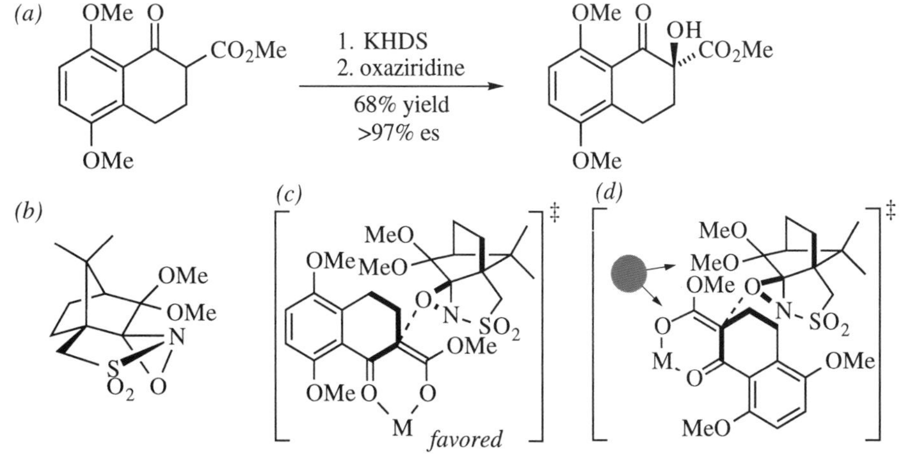

(a)

(b)

(c)

(d)

Scheme 8.23. *(a)* Application of enolate oxidation reactions of a chiral oxaziridine to the synthesis of an AB ring synthon of daunomycin [100]. *(b)* Structure of the oxaziridine used. Proposed *(c)* favored and *(d)* disfavored transition structures (see also ref. [101]).

An indirect route to α-hydroxy carbonyl compounds uses enol ethers as substrates for dihydroxylations (Scheme 8.24). The primary product is a vicinal hydroxy-hemiacetal which fragments to afford an α-hydroxyketone, rendering the overall route a two-step conversion of ketone to α-hydroxy ketone. The stereochemically important step can use a chiral auxiliary or enantioselective catalysis [64]. The sense of asymmetric induction found in Oppolzer's sulfonamide, shown in Scheme 8.24a [102], deserves comment, since this auxiliary is one seen previously in the Diels-Alder reaction (Figure 6.12f) , and the topicity is not obvious [103]. The illustrated conformation is the most stable (*cf.* Figure 4.21, Figure 6.23, and the accompanying discussion). In this conformation, the sulfonamide shields the *Re* face of C-2 (toward the viewer), so that the lead then adds to the *Si* face. Opening of the plumbacycle with acetate affords an intermediate that suffers fragmentation and acetate migration to give the *R* isomer by the mechanism shown [102]. The Sharpless asymmetric dihydroxylation shown in Scheme 8.24b follows the topicity suggested by the mneumonic of Scheme 8.17, with the aromatic moiety playing the part of the large (R$_L$) substituent.

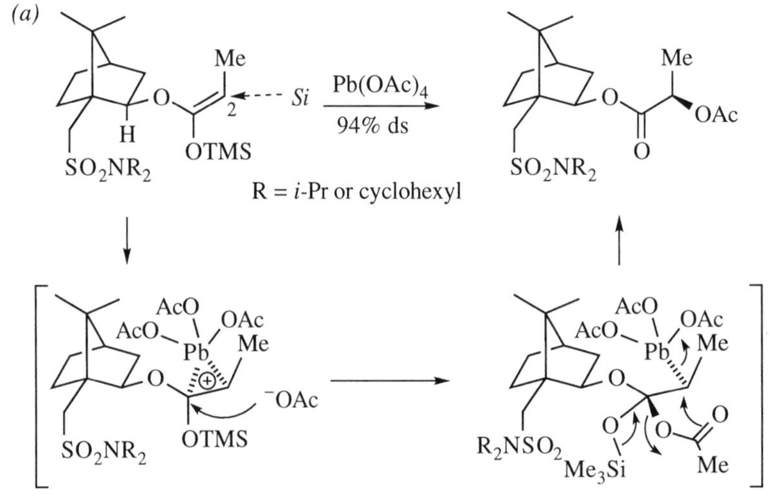

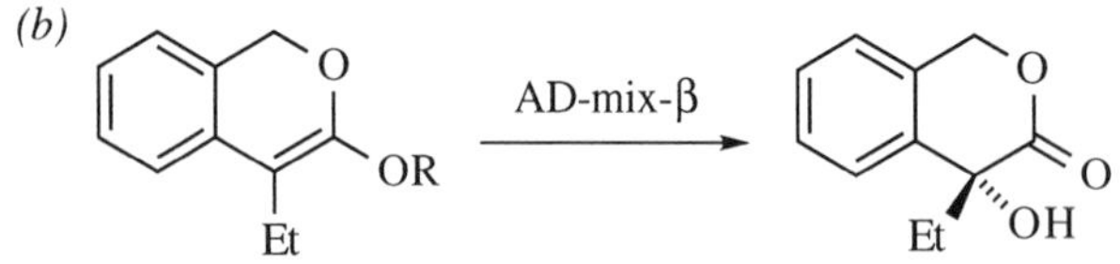

Scheme 8.24. Dihydroxylation reactions used for the synthesis of α-hydroxy carbonyl compounds. *(a)* A chiral auxiliary approach [102]. *(b)* Application of the Sharpless AD procedure to an intermediate for the synthesis of camptothecin [104].

8.4.2 Aminations

Amination reactions of carbonyl compounds provide access to useful building blocks for nitrogen-containing compounds, the conversion of esters to amino acid derivatives being particularly important [105]. In 1986, the groups of Gennari, Evans, and Vederas simultaneously published routes to α-hydrazino ester derivatives by the addition of the electrophilic reagent di(*tert*-butyl)azodicarboxylate (DBAD) to enolates or trimethylsilyl ketene acetals (Scheme 8.25) [106-109]. Excellent yields were obtained, and the products were formed in accord with the diastereofacial selectivity of the nucleophiles in alkylation or aldol reactions (Chapters 3 and 5).

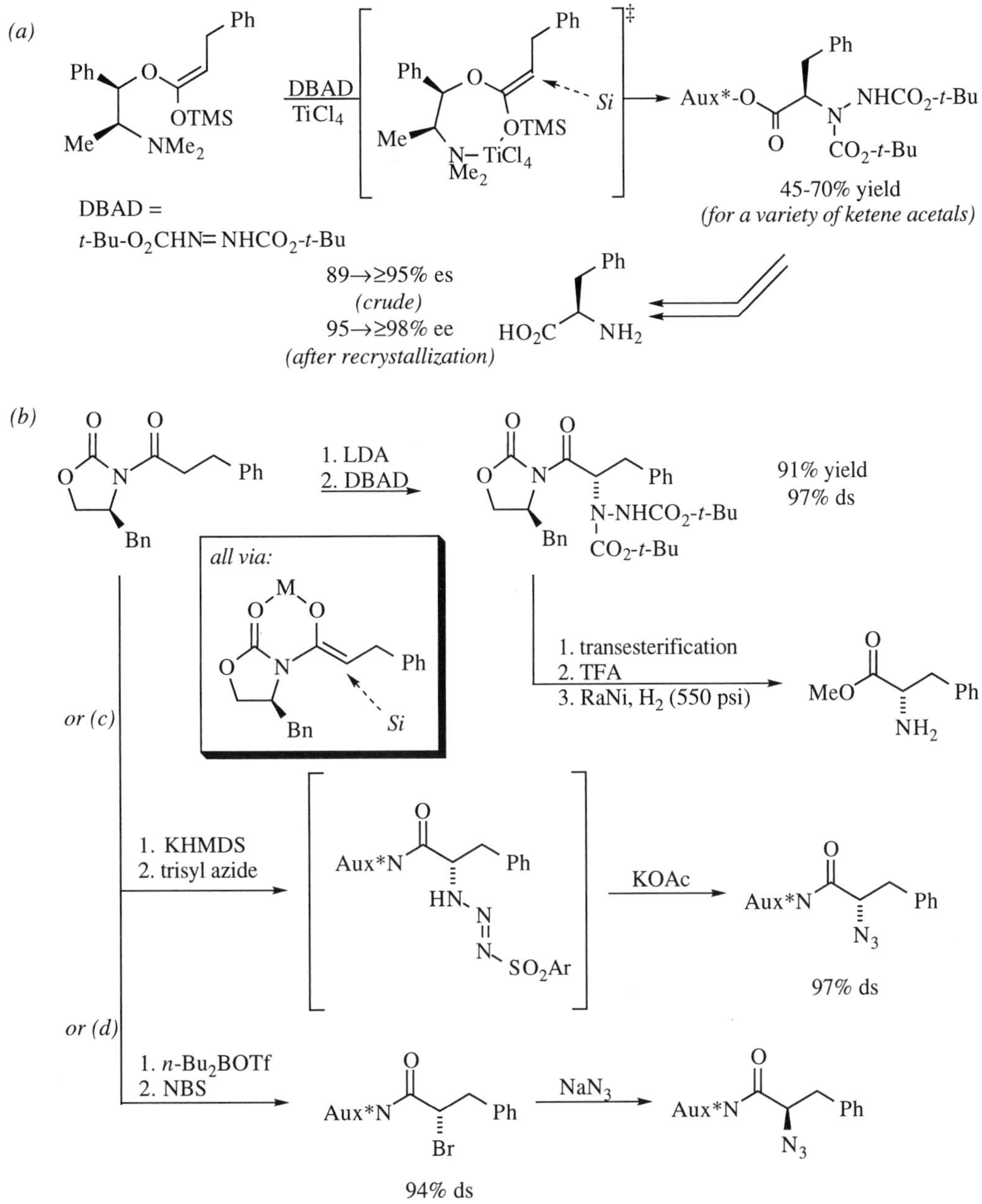

Scheme 8.25. α-Amidation of chiral ester enolates using di(*tert*-butyl)azodicarboxylate and *(a)* *N*-methylephedrine [106] or *(b)* oxazolidinone chiral auxiliaries [107]. Azidation of a chiral enolate *(c)* directly or *(d)* via bromination/azidation [110].

Unfortunately, the hydrazino esters or amides required inconveniently high pressures for their hydrogenolysis (500 psi; Schemes 8.25a and b). An improvement involved the direct azidation of the same enolates using arylazide derivatives, which were found to undergo reactions with enolate nucleophiles to provide a *C*-sulfonyltriazene intermediate which could be decomposed to the α-azido ester (Scheme 8.25c) [110]. Alternatively, azides may be obtained by enolate bromination followed by S$_N$2 azide displacement; note that these techniques are stereochemically complementary. Similar chemistry has also been accomplished using 10-sulfonamidoisobornyl chiral auxiliaries (Scheme 8.25d) [111].

A different approach utilizes achiral enolates and a chiral amination reagent (interligand asymmetric induction) [112]. α-Chloro-α-nitrosocyclohexane had previously been used as an aminating reagent with chiral enolates, providing nitrones as the primary product [113]. The adaptation of this chemistry to chiral aminating agents gave the nitrones with high diastereoselectivity (Scheme 8.26), which could be hydrolyzed to give α-hydroxylamino ketones. These were further reduced to the amino alcohols using borohydride reagents and zinc/HCl. The reactions were proposed to proceed through a Zimmerman-Traxler-type transition structure (Scheme 5.1) in which the *Z(O)*-enolate of the ketone was coordinated via zinc to the nitroso group and the whole ensemble oriented to avoid steric interactions between the incoming nucleophile and the sulfonamide group.

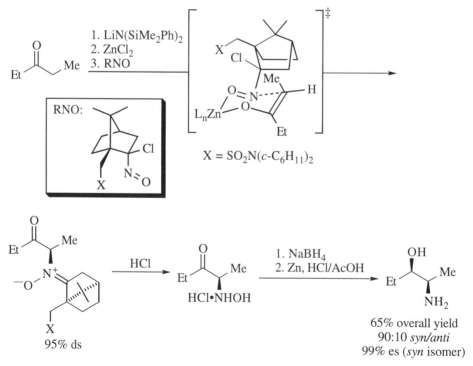

Scheme 8.26. Reaction of an achiral enolate with a chiral α-chloro-α-nitroso reagent [113].

8.5 Miscellaneous oxidations

8.5.1 Oxidation of sulfides

A great deal of effort has been expended in the development of ways to carry out the asymmetric oxidation of sulfides to sulfoxides; progress in this field has been exhaustively reviewed [114]. This is interesting from both a theoretical viewpoint and from the utility of certain chiral sulfoxides as reagents in asymmetric synthesis [115]. Some natural products also contain sulfoxide stereogenic centers.

The most common method for obtaining chiral sulfur compounds is the Andersen synthesis, which utilizes a chiral sulfinate ester such as that derived from menthol by diastereoselective oxidation and isomeric enrichment via epimerization and recrystallization [115,116]. Scheme 8.27a shows a simple example; note that the Grignard reaction occurs with inversion of configuration at the sulfur atom. More recently, substantial advances have been made in the application of chiral reagents (*i.e.*, chiral *N*-sulfonyloxaziridines [117]) and catalytic systems to this problem. A promising route uses a Sharpless-style titanium-diethyl tartrate system (Scheme 8.27b and c); this reaction can proceed with high enantioselectivity when appropriate substrates such as aryl sulfides are used [114,118]. The reaction is highly dependent on the stoichiometry of Ti(O-*i*-Pr)$_4$/DET/added water, with the proportion of 1:2:1 being critical.

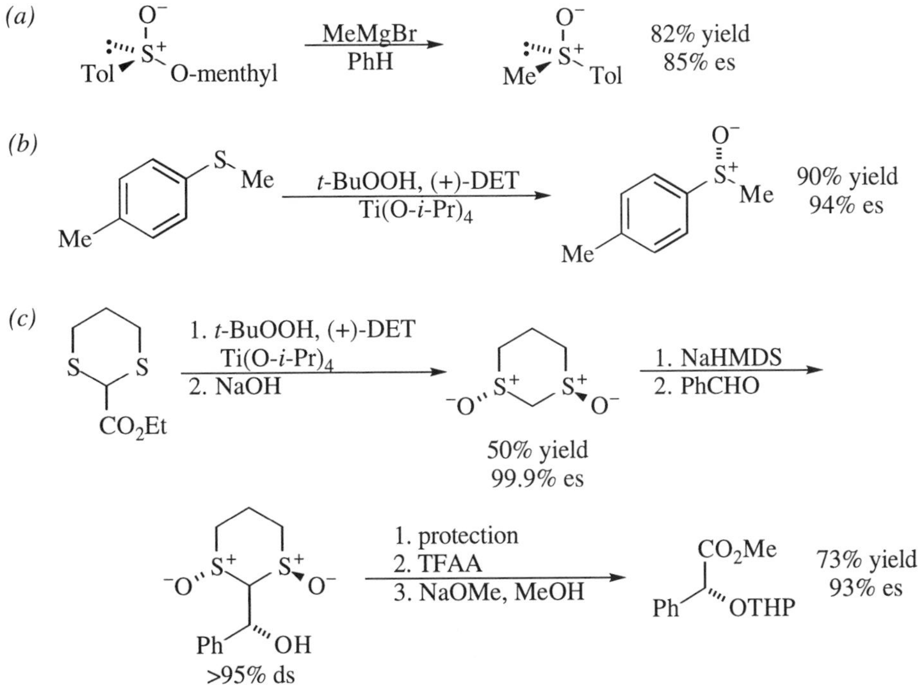

Scheme 8.27. (*a*) An example of the Andersen synthesis of chiral sulfoxides [119]. (*b*) Catalytic oxidation of an aromatic sulfide using a chiral titanium complex [118]. (*c*) Synthesis of a C$_2$-symmetrical *trans*-1,3-dithiane-1,3-dioxide and its use as an asymmetric acyl anion equivalent [120,121].

An interesting application of this chemistry has been the subject of considerable work by the Aggarwal group. These workers prepared the *trans* isomer of 1,3-dithiane-1,3-dioxide in high ee; note the use of a temporary carbomethoxy group, which proved necessary for high enantioselectivity (Scheme 8.27c) [120]. Deprotonation gave an acyl anion equivalent which reacted with aromatic aldehydes with high diastereoselectivity [121]. Pummerer removal of the heterocycle followed by basic transesterification met with some isomerization and loss of enantiomeric purity, although this problem could be mitigated by a multistep procedure involving intermediate thioesters.

The reader is referred to the review literature for other applications of chiral sulfoxides in diastereoselective synthesis, including their use as directing groups for a wide variety of organometallic reactions [122,123] and the related chemistry of chiral sulfoxonium salts [115]. Although the vast majority of these efforts have to date obtained the stereogenic sulfur atom through the Andersen procedure or a variant thereof, one expects chiral catalysis to play an increasing role in the future.

8.5.2 *Group-selective oxidation of C-H and C-C σ bonds*

The oxidation of enantiotopic alkyl groups is only rarely accomplished through chemical means, although group selectivity is very commonly utilized in enzymatic reactions. This is partly because the conceptually simplest kind of group-selective reaction involves the formal cleavage of C-H or C-C σ bonds, which is still difficult to do with useful levels of chemo- or regioselectivity, much less enantioselectivity. Due to the significant potential of this strategy in asymmetric synthesis, a few early results are presented below.

A few non-enzymatic methods that permit the formal group-selective insertion of oxygen into enantiotopic C-H bonds are shown in Scheme 8.28. The asymmetric Kharash reaction in Scheme 8.28a uses a chiral bisoxazoline complex of copper similar to that employed by Evans in his approach to asymmetric aziridination (*cf.* Scheme 8.12) [124]. Without going too deeply into the details of the mechanism, note that although the overall reaction selectively replaces one of the enantiotopic hydrogen atoms with a benzyloxy group, the actual enantioselective step entails the face-selective recombination of the allylic radical with the copper(II)-benzoate salt! According to this hypothesis, the oxygen is delivered suprafacially and intra-molecularly with expulsion of the original catalyst. Note that this example proceeds in low yield. Should more reactive catalyst systems that proceed with higher turnovers be found, this might provide a useful route to allylic oxidation.

The benzylic oxidation depicted in the second step of Scheme 8.28b is actually a formal group-selective differentiation of diastereotopic C-H bonds, since asymmetric epoxidation occurs prior to hydroxylation [125]. However, the reaction is an interesting example of a kinetic resolution that depends on the fact that the catalyst used reacts with the two epoxides at different rates, apparently because the chiral catalyst system:

1. Prefers to remove a hydrogen atom from the same face as the epoxide oxygen, possibly because of conformational effects.
2. Is selective for the removal of the H_{Si} hydrogen.

Since the salen(Mn) catalyst used promotes both epoxidation and hydroxylation reactions (although not with the same efficiency), the net effect is to stereoselectively oxidize the *S,R*-monoepoxide in order to increase the enantiomeric purity of the *R,S*-epoxide.

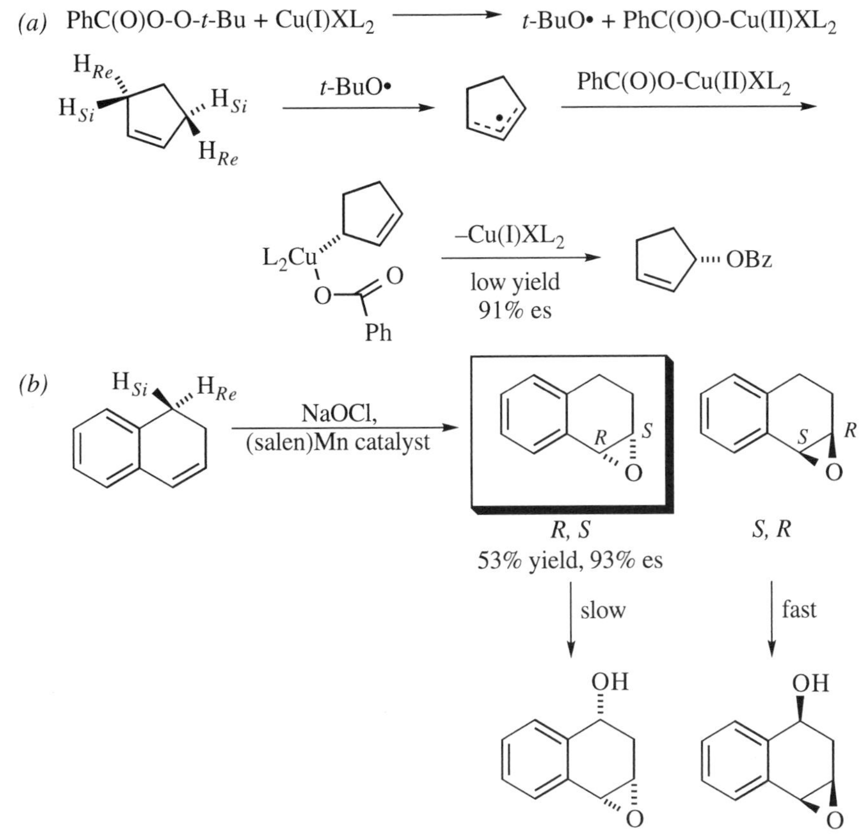

(a) $PhC(O)O\text{-}O\text{-}t\text{-}Bu + Cu(I)XL_2 \longrightarrow t\text{-}BuO\bullet + PhC(O)O\text{-}Cu(II)XL_2$

Scheme 8.28. Formally group-selection insertion of oxygen into enantiotopic C-H bonds. *(a)* An asymmetric Kharasch reaction [124]. The catalyst is similar to that shown in Scheme 8.12, except that each oxazoline bears two methyl substituents at C-5. *(b)* Kinetic resolution of dihydronaphthalenes [125]. The reaction uses a Jacobsen epoxidation catalyst (Scheme 8.6, type A).

The group-selective stereodifferentiation of C-C bonds has been investigated in the context of ring-expansion chemistry. In a symmetrical ketone like 4-*tert*-butylcyclohexanone, the two methylene groups adjacent to the ketone are enantiotopic. Several groups interested in the overall "oxidation" of these bonds have developed asymmetric versions of the classical Baeyer-Villiger and Beckmann reactions (Scheme 8.29a).

Although highly efficient enzymatic Baeyer-Villiger reactions have been accomplished using cyclohexanone oxygenase [126,127], only one method using an abiotic method has been reported [128]. A chiral copper complex is used to activate the oxidizing agent, this time molecular oxygen itself. In the example shown, a kinetic resolution of a 2-substituted cyclohexanone is carried out to afford both product lactone and the starting cyclohexanone with reasonable enantioselectivities.

Such reactions have not yet found their way into the arsenals of synthetic organic chemists; instead, multistep sequences such as the chiral base-mediated method shown in Scheme 8.29c are usually used to achieve the net result of enantioselective oxygen insertion [129].

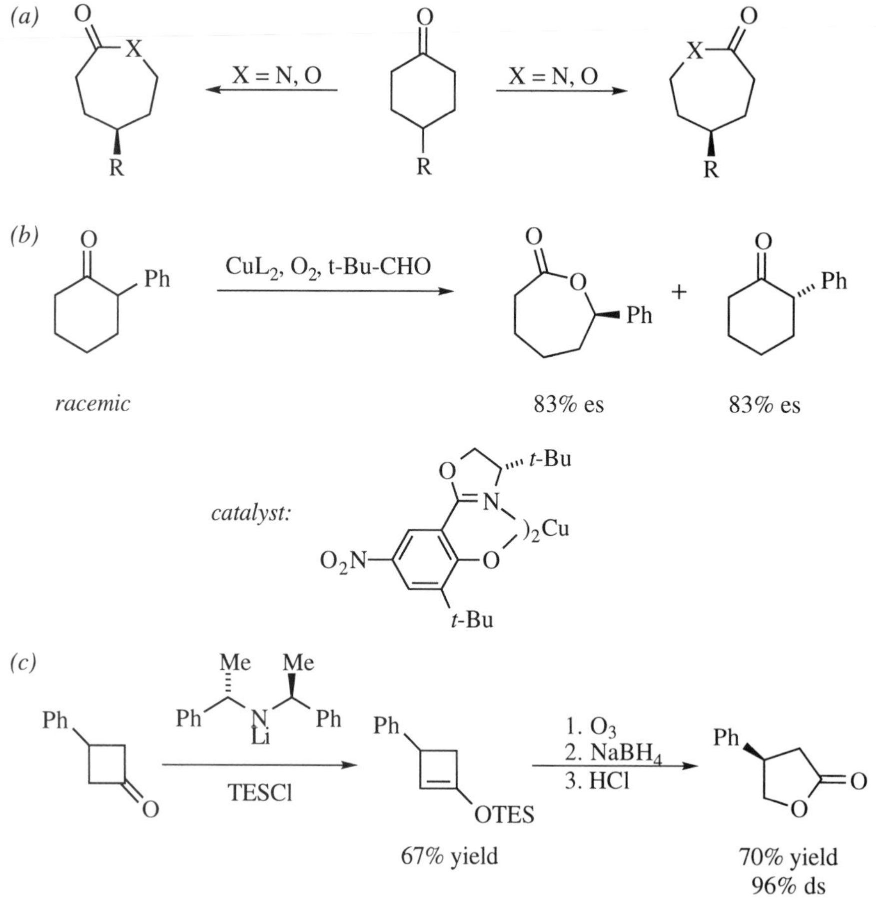

Scheme 8.29. *(a)* A generic asymmetric ring-expansion reaction. *(b)* A kinetic resolution using an asymmetric Baeyer-Villiger catalyst [128]. *(c)* A synthetic equivalent of an asymmetric Baeyer-Villiger reaction [129].

In contrast, there is no known enzymatic version of an asymmetric nitrogen insertion process. Rather, there are two methods available that utilize variations on known ring-expansion processes (Scheme 8.30). The first utilizes oxaziridines as the first isolated intermediate in a three-step overall sequence [130]. Axially dissymmetric spirocyclic oxaziridines are available by the oxidation of imines derived from the starting ketone and α-methylbenzylamine. The reaction utilizes one element of diastereofacial selectivity (interpreted here as equatorial attack of the peracid oxidizing agent) and an interesting kind of selectivity whereby intramolecular attack of the now-secondary nitrogen causes ejection of a carboxylic acid; in this latter reaction, the stereogenic nitrogen atom of the oxaziridine [131] is formed with good diastereoselectivity. The oxaziridine is then photolyzed, which causes the molecule to undergo bond reorganization to give the lactam. This reaction takes advantage of the known (but not well-understood) tendency of the oxaziridine to react with regioselective migration of the bond antiperiplanar to the lone pair on the nitrogen atom (emboldened in the scheme) [132]. Reductive removal of the chiral substituent on nitrogen then finishes off the overall ring-expansion protocol.

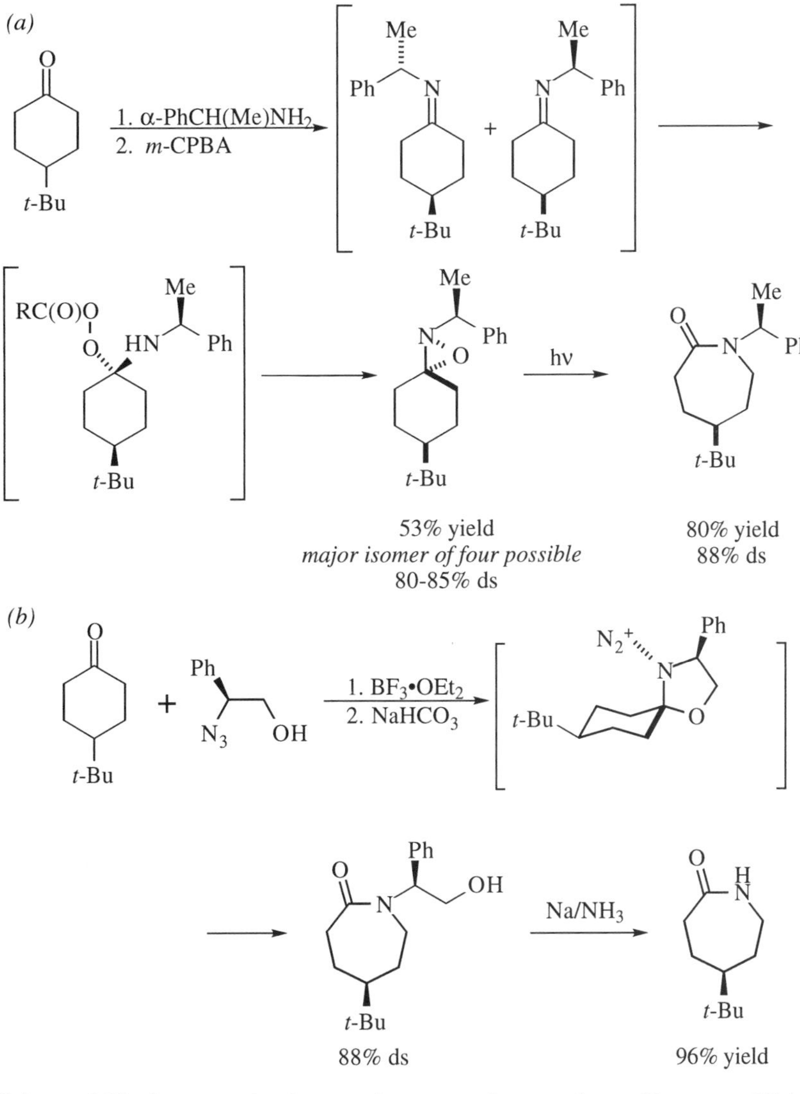

Scheme 8.30. Asymmetric nitrogen ring-expansion reactions of ketones utilizing (*a*) oxaziridine synthesis and photolysis [130] and (*b*) an azide-Schmidt reaction [133].

A few examples of a similar conversion utilizing an azide-based variant of the Schmidt reaction have also appeared [133]. The reaction is thought to entail the formation of a hemiketal between the hydroxyl group of the reagent and the ketone; dehydration of the hemiketal leads to an oxonium ion that is subject to attack by the now-tethered azido group. Migration of the bond antiperiplanar to the departing N_2 substituent, possibly through a conformation such as that depicted, was proposed to lead to the observed lactam. Again, removal of the chiral substituent on nitrogen afforded the formal asymmetric Schmidt reaction product.

8.6 References

1.   J. D. Morrison; H. S. Mosher, In *Asymmetric Organic Reactions*; American Chemical Society: Washington, D.C., 1976, pp 258-262.
2.   S. W. Henbest; R. A. L. Wilson *J. Chem. Soc.* **1957**, 1968-1965.
3.   C. G. Chavdarian; C. H. Heathcock *Synth. Commun.* **1976**, *6*, 277-280.
4.   P. Chamberlain; M. L. Roberts; G. H. Whitham *J. Chem. Soc. B* **1970**, 1374-1381.
5.   K. B. Sharpless; T. R. Verhoeven *Aldrichimica Acta* **1979**, *12*, 63-74.
6.   P. Kocovsky; I. Stary *J. Org. Chem.* **1990**, *55*, 3236-3243.
7.   A. H. Hoveyda; D. A. Evans; G. C. Fu *Chem. Rev.* **1993**, *93*, 1307-1370.
8.   K. B. Sharpless; R. C. Michaelson *J. Am. Chem. Soc.* **1973**, *95*, 6136-6137.
9.   B. E. Rossiter; T. R. Verhoeven; K. B. Sharpless *Tetrahedron Lett.* **1979**, 4733-4736.
10. T. Katsuki; K. B. Sharpless *J. Am. Chem. Soc.* **1980**, *102*, 5974-5976.
11. M. G. Finn; K. B. Sharpless, In *Asymmetric Synthesis*; J. D. Morrison, Ed.; Academic: Orlando, 1985; Vol. 5, pp 247-308.
12. B. E. Rossiter, In *Asymmetric Synthesis*; J. D. Morrison, Ed.; Academic: Orlando, 1985; Vol. 5, pp 194-246.
13. R. A. Johnson; K. B. Sharpless, In *Catalytic Asymmetric Synthesis*; I. Ojima, Ed.; VCH: New York, 1993, pp 103-158.
14. T. Katsuki; V. S. Martin *Org. React.* **1996**, *48*, 1-299.
15. S. S. Woodard; M. G. Finn; K. B. Sharpless *J. Am. Chem. Soc.* **1991**, *113*, 106-113.
16. M. G. Finn; K. B. Sharpless *J. Am. Chem. Soc.* **1991**, *113*, 113-126.
17. E. J. Corey *J. Org. Chem.* **1990**, *55*, 1693-1694.
18. I. D. Williams; S. F. Pederson; K. B. Sharpless; S. J. Lippard *J. Am. Chem. Soc.* **1984**, *106*, 6430-6431.
19. E. N. Jacobsen, In *Catalytic Asymmetric Reactions*; I. Ojima, Ed.; VCH: New York, 1993, pp 159-202.
20. W. Zhang; J. L. Loebach; S. R. Wilson; E. N. Jacobsen *J. Am. Chem. Soc.* **1990**, *112*, 2801-2803.
21. Y. Gao; R. Hanson; J. M. Klunder; S. Y. Ko; H. Masamune; K. B. Sharpless *J. Am. Chem. Soc.* **1987**, *109*, 5765-5780.
22. R. D. Wood; B. Ganem *Tetrahedron Lett.* **1982**, *23*, 707-710.
23. P. Garner; J. M. Park; V. Rotello *Tetrahedron Lett.* **1985**, *26*, 3299-3302.
24. D. P. G. Hamon; N. J. Shirley *J. Chem. Soc., Chem. Commun.* **1988**, 425-426.
25. E. N. Jacobsen; W. Zhang; A. Muci; J. R. Ecker; L. Deng *J. Am. Chem. Soc.* **1991**, *113*, 7063-7064.
26. S. Chang; J. M. Galvin; E. N. Jacobsen *J. Am. Chem. Soc.* **1994**, *116*, 6937-6938.
27. M. Palucki; P. J. Pospisil; W. Zhang; E. N. Jacobsen *J. Am. Chem. Soc.* **1994**, *116*, 9333-9334.
28. B. D. Brandes; E. N. Jacobsen *J. Org. Chem.* **1994**, *59*, 4378-4380.
29. B. Brandes; E. N. Jacobsen *Tetrahedron Lett.* **1995**, *36*, 5123-5126.
30. N. H. Lee; E. N. Jacobsen *Tetrahedron Lett.* **1991**, *32*, 6533-6536.

31. S. Chang; R. M. Heid; E. N. Jacobsen *Tetrahedron Lett.* **1994**, *1994*, 669-672.

32. R. Helder; J. C. Hummelen; R. W. P. M. Laane; J. S. Wiering; H. Wynberg *Tetrahedron Lett.* **1976**, 1831-1834.

33. S. Colonna; H. Molinari; S. Banfi; S. Juliá; J. Masana; A. Alvarez *Tetrahedron* **1983**, *39*, 1635-1641.

34. J. R. Flisak; K. J. Gombatz; M. M. Holmes; A. A. Jarmas; I. Lantos; W. L. Mendelson; V. J. Novack; J. J. Remich; L. Snyder *J. Org. Chem.* **1993**, *58*, 6247-6254.

35. C. Baccin; A. Gusso; F. Pinna; G. Strukul *Organometallics* **1995**, *14*, 1161-1167.

36. V. S. Martin; S. S. Woodard; T. Katsuki; Y. Yamada; M. Ikeda; K. B. Sharpless *J. Am. Chem. Soc.* **1981**, *103*, 6237-6240.

37. B. E. Rossiter; K. B. Sharpless *J. Org. Chem.* **1984**, *49*, 3707-3711.

38. K. B. Sharpless; C. H. Behrens; T. Katsuki; A. W. M. Lee; V. S. Martin; M. Takatani; S. M. Viti; F. J. Walker; S. S. Woodard *Pure Appl. Chem.* **1983**, *55*, 589-604.

39. M. Kusakabe; F. Sato *J. Org. Chem.* **1989**, *54*, 3486-3487.

40. S. Miyano; L. D.-L. Lu; S. M. Viti; K. B. Sharpless *J. Org. Chem.* **1985**, *40*, 4350-4360.

41. J. A. Marshall; K. E. Flynn *J. Am. Chem. Soc.* **1983**, *105*, 3360-3362.

42. S. Y. Ko; A. W. M. Lee; S. Masamune; L. A. Reed, III; K. B. Sharpless; F. J. Walker *Science* **1983**, *220*, 949-951.

43. S. Y. Ko; A. W. M. Lee; S. Masamume; I. Reed, L.A.; K. B. Sharpless; F. J. Walker *Tetrahedron* **1990**, *46*, 245-264.

44. M. J. Chong; K. B. Sharpless *J. Org. Chem.* **1985**, *50*, 1560-1563.

45. C. H. Behrens; S. Y. Ko; K. B. Sharpless; F. J. Walker *J. Org. Chem.* **1985**, *50*, 5687-5696.

46. G. Stork; S. D. Rychnovsky *Pure Appl. Chem.* **1987**, *59*, 345.

47. S. Masamune; W. Choy *Aldrichimica Acta* **1982**, *15*, 47-63.

48. S. Takano; K. Sakurai; S. Hatakeyama *J. Chem. Soc., Chem. Commun.* **1985**, 1759-1761.

49. R. E. Babine *Tetrahedron Lett.* **1986**, *27*, 5791-5794.

50. B. Häfele; D. Schröter; V. Jäger *Angew. Chem., Int. Ed. Engl.* **1986**, *25*, 87-89.

51. S. L. Schreiber; T. S. Schreiber; D. B. Smith *J. Am. Chem. Soc.* **1987**, *109*, 1525-1529.

52. D. B. Smith; Z. Wang; S. L. Schreiber *Tetrahedron* **1990**, *46*, 4793-4808.

53. Y.-F. Wang; C.-S. Chen; G. Girdaukas; C. J. Sih *J. Am. Chem. Soc.* **1984**, *106*, 3695-3696.

54. S. L. Schreiber *Chem. Scr.* **1987**, *27*, 563-566.

55. L. E. Martinez; J. L. Leighton; D. H. Carsten; E. N. Jacobsen *J. Am. Chem. Soc.* **1995**, *117*, 5897-5898.   See this paper for references to earlier approaches to this problem.

56. L. E. Overman; S. Sugai *J. Org. Chem.* **1985**, *50*, 4154-4155.

57. W. A. Nugent *J. Am. Chem. Soc.* **1992**, *114*, 2768-2769.

58. D. A. Evans; K. A. Woerpel; M. M. Hinman; M. M. Faul *J. Am. Chem. Soc.* **1991**, *113*, 726-728.

59. K. J. O'Connor; S.-J. Wey; C. J. Burrows *Tetrahedron Lett.* **1992**, *33*, 1001-1004.

60. Z. Li; K. R. Conser; E. N. Jacobsen *J. Am. Chem. Soc.* **1993**, *115*, 5326-5327.

61. Z. Li; R. W. Quan; E. N. Jacobsen *J. Am. Chem. Soc.* **1995**, *117*, 5889-5890.

62. B. B. Lohray *Tetrahedron: Asymm.* **1992**, *3*, 1317-1349.

63. R. A. Johnson; K. B. Sharpless, In *Catalytic Asymmetric Synthesis*; I. Ojima, Ed.; VCH: New York, 1993, pp 227-272.

64. H. C. Kolb; M. S. VanNieuwenhze; K. B. Sharpless *Chem. Rev.* **1994**, *94*, 2483-2547.

65. M. Schröder *Chem. Rev.* **1980**, *80*, 187-213.

66. K. Hofmann *Chem. Ber.* **1912**, *45*, 3329-3336.

67. V. Van Rheenen; R. C. Kelly; D. Y. Cha *Tetrahedron Lett.* **1976**, 1973-1976.

68. R. Criegee; B. Marchand; H. Wannowius *Liebigs Ann. Chem.* **1942**, *550*, 99-133.

69. S. Hanessian; P. Meffre; M. Girard; S. Beaudoin; J.-Y. Sancéau; Y. Bennani *J. Org. Chem.* **1993**, *58*, 1991-1993.

70. K. Tomioka; M. Nakajima; K. Koga *J. Am. Chem. Soc.* **1987**, *109*, 6213-6215.

71. E. N. Jacobsen; I. Markó; W. S. Mungall; G. Schröder; K. B. Sharpless *J. Am. Chem. Soc.* **1988**, *110*, 1968-1970.

72. J. S. M. Wai; I. Markó; M. B. France; J. S. Svendsen; K. B. Sharpless *J. Am. Chem. Soc.* **1989**, *111*, 1123-1125.

73. H. Kwong; C. Sorato; Y. Ogino; H. Chen; K. B. Sharpless *Tetrahedron Lett.* **1990**, *31*, 2999-3002.

74. K. B. Sharpless; W. Amberg; Y. L. Bennani; G. A. Crispino; J. Hartung, Jeong, Kyu-Sung; H.-L. Kwong; K. Morikawa; Z.-M. Wang; D. Xu; X.-L. Zhang *J. Org. Chem.* **1992**, *57*, 2768-2771.

75. W. Amberg; Y. L. Bennani; R. K. Chadha; G. A. Crispino; W. D. Davis; J. Hartung; K.-S. Jeong; Y. Ogino; T. Shibata; K. B. Sharpless *J. Org. Chem.* **1993**, *58*, 844-849.

76. G. Crispino; K.-S. Jeong; H. C. Kolb; Z.-M. Wang; D. Xu; K. B. Sharpless *J. Org. Chem.* **1993**, *58*, 3785-3786.

77. K. Morikawa; J. Park; P. G. Andersson; T. Hashiyama; K. B. Sharpless *J. Am. Chem. Soc.* **1993**, *115*, 8463-8464.

78. L. Wang; K. B. Sharpless *J. Am. Chem. Soc.* **1992**, *114*, 7568-7570.

79. H. C. Kolb; P. G. Andersson; K. B. Sharpless *J. Am. Chem. Soc.* **1994**, *116*, 1278-1291.

80. P.-O. Norrby; H. C. Kolb; K. B. Sharpless *J. Am. Chem. Soc.* **1995**, *13*.

81. E. J. Corey; M. C. Noe *J. Am. Chem. Soc.* **1993**, *115*, 12579-12580.

82. E. J. Corey; M. C. Noe *J. Am. Chem. Soc.* **1996**, *118*, 319-319.

83. Z.-M. Wang; X.-L. Zhang; K. B. Sharpless *Tetrahedron Lett.* **1993**, *34*, 2267-2270.

84. Z. M. Wang; K. B. Sharpless *Synlett* **1993**, *8*, 603-604.

85. J. Jurczak; S. Pikul; T. Bauer *Tetrahedron* **1986**, *42*, 447-488.

86. R. Oi; K. B. Sharpless *Tetrahedron Lett.* **1992**, *33*, 2095-2098.

87. Y. Gao; K. B. Sharpless *J. Am. Chem. Soc.* **1988**, *110*, 7538-7539.

88. S. Y. Ko; M. Malik *Tetrahedron Lett.* **1993**, *34*, 4675-4678.

89. S. Masamune; W. Choy; J. S. Petersen; L. R. Sita *Angew. Chem., Int. Ed. Engl.* **1985**, *24*, 1-30.

90. J. K. Cha; W. C. Christ; Y. Kishi *Tetrahedron* **1984**, *40*, 2247-2255.

91. G. Stork; M. Kahn *Tetrahedron Lett.* **1983**, *24*, 3951-3954.

92. K. Morikawa; K. B. Sharpless *Tetrahedron Lett.* **1993**, *34*, 5575-5578.

93. V. J. Rawal; A. S. Florjancic; S. P. Singh *Tetrahedron Lett.* **1994**, *35*, 8985-8988.

94. K. Tomioka; K. Ando; Y. Takemasa; K. Koga *J. Am. Chem. Soc.* **1984**, *106*, 2718-2719.

95. J. Lee; S. Oya; J. K. Snyder *Tetrahedron Lett.* **1991**, *32*, 5899-5902.

96. D. A. Evans; M. M. Morrissey; R. L. Dorow *J. Am. Chem. Soc.* **1985**, *107*, 4346-4348.

97. D. Enders; V. Bhushan *Tetrahedron Lett.* **1988**, *29*, 2437-2440.

98. F. A. Davis; B.-C. Chen *Chem. Rev.* **1992**, *92*, 919-934.

99. F. A. Davis; B.-C. Chen *J. Org. Chem.* **1993**, *58*, 1751-1753.

100. F. A. Davis; C. Clark; A. Kumar; B.-C. Chen *J. Org. Chem.* **1994**, *59*, 1184-1190.

101. F. A. Davis; M. C. Weismiller; C. K. Murphy; R. T. Reddy; B.-C. Chen *J. Org. Chem.* **1992**, *57*, 7274-7285.

102. W. Oppolzer; P. Dudfield *Helv. Chim. Acta* **1985**, *68*, 216-219.

103. W. Oppolzer; C. Chapuis; G. Bernardinelli *Tetrahedron Lett.* **1984**, *25*, 5885-5888.

104. D. P. Curran; S.-B. Ko *J. Org. Chem.* **1994**, *59*, 6139-6141.

105. R. M. Williams *Synthesis of Optically Active α-Amino Acids*; Pergamon: Oxford, 1989.

106. C. Gennari; L. Colombo; G. Bertolini *J. Am. Chem. Soc.* **1986**, *108*, 6394-6395.

107. D. A. Evans; T. C. Britton; J. A. Ellman; R. L. Dorow *J. Am. Chem. Soc.* **1986**, *108*, 6395-6397.

108. L. A. Trimble; J. A. Vederas *J. Am. Chem. Soc.* **1986**, *108*, 6397-6399.

109. D. A. Evans; T. C. Britton; R. L. Dorow; J. F. Dellaria *Tetrahedron* **1988**, *44*, 5525-5540.

110. D. A. Evans; T. C. Britton; J. A. Ellman; R. L. Dorow *J. Am. Chem. Soc.* **1990**, *112*, 4011-4030.

111. W. Oppolzer; R. Moretti *Tetrahedron* **1988**, *44*, 5541-5552.

112. W. Oppolzer; O. Tamura; H. Sundarababu; M. Signer *J. Am. Chem. Soc.* **1992**, *114*, 5900-5902.

113. W. Oppolzer; O. Tamura *Tetrahedron Lett.* **1990**, *31*, 991-994.

114. H. B. Kagan, In *Catalytic Asymmetric Synthesis*; I. Ojima, Ed.; VCH: New York, 1993, pp 203-226.

115. M. R. Barbachyn; C. R. Johnson, In *Asymmetric Synthesis*; J. D. Morrison and J. W. Scott, Ed.; Academic: Orlando, 1984; Vol. 4, pp 227-261.

116. K. K. Andersen *Tetrahedron Lett.* **1962**, 93-95.

117. F. A. Davis; A. C. Sheppard *Tetrahedron* **1989**, *45*, 5703-5742.

118. P. Pritchen; M. Deshmukh; E. Dunach; H. B. Kagan *J. Am. Chem. Soc.* **1984**, *106*, 8188-8193.

119. J. Drabowicz; B. Buknicki; M. Mikolajczyk *J. Org. Chem.* **1982**, *47*, 3325-3327.

120. V. K. Aggerwal; G. Evans; E. Moya; J. Dowden *J. Org. Chem.* **1992**, *57*, 6390-6391.

121. V. K. Aggarwal; A. Thomas; R. J. Franklin *J. Chem. Soc., Chem. Commun.* **1994**, 1653-1655.

122. G. Solladié *Synthesis* **1981**, 185-196.

123. J. Aubé, In *Comprehensive Organic Synthesis*; B. M. Trost and I. Fleming, Ed.; Pergamon: Oxford, 1991; Vol. 1, pp 819-842.

124. M. B. Andrus; A. B. Argade; X. Chen; M. G. Pamment *Tetrahedron Lett.* **1995**, *36*, 2945-2948.

125. J. F. Larrow; E. N. Jacobsen *J. Am. Chem. Soc.* **1994**, *116*, 12129-12130.

126. M. J. Taschner; D. J. Black *J. Am. Chem. Soc.* **1988**, *110*, 6892-6893.

127. M. J. Taschner; D. J. Black; Q.-Z. Chen *Tetrahedron: Asymm.* **1993**, *4*, 1387-1390.

128. C. Bolm; G. Schlingloff; K. Weickhardt *Angew. Chem., Int. Ed. Engl.* **1994**, *33*, 1848-1849.

129. T. Honda; N. Kimura; M. Tsubuki *Tetrahedron: Asymm.* **1993**, *4*, 1475-1478.

130. J. Aubé; Y. Wang; M. Hammond; M. Tanol; F. Takusagawa; D. Vander Velde *J. Am. Chem. Soc.* **1990**, *112*, 4879-4891.

131. F. A. Davis; R. H. Jenkins, Jr., In *Asymmetric Synthesis*; J. D. Morrison and J. W. Scott, Ed.; Academic: Orlando, 1984; Vol. 4, pp 313-353.

132. A. Lattes; E. Oliveros; M. Rivière; C. Belzecki; D. Mostowicz; W. Abramskj; C. Piccinni-Leopardi; G. Germain; M. Van Meerssche *J. Am. Chem. Soc.* **1982**, *104*, 3929-3934.

133. V. Gracias; G. L. Milligan; J. Aubé *J. Am. Chem. Soc.* **1995**, *117*, 8047-8048.

Index

Entries in italics refer to definitions of terms

DATE DUE			

GAYLORD No. 2333   PRINTED IN U.S.A.